DIAMAGNETISM

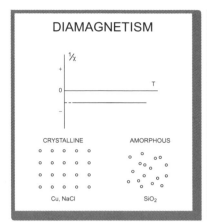

CRYSTALLINE   AMORPHOUS

Cu, NaCl   SiO$_2$

IDEAL FERROMAGNETISM

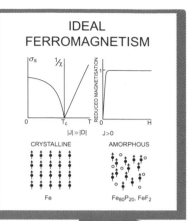

CRYSTALLINE   AMORPHOUS

Fe   Fe$_{80}$P$_{20}$, FeF$_2$

ANTIFERROMAGNETISM

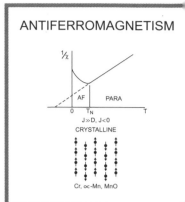

J≫D, J<0
CRYSTALLINE

Cr, ∝-Mn, MnO

INCIPIENT FERROMAGNETISM

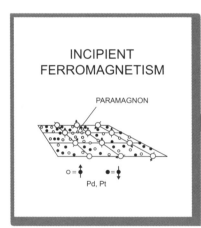

PARAMAGNON

o = ⬆ ●= ⬇
Pd, Pt

METAMAGNETISM
FIELD-INDUCED TRANSITIONS

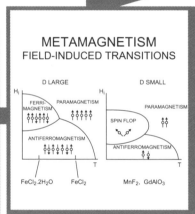

D LARGE   D SMALL

FeCl$_2$.2H$_2$O   FeCl$_2$ MnF$_2$, GdAlO$_3$

SUPERPARAMAGNETISM

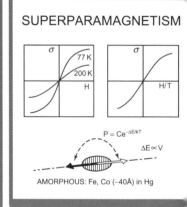

$P = Ce^{-\Delta E/kT}$

$\Delta E \propto V$

AMORPHOUS: Fe, Co (~40Å) in Hg

Family Tree of Magnetism

"The Taxonomy of Magnetism", courtesy C. M. Hurd.
See also C. M. Hurd: *Contemp. Phys.* **23**, 469 (1982).

MICTOMAGNETISM
CLUSTER GLASS

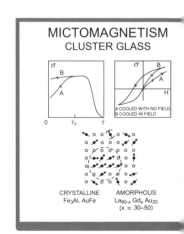

CRYSTALLINE   AMORPHOUS
Fe$_3$Al, AuFe   La$_{80-x}$ Gd$_x$ Au$_{20}$
(x ≈ 30–50)

IDEAL PARAMAGNETISM

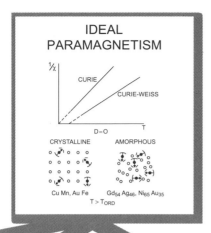

CURIE

CURIE-WEISS

D=O

CRYSTALLINE   AMORPHOUS

Cu Mn, Au Fe $Gd_{54} Ag_{46}$, $Ni_{65} Au_{35}$

$T > T_{ORD}$

FERRIMAGNETISM

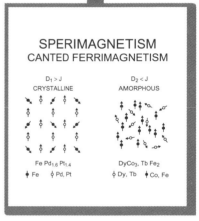

σ_s $1/\chi$

T_c

CRYSTALLINE   AMORPHOUS

Spinel : $XOFe_2O_3$ $REFe_2$, $Gd_{30}Co_{70}$

SPEROMAGNETISM

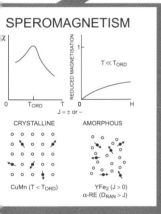

$T \ll T_{ORD}$

T_{ORD} T   H

$J = \pm$ or $-$

CRYSTALLINE   AMORPHOUS

CuMn ($T < T_{ORD}$) YFe_2 ($J > 0$)
α-RE ($D_{RAN} > J$)

ASPEROMAGNETISM

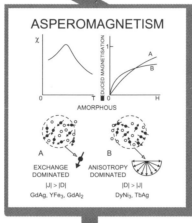

T   H

AMORPHOUS

A   B
EXCHANGE   ANISOTROPY
DOMINATED   DOMINATED
$|J| > |D|$ $|D| > |J|$
GdAg, YFe_3, $GdAl_2$ $DyNi_3$, TbAg

SPERIMAGNETISM
CANTED FERRIMAGNETISM

$D_1 > J$ $D_2 < J$
CRYSTALLINE   AMORPHOUS

Fe $Pd_{1.6}$ $Pt_{1.4}$ $DyCo_3$, Tb Fe_2
Fe Pd, Pt   Dy, Tb Co, Fe

IDEAL SPIN GLASS
$J = \pm$ or $-$

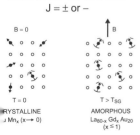

B = 0   B

T = 0 $T > T_{SG}$
CRYSTALLINE   AMORPHOUS
Mn$_x$ ($x \rightarrow 0$) La_{80-x} Gd_x Au_{20}
($x \lesssim 1$)

HELIMAGNETISM
CRYSTALLINE ASPEROMAGNETISM

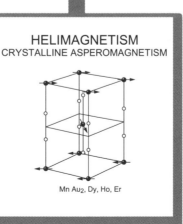

Mn Au_2, Dy, Ho, Er

THE THEORY OF MAGNETISM MADE SIMPLE

An Introduction to Physical Concepts
and to Some Useful Mathematical Methods

Other World Scientific titles by Prof. D. C. Mattis

The Many-Body Problem
An Encyclopedia of Exactly Solved Models in One Dimension

Statistical Mechanics
A Guide for Students and Researchers

THE THEORY OF MAGNETISM MADE SIMPLE

An Introduction to Physical Concepts and to Some Useful Mathematical Methods

Daniel C. Mattis

Department of Physics, University of Utah

World Scientific

NEW JERSEY • LONDON • SINGAPORE • BEIJING • SHANGHAI • HONG KONG • TAIPEI • CHENNAI

Published by

World Scientific Publishing Co. Pte. Ltd.

5 Toh Tuck Link, Singapore 596224

USA office: 27 Warren Street, Suite 401-402, Hackensack, NJ 07601

UK office: 57 Shelton Street, Covent Garden, London WC2H 9HE

Library of Congress Cataloging-in-Publication Data
Mattis, Daniel Charles, 1932–
    The theory of magnetism made simple : an introduction to physical concepts and to
some useful mathematical methods / Daniel C. Mattis.
        p. cm.
    Includes bibliographical references and index.
    ISBN 981-238-579-7 -- ISBN 981-238-671-8 (pbk.)
    1. Magnetism. 2. Mathematical physics. I. Title.

    QC753.2.M38 2006
    538--dc22
                                                            2005057871

British Library Cataloguing-in-Publication Data
A catalogue record for this book is available from the British Library.

Printed by FuIsland Offset Printing (S) Pte Ltd, Singapore

Prologue

Research into magnetism and magnetic media is undergoing a sea change. Consider the following points.

- In the attempt to satisfy an ever-increasing need for higher density recording materials ultra-small *mesoscopic* and even *nanoscopic* magnetic materials (magnetic *dots* or *grains*) are being made in the laboratory while their electrical and magnetic properties are undergoing intense scrutiny. Giant — even *colossal* — magnetoresistance are newly identified physical phenomena already being exploited for the latest in high-density media.
- Quantum tunneling of spins, virtually unknown a decade ago, has suddenly become an important and popular topic in its theory and in its practice; this new field has been named *spintronics*.
- "Quantum computation" based on quantum entanglement of spins vies with its electronic counterparts for a role in the information processing units of the future.
- "Quantum fluctuations" drive *long-range order* in a phenomenon colorfully denoted "order from disorder."
- Artificially engineered metallic alloys with fully spin-polarized conduction electrons have been dubbed "half-metals."
- A purely magnetic phenomenon (i.e., *antiferromagnetism*) is reputedly at the basis of the extraordinary discovery, high-temperature superconductivity — although the facts are either not all yet known or are insufficiently understood at the time of writing.

At the dawn of this new millennium it becomes apparent that the study of magnetism — arguably already dating back several millennia — has acquired a new urgency along with its new vocabulary. In a previous version of this book (*The Theory of Magnetism*, vols. I and II, Springer, Berlin, 1981 and 1985), the statics and dynamics of magnetism and its thermodynamic properties were examined. Insofar as they dealt with basic theory and documented the history of the science, these texts should still serve regardless of their flaws. However in other aspects the earlier works are now as hopelessly out of date as is the original 1965 edition, based on a set of lecture notes for a Summer course.

But even when applications evolve rapidly in time they are still secondary to the immutable basic physical principles that empower them. The purpose of the present text is to expound the basics of quantum magnetism to a new audience, whether undergraduate, graduate, or professional. The presentation is intended to be quickly and easily grasped by the reader — preferably with an instructor's help. The field is vast and not all the important topics could be covered. But most concepts that *are* introduced in these pages are worked out in minute detail. A lot of redundancy should make it easier to master some technical aspects. The mathematics is held at first to the minimum required for the formulation of classical and then quantum phenomena, but is then further developed to whatever extent needed as the *many-body* aspects that are at the heart of the theory are progressively allowed to unfold. There is a deliberate attempt to use similar tools wherever possible so that the reader can hone newly acquired skills on various topics; additionally, references to more advanced methods abound.

Following a historical Chapter 1 (co-authored with Dr. Noémi P. Mattis), Chapter 2 provides a quick overview of the subject matter and touches upon all the topics that follow. The study of the theory of magnetism provides more than just an entry into today's hot topics: it gives insight into all of theoretical physics, of which it is an integral part. A half-century ago when this author was still a graduate student, one of his instructors joked that, "all of mathematics is a special case of quantum mechanics." In the spirit of this witty but profound observation, let us see whether all of theoretical physics might not be "a special case of the theory of magnetism!" And there are several points to buttress this claim.

- The study of spin glasses, first undertaken thirty years ago, was intended as a first step to understanding amorphous SiO_2 (window glass) and random media in general rather than as an end in itself. One of the purposes was to understand whether a *phase transition* from a disordered high-temperature phase into a disordered low-temperature phase was even conceptually possible. The question was asked, does the "glassy" temperature signal a true discontinuity in thermodynamic properties, or merely an enhanced viscosity? Is the spin glass a genuine thermodynamic phase? This question is answered in the affirmative by some exactly solvable examples. Actually, this is just a special case of a larger class of so-called "critical phenomena." The very creation of ferromagnetism at or near the Curie temperature T_c, provides us with the original example of what is now classified as a *second-order phase transition*. This phenomenon is now known to be common to all physical sciences. Historically, the first serious theory purporting to explain how a paramagnet attains a macroscopic magnetic moment and long-range order when the temperature drops below T_c was devised one hundred years ago by Pierre Weiss, inspired by van der Waals' seminal theory of vapor-to-liquid phase transitions.

- All mean-field approximations in current use are outgrowths of Weiss' "molecular field" theory of ferromagnetism. Spin waves (magnons) are the simplest realization of Goldstone's theorem in quantum field theory. The interacting quantum spins in the Heisenberg model are the very prototype of a nonlinear quantum field theory. The first exact solution of *any* quantum "many-body problem" was "Bethe's ansatz," which revealed the eigenstates and eigenvalues of the one-dimensional Heisenberg model. And so it goes ...

Let us now mention up front what was purposely left out of this book for want of space, expertise or time. At the top of this list is the now famous "Quantum Hall Effect," whereby the two-dimensional electron gas became a laboratory for particles of fractional spin and charge. There is also little discussion of latter-day uses of Bethe's ansatz — as in the Lieb–Wu solution of the one-dimensional Hubbard model or the Andrei–Wiegman analysis of the Kondo effect. There is no development of the renormalization group (the numerical technique that first allowed insight into the second-order phase transition) nor of conformal field theory, central to the characterization of phase transitions in 2D, nor of Baxter's IRF transfer matrices and 8-Vertex models. The Replica Method for spin-glasses and the replica-breaking solutions of Parisi are also omitted.

The reason for all these omissions is obvious. These are all "advanced" topics beyond the scope of this book, in the sense that each requires a distinct approach and mind-set and grasp of mathematical principles. These are the domains of specialists. What gave this author the impetus to omit or skirt these important topics? By now, each of these topics has become so popular that one or more reviews, conference proceedings and textbooks have been written on it, sometimes even by the greatest experts: the discoverers or inventors, their close collaborators or their students. It is not for an introductory treatise to reinterpret what has been authoritatively stated elsewhere. Instead, we stick to a few of the author's favorite themes and exploit them from various points of view.

The goal is to prepare the reader, as best as possible, for further adventures into advanced research and for unforeseeable discoveries to come, all surely requiring a good background and a consistent methodology. Where this could not be done conveniently references to the best available tutorials are given, in keeping with another goal: to keep it *simple*, as in the title.

The author is grateful to Prof. E. H. Lieb at Princeton, Prof. J. Hirsch of UCSD and Dr. G. Ortiz of Los Alamos National Laboratory for discussions of material pertinent to their work and other helpful comments. I thank the many others who have altruistically allowed their results to be copied or quoted in this text. Thanks also to my dear wife for the historical research she performed in 1963 incorporated in Chapter 1; forty-two years later this part of the text remains fresh as new!

I also thank my editors — Dr. K. K. Phua and Kim Tan at World Scientific Publishing for their boundless patience with this procrastinating author.

Salt Lake City, *July 2005*

Contents

Chapter 5.   From Magnons to Solitons: Spin Dynamics   175

Chapter 6.   Magnetism in Metals   257

Chapter 1

History of Magnetism

Western scientists and historians generally believe it was the Greeks who first reflected upon the wondrous properties of magnetite, the magnetic iron ore $FeO–Fe_2O_3$ and famed lodestone (leading stone, or compass). This mineral, which even in the natural state often has a powerful attraction for iron and steel, was mined in the province of Magnesia.

> The magnet's name the observing Grecians drew
> From the magnetick region where it grew.[1]

This origin is not incontrovertible. According to Pliny's account the magnet stone was named after its discoverer, the shepherd Magnes, "the nails of whose shoes and the tip of whose staff stuck fast in a magnetick field while he pastured his flocks."[1]

But more than likely, according to Chinese writings dating back to 4000 B.C. that mention magnetite, the original discoveries were made in China.[2] Meteoric iron was disovered and utilized in the period 3000–2500 B.C. A primitive compass illustrated below dates from that period.

---

[1] Lucretius Carus, *De Rerum Natura*, 1st century B.C. References are to *vv.* 906 ff., in the translation by Th. Creech, London, 1714. Pliny, quoted in W. Gilbert, *De Magnete*, trans., Gilbert Club, London, 1900, rev. ed., Basic Books, New York, 1958, p. 8.

[2] The text and illustrations relating to China are adapted from Yu-qing Yang's paper "*Magnetic Materials in China*," presented in Poland at the 3rd International Conference on *Physics of Magnetic Materials*, (9–14 Sept. 1986), W. Gorzkowski, H. Lachowicz and H. Szymczak, eds., World Scientific, 1987.

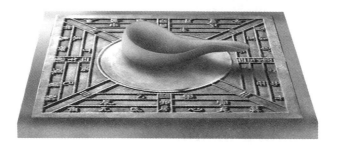

It consists of a spoon-shaped magnetite object with a smooth bottom, set on a polished copper surface. When pushed it rotated freely and usually came to rest with the handle pointing to the South.

The head of a floating "fish" made of magnetized iron, illustrated in a military manual of 1044 A.D., also pointed South.

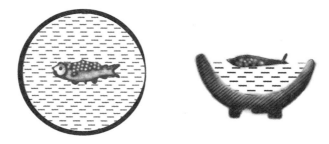

Dated only a couple of decades later, the book "Notes by the Dream Brook" describes the making of a magnetic steel needle in the remarkably modern needle-and-pivot compass shown here.

1.1. Physics and Metaphysics

The lodestone also appeared in Greek writings by the year 800 B.C., and Greek thought and philosophy dominated all thinking on the subject for some 23 centuries following this. A characteristic of Greek philosophy was that it did not seek so much to explain and predict the wonders of nature as to force them to fit within a preconceived scheme of things. It might be argued that this seems to be precisely the objective of modern physics as well, but the analogy does not bear close scrutiny. To understand the distinction between modern and classical thought on this subject, suffice it to note the separate meanings of the modern word *science* and of its closest Greek equivalent, $\epsilon\pi\iota\sigma\tau\acute{\eta}\mu\eta$. We conceive science as a specific activity pursued for its own sake, one which we endeavor to keep free from "alien" metaphysical beliefs. Whereas, $\epsilon\pi\iota\sigma\tau\acute{\eta}\mu\eta$ meant *knowledge* for the Greeks, with aims and methods undifferentiated from those of philosophy.

The exponents of one important school of philosophy, the *animists*, took cognizance of the extraordinary properties of the lodestone by ascribing to it a divine origin. Thales, then later Anaxagoras and others, believed the lodestone to possess a soul. We shall find this idea echoed into the seventeenth century A.D.

The school of the *mechanistic*, or atomistic, philosophers should not be misconstrued as being more scientific than were the animists, for their theories were similarly deductions from general metaphysical conceptions, with little relation to what we would now consider "the facts." Diogenes of Apollonia (about 460 B.C.), a contemporary of Anaxagoras, says there is humidity in iron which the dryness of the magnet feeds upon. The idea that magnets feed upon iron was also a long lived superstition. Still trying to check on it, John Baptista Porta, in the sixteenth century, reported as follows:

> I took a Loadstone of a certain weight, and I buried it in a heap of Iron-filings, that I knew what they weighed; and when I had left it there many months, I found my stone to be heavier, and the Iron-filings lighter: but the difference was so small, that in one pound I could finde no sensible declination; the stone being great, and the filings many: so that I am doubtful of the truth.[3]

---

[3] John Baptista Porta, *Natural Magick*, Naples, 1589, reprint of 1st English ed., Basic Books, New York, 1957, p. 212.

But the more sophisticated theories in this category involved effluvia, which were invisible emanations or a sort of dynamical field. The earliest of these is due to Empedocles, later versions to Epicurus and Democritus. We quote a charming accounting by the Roman poet Lucretius Carus showing that in the four centuries since Empedocles, in an era of high civilization, the theory had not progressed:

> Now sing my muse, for 'tis a weighty cause.
> Explain the Magnet, why it strongly draws,
> And brings rough Iron to its fond embrace.
> This Men admire; for they have often seen
> Small Rings of Iron, six, or eight, or ten,
> Compose a subtile chain, no Tye between;
> But, held by this, they seem to hang in air,
> One to another sticks and wantons there;
> So great the Loadstone's force, so strong to bear! ...
> First, from the Magnet num'rous Parts arise.
> And swiftly move; the Stone gives vast supplies;
> Which, springing still in Constant Stream, displace
> The neighb'ring air and make an empty Space;
> So when the Steel comes there, some Parts begin
> To leap on through the Void and enter in ...
> The Steel will move to seek the Stone's embrace,
> Or up or down, or t'any other place
> Which way soever lies the Empty Space.[1]

The first stanza is a vivid enough description of magnetic induction, the power of magnetized iron to attrace other pieces of iron. Although this fact was already known to Plato, Lucretius was perhaps among the first to notice, by accident, that magnetic materials could also repel. The phenomenon awaited the discovery of the existence of two types of magnetic poles for an explanation.

There followed many centuries without further progress at a time when only monks were literate and research was limited to theological considerations.

The date of the first magnetic technological invention, the compass, and the place of its birth are still subjects of dispute among historians. Considerable weight of opinion places this in China at some time between 2637 B.C.

and 1100 A.D., as previously noted. Less reliable sources have it that the compass was introduced into China only in the thirteenth century A.D. and owed its prior invention to Italian or Arab origin. In any event, the compass was certainly known in western Europe by the twelfth century A.D. It was an instrument of marvelous utility and fascinating properties. Einstein has written in his autobiography of its instinctive appeal:

> A wonder ... I experienced as a child of 4 or 5 years, when my father showed me a compass. That this needle behaved in such a determined way did not at all fit into the nature of events, which could find a place in the unconscious world of concepts (effects connected with direct "touch"). I can still remember — or at least believe I can remember — that this experience made a deep and lasting impression upon me.[4]

Many authors in the middle ages advanced metaphysical explanations of the phenomenon. However, the Renaissance scientist William Gilbert[5] said of these writers:

> ... they have lost their oil and their pains; for, not being practised in the subjects of Nature, and being misled by certain false physical systems, they adopted as theirs, from books only, without magnetical experiments, certain inferences based on vain opinions, and many things that are not, dreaming old wives' tales.

Doubtless his condemnation was too severe. Before Gilbert and the sixteenth century, there had been some attempts at experimental science, although not numerous. The first and most important was due to Pierre Pélerin de Maricourt, better known under the Latin nom de plume Petrus Peregrinus. His "Epistola Petri Peregrini de Maricourt ad Sygerum de Foucaucourt Militem de Magnete," dated 1269 A.D., is the earliest known treatise of experimental physics. Peregrinus experimented with a spherical lodestone which he called *terrella*. Placing on it an oblong piece of iron at various spots, he traced lines in the direction it assumed and thus found these lines to circle the lodestone the way meridians gird the earth, crossing

---

[4] P. A. Schilp (ed.), *Albert Einstein: Philosopher-Scientist*, Vol. 1, Harper & Row, New York, 1959, p. 9.

[5] W. Gilbert, *De Magnete*, trans., Gilbert Club, London 1900, rev. ed., Basic Books, New York, 1958, p. 3.

at two points. These he called the *poles* of the magnet, by analogy with the
poles of the earth.

1.2.   Gilbert and Descartes

Of the early natural philosophers who studied magnetism the most famous
is William Gilbert of Colchester, the "father of magnetism."

> Gilbert shall live till loadstones cease to draw
> Or British fleets the boundless ocean awe.[6]

The times were ripe for him. Gilbert was born in 1544, after Copernicus
and before Galileo, and lived in the bloom of the Elizabethan Renaissance.
Physics was his hobby, and medicine his profession. Eminent in both, he
became Queen Elizabeth's private physician and president of the Royal
College of Physicians. It is said that when the Queen died, her only per-
sonal legacy was a research grant to Gilbert. But this he had no time to
enjoy, for he died a few months after her, carried off by the plague in 1603.

Some 20 years before Sir Francis Bacon, he as a firm believer in what we
now call the experimental method. Realizing that "it is very easy for men
of acute intellect, apart from experiment and practice, to slip and err," he
resolved to trust no fact that he could not prove by his own experience. *De
Magnete* was Gilbert's masterpiece, 17 years in the writing and containing
almost all his results prior to the date of publication in 1600. There he assem-
bled all the trustworthy knowledge of his time on magnetism, together with
his own major contributions. Among other experiments, he reproduced those
performed three centuries earlier by Peregrinus with the terrella; but Gilbert
realized that his terrella was an actual model of the earth and thus was the
first to state specifically that the earth is itself a magnet, "which opinion of
his was no sooner broached that it was embraced and wel-commed by many
prime wits as well English as Forraine"[7] Gilbert's theory of magnetic fields
went as follows: "Rays of magnetick virtue spread out in every direction in
an orbe; the center of this orbe is not at the pole (as Porta reckons) but in
the center of the stone and of the terrella" [Gilbert[5], p. 95].

---

[6] J. Dryden, from "epistle to Doctor Walter Charleton, physician in ordinary to King
Charles I."

[7] Nathaniel Carpenter, Dean of Ireland, quoted in P. F. Mottelay, *A Bibliographical
History of Electricity and Magnetism*, London, 1922, p. 107.

Gilbert dispelled superstitions surrounding the lodestone, of which some dated from antiquity, such as "if a loadstone be anointed with garlic, or if a diamond be near, it does not attract iron." Some of these had already been disproved by Peregrinus in 1269, and even nearer to Gilbert's time, by the Italian scientist Porta, founder of one of the earliest scientific academies. Let Porta recount this:

> It is common Opinion amongst Sea-men, that Onyons and Garlick are at odds with the Loadstone: and Steersmen, and such as tend the Mariners Card are forbidden to eat Onyons or Garlick, lest they make the Index of the Poles drunk. But when I tried all these things, I found them to be false: for not only breathing and beleching upon the Loadstone after eating of Garlick, did not stop its Virtues: but when it was all anoynted over with the juice of Garlick, it did perform its office as well as if it had never been touched with it: and I could observe almost not the least difference, lest I should seem to make void the endeavours of the Ancients. And again, When I enquired of the Mariners, whether it were so, that they were forbid to eat Onyons and Garlick for that reason; they said, They were old Wives fables, and things ridiculous; and that Sea-men would sooner lose their lives, than abstain from eating Onyons and Garlick [Porta[3], p. 211].

But the superstitions survived the disproofs of Peregrinus, Porta, and Gilbert, and have left their vestiges in our own time and in common language. Between superstition and fraud there is but a thin line, and Galileo recounts how his natural skepticism protected him in one instance from a premature Marconi:

> ... a man offered to sell me a secret for permitting one to speak, through the attraction of a certain magnet needle, to someone distant two or three thousand miles, and I said to him that I would be willing to purchase it, but that I would like to witness a trial of it, and that it would please me to test it, I being in one room and he being in another. He told me that, at such short distance, the action could not be witnessed to advantage; so I sent him away, and said that I could not just then go to Egypt or Muscovy to see his experiment, but that if he would go there himself I would stay and attend to the rest in Venice.[8]

---

[8] Spoken by Sagredus in Galileo's, *Dialogo sopra i due massimi sistemi del mondo Tolemaico e Copernicano*, 1632.

Medical healers of all times have been prompt to invoke magnetism. For example, mesmerism, or animal magnetism, was just another instance of the magnetic fluids, which we shall discuss shortly, invading the human body. Still seeking to disprove such hypotheses, it was in the interest of science that Thomas Alva Edison, as late as 1892, subjected himself "together with some of his collaborators and one dog" to very strong magnetic fields without, however, sensing any effects.

But how could experiments disprove metaphysics? In spite of their own extensive investigations, Gilbert and Porta were themselves believers in an animistic philosophy, and such were their theory and explanations of the phenomena which they had studied. Note in what sensuous terms Porta describes magnetic attraction:

> ... iron is drawn by the Loadstone, as a bride after the bridegroom, to be embraced; and the iron is so desirous to joyn with it as her husband, and is so sollicitious to meet the Loadstone: when it is hindred by its weight, yet it will stand an end, as if it held up its hands to beg of the stone, and flattering of it, ... and shews that it is not content with its condition: but if it once kist the Loadstone, as if the desire were satisfied, it is then at rest; and they are so mutually in love, that if one cannot come to the other it will hang pendulous in the air ... [Porta[3], p. 201].

His explanation, or theory, for this phenomenon is no less anthropomorphic:

> I think the Loadstone is a mixture of stone and iron ... whilst one labors to get the victory of the other, the attraction is made by the combat between them. In that body, there is more of the stone than of the iron; and therefore the iron, that it may not be subdued by the stone, desires the force and company of iron; that being not able to resist alone, it may be able by more help to defend itself. For all creatures defend their being. [*ibid.*, p. 191].

To which Gilbert, picking up the dialogue 40 years later, retorts:

> As if in the Loadstone the iron were a distinct body and not mixed up as the other metals in their ores! And that these, being so mixed up, should fight with one another, and should extend their quarrel, and that in consequence of the battle auxilliary forces should be called in, is indeed absurd. But iron itself, when excited by the Load-

stone, seizes iron no less strongly than the Loadstone. Therefore those
fights, seditions, and conspiracies in the stone ... are the ravings of
a babbling old woman, not the inventions of a distinguished mage.
[Gilbert[5], p. 63].

Gilbert's own ideas are themselves a curious blend of science and myth.
On the one hand he dismisses the effluvia theory of magnetism with cogent
reasoning, although he admits this concept might apply to electricity. His
arguments are pithy: magnetic force can penetrate objects and the lodestone attracts iron through solid materials other than air, which should act
as deterrents to any sort of effluvium. Electricity, on the other hand, is
strongly affected by all sorts of materials. But when he comes to give his
own explanation for magnetic attraction, he states it arises because "the
Loadstone hath a soul." He believed the earth to have one, and therefore the
loadstone also, it "being a part and choice offspring of its animate mother
the earth." [*ibid.*, p. 210].

Notwithstanding the shortcomings of his theory, Gilbert had indeed in-
agurated the experimental method. At the other pole stands René Descartes
(1596–1650). Here was a philosopher who ignored the facts, but whose merit
it was to exorcise the soul out of the lodestone, laying the foundations of
a rational theory. Descartes is the author of the first extensive theory of
magnetism, elaborated in his *Principia*, Part IV, sections 133–183.

Since Descartes was among the "prime forraine wits" to embrace Gilbert's
hypothesis linking the lodestone to the earth, his theory of ferromagnetism is
accessory to his theory of geomagnetism. Both can be summarized as follows:
The prime imponderables were not specifically denoted "effluvia" but rather
"threaded parts" (*parties cannelées*). These were channeled in one-way ducts
through the earth, entering through pores in one pole while leaving through
pores in the other. Two kinds of parts were distinguised: those that could
only enter the North Pole and leave by the South, and the ones that made
the inverse voyage. The return trip was, in either case, by air. The parts
find this a disagreeable mode of travel, and seize upon the opportunity to
cross any lodestone in the way. So much so, indeed, that if they chance to
meet a loadstone they will even abandon their ultimate destination and stay
with it, crossing it over and over again. This is shown in Fig. 1.1. Vortices
are thus created in and around the material. Lodestone, iron, and steel are
the only materials having the proper channels to accommodate the parts
because of their origin in the inner earth. Of these, lodestone ducts are best,

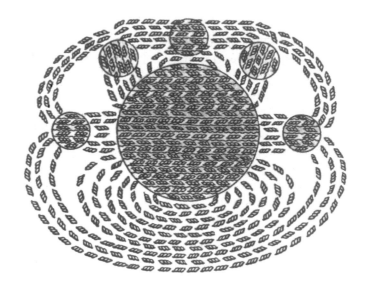

Fig. 1.1.   The first theory: Descartes' threaded parts are shown going through the earth (center sphere) and in and around other magnetized bodies.

whereas iron is malleable, and therefore the furlike cillia which cause the ducts to be one-way are disturbed in the process of mining. The threaded parts — throwing themselves upon the iron with great speed — can restore the position of the cillia, and thus magnetize the metal. Steel, being harder, retains magnetization better.

With this theory, Descartes claimed to be able to interpret all magnetic phenomena known to his time. From today's perspective it is hard to see how he even met Gilbert's objection to the theory of effluvia, stated a generation earlier, nor how this theory could answer the practical questions which arise in the mind of anyone working directly with magnetism. However, such was Descartes' reputation that this theory came to be accepted as fact, and influenced all thinking on the subject throughout this century, and much of the eighteenth. Two of his more prominent disciples in the eighteenth century were the famed Swiss mathematician Léonard Euler and the Swedish mystic and physicist Emmanuel Swedenborg.

Descartes, in his physics and in his philosophy, marks the transition between metaphysical and scientific thought. First, he re-established confidence in the power of reason, which was an absolute necessity for the birth of theoretical science. Second, he postulated a dichotomy of the soul and of the body, which opened the door to the study of nature on her own terms. In that, he was not alone. The beginning of the seventeenth century had witnessed a widespread mechanistic revolution in the sciences, led by such men as Gassendi, Mersenne, Hobbes, Pascal, Huygens and others. These people often went much further than Descartes in divorcing physics from metaphysics. Descartes still believed that physics could be deduced from unprovable first principles, and this mechanism was thus close to the Greeks'. He invoked metaphysics to ascertain his scientific assumptions: his argument was that, since God created both nature and our reason, we can trust that the certitudes He has instilled in us correspond to Truth. The other mechanists, however, were content with a more humble approach to nature: let us describe the phenomena, they would say, and not mind the deeper essence of things.

Probably the most important contribution of mechanism to modern science is the adoption of a separate language to describe nature, that of mathematics. At first, there is an intuition. Galileo had already said in 1590: "Philosophy is written in a great book which is always open in front of our eyes (I mean: the universe), but one cannot understand it without first applying himself to understanding its language and knowing the characters in which it is written. It is written in the *mathematical* language".[9] Descartes, almost 30 years later, receives the same idea as an illumination, though he fails to apply it successfully. The new language having been adopted, physics will receive an impetus from the invention of calculus (by Newton and Leibnitz, in the 1680's) and (in the words of twentieth century physicist and Nobel Prize winner, P. A. M. Dirac) from a "mathematics that continually shifts its foundations and gets more abstract ... It seems ... that advance in physics is to be associated with continual modification and generalisation of the axioms at the base of mathematics rather than ... any one ... scheme". [see footnote 3 in Chap. 3.] In this way, mathematics has come to replace metaphysics.

In magnetism, it is the French monk Mersenne, a friend of Descartes', who is the first, in 1644, to quantify many of Gilbert's observations.[10] Progress in theory remains slow. The mechanists are reluctant to speculate about deeper causes, and the field is left to the neocartesians. Even so, by the year 1700 there is one (as later there will be many) dissident voice singing a new tune. It belongs to John Keill, Savilian professor of astronomy at Oxford, who in

---

[9] *Opera di Galileo Galilei*, Ed. Nazionale, Firenze, 1890–1909, Vol. V, p. 232.
[10] Marin Mersenne, "Cogitata physico-mathematica" (1644), the part entitled "Tractatus de Magnetis proprietatibus."

his eighth lecture of that year observes:

> It is certain that the magnetic attractions and directions arise from
> the structure of parts; for if a loadstone be struck hard enough, so
> that the position of its internal parts be changed, the loadstone will
> also be changed. And if a loadstone be put into the fire, insomuch as
> the internal structure of the parts be changed or wholly destroyed,
> then it will lose also its former virtue and will scarce differ from
> other stones ... And what some generally boast of, concerning efflu-
> via, a subtile matter, particles adapted to the pores of the loadstone,
> etc. ... does not in the least lead us to a clear and distinct explication
> of those operations; but notwithstanding all these things, the mag-
> netick virtues must still be reckoned amongst the occult qualities.[11]

1.3.   Rise of Modern Science

It is not until the second half of the eighteenth century that we see the
beginnings of a modern scientific attack on the problems of magnetism,
characterized by a flexible interplay between theory and experiment and
founded in rational hypotheses. For a while, theory becomes variations on
the theme of *fluids*. Even Maxwell was to swim in this hypothesis, and it was
not until the discovery of the electron that magnetic theory could be placed
on more solid ground.

The fluid theory was originally proposed as an explanation for electricity
after the discovery by Stephen Gray in 1729 that electricity could be con-
veyed from one body to another. This was through the medium of metals
or other "nonelectric" substances, i.e., substances which conduct electricity
and do not lend themselves readily to the accumulation of static charge. An
early, and eminent proponent of the "one-fluid" hypothesis was Benjamin
Franklin. He interpreted static charge as the lack, or excess, of the electric
fluid. It was said of him: "Eripuit caelo fulmen, sceptrumque tyrannis." (He
snatched lightning from the sky and the scepter from tyrants.)

One of his scientific disciples, a German *émigré* to St. Petersburg by the
name of Franz Maria Aepinus (1724–1802), applied the one-fluid theory to
magnetism. His theories of electricity and magnetism appear in *Tentament*

---

[11] J. Keil, *Introductio ad Veram Physicam*, 1705 (transl., 1776).

Theoriae Electricitatis et Magnetismi, published in St. Petersburg, 1759.[12]
This is an important work for it brought many ideas of Franklin into sharp
focus and gave the theories of effluvia the *coup de grâce* by dint of mathematical and experimental reasoning.

In 1733, Charles François duFay, superintendent of the French Royal
Botanical Gardens, discovered that there were two types of electricity. These
he denoted *vitreous* and *resinous*, each of which attracted its opposite and
repelled its own kind. As this idea came into competition with the Franklin
one-fluid theory, much thought and many experiments were expended on
proving one at the expense of the other.

Some years after electricity was granted a second fluid, so was magnetism.
The two fluids were denoted *austral* and *boreal* in correspondence with the
two poles. It was said that in the natural nonmagnetic state the two fluids
saturate iron equally, but that magnetization parts them and leads them
slightly towards their respective poles, where they accumulate. The Swede
Johan C. Wilcke, a former student and collaborator of Aepinus, and the
Dutchman Anton Brugmans presented this two-fluid hypothesis independently in 1778.

The best-known proponent of the two-fluid theory was Charles Augustin
Coulomb (1736–1806), who proposed an important modification in the
theory, and whose experiments have immortalized his name as the unit of
charge and in connection with the law of force.[13] It was by means of a
torsion balance of his invention that he established with some precision the
law which bears his name: that infinitesimals of either fluid, in electricity as
in magnetism, attract or repel in the ratio of the inverse of the square of their
distance. After establishing this in 1785, he performed many experiments on
the thermal properties of magnets. On the theoretical side, his principal contribution was the realization that the magnetic fluids could not be free to flow
like their electrical counterpart, but perforce were bound to the individual
molecules. Thus, he supposed each molecule to become somewhat polarized
in the process of magnetization. In this manner it could be explained why
the analogue of Gray's effect had never been found in magnetism, and why
two new poles always appeared when a magnet was cut in twain. Coulomb

---

[12] Although this work has never been translated from the Latin, a good account of it is
given by Père René Just Haüy in his "Exposition raisonnée de la théorie de l'électricité et
du magnétisme," 1787.
[13] Almost all of Coulomb's memoirs are collected in a single work published by the *Société
française de physique* in 1884, *Collection des Mémoires Relatifs à la Physique*, Vol. 1.

was also aware that the laws of force he had discovered were *not* applicable
on an atomic scale, that a solid body was not in stable equilibrium under
the effect of the inverse square law of force alone. However, concerning the
unknown laws of molecular repulsion, attraction, and cohesion, he was to
write, "It is almost always more curious than useful to seek to know their
causes,"[14] a correct (if defeatist) attitude considering the nearly complete
ignorance of atomic structure in his day.

Siméon Denis Poisson (1781–1840) is the man who eventually became
the best interpreter of the physical constructs which Coulomb discovered.
Here was a brilliant mathematician whose scientific career appears to have
been predestined from his early studies, whose school-teacher had predicted,
punning on the verse of La Fontaine:

Petit Poisson deviendra grand
Pourvu que Dieu lui prête vie.[15]

To magnetism, Poisson brought the concept of the static potential, with
which he had been so successful in solving the problems of static electricity.
And having invented the mathematical theory of magnetostatics, he proceeded by solving a considerable number of problems in that field. This
work dates from 1824 onwards, a very exciting time in physics as we shall
see in the following section. But Poisson ignored all developments that came
after the experiments of Coulomb, and while he gave the full theory of all
the discoveries by the master of the torsion balance, he did not participate
in the exciting movement that followed and which was to lead science into
a totally new direction.

We understand the magnetical work of Poisson today mostly in the manner in which it was extended and interpreted by George Green (1793–1841).
Poisson's "equivalent volume and surface distributions of magnetization" are
but a special case of one of the later Green theorems. Poisson found it conve-
nient to consider instead of $\mathbf{H}(\mathbf{r})$ the scalar potential $V(\mathbf{r})$ with the property
that

$$\nabla V(\mathbf{r}) = \mathbf{H}(\mathbf{r}). \tag{1.1}$$

---

[14] *Mémoire lu à l'Institute le 26 prairial, an 7, par le citoyen Coulomb*, Mémoires de
l'Institut, Vol. III, p. 176.
[15] Little Fish will become great/if God allows him but to live.

The contribution to $V(\mathbf{r})$ from the point $\mathbf{r}'$ in a magnetic material is

$$\left(m_x \frac{\partial}{\partial x'} \frac{1}{|\mathbf{r} - \mathbf{r}'|} + \cdots \right) d_3 r' = \mathbf{m} \cdot \nabla' \frac{1}{|\mathbf{r} - \mathbf{r}'|} d_3 r' \tag{1.2}$$

and the potential from the entire sample is therefore,

$$V(\mathbf{r}) = \int_{\text{vol}} \mathbf{m}(\mathbf{r}') \cdot \nabla' \frac{1}{|\mathbf{r} - \mathbf{r}'|} d^3 \mathbf{r}' . \tag{1.3}$$

By partial integration, this is transformed into the sum of a surface integral and of a volume integral,[16]

$$V(\mathbf{r}) = \int_{\text{sur}} \frac{1}{|\mathbf{r} - \mathbf{r}'|} \mathbf{m} \cdot d\mathbf{S}' - \int_{\text{vol}} \frac{1}{|\mathbf{r} - \mathbf{r}'|} \nabla' \cdot \mathbf{m}(\mathbf{r}') d^3 \mathbf{r}' \tag{1.4}$$

the latter vanishing for the important special case of constant magnetization. In that special case, the sources behaved *as if* they were all on the surface of the north and south poles of the magnet, and from this one might suspect that experiments concerning the nature of magnetic fields surrounding various substances would *never* reveal the slightest information about the mechanism within the material. Therefore, the fluid hypothesis seemed as good a working model as any. Poisson also extended the theory in several directions. For example, by means of certain assumptions regarding the susceptibility of magnetic fluids to applied fields he obtained a law of induced magnetization, thereby explaining the phenomenon of which Lucretius had sung.

Poisson, like Coulomb, refused to become excited about any speculation concerning the *nature* of the sources of the field, which is to say, the fluids. This reluctance to discuss the profound nature of things was, of course, an extreme swing of the pendulum away from metaphysics. That attitude was itself developed into an all-embracing philosophy by Auguste Comte. *Positivism*, as it was called, holds that in every field of knowledge general laws can only be induced from the accumulated facts and that it is possible to arrive, in this way, at fundamental truths. Comte believed this to be the only scientific attitude. But, after Poisson, the main theoretical advances will be made by those who ask *why* as well as *how*. That is, by physicists who will make *hypotheses* that are vaster, simpler, and more speculative *than the mere facts allow*. The specialization to the facts at hand answers the *how*; the vaster theory, the *why*.

---

[16] S. D. Poisson, *Mémoire sur la Théorie du Magnétisme*, Mémoires de l'Académie, Vol. V, p. 247.

1.4. Electrodynamics

As early as the seventeenth century there was reason to connect the effects of electricity and magnetism. For example, "in 1681, a ship bound for Boston was struck by lightning. Observation of the stars showed that 'the compasses were changed'; 'the north point was turn'd clear south.' The ship was steered to Boston with the compass reversed".[17] The fluids had proliferated and this now made it desirable to seek a relationship among them. The invention of the voltaic pile about 1800 was to stimulate a series of discoveries which would bring relative order out of this chaos.

During the time that Poisson, undisturbed, was bringing mathematical refinements to the theory of fluids, an exciting discovery opened the view to a new science of electrodynamics. In April, 1820, a Danish physicist, Hans Christian Oersted (1777–1851) came upon the long-sought connection between electricity and magnetism. Oersted himself had been seeking to find such a connection since the year 1807, but always unfruitfully, until the fateful day when he directed his assistant Hansteen to try the effect of a current on a delicately suspended magnetic needle nearby. *The needle moved.* The theory which had guided his earlier research had indicated to him that the relation should manifest itself most favorably under open-circuit conditions and not when the electric fluid was allowed to leak away, and one can easily imagine his stupefaction when the unexpected occurred. On July 21, 1820, he published a memoir in Latin (then a more universal tongue than Danish), which was sent to scientists and scientific societies around the world.[18] Translations of his paper were published in the languages and journals of every civilized country.

The reaction was feverish; immediately work started, checking and extending the basic facts of electromagnetism. French and British scientists led the initial competition for discoveries. At first the French came in ahead. The French Academy of Science of that time was a star-studded assembly, and besides Poisson it included such personalities as Laplace, Fresnel, Fourier, and more particularly active in this new field, Biot, Savart, Arago, and Ampère.

Dominique F. J. Arago (1786–1853) was the first to report upon the news of Oersted's discovery, on September 11, 1820. Here was a most remarkable

---

[17] F. Cajori, *A History of Physics*, Dover, New York, 1962, p. 102.

[18] This memoir refers to the experiment performed "last year," by which Oersted meant "last academic year," viz., "last April." This has often been misinterpreted, and the date 1819 incorrectly given for the discovery.

scientist, who had been elected to the Academy 12 years earlier at the age of 23 as a reward for "adventurous conduct in the cause of science." The story of his dedication is worth retelling. In 1806, Arago and Biot had been commissioned to conduct a geodetic survey of some coastal islands of Spain. This was the period of Napoleon's invasion of that country, and the populace took them for spies. After escaping from a prison, Arago escaped to Algiers, whence he took a boat back to Marseilles. This was captured by a Spanish man-of-war almost within sight of port! After several years of imprisonment and wanderings about North Africa, he finally made his way back to Paris in the Summer of 1809, and forthwith deposited the precious records of his survey in the *Bureau des Longitudes*, having preserved them intact throughout his vicissitudes. In this, he followed in a great tradition, and today's armchair scientists may well find their predecessors an adventurous lot. For example, earlier, in 1753, G. W. Richmann of St. Petersburg, had been struck dead by lightning while verifying the experiments of Franklin. The effects of the electricity on his various organs were published in leading scientific journals, and Priestley wrote: "It is not given to every electrician to die in so glorious manner as the justly envied Richmann".[19]

Shortly after his initial report to the Academy, Arago performed experiments of his own and established that a current acts like an ordinary magnet, both in attracting iron filings and in its ability to induce permanent magnetism in iron needles.

Seven days after Arago's report, André Marie Ampère (1775–1836) read a paper before the Academy, in which he suggested that internal electrical currents were responsible for the existence of ferromagnetism, and that these currents flowed perpendicular to the axis of the magnet. By analogy, might not steel needles magnetized in a solenoid show a stronger degree of magnetization than those exposed to a single current-carrying wire? Ampère proposed this idea to Arago, and they jointly performed the successful experiment on which Arago reported to the Academy on November 6, 1820.

The English were not far behind. It took Sir Humphrey Davy (1778–1829) until November 16 of that year to report on his similar experiments. Everyone took particular pains to witness and record important experiments and the dates thereof, and thus establish priority. No scientist on either side of the Channel underestimated the importance of Oersted's discovery nor of its consequences.

---

[19] J. B. Priestley, *History of Electricity*, London, 1775, p. 86.

After Ampère's death, correspondence to him from Fresnel (one letter undated, the other dated 5 June, 1821) was found among his papers, containing the suggestion that the "Amperian currents" causing ferromagnetism should be molecular rather than macroscopic in dimension. Ampère had hesitated on this point. Fresnel wrote that the lack of (Joule) heating, and arguments similar to those advanced above in connection with Coulomb's theory, suggested the existence of elementary, atomic or molecular currents.[20] This must have accorded well with Ampère's own ideas, and he made some (unpublished) calculations on the basis of such a model. This work was carried on subsequently by W. E. Weber (1804–1891), who assumed the molecules of iron or steel to be capable of movement around their fixed centers. These molecules, in unmagnetized iron or steel, lie in various directions such as to neutralize each other's field; but under the application of an external force, they turn around so that their axes lie favorably oriented with respect to this external field. Precisely the same concept had been stated by Ampère, in a letter addressed to Faraday dated 10 July, 1822, and it proved superior to Poisson's theory in explaining the saturation of magnetization, for example. The final evolution of this idea may be traced to J. A. Ewing (1855–1935), who pivoted tiny magnets arranged in geometric arrays so that they might be free to turn and assume various magnetic configurations. If these magnets could be assumed each to represent a magnetic molecule, and the experimental distance scaled to molecular size, the experiments of Ewing might have been expected to yield quantitative as well as qualitative information about ferro magnets. As Digby had written long before, in connection with Gilbert's terrella, "any man that hath an ayme to advance much in naturall science, must endeavour to draw the matter he inquireth of, into some such modell, or some kind of manageable methode; which he may turne and winde as he pleaseth ..." But for the unfortunate outcome, see ahead.

The fact that current loops had been found by Ampère to behave in every manner like elementary magnets did not logically prove that ferromagnetism is caused by internal electric currents. Nevertheless, this was the hypothesis most economical in concepts that could be put forward, and as it turned out, the most fruitful in stimulating new discoveries and in creating "insight" into the "physics" of magnetism. It was Ampère and his followers, rather than

---

[20] *Collection des Mémoires Relatifs à la Physique, Soc. Franc. de Phys.*, 1884, Vol. II, pp. 141, 144.

Poisson and his school, who acted in the modern style, the harmonious union of theory and experiment, economical in hypotheses.

The modern scientific method, which found in electrodynamics one of its early applications, seeks to imbed every phenomenon into a vaster mathematical and conceptual framework and, within this context, to answer the question of why it occurs, and not merely of how it happens. The long list of discoveries by the electrodynamicists is sufficient tribute to the efficacy of this method, even though their explanations of the cause of magnetism were to be proved wrong. If we dwell on the subject, it is because of the popular misconception that modern science is positivistic, an idea which is not justified by the methods with which science is carried forward today. Einstein wrote: "There is no inductive method which could lead to the fundamental concepts of physics ... in error are those theorists who believe that theory comes inductively from experience."[21]

The nineteenth century was so rich in interrelated theories and discoveries in the fields of atomic structure, thermodynamics, electricity, and magnetism, that it is quite difficult to disentangle them and neatly pursue our history of the theory of magnetism. Fortunately many excellent and general accounts exist of the scientific progress made in that epoch, in which magnetism is discussed in the proper perspective, as one of the many areas of investigation. Here we concentrate only on the conceptual progress and distinguish between progress in the physical theory of magnetism and progress in the understanding of the nature of the magnetic forces.

Much is owed to the insight of Michael Faraday (1791–1867), the humble scientist who is often called the greatest experimental genius of his century. Carrying forward the experiments of the Dutchman Brugmans, who had discovered that (paramagnetic) cobalt is attracted, whereas (diamagnetic) bismuth and antimony are repelled from the single pole of a magnet, Faraday studied the magnetic properties of a host of ordinary materials and found that all matter has one magnetic property or the other, although usually only to a very small degree. It is in describing an experiment with an electromagnet, in which a diamagnetic substance set its longer axis at right angles to the magnetic flux, that Faraday first used the term, "magnetic field" (December, 1845). He was not theoretically minded and never wrote an equation in his life. Nevertheless, his experiments led him unambiguously to the belief that magnetic substances acted upon one another by means of

---

[21] A. Einstein, *The Method of Theoretical Physics*, Oxford, 1933.

intermediary fields and not by "action at a distance." This was best explained by Maxwell, who

> ... resolved to read no mathematics on the subject till I had first read through Faraday's *Experimental Researches in Electricity*. I was aware that there was supposed to be a difference between Faraday's way of conceiving phenomena and that of the mathematicians .... As I proceeded with the study of Faraday, I perceived that his method ... [was] capable of being expressed in ordinary mathematical forms .... For instance, Faraday, in his mind's eye, saw lines of force traversing all space where the mathematicians saw centres of force attracting at a distance: Faraday saw a medium where they saw nothing but distance .... I also found that several of the most fertile methods of research discovered by the mathematicians could be expressed much better in terms of ideas derived from Faraday than in their original form.[22]

Faraday's concept of fields led him to expect that they would influence light, and after many unfruitful experiments he finally discovered in 1845 the effect which bears his name. This was a rotation of plane-polarized light upon passing through a medium in a direction parallel to the magnetization. Beside the Faraday effect, several other magneto-optic phenomena have been found: Kerr's magneto-optic effect is the analogue of the above, in the case of light reflected off a magnetic or magnetized material. Magnetic double refraction, of which an extreme example is the Cotton-Mouton effect, is double refraction of light passing perpendicular to the magnetization. But the effect which was to have the greatest theoretical implications was that discovered by Zeeman, which we shall discuss subsequently.

By virtue of the physicomathematical predictions to which it led, the hypothesis of fields acquired a reality that it has never since lost. Henry Adams wrote *ca.* 1900:

> For a historian, the story of Faraday's experiments and the invention of the dynamo passed belief; it revealed a condition of human ignorance and helplessness before the commonest forces, such as his mind refused to credit. He could not conceive but that someone,

---

[22] J. C. Maxwell, *A Treatise on Electricity and Magnetism*, 1873, reprinted by Dover, New York, 1954.

somewhere, could tell him all about the magnet, if one could but find the book ... .[23]

But of course there was such a book: James Clerk Maxwell (1831–1879) had summarized Faraday's researches, his own equations, and all that was known about the properties of electromagnetic fields and their interactions with ponderable matter.[22]

With Faraday, Maxwell believed the electric field to represent a real, physical stress in the ether (vacuum). Because electrodynamics had shown the flow of electricity to be responsible for magnetic fields, the latter must therefore have a physical representation as rates of change in the stress fields. Thus the energy density $E^2/8\pi$ stored in the electric field was of necessity potential energy; and the energy density $H^2/8\pi$ stored in the magnetic field was necessarily kinetic energy of the field. This interpretation given by Maxwell showed how early he anticipated the Hamiltonian mechanics which was later to provide the foundations of quantum electrodynamics, and how modern his outlook was.

The harmonic solutions of Maxwell's equations were calculated to travel with a velocity close to that of light. When the values of the magnetic and electrical constants were precisely determined, it was found by H. R. Hertz in 1888 that these waves were *precisely* those of light, radio, and those other disturbances that we now commonly call *electromagnetic waves*. Later, the special theory of relativity was invented by Einstein with the principal purpose of giving to the material sources of the field the same beautiful properties of invariance that Maxwell had bestowed on the fields alone. One formulation of Maxwell's equations that has turned out most useful introduces the vector potential $\mathbf{A}(\mathbf{r}, t)$ in terms of the magnetic field $\mathbf{H}(\mathbf{r}, t)$ and the magnetization $\mathbf{M}(\mathbf{r}, t)$ as

$$\nabla \times \mathbf{A} = \mathbf{H} + 4\pi\mathbf{M} = \mathbf{B}. \tag{1.5}$$

The vector $\mathbf{B}$ is defined as the curl of $\mathbf{A}$, and is therefore solenoidal by definition; that is, $\nabla\cdot\mathbf{B} = 0$, as is instantly verified. The next equation introduces the electric field in terms of the potentials $\mathbf{A}(\mathbf{r}, t)$ and the scalar potential $U(\mathbf{r})$ without the necessity of describing the sources. More precisely,

$$\mathbf{E} = -\nabla U - \frac{1}{c}\frac{\partial \mathbf{A}}{\partial t}. \tag{1.6}$$

---

[23] H. Adams, *The Education of Henry Adams*, Random House, New York, 1931.

This is but the equation of the dynamo: the motion or rate of change of the magnetic field is equivalent to an electric field, and if this is made to occur in a wire, a current will flow. The familiar differential form of this equation,

$$\nabla \times \mathbf{E} = -\frac{1}{c}\frac{\partial \mathbf{B}}{\partial t} \tag{1.7}$$

is obtained by taking the curl of both sides. Next, the results of Oersted, Ampère, Arago, and their colleagues almost all were summarized in the compact equation

$$\nabla \times \mathbf{H} = \frac{4\pi}{c}\mathbf{j} + \frac{1}{c}\frac{\partial \mathbf{D}}{\partial t} \tag{1.8}$$

where $\mathbf{j}$ is the real current density, and

$$\mathbf{D} = \mathbf{E} + 4\pi\mathbf{P} \tag{1.9}$$

relates the electric displacement vector $\mathbf{D}$ to the electric field $\mathbf{E}$ and the polarization of material substances, $\mathbf{P}$. Equation (1.8) can be transformed into more meaningful form by using the definitions of $\mathbf{H}$ and $\mathbf{D}$ in terms of $\mathbf{A}$, $\mathbf{P}$, and $\mathbf{M}$, and assuming a gauge $\nabla\cdot\mathbf{A} = 0$:

$$\left(-\nabla^2 + \frac{1}{c^2}\frac{\partial^2}{\partial t^2}\right)\mathbf{A}(\mathbf{r}, t) = 4\pi\left(\frac{1}{c}\mathbf{j} + \frac{1}{c}\frac{\partial \mathbf{P}}{\partial t} + \nabla \times \mathbf{M}\right). \tag{1.10}$$

This equation shows that in regions characterized by the absence of all material sources $\mathbf{j}$, $\mathbf{P}$, and $\mathbf{M}$, the vector potential obeys a wave equation, the solutions of which propagate with c = speed of light. The right-hand side of this equation provides a ready explanation for Ampère's famous theorem, that every current element behaved, insofar as its magnetic properties were concerned, precisely like a fictitious magnetic shell which would contain it. For if we replace all constant currents $\mathbf{j}$ by a fictitious magnetization $\mathbf{M} = \mathbf{r} \times \mathbf{j}/2c$ the fields would be unaffected. Nevertheless, for the sake of definiteness it will be useful to assume that $\mathbf{j}$ always refers to real free currents, $\mathbf{P}$ to bound or quasibound charges, and $\mathbf{M}$ to real, permanent magnetic moments.[24]

---

[24] For the latest on Maxwell equations, see L. L. Hirst, "The microscopic magnetization; concept and application," *Rev. Mod. Phys.* **69**, 607–628 (1997). For a comprehensive history, see O. Darrigol's *Electrodynamics from Ampère to Einstein*, Oxford Univ. Press, New York, 2000.

1.5. The Electron

Once these equations were formulated, permitting a conceptual separation of cause and effect, of the fields and of their sources, progress had to be made in understanding the *sources*. These related to the nature of matter itself, a study which at long last became "more useful than curious," to turn about the words of Coulomb. A giant step was taken in this direction by the discovery of the electron, one of the greatest scientific legacies of the nineteenth century to our own.

While Faraday, Maxwell, and many others had noted the likelihood that charge existed in discrete units only, this idea did not immediately make headway into chemistry, and Mendeléev's 1869 atomic table was based on atomic weights rather than atomic numbers. The first concrete suggestion was made in 1874 by G. Johnstone Stoney, the man who was to give the particle its name in 1891.[25] We know the electron as the fundamental particle, carrier of $e = 1.602 \times 10^{-19}$ Coulomb unit of charge, and $m = 9.11 \times 10^{-28}$ gram of rest mass, building block of atoms, molecules, solid and liquid matter. But it is amusing to us to recall that it was first isolated far from its native habitat, streaming from the cathode of the gas discharge tubes of the 1870's. Certainly it was of mixed parentage. To mention but two of the greatest contributors to its "discovery," Jean Perrin found in his 1895 thesis work that the cathode rays consisted of negatively charged particles, and Thomson had obtained the ratio e/m to good precision by 1897 (curvature in a magnetic field). Its existence was consecrated at the Paris International Congress of Physics held in 1900 inaugurating the twentieth century with a survey of the problems which the discovery of the existence of the electron had finally solved, and of those which it now raised.[26]

By this time there already was great interest in the spectral lines emitted by incandescent gases, for their discrete nature suggested that the fluids that constituted the atoms were only capable of sustaining certain well-defined vibrational frequencies. In 1896 Zeeman had shown that the spectral lines could be decomposed into sets of lines, *multiplets*, if the radiating atoms were subjected to intense magnetic fields. This experiment had disastrous consequences on various hypotheses of atomic structure which had hitherto appeared to be in accord with experimental observations, for example, that

---

[25] G. Johnstone Stoney, *Trans. Roy. Dub. Soc.* **4**, 583 (1891).
[26] For a more complete account of the birth of the electron, see D. L. Anderson, *The Discovery of the Electron*, Van Nostrand, Princeton, N. J., 1964.

of Kelvin. Concerning his "gyrostatic" model of the atom as an electrified ring, Kelvin himself, in 1899, was to write its epitaph:

> No simplifying suppositions as to the character of the molecule, such as the symmetry of forces and moments of inertia round the axis of the ring, can possibly give Zeeman's normal results of the splitting of a bright line into two sharp lines circularly polarized in opposite directions, when the light is viewed (in a spectrograph) from a direction parallel to the lines of magnetic force; and the dividing of each bright line into three, each plane polarized, when the light is viewed from a direction perpendicular to the lines of force. Hence, although from 1856 till quite lately I felt quite satisfied in knowing that it sufficed to explain Faraday's magnetooptic discovery, I now, in the light of Zeeman's recent discovery, discard my old tempting gyrostatic hypothesis for an irrefragable reason .... .[27]

It was Zeeman's teacher, the Dutch theoretician Hendrik Antoon Lorentz (1853–1928), who provided the first reasonable theory of the phenomenon, and he based it on the electron theory. Later he extended his electron theory to give a physical basis for all of electrodynamics. "The judgment exhibited by him here is remarkable," wrote Sommerfeld in a now classic textbook,[28] "he introduced only concepts which retained their substance in the later theory of relativity." Consider Lorentz' formulation of the force exerted by electric and magnetic fields on a (nonrelativistic) particle of charge e and velocity $\mathbf{v}$

$$\mathbf{F} = e\left(\mathbf{E} + \frac{\mathbf{v}}{c} \times \mathbf{B}\right). \tag{1.11}$$

In the Lorentz theory of the Zeeman effect, one supposes the electron to be held to the atom by a spring (of strength K) and a weak magnetic field $\mathbf{H}(=\mathbf{B})$ applied along the z direction. Thus, by the laws of Newton and Lorentz,

$$m\ddot{\mathbf{r}} = \mathbf{F} = -K\mathbf{r} + \frac{e}{c}\dot{\mathbf{r}} \times \mathbf{H}. \tag{1.12}$$

---

[27] "Mathematical and Physical Papers of W. Thomson, Lord Kelvin," Vol. V, Cambridge, 1911.
[28] A. Sommerfeld, *Electrodynamics*, Academic Press, New York, 1952, p. 236.

The parallel motion $z(t)$ proceeds at the unperturbed frequency

$$z(t) = z(0) \cos \omega_0 t \quad \text{with} \quad \omega_0 = \sqrt{\frac{K}{m}} \tag{1.13}$$

but in the perpendicular direction there is a frequency shift in the amount

$$\Delta\omega \cong \pm\frac{1}{2}\frac{eH}{mc} + O(H^2)(\text{assuming } \Delta\omega \ll \omega_0) \tag{1.14}$$

according to whether the angular motion is clockwise or counterclockwise in the plane perpendicular to the magnetic field. If this is equated to the width of the splittings observed, it yields a value for e/m within a factor of 2 from that which had been established for the cathode rays. This factor was not to be explained for another quarter century and it required both relativity and quantum theory for its explanation.

The thermal properties of magnetic substances were first investigated in a systematic manner by Pierre Curie (1859–1906), who found $\mathcal{M}$ to be proportional to the applied field H. He studied χ, the constant of proportionality, known as the magnetic susceptibility, finding for paramagnetic substances:

$$\chi = \lim_{H \to 0} \frac{\mathcal{M}}{H} = \frac{C}{T}. \tag{1.15}$$

Curie's constant C assumed different values depending on the materials, with T the temperature measured from the absolute zero. In diamagnetic substances χ is negative and varies little with T. In all ferromagnetic materials, he found a relatively rapid decrease of the spontaneous magnetization $\mathcal{M}$ as the temperature was raised to a critical value, now known as the Curie temperature; above this temperature, the ferromagnets behaved much like ordinary paramagnetic substances.[29] While many of these results had been known for one or another material, the scope of his investigations and the enunciation of these general laws gave particular importance to Curie's research.

The diamagnetism was explained a decade after Curie's experiments, in a famous paper[30] by Langevin (1872–1946), as a natural development of Lorentz's electronic theory of the Zeeman effect. As far as Langevin could see, diamagnetism was but another aspect of the Zeeman effect. Without entering into the details of his calculation, we may merely observe that this

---

[29] P. Curie, *Ann. Chim. Phys.* **5** (7), 289 (1895), and *Oeuvres*, Paris, 1908.
[30] P. Langevin, *Ann. Chim. Phys.* **5** (8), 70 (1905), and *J. Phys.* **4** (4), 678 (1905).

phenomenon is apparently already contained in one of the Maxwell equations: Eqs. (1.6, 1.7) indicate that when the magnetic field is turned on, an electric field will result; this accelerates the electron, producing an incremental current loop, which in turn is equivalent to a magnetization opposed to the applied field. Thus *Lenz' Law*, as this is called in the case of circuits, was supposed valid on an atomic scale, and held responsible for the universal *diamagnetism* of materials. *Paramagnetism* was explained by Langevin as existing only in those atoms which possessed a *permanent* magnetic moment. The applied magnetic field succeeded in aligning them against thermal fluctuations. Standard thermodynamic reasoning led Langevin to the relationship

$$\mathscr{M} = f\left(\frac{H}{T}\right) \tag{1.16}$$

where f = odd function of its argument. Then, in weak fields the leading term in a Taylor series expansion of the right-hand side yields Curie's law, without further ado and associated ferromagnetism with the existence of permanent magnetic moments within the material.

1.6.   The Demise of Classical Physics

As we know, the great development of physics in the 20th century was indebted mainly to the invention of quantum mechanics. Progress in the modern understanding of magnetism has depended, to a great extent, on progress in quantum theory. Conversely, the greatest contributions of the theory of magnetism to general physics have been in the field of quantum statistical mechanics and thermodynamics. Whereas understanding in this important branch was limited in the nineteenth century to the theory of gases, the study of *magnetism as a cooperative phenomenon* has been responsible for the most significant advances in the theory of thermodynamic phase transitions. This has transformed statistical mechanics into one of the sharpest and most significant tools for the study of solid matter.

The initial, and perhaps the greatest, step in this direction was taken in 1907 when Pierre Weiss (1865–1940) gave the first modern theory of magnetism.[31] Long before, Coulomb already knew that the ordinary laws of electrostatics and magnetostatics could not be valid on the atomic scale;

---

[31] P. Weiss, *J. de Phys.* **6** (4), 661 (1907).

and neither did Weiss presume to guess what the microscopic laws might be. He merely assumed that the *interactions* between magnetic molecules could be described empirically by what he called a "molecular field." This molecular field H_m would act on each molecule just as an external field did, and its magnitude would be proportional to the magnetization and to a parameter $\mathbb{N}$ which would be a constant physical property of the material. Thus, Weiss' modification of the Langevin formula was

$$\mathscr{M} = f\left(\frac{H + \mathbb{N}\mathscr{M}}{T}\right).$$

(1.17)

If the molecular field were due to the demagnetizing field caused by the free north and south poles on the surface of a spherical ferromagnet, the Weiss constant would be $\mathbb{N} \approx -4\pi/3$ in some appropriate units. In fact, experiments gave to $\mathbb{N}$ the value $\approx +10^4$ for iron, cobalt, or nickel, as could be determined by solving the equation above for $\mathscr{M}(\mathbb{N}, T)$ and fitting $\mathbb{N}$ to secure best agreement with experiment. This is easy to do at high temperature, where f can be replaced by the leading term in the Taylor series development. That is,

$$\mathscr{M} \doteq \frac{C}{T}(H + \mathbb{N}\mathscr{M})$$

(1.18)

predicting a Curie temperature (where $\mathscr{M}/H = \infty$) at

$$T_c = C\mathbb{N}.$$

(1.19)

Above the Curie temperature, the combination of the two equations above yields

$$\chi = \frac{\mathscr{M}}{H} = \frac{C}{T - T_c}$$

(1.20)

the famous Curie-Weiss law which is nearly, if not perfectly, obeyed by all ferromagnets. This agreement with experiment was perhaps unfortunate, for it meant that the gross features of magnetism could be explained without appeal to any particular mechanism and with the simplest of quasi-thermodynamic arguments. Therefore *any* model that answered to those few requirements would explain the gross facts, and a correct theory could be tested only on the basis of its predictions for the small deviations away from the laws of Weiss. It was to be some time before the deviations were systematically measured and interpreted, but the large value of Weiss' constant $\mathbb{N}$ was intriguing enough in his day and pointed to the existence of new phenomena on the molecular scale.

To the mystery of the anomalously large molecular fields was added that of the *anomalous Zeeman effect,* such as that observed when the sodium *D* lines were resolved in a strong magnetic field. Unlike the spectrum previously described, these lines split into quartets and even larger multiplets, which could no more be explained by Lorentz' calculation than the normal Zeeman effect could be explained by Kelvin's structures. Combined with many other perplexing facts of atomic, molecular, and solid-state structure, these were truly mysteries.

A development now took place that should have effected a complete and final overthrow of the Langevin-Lorentz theory of magnetism. It was the discovery of an important theorem in statistical mechanics by Niels Bohr (1885–1962), contained in his doctoral thesis of 1911. In view of the traditional obscurity of such documents, it is not surprising that this theorem should have been rediscovered by others, in particular by Miss J. H. van Leeuwen in her thesis work in Leyden, 8 years later. In any event, with so many developments in quantum theory occurring in that period, the Bohr-van Leeuwen theorem was not universally recognized as the significant landmark it was, until so pointed out by Van Vleck in 1932.[32],[33] Consider the following paraphrase of it, for classical nonrelativistic electrons: *At any finite temperature, and in all finite applied electrical or magnetical fields, the net magnetization of a collection of electrons in thermal equilibrium vanishes identically.* Thus, this theorem demonstrates the lack of relevance of classical theory and the need for a quantum theory. We now proceed to a proof of it.

The Maxwell-Boltzmann thermal distribution function gives the probability that the *n*th particle have momentum $\mathbf{p}_n$ and coordinate $\mathbf{r}_n$, as the following function:

$$dP(\mathbf{p}_1, \ldots, \mathbf{p}_N; \mathbf{r}_1, \ldots, \mathbf{r}_N)$$

$$= \exp\left[-\frac{1}{kT}\mathscr{H}(\mathbf{p}_1, \ldots; \ldots, \mathbf{r}_N)\right] d\mathbf{p}_1 \cdots d\mathbf{r}_N \qquad (1.21)$$

where k = Boltzmann's constant,
$\quad T$ = temperature,
$\quad \mathscr{H}$ = Hamilton's function, total energy of the system.

---

[32] J. H. Van Vleck, *The Theory of Electric and Magnetic Susceptibilities,* Oxford, 1932, p. 104.

[33] Summarizing the consequences of this theorem, Van Vleck quips: "... when one attempts to apply classical statistics to electronic motions within the atom, the less said the better ...."

The thermal average (TA) of any function $F(\mathbf{p}_1, \ldots; \ldots \mathbf{r}_N)$ of these generalized coordinates is then simply

$$\langle F \rangle_{\text{TA}} = \frac{\int F \, dP}{\int dP} \tag{1.22}$$

with the integration carried out over all the generalized coordinates, or "phase space."

Next we consider the solution of Maxwell's equation, (1.8), giving the magnetic field in terms of the currents flowing. As the current density created by the motion of a single charge e_n is $\mathbf{j}_n = e_n \mathbf{v}_n$, the integral solution of (1.8) becomes

$$\mathbf{H}(\mathbf{r}) = \sum_{n=1}^{N} e_n \frac{\mathbf{v}_n \times \mathbf{R}_n}{c R_n^3} \tag{1.23}$$

where $\mathbf{R}_n = \mathbf{r} - \mathbf{r}_n$. This is the law of Biot-Savart in electrodynamics. The quantity $\mathbf{v}_n \times \mathbf{R}_n$ which appears above is closely related to the *angular momentum*, to be discussed in Chapter 3. To calculate the expected magnetic field caused by the motion of charges in a given body, it suffices to take the thermal average of $\mathbf{H}(\mathbf{r})$ over the thermal ensemble dP characteristic of that body. For this purpose it is necessary to know something about Hamilton's function $\mathscr{H}$, also denoted the *Hamiltonian*, for particles in electric and magnetic fields. In the absence of magnetic fields, this function is

$$\mathscr{H} = \sum_n \frac{1}{2} m_n v_n^2 + U(\mathbf{r}_1, \ldots, \mathbf{r}_n) \tag{1.24}$$

where $\mathbf{v}_n = \mathbf{p}_n/m_n = d\mathbf{r}_n/dt$ as a consequence of Lagrange's equations; the first term being the kinetic energy of motion and the second the potential energy due to the interactions of the particles amongst themselves and with any fixed potentials not necessarily restricted to Coulomb law forces. We observe from Maxwell's equation, (1.6), that Newton's second law $e_n \mathbf{E} = \dot{\mathbf{p}}_n$ assumes once more the form in which forces are derivable from potentials if we incorporate the magnetic field by the substitution $\mathbf{p}_n \to \mathbf{p}_n + e_n \mathbf{A}(\mathbf{r}_n, t)/c$. This is in fact correct, and the appropriate velocity for a particle under the effects of a vector potentials is

$$m_n \mathbf{v}_n = \mathbf{p}_n + \frac{e_n}{c} \mathbf{A}(\mathbf{r}_n, t) \,. \tag{1.25}$$

For a more rigorous derivation, see the text by Goldstein.[34] Substitution of (1.25) into (1.24) yields the correct Hamiltonian in a field.

It shall be assumed that a magnetic field as described by a vector potential $\mathbf{A}(\mathbf{r}, t)$, is applied to the particles constituting a given substance, and the resultant motions produce in turn a magnetic field $\mathbf{H}(\mathbf{r})$ as given by (1.23). The thermal average of this field can be computed with three possible outcomes:

(1) $\langle\mathbf{H}\rangle_{\mathrm{TA}}$ has a finite value that is independent of $\mathbf{A}$ in the limit $\mathbf{A} \to 0$. In this case, the substance is evidently a *ferromagnet*.
(2) $\langle\mathbf{H}\rangle_{\mathrm{TA}}$ is parallel to the applied magnetic field $\nabla \times \mathbf{A}$, but its magnitude is proportional to the applied field and vanishes when the latter is turned off. This is the description of a *paramagnetic* substance.
(3) $\langle\mathbf{H}\rangle_{\mathrm{TA}}$ is proportional but *antiparallel* to the applied field. Such a substance is repelled by a magnetic pole and is a *diamagnet*.

The actual calculation is simplicity itself. In calculating the thermal average of $\mathbf{H}(\mathbf{r})$, we note that the convenient variables of integration are the $\mathbf{v}_n$ and not the $\mathbf{p}_n$. A transformation of variables in the integrals (1.22) is then made from the $\mathbf{p}_n$ to the $\mathbf{v}_n$, and according to the standard rules for the transformation of integrals, the integrands must be divided by the Jacobian of the transformation

$$J = \det \left\| \frac{m_n \partial v_n}{\partial p_m} \right\| = 1 \,. \tag{1.26}$$

As is seen, the Jacobian equals unity, and therefore $\mathbf{A}(\mathbf{r}, t)$ simply disappears from the integrals, both from the Maxwell-Boltzmann factor and from the quantity $\mathbf{H}(\mathbf{r})$, which vanishes because it is odd under the inversion $\mathbf{v}_n \to -\mathbf{v}_n$. This completes the proof that the currents and associated magnetic moments induced by an external field all vanish identically.

Therefore, by actual calculation the classical statistical mechanics of charged particles resulted in none of the three categories of substances described by Faraday, but solely in a *fourth* category, viz., substances carrying no currents and having *no* magnetic properties whatever. This stood in stark conflict with experiment.

---

[34] H. Goldstein, *Classical Mechanics*, Addison-Wesley, Reading, Mass., 1951.

1.7. Quantum Theory

In the period between 1913 and 1925 the "old quantum theory" was in its prime and held sway. Bohr had quantized Rutherford's atom, and the structure of matter was becoming well understood. Spatial quantization was interpretable on the basis of the old quantum theory. A famous experiment by Stern and Gerlach[35] allowed the determination of the angular momentum quantum number and magnetic moment of atoms and molecules. The procedure was to pass an atomic beam through an inhomogeneous magnetic field, which split it into a discrete number of divergent beams. In 1911 the suggestion was made that all elementary magnetic moments should be an integer multiple of what later came to be called the *Weiss magneton*, in honor of its inventor. Although the initial ideas were incorrect, this suggestion was taken up again by Pauli in 1920, the unit magnetic moment given a physical interpretation in terms of the Bohr atom, acquiring a new magnitude some five times larger

$$\mu_{\rm B} = \frac{e\hbar}{2mc} = 0.927 \times 10^{-20} \text{ erg/G} \tag{1.27}$$

and renamed the *Bohr magneton*. In 1921, Compton[36] proposed that the electron possesses an *intrinsic* spin and magnetic moment, in addition to any orbital angular momentum and magnetization. This was later proven by Goudsmit and Uhlenbeck, then students of Ehrenfest. In a famous paper in 1925 they demonstrated that the available evidence established the spin of the electron as $\hbar/2$ beyond any doubt.[37] The magnetic moment was assigned the value in (1.27), i.e., twice larger than expected for a charge rotating with the given value of angular momentum. The reason for this was yet purely empirical: the study of the anomalous Zeeman effect had led Landé some two years earlier to his well-known formula for the "gyro magnetic" g factor, the interpretation of which in terms of the spin quantum number obliged Goudsmit and Uhlenbeck to assign the anomalous factor $g = 2$ to the spin of the electron. But a satisfactory explanation of this was to appear within a few years, when Dirac wedded the theory of relativity to quantum mechanics.

Developments in a new quantum mechanics were then proceeding at an explosive rate and it is impossible in this narrative to give any detailed

---

[35] W. Gerlach and O. Stern, *Z. Phys.* **9**, 349 (1922).
[36] A. H. Compton, *J. Franklin Inst.* **192**, 144 (1921).
[37] G. Uhlenbeck and S. Goudsmit, *Naturwiss.* **13**, 953 (1925).

accounting of them. In 1923 De Broglie made the first suggestion of wave mechanics,[38] and by 1926 this was translated into the wave equation through his own work and particularly that of Schrödinger. Meantime, Heisenberg's work with Kramers on the quantum theory of dispersion, in which the radiation field formed the "virtual orchestra" of harmonic oscillators, suggested the power of noncommutative matrix mechanics. This was partly worked out, and transformation scheme for the solution of general quantum-mechanical problem given, in a paper by Heisenberg et al.[39] Born and Wiener[40] then collaborated in establishing the general principle that to every physical quantity there corresponds an operator. From this it followed that Schrödinger's equation and that of matrix mechanics were identical, as was actually shown by Schrödinger[41] also in 1926. It is the time-independent formulation of his equation which forms the basis for much of the succeeding work in solid-state physics and in quantum statistical mechanics, when only equilibrium situations are contemplated, a generalization of $\mathscr{H} = E$ of classical dynamics:

$$\mathscr{H}\left(\frac{\hbar}{i}\nabla_n, \mathbf{r}_n\right)\psi = E\psi \tag{1.28}$$

with the operator $\hbar\nabla_n/i$ replacing the momentum $\mathbf{p}_n$ in Hamilton's function. By radiation or otherwise, a system will always end up with the lowest possible, or ground state, energy eigenvalue E_0. When thermal fluctuations are important, however, the proper balance between maximizing the entropy (i.e., the logarithm of the probability) and minimizing the energy is achieved by minimizing an appropriate combination of these, the *free energy*. This important connection between quantum theory and the contemporary postulational statistical mechanics of J. W. Gibbs was forged from the very beginnings, in the introduction of quantization to light and heat by Planck in 1900, through Einstein's theory of radiation (1907), its adaptation by Debye to the specific heat of solids (1912), and most firmly in the new quantum

---

[38] See *Compt. Rend.*, September, 1923, and also De Broglie's thesis. For this work, he was awarded the 1929 Nobel prize in physics.

[39] W. Heisenberg, M. Born and P. Jordan, *Z. Phys.* **35**, 557 (1926).

[40] M. Born and N. Wiener, *J. Math. Phys. (M.I.T.)* **5**, 84 (1926).

[41] E. Schrödinger, *Ann. Phys.* **79** (4), 734 (1926); C. Eckart, *Phys. Rev.* **28**, 711 (1926). In these references, and in much of the narrative in the text, we follow a truly fascinating account given by Sir E. Whittaker: *A History of the Theories of Aether and Electricity*, Vol. II, Harper & Row, New York, 1960.

mechanics. For Heisenberg's uncertainty principle,

$$\Delta p_n \Delta r_n \gtrsim \hbar \tag{1.29}$$

which expressed the lack of commutativity of $\partial/\partial x_n$ with x_n, gave a natural size to the cells in phase space in terms of Planck's constant $2\pi\hbar$. More pertinently perhaps, the interpretation of processes in quantum mechanics as statistical, aleatory events seemed an important necessary step in applying this science to nature. This required unravelling the properties of ψ.

The importance of the energy eigenfunction ψ in the subsequent development of quantum theory was overriding, and is discussed in some detail in the later chapters of this volume. Here a bare outline will suffice. In 1927 Pauli invented his spin matrices; the next year Heisenberg and (almost simultaneously) Dirac explained ferromagnetism by means of *exchange*, the mysterious combination of the Pauli exclusion principle and physical overlap of the electronic wavefunctions. The next development was to consider the wavefunction itself as a field operator, i.e., *second quantization*. The importance of this was perhaps not immediately recognized. But the quantization of the electromagnetic field by Planck, leading to the statistics of Bose and Einstein, had here its precise analogue: the quantization of the electronic field by Jordan and Wigner[42] in 1928. This incorporated Pauli's exclusion principle and led to the statistics of Fermi and Dirac and to the modern developments in field theory and statistical mechanics. In 1928 also, Dirac[43] incorporated relativity into Hamiltonian wave mechanics, and the first and immediate success of his theory was the prediction of electron spin in precisely the form postulated by Pauli by means of his 2×2 matrix formulation, and an explanation of the anomalous factor $g = 2$ for the spin. The anomalous Zeeman effect finally stood stripped of all mystery. The Jordan-Wigner field theory permitted one loophole to be eliminated from the theory, viz., the existence of negative energy states. These were simply postulated to be filled.[44] The striking confirmation came with the 1932 discovery by C. D. Anderson of the positron — a *hole* in the negative energy Dirac sea.

Side by side with these fundamental developments, Hartree, Fock, Heitler, London, and many others were performing atomic and molecular calculations that were the direct applications of the new theory of the quantum

[42] P. Jordan and E. Wigner, *Z. Phys.* **47**, 631 (1928).
[43] P. A. M. Dirac, *Proc. Roy. Soc.* **117A**, 610 (1928).
[44] P. A. M. Dirac, *Proc. Roy. Soc.* **126A**, 360 (1930).

electron. In 1929 Slater[45] showed that a single determinant, with entries which are individual electronic wavefunctions of space *and* spin, provides a many-electron wavefunction that has the correct symmetries to be useful in problems of atomic and molecular structure. Heisenberg, Dirac, Van Vleck, Frenkel, Slater, and many others contributed to the notions of exchange, and to setting up the operator formalism to deal with it. And so, without further trying to list individual contributions, we can note that by 1930, after four years of the most exciting and brilliant set of discoveries in the history of theoretical physics, the foundations of the modern electronic theory of matter were definitively laid and an epoch of consolidation and calculation based thereupon was started in which we yet find ourselves at the date of this writing. We note that 1930 was the year of a Solvay conference devoted to magnetism; it was time to pause and review the progress which had been made, time to see if "someone ... could tell all about the magnet ... ."

It is not out of place to note the youth of the creators of this scientific revolution, most of them under 30 years of age. None of them had ever known a world without electrons or without the periodic table of the atoms. On the one hand their elders saw determinism and causality crumbling: "God does not throw dice!" Einstein was to complain.[46] But determinism was just the scientific substitute for God's prescience[47] and the elders forgot that long before, in the very foundations of statistical mechanics and thermodynamics, such as had been given by J. Willard Gibbs and Einstein himself, determinism had already been washed out to sea. The alarm about causality was equally unwarranted, for it has retained the same status in quantum theory as in classical theory in spite of all the assults upon it. On the other hand, the young workers only saw in the new theory a chance finally to understand *why* Bohr's theory, *why* the classical theory worked when they did — and also why they failed when they did. They saw in quantum theory a greater framework with which to build the universe, no less than in the theory of relativity; and being devoid of severe metaphysical bias, they did not interpret this as philosophical retrogression. There is a psychological truth in this which had not escaped the perspicacious Henry Adams: "Truly the animal that is to be trained to unity must be caught young. Unity is vision; it must have been part of the process of *learning to see*"[23] (italics ours).

---

[45] J. C. Slater, *Phys. Rev.* **34**, 1293 (1929).

[46] Quoted by J. H. Van Vleck in a talk, "American Physics Becomes of Age," *Phys. Today* **17**, 21 (1964).

[47] We thank Ch. Perelman for this insightful observation.

1.8.   Modern Foundations

At the same time as the theoretical foundations were becoming firmer, the
experimental puzzles concerning magnetism were becoming numerous. Let
us recount these with utmost brevity. Foremost was the question, why is
not iron spontaneously ferromagnetic? Weiss proposed that his molecular
field had various directions in various elementary crystals forming a solid;
thus the magnetic circuits are all closed, minimizing magnetostatic energy.
Spectacular evidence was provided by Barkhausen in 1919, who by means
of newly developed electronic amplifiers, heard distinct *clicks* as an applied
field aligned the various Weiss domains. This irreversible behavior also explained hysteresis phenomena. Measurements of the gyromagnetic ratio gave
$g \cong 2$ for most ferromagnetic substances, showing that unlike in atomic or
molecular magnetism only the spins participated in the magnetic properties
of the solids. In atoms, the mounting spectroscopic data permitted Stoner
to assign the correct number of equivalent electrons to each atomic shell and
Hund to enunciate his rules concerning the spontaneous magnetic moment
of a free atom or ion.

In the study of metals it was found that alloying magnetic metals with
non-magnetic ones resulted in a wide spectrum of technical properties. In
metals, unlike insulators, it was also found that the number of magnetic electrons *per* atom was not, in general, an integer. In many respects, knowledge
of the magnetic properties of many classes of solids was fairly definitive by
1930. Only the class of solids which are magnetic, but non*ferro*magnetic,
ordered structures remained to be discovered, and only two crucial tools of
investigation were lacking: neutron diffraction, and magnetic resonance, each
of which has permitted modern investigators to study solids from within.

The progress of the theory of magnetism in the first third of the twentieth
century can be followed in the proceedings of the Sixth Solvay Conference
of 1930, which was entirely devoted to this topic, and in two important
books: Van Vleck's *The Theory of Electric and Magnetic Susceptibilities*,
Oxford, 1932, and E. C. Stoner's *Magnetism and Matter*, Methuen, London,
1934. In the first of these but one chapter out of thirteen is devoted to
the study of magnetism as a cooperative phenomenon. In the other, barely
more emphasis is given to this field of study; after all, Heisenberg had written "it seems that till now, Weiss' theory is a sufficient basis, even for
the deduction of second-order effects," in his contribution to the Solvay
conference. Nevertheless, some attempts at understanding magnetism as a

collective phenomenon dating from this time were to lay the groundwork for our present understanding of the subject.

First came the Lenz-Ising formulation[48] of the problem of ferromagnetism; spins were disposed at regular intervals along the length of a *one-dimensional* chain. Each spin was allowed to take on the values ±1, in accordance with the laws of Goundsmit and Uhlenbeck. This model could be solved exactly (*vide-infra*), and as long as each spin interacted with only a finite number of neighbors, the Curie temperature could be shown to vanish identically. Did this signify that forces of indefinitely large range were required to explain ferromagnetism?

The introduction by Pauli of his spin matrices established spin as a vector quantity; the requirements that the interaction between two spins be an isotropic scalar and the theory of permutations led Dirac in 1929 to the explicit formulation of the "exchange operator", a scalar quantity that measures the parallelism of neighboring spins, one pair at a time:

$$\mathbf{S}_i{\cdot}\mathbf{S}_j \tag{1.30}$$

as the essential ingredient in the magnetic interaction. The coefficient (usually denoted J) of this operator was a function of the electrostatic force between the electrons, a force so large that the magnetic dipole–dipole interactions (the Amperian forces) could even be neglected to a first approximation. *At this juncture the theory of Weber and Ewing was rendered obsolete.* Quoting Stoner:

> ... As soon as the elementary magnets are interpreted as corresponding to electrons in orbits, or as electron spins, it is found that the magnetic forces between neighbouring molecules are far too small to give rise to constraints of the magnitude required in the Ewing treatment of ferromagnetism. Atoms or molecules may in certain cases have some of the characteristics of small bar magnets — in possessing a magnetic moment; but an interpretation of the properties of a bar of iron as consisting of an aggregate of atomic bar magnets ceases to be of value, whatever its superficial success, once it is known that the analogy between atoms and bar magnets breaks down just at those points which are essential to the interpretation.[49]

---

[48] E. Ising, *Z. Phys.* **31**, 253 (1925).
[49] E. C. Stoner, *Magnetism and Matter*, Methuen, London, 1934, p. 100.

With (1.30) or its equivalent as the basic ingredient in the magnetic Hamiltonian, Bloch[50] and Slater[51] discovered that *spin waves* were the elementary excitations. Assigning to them Bose-Einstein statistics, Bloch showed that an indefinitely large number of them would be thermally excited at any finite positive temperature, no matter how small, in one or two dimensions; but that a three-dimensional ferromagnet possessed a finite Curie temperature. For the magnetization in three dimensions, Bloch derived his "three-halves' power law"

$$\mathscr{M}(T) = \mathscr{M}(0) \left[1 - \left(\frac{T}{T_c} \right)^{3/2} \right] \tag{1.31}$$

with T_c a measurable quantity that also was calculable from the basic interaction parameters. The same calculation leads to $T_c \equiv 0$ in 1D. *Now*, at last, Ising's result, $T_c = 0$, could be understood to have resulted primarily from the one-dimensionality of his array and not at all from the old-quantum-theoretical formulation of the spins.

The properties of metallic conduction electrons were fairly well understood by the 1930's. In 1926 Pauli had calculated the spin paramagnetism of conduction electrons obeying Fermi statistics; and not much later Landau obtained their motional diamagnetism; one of the primary differences between classical and quantum-mechanical charged particles arose in this violation of the Bohr-van Leeuwen theorem. Bloch then considered the Coulomb repulsion among the carriers in a very dilute gas of conduction electrons in a monovalent metal. In the Hartree-Fock approximation, he found that for sufficiently low concentrations (or sufficiently large effective mass, we would add today) this approximation gave lowest energy to the ferromagnetic configuration, in even greater contrast to classical theory. However, Pauli's criticized the calculation:

> Under those conditions (of low concentration, etc.), in fact, the approximation used by Bloch is rather bad; one must, however, consider as proved his more general result, which is, that ferromagnetism is possible under circumstances very different from those in which the Heitler-London method is applicable; and that it is not sufficient, in general, to consider merely the signs of the exchange integrals.[52]

---

[50] F. Bloch, *Zeit. Phys.* **61**, 206 (1930).
[51] J. C. Slater, *Phys. Rev.* **35**, 509 (1930).
[52] W. Pauli, in *Le Magnétisme*, 6th Solvay Conf., Gauthier-Villars, Paris, 1932, p. 212.

Subsequent theories have borne out Pauli's skepticism, but only partly. For example, it is now widely believed that the charged electrons of a metal behave, in most respects, as ideal, noninteracting fermions. However, the *dilute* Coulomb gas of electrons is now known to be ferromagnetic, for other complex reasons. Nevertheless, many detailed studies are still anchored on the simplistic Hartree-Fock ferromagnetism due to the natural reluctance of physicists to abandom the only model of ferromagnetism other than the Heitler-London theory, which could be characterized as truly *simple*. Modern many-body theory took another 20 years to develop (see Chapters 4–9).

In 1932, Néel[53] put forward the idea of *anti*ferromagnetism to explain the temperature independent paramagnetic susceptibility of such metals as Cr and Mn, too large to be explained by Pauli's theory. He proposed the idea of two compensating sublattices undergoing negative exchange interactions, resulting in (1.20) with a *negative* Curie temperature — now known as the Néel temperature, T_N.

In 1936, Slater and Wannier both found themselves at Princeton, one at the Institute for Advanced Studies, and the other at the University. This overlap must have had positive results, for in an issue of the *Physical Review* of the succeeding year there are two consecutive papers of some importance to the theory of magnetic solids. In the first of these, Wannier[54] introduced the set of orthogonal functions bearing his name, of which we shall have much more to say in later chapters. In the second of these, Slater[55] gave a nontrivial theory of ferromagnetism of metals partly based on the use of Wannier functions. He discussed the case of a ferromagnetic metal with a half-filled band which, while least favorable to ferromagnetism, is amenable to analysis (one electron + one hole is *exactly* soluble):

> Starting with the theory of energy bands we have set up the perturbation problem and solved approximately the case of a band containing half enough electrons to fill it, all having parallel spins but one. This problem is a test for ferromagnetism: if the lowest energy of the problem is lower than the energy when all have parallel spins, the system will tend to reduce its spin, and will not be ferromagnetic; whereas if all energies of the problem are higher ... we shall have ferromagnetism.[55]

[53] L. Néel, *Ann. Phys.* (Paris) **17**, 64 (1932); *J. Phys. Radium* **3**, 160 (1932).
[54] G. H. Wannier, *Phys. Rev.* **52**, 191 (1937).
[55] J. C. Slater, *Phys. Rev.* **52**, 198 (1937). It should also be noted that the very first such criteria dated back to Y. Frenkel, *Z. Physik* **49**, 31 (1928).

This followed hard upon his band theory of the ferromagnetism of nickel.[56]

The study of nickel-copper alloys was strong evidence for the band theory of ferromagnetism in metals. A single copper atom dissolved in nickel does not succeed in binding the extra electron which it brings into the metal by virtue of its higher valency. This electron finds its way into the lowest unfilled band, which belongs to the minority spins; and therefore decreases the magnetization of the entire crystal in the amount of one Bohr magneton, homogeneously distributed. As pure nickel possesses 0.6 Bohr magnetons per atom, the magnetization should decrease linearly with copper concentration, extrapolating to zero at 60 percent concentration. The experiments accorded beautifully with this hypothesis.[57]

Continuing study of Weiss magnetic domains soon indicated that they did not necessarily coincide with the physical crystals, that their size was determined by such effects as magnetostriction, quantum-mechanical exchange energy, etc., as well as the magnetostatic energy. The first complete theory (for Co) was sketched by Heisenberg in 1931, and subsequent work by Bloch, Landau, Bozorth, Becker, and many others have laid the experimental and theoretical framework of a theory of domains. Recent formulations by Brown and collaborators, Landau and Ginzburg, and others, have produced the interesting differential equations that regulate the slow distortions in space and time associated with domains, spinwaves, and the like. This field of studies[58] has even acquired a name: micromagnetics, and spawned new industries: magnetic "bubbles," "magnetic disks," etc.

1.9. Magnetic Bubbles

In the 1960's a number of experimental and theoretical studies of strip- and cylindrical domains[59] were followed in 1967 by Bobeck's suggestion of the device potential of the cylindrical domains.[60] This invention by Bobeck, Shockley, Sherwood and Gianola has become known as *magnetic*

---

[56] J. C. Slater, *Phys. Rev.* **49**, 537, 931 (1936); R. Bozorth, *Bell Syst. Tech. J.* **19**, 1 (1940).

[57] E. C. Stoner, *Proc. Roy. Soc.* **169A**, 339 (1939).

[58] A. Hubert, *Theorie der Domänewände in geordneten Medien*, Lecture Notes in Physics, Vol. 26, Springer, Berlin, Heidelberg, New York, 1974, is a recent review. Quantum aspects are first treated in A. Antoulas, R. Schilling and W. Baltensperger, *Solid State Commun.* **18**, 1435 (1976); and R. Schilling, *Phys. Rev.* **B15**, 2700 (1977).

[59] C. Kooy and U. Enz, *Philips Res. Report* **15**, 7 (1960).

[60] A. H. Bobeck, *Bell Syst. Tech. J.* **46**, 1901 (1967).

bubbles; they have been ubiquitous, with applications in telephone switching, computer memory banks — that is wherever a high density, low energy rapidly switchable information storage is required. Today they are mostly obsolete, supplanted by modern electronics. But in the future? As a simple demonstration of micromagnetics we outline some of the basic facts and theory and refer the reader to the monograph by Eschenfelder[61] for supplementary information.

Magnetic bubbles are small, mobile cylindrical domains in thin films magnetized perpendicular to the film surface. Bubble memory chips with capacities of 10^6 bits and data rates of the order of megabits/s are now technological commonplace, with other applications in the offing. Yet the field is so specialized that two otherwise excellent books on thin magnetic films[62] have no mention of this phenomenon at all! Magnetic thin films date to Blois' fabrication of 1000 Å thick permalloy films (80% Ni, 20% Fe), chosen for the low value of magnetocrystalline anisotropy.[63] Magnetostatic considerations favor the magnetization in the plane of the film, and the early literature[62] discussed mainly applications of this type. But already by 1960 there were photographs of bubble domains[64] and attempts at a theory.[59] By 1967 Bobeck and colleagues[60] had invented a bubble device (capable of generating, storing and counting the units) and shortly thereafter a complete theory was developed by *Thiele*.[65] Such a theory must predict the range of parameters (film parameters and thickness, external field parameter) over which bubbles are stable, and take into account the factors affecting their motion when subjected to magnetic field gradients. So first, what are the materials and parameters?

Magnetic rare earth garnets — orthoferrites such as $TmFeO_3$ or $DyFeO_3$ — oriented with c axis perpendicular to the film have very small barriers (as small as 0.01 Oe) against the motion of a domain wall and have been found eminently suitable. Other materials, including amorphous magnetic metal alloys, have been used.

Figure 1.2 shows a demagnetized sample, with contrasting shades indicating domains of opposite magnetization (into or out of the film). In such

---

[61] A. H. Eschenfelder, *Magnetic Bubble Technology*, Springer Series in Solid State Sciences, Vol. 14, Springer, Berlin, Heidelberg, New York, 1980.

[62] M. Prutton, *Thin Ferromagnetic Films*, Butterworth, Washington, 1964; and R. Soohoo, *Magnetic Thin Films*, Harper & Row, New York, 1965.

[63] M. Blois, *J. Appl. Phys.* **26**, 975 (1955).

[64] R. Sherwood et al., *J. Appl. Phys.* **30**, 217 (1959).

[65] A. Thiele, *Bell. Syst. Tech. J.* **48**, 3287 (1969); A. Thiele et al., *Bell Syst. Tech. J.* **50**, 711, 725 (1971).

Fig. 1.2. Magnetic domains in a thin platelet of orthoferrite $TmFeO_3$ in the absence of any magnetic field. A lone "bubble" domain is seen — the strip domains are in stable equilibrium.

a configuration, strip domains are most stable, the sole bubble being the exception. An isolated bubble is always unstable and will grow indefinitely unless an external magnetic field is present, in such a direction as to collapse the bubble. In Fig. 1.2, the bubble is stabilized by the demagnetizing field of the neighboring domains. An external field applied to the demagnetized sample causes the domains in the direction of the field to grow and the others to shrink until they assume the circular shape denoted as bubbles. Increasing the field causes the bubbles to shrink until at a second critical field they disappear totally. This is illustrated in Fig. 1.3 [(a) is no bias,

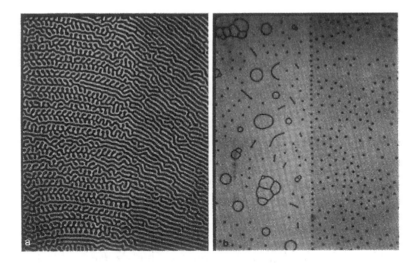

Fig. 1.3. The left-hand sides of (a) and (b) are as grown 6 μm $(YGdTm)_3$ $(GaFe)_5O_{12}$ garnet film, the right-hand sides have been treated by superficial ($\frac{1}{2}$ μm) hydrogen ion implantation to improve the quality of the domains. (a) is in zero bias — note the total absence of bubble domains; (b) is in 170 Oe applied magnetic field, sufficient to produce *only* bubble domains.[60]

(b) is for 170 Oe external field)], which also shows the importance of the proper choice of materials. The material is 6 μm $(YGdTm)_3$ $(GaFe)_5O_{12}$ garnet film, untreated in the left-hand side of the pictures and after hydrogen ion implantation on the right-hand sides.

Thiele's theory was subsequently nicely simplified by Callen and Josephs[66] and it is this version we now present starting with the conditions for static equilibrium.

We consider a film of magnetic material of thickness $h = 0.1$ mm or less. The bubble radius r will turn out to be of approximately the same magnitude as h and both are related to a characteristic length $l \equiv \sigma_w/4\pi M^2$. The magnetization has the magnitude $|M|$ everywhere, pointing in the $+z$ direction (normal to the film) within the domain, and $-z$ direction elsewhere. An applied field $\mathbf{H}$ stabilizes the domain.

The total energy $E_T = E_w + E_H - E_D$ comprises the following: the wall energy,

$$E_w = 2\pi r h \sigma_w \tag{1.32}$$

in which the important parameter is σ_w, the wall energy per unit surface; the interaction with the external field

$$E_H = 2MH\pi r^2 h, \tag{1.33}$$

and the demagnetizing energy of a single cylindrical domain in an infinite plate E_D, given by Thiele[65] in terms of elliptic functions, for which Callen and Josephs found an excellent — and simple — approximation. Before proceeding with the calculation, it is convenient to express all the quantities in dimensionless form: a dimensionless energy $\mathscr{E} \equiv E/(16\pi^2 M^2 h^3)$, a dimensionless field $\mathscr{H} = H/4\pi M$ a dimensionless radius $x = r/h$, and a dimensionless characteristic length $\lambda = l/h = \sigma_w/4\pi M^2 h$.

Then, $d\mathscr{E}_T/dx = 0$, the condition for bubble stability, yields

$$\frac{1}{2}\lambda + \mathscr{H}x - x \bigg/ \left(1 + \frac{3}{2}x\right) = 0 \tag{1.34}$$

where the Callen-Josephs approximation to Thiele's E_D yields the simple term, $-x/(1+3x/2)$. The equation is solved graphically as shown in Fig. 1.4 with the regions of stability being indicated, as derived from the sign of $d^2\mathscr{E}_T/dx^2$. Above a critical field $\mathscr{H}_0 = 1 + 3\lambda/4 - (3\lambda)^{1/2}$ *no* bubble domain

---

[66] H. Callen and R. M. Josephs, *J. Appl. Phys.* **42**, 1977 (1971).

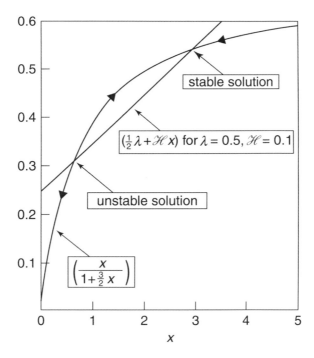

Fig. 1.4. Graphical solution of (1.34). Tilt of straight-line segment depends on external field. Stable and unstable solutions are shown, with the direction of bubble growth or shrinkage shown by arrows.[65]

is stable (the straight line is tangent to the curve). It is similarly possible to calculate a criterion for the transformation of bubbles into strip domains.[65]

In bubble dynamics, one assumes that the rate of change of the energy in the bubble is balanced by drag (the friction force is proportional to the velocity so the frictional loss is proportional to the *square* of $\dot{x}$) and by a second mechanism dependent only on the distance traversed per unit time (i.e., proportional to $|\dot{x}|$). Writing the energy balance per unit length of wall:

$$\left[\frac{\lambda}{2x} + \mathscr{H} - \left(1 + \frac{3}{2}x\right)^{-1}\right] x\dot{x} = \eta^{-1}x\dot{x}^2 - \mathscr{H}_{\mathrm{c}}x|\dot{x}| \tag{1.35}$$

$\mathscr{H}_{\mathrm{c}}$ is the coercive field for wall motion, and η is the wall mobility parameter (in our units of $h/4\pi M$). The equation of bubble dynamics is then

$$\dot{x} = -\eta\left[\left(\frac{\lambda}{2x}\right) - \left(1 + \frac{3}{2}x\right)^{-1} + \mathscr{H} + \mathscr{H}_{\mathrm{c}}\,\mathrm{sgn}(\dot{x})\right] \tag{1.36}$$

determining the rates at which the bubbles can be grown or shrunk.

At the time of writing (2005) Seagate Technology is introducing magnetic disk memories able to store vastly more information than in conventional magnetic disks. They are "novel" in that the individual moments are aligned perpendicular to, rather than parallel to, the surface. This is a case of magnetic bubbles *redux*, in a contemporary context.

For readers anxious to pursue this topic, micromagnetics is studied from a more fundamental point of view in a recent text a well-known theoretical physicist.[67]

1.10.   Ultimate Thin Films

The search for the ultimate thin film led Pomerantz and collaborators to the fabrication of literally two-dimensional ferromagnets.[68] Two types are shown in Fig. 1.5; in either case, the magnetic stratum consists of Mn^{2+} ions.

Despite this *tour de force*, two-dimensional systems stand in peculiar relation to the rest of physics. Difficult of resolution, they seem capable of exhibiting some of the properties of the "real" three-dimensional world, but not all. In any event, they have been of some interest for many years now. In 1942, the chemical engineer, turned theoretical chemist/physicist and mathematician, Lars Onsager developed an exact solution to the problem of Ising spins in a plane, the "two-dimensional Ising model".[69] This work stands as a pinnacle of the achievements of theoretical physics in mid-twentieth century. Onsager's solution yielded the thermodynamic properties of the interacting system, and demonstrated the phase transition at T_c but in a form quite unlike that of Curie-Weiss. In particular, the infinite specific-heat anomaly at T_c is a challenge for approximate, simpler theories to reproduce. Onsager's discovery was not without an amusing sequel. The original solution was given by Onsager as a discussion remark, following a paper presented to the New York Academy of Science in 1942 by Gregory Wannier, but the paper, based on an application of Lie algebras, only appeared two years later.[70] However, his notorious formula for the spontaneous magnetization

---

[67] A. Aharoni, *Introduction to the Theory of Ferromagnetism*, Oxford Univ. Press, Oxford, 2000.

[68] M. Pomerantz, F. Dacol and A. Segmuller, *Phys. Rev. Lett.* **40**, 246 (1978); *Physics Today* **34**, 20 (1981).

[69] L. Onsager, *Phys. Rev.* **65**, 117 (1944).

[70] L. Onsager, *Phys. Rev.* **65**, 117 (1944).

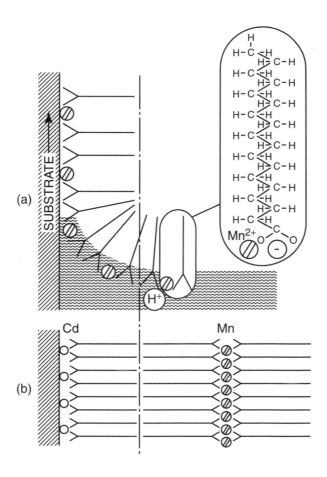

Fig. 1.5.   (a) Schematic diagram of the deposition of a literally 2-D magnet by the Langmuir-Blodgett technique. The hydrophilic substrate (e.g., quartz) is pulled through the water surface. A monolayer of manganese stearate that was floating on the water adheres to the quartz to produce a "Structure I" film. The inset shows the structure of a stearate anion, and a Mn^{2+} ion. (b) Schematic cross section of a 2-D magnetic of structure II. The Mn ions are shown in a plane, but the actual structure may be nonplanar.[68]

below T_c, which requires substantial additional analysis,

$$\mathcal{M} = (1 - x^{-2})^{1/8}, \qquad x = \sinh \frac{2J_1}{kT} \sinh \frac{2J_2}{kT} \qquad (1.37)$$

was *never* published by him, but merely "disclosed". It

>   ... required four years for its decipherment. It was first exposed to
>   the public on 23 August 1948 on a blackboard at Cornell University

on the occasion of a conference on phase transitions. Laslo Tisza
had just presented a paper on The General Theory of Phase Transitions. Gregory Wannier opened the discussion with a question concerning the compatibility of the theory with some properties of the
Ising model. Onsager continued this discussion and then remarked
that — incidentally, the formula for the spontaneous magnetization
of the two-dimensional model is just that (given above). To tease
a wider audience, the formula was again exhibited during the discussion which followed a paper by Rushbrooke at the first postwar
IUPAP (International Union of Pure and Applied Physics) statistical
mechanics meeting in Florence in 1948; it finally appeared in print
as a discussion remark ... [cf. *Nuovo Cimento Suppl.* **6**, 261 (1949).]
However, Onsager never published his derivation. The puzzle was finally solved by C. N. Yang and its solution published in 1952. Yang's
analysis is very complicated ... .[71]

Of course the Ising model is peculiar in that it does not accommodate
spin waves. Bloch had already established that in two dimensions, as well as
in one, the number of spin waves excited at finite temperature is sufficient
to destroy the long range order. In 1966, Mermin and Wagner[72] succeeded
in proving rigorously the absence of long-range order at finite T in both the
Heisenberg ferromagnet and antiferromagnet in one and two dimensions,
i.e., $\mathscr{M} = 0$ at all temperatures. This by itself is, however, not sufficient to
rule out a phase transition at finite temperature T_c! This interesting possibility came out as a result of numerical work by Stanley and Kaplan,[73]
later confirmed by a remarkable theory of the X-Y model due to Berezinskii
and Kosterlitz and Thouless,[74] in which vortices — a topological disorder
— were seen to be an important ingredient, an adjunct to the spin waves
which are the more common excitations of a magnetic substance. The two-
dimensional phase transition which they obtained was quite different from
the usual. Starting from this discovery an entire field of physics has now

---

[71] Quoted from E. Montroll, R. Potts and J. Ward, "Correlations and Spontaneous Magnetization of the Two-Dimensional Ising Model," in Onsager celebration issue of *J. Math.
Phys.* **4**, 308 (1963).

[72] N. Mermin and H. Wagner, *Phys. Rev. Lett.* **17**, 1133 (1966). At $T = 0$, the two-
dimensional ferromagnet has obvious long-range orbit, but this is not so obvious in the
*anti*ferromagnet. Nevertheless, it does; according to H. Q. Lin *et al.*, *Phys. Rev.* **B50**,
12,702 (1994).

[73] H. E. Stanley and T. A. Kaplan, *Phys. Rev. Lett.* **17**, 913 (1966).

[74] J. Kosterlitz and D. Thouless, *J. Phys.* **C6**, 1181 (1973).

developed around the properties of low-dimensional system, principally two-
and one-dimensional (fractional dimensionality being also admitted).

1.11.   Dilute Magnetic Alloys

At first sight, no topic could appear less intellectually challenging than that
of dilute magnetic alloys. Yet some of the great recent conceptual advances of
theoretical physics have had their inception in this seemingly modest subject!

The original studies by Friedel[75] centered about the question of formation
of a local magnetic moment — how do iron or manganese maintain their
atomic magnetic moments when imbedded in a nonmagnetic host matrix
such as copper metal? Anderson then suggested a model[76] with the intention
of resolving or closing this question in a semiquantitative way. Far from it!
Anderson's model, which was not rigorously soluble either, opened an entire
new field of investigation, seeking to explain or predict the occurrence and
non-occurrence of local moments, as in the variety of examples in Fig. 1.6
and Table 1.1.

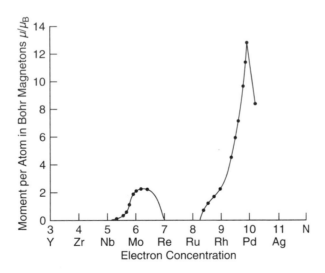

Fig. 1.6.   Magnetic moment on Fe impurity ion as function of electron number of
the host. (From Clogston[77]).

[75] J. Friedel, *Canad. J. Phys.* **34**, 1190 (1956); *J. Phys. Rad.* **19**, 573 (1958).
[76] P. W. Anderson, *Phys. Rev.* **124**, 41 (1961).
[77] A. Clogston, *Phys. Rev.* **125**, 541 (1962).

Table 1.1. Which "magnetic" atoms retain their moments in dilute alloys (adapted from Heeger[78]).

Magnetic	Host Metal			
Species	Au	Cu	Ag	Al
Ti	no	?	?	no
V	?	?	?	no
Cr	yes	yes	yes	no
Mn	yes	yes	yes	no
Fe	yes	yes	?	no
Co	?	?	?	no
Ni	no	no	?	no

If the impurity atom does possess a magnetic moment this polarizes the conduction electrons in its vicinity by means of the exchange interaction and thereby influences the spin orientation of a second magnetic atom at some distance. Owing to quantum oscillations in the conduction electrons' spin polarization the resulting effective interaction between two magnetic impurities at some distance apart can be ferromagnetic (tending to align their spins) or antiferromagnetic (tending to align them in opposite directions). Thus a given magnetic impurity is subject to a variety of ferromagnetic and antiferromagnetic interactions with the various neighboring impurities. What is the state of lowest energy of such a system? This is the topic of an active field of studies entitled "spin glasses," the magnetic analog to an amorphous solid.

When the impurity atoms are very few, in the *very* dilute magnetic alloys, each impurity atom correlates with just the conduction electrons in its own vicinity. This has a surprising consequence: an anomalous resistivity of *high-purity* gold and copper samples has been remarked, in which the resistance *increased* with decreasing temperature. This contradicts the usual behavior, for generally thermal disorder decreases as the temperature is lowered. These anomalies were finally found to be caused by the trace impurities of iron deposited by contact with the steel used in the fabrication.

The resistance mystery was ultimately pierced by J. Kondo in the early 1960's, with his explanation of what has come to be called the "Kondo effect".[79] It occurs only when the magnetic species is *anti*ferromagnetically

---

[78] A. Heeger, *Solid State Phys.* **23**, 283 (1969).

[79] J. Kondo, *Prog. Theor. Phys.* **28**, 846 (1962); *ibid.* **32**, 37 (1964).

coupled with the conduction electrons, causing a scattering resonance at the Fermi level which becomes the sharper the lower the temperature. A completely satisfactory theory of this many-body effect is not yet known, but simple calculations of the ground state and thermodynamic properties have led us to believe that an explanation of the transport anomalies is firmly in place. First, the thermodynamics has been obtained by analytical, numerical and renormalization-group type methods.[80] The symmetry of the ground state — in many cases, a nondescript *singlet* — is known.[81] The thermodynamic and transport properties have been proved, rigorously, to be analytic functions of the temperature by Hepp,[82] despite indications, from the initial calculations of Kondo and his followers, that logarithmic singularities of the form $\log(T/T_K)$ were the *signature* of this phenomenon down to the lowest temperatures.

There have also been breakthroughs. Within months or weeks of each other, in 1980 Andrei[83] in the United States and Wiegmann[84] of the Soviet Union found, independently, almost identical solutions to the Kondo Hamiltonian, which they believe to be exact. These solution are broadly based on a method first proposed by Hans Bethe in 1932 to solve the one-dimensional Heisenberg antiferromagnet, now known as *Bethe ansatz*. Other, more straightforward calculations, are used as examples throughout the present book.

1.12.   New Directions

To quote Dryden, on the Canterbury Pilgrims: "... But enough of this; there is such a variety of game springing up before me, that I am distracted in my choice and know not which to follow ..."

The world of magnetism has been expanding. In biological systems, the minute presence of magnetite — the loadstone — has been found to explain such remarkable properties as the orientational ability of homing pigeons. Magnetostatic *bacteria* are commonly found in pools, marshes and other bodies of water throughout the world. Their orientational mechanism consists of single-domain, nanometersized, magnetite (Fe_3O_4) or greigite (Fe_3S_4),

[80] K. G. Wilson, The renormalization group: Critical phenomena and the Kondo problem, *Rev. Mod. Phys.* **47**, 773 (1975).
[81] D. Mattis, *Phys. Rev. Lett.* **19**, 1478 (1967).
[82] K. Hepp, *Solid State Commun.* **8**, 2087 (1970).
[83] N. Andrei, *Phys. Rev. Lett.* **45**, 379 (1980).
[84] P. B. Wiegmann et al., *Phys. Lett.* **81A**, 175, 179 (1981).

enveloped in protective membranes. This allows them to orient and propel themselves along the earth's magnetic lines of force ($\sim$1 G), the better to seek out regions high in nutrients.[85]

The production and detection of minute magnetic fields has become an industry, thanks to a remarkable development in electrical science, viz., superconductivity. The theory of superconductivity, jointly invented by J. Bardeen, J. R. Schrieffer, and L. N. Cooper in 1957 explained how and why it might be possible to produce infinitesimal magnetic fields in amperian current loops, and devices acronymed SQUIDS and based on the "Josephson effect" (pair tunneling) are now available for this purpose. This many-body theory also introduced the use of field-theoretic methods into condensed matter physics.

Nowadays, we find the traffic of information is sometimes reversed; quantum statistical methods specifically designed for the solution of magnetic problems have become the prototypes even in field theory and particle physics.[86] Nobel prizes have included the above-named theorists of superconductivity, as well as B. Josephson, L. Onsager, A. A. Abrikosov, P. W. Anderson and J. H. Van Vleck, all Nobelists conducting research into the fundaments of magnetism.

The latest discoveries to incorporate magnetism as an integral part involve *high-temperature superconductivity*. Discovered by Bednorz and Müller in 1987 this phenomenon swept through the physics community like a firestorm for a few years,[87] had its own "Woodstock" (the March, 1987 meeting of the American Physical Society), and won its discoverers the Nobel prize. It is now fairly quiescent, but after almost two decades and a few false starts, theory and experiment are starting to come together. Figure 1.7 which follows, summarizes observations by an international group led by Tranquada. These show spin and charge separation and the signature of one-dimensional *antiferromagnets*, the magnon spectrum of magnetic *stripes*, in a high-temperature superconductor.

In the last half of the twentieth century, improvements in alloying magnetic materials (take, for example, superpermalloy; developed in 1947, it

---

[85] M. Posfai *et al.*, *Science* **280**, 880 (1998), R. Dunin-Borkowski *et al.*, *Science* **282**, 1868 (1998); D. A. Bazylinski and R. B. Frankel, *Nature Rev. Microbiol.* **2**, 217 (2004).
[86] J. Kogut, *Rev. Mod. Phys.* **51**, 659 (1979).
[87] For a stimulating account of this social phenomenon see Chapter 9 in G. Vitali, *Superconductivity, The Next Revolution?*, Cambridge Univ. Press, 1993. Note that "high temperature" is a relative term: the phenomenon has not yet reached room temperature . . . .

Cartoons of charge and spin stripes within the Cu-O planes. Only copper sites are represented. Arrows indicate orientations of magnetic moments on copper atoms. Empty circles represent copper sites within the charge stripes. (a) and (b) represent equivalent domains.

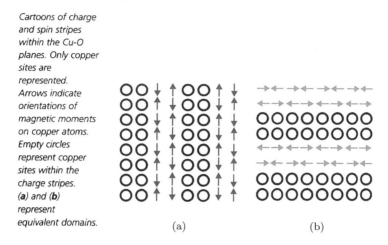

(a) (b)

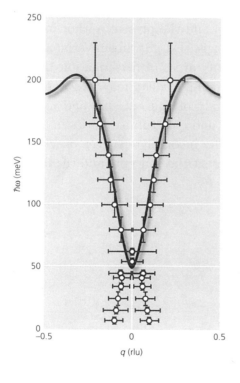

Dispersion of magnetic scattering along a diagonal direction in fig. 2. Symbols indicate experimental results, with bars representing peak widths. The red curve represents the theoretical dispersion of singlet-triplet excitations for an isolated two-leg antiferromagnetic ladder.

Fig. 1.7. Some Magnetic Aspects of a High-Temperature Superconductor.
[Source: Isis 2004 Science Highlights. See J. M. Tranquada et al., *Nature* **429**, 534 (2004).]

contains iron, nickel, molybdenum, etc. in well defined proportions and has a permeability 1000 times that of iron) has been matched by recent progress in the theory and in computational ability. The density-functional theory of Walter Kohn (also a recent Nobelist) and his collaborators, when combined with a local-density approximation (LDA), has made it somewhat simpler to calculate and understand the quantum mechanics of solids from "first principles." One very practical application of these theories is to find a convincing explanation of why certain materials are magnetic and others are not. Finally, a topic that was once in vogue and then disappeared is now back at the cutting edge: magnetic semiconductors. These are the designated engines to drive "spintronics" and other future applications of magnetism in nanotechnology. It is appropriate to recall the first work dedicated to this topic, a book-length article written almost 40 years ago in which the term "giant magnetoresistance" was introduced and in which the "magnetic polaron" was first named and solved in closed form. I am referring, of course, to *Magnetic Semiconductors* by S. Methfessel and the present author in Vol. XVIII/1 of the *Handbuch der Physik*, S. Flügge, Ed., Springer-Verlag, Berlin.

In the next chapter we identify areas of investigation that appear to be fruitful and observe why some simple ideas may not qualify. This analysis naturally leads us to the study of spin, atomic and molecular physics, many-body quantum mechanics and quantum statistics in subsequent chapters. While forming an admittedly incomplete survey of what we know, the chapters that follow serve to present the concepts and the mathematics that permeate the field.

Chapter 2

Currents or Spins?

We bring up in the simplest possible terms some important concepts in magnetic phenomena, while seeking to dispel a few commonly held misconceptions. The topics broached here are treated more thoroughly in subsequent chapters.

2.1. Charge Currents or Spins?

An *insulator* with magnetic properties is a solid (or exceptionally, a liquid) in which the significant internal degrees of freedom are those of quantized atomic or ionic *spins*. A century ago, observation of macroscopic magnetic fields emanating from a solid would have been first-hand evidence of the presence of elementry Amperian charged current loops flowing within the material. Today, one turns to quantum mechanics for a correct explanation. For some insight into this change of perspective, consider the classical current loop illustrated below, in which a current I circulates around an area A.

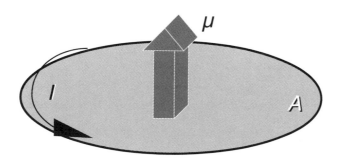

The effective magnetic dipole strength is $\mu = IA/c$. In the limit $A \to 0$ and $I \to \infty$ with the product held constant, all the higher multipole fields vanish, as do the near-field corrections, leaving only the magnetic field of a point-dipole. The current associated with a charge e rotating at velocity $\mathbf{v}$ around a circular track of radius a is $I = \frac{e|\mathbf{v}|}{2\pi a}$, hence the dipole moment is $\mu = I\pi a^2/c = \frac{e}{2mc}L$, where m is the mass, $e/2mc$ is the *classical gyromagnetic* ratio and L is the angular momentum $m|\mathbf{v}|a$.

At first thought, an array of Amperian current loops is indistinguishable from a similar array of permanent magnetic dipoles. It is tempting to construct a theory of ferromagnetism based on a distribution of currents within a solid. However, on further reflection this attribution becomes untenable. "Permanent" implies immutable, whereas array of current loops *must* and *will* wind themselves down to zero — given that in equilibrium, $|\mathbf{v}| \to 0$. The decay of magnetic moments can only be prevented if the individual angular momenta L are, somehow, *quantized* and $\neq 0$.

Indeed, while the concept of an Amperian current flow may be applicable to an induction coil or to an electromagnet, it does not address the source of magnetic fields in "permanent" magnetic materials, many of which are also electrical insulators. How could one justify a persistent *flow of current* in an *insulator*? Even if the material were doped and thereby acquired some free charges, one of the tenets of classical statistical mechanics is that *charge currents do not flow* spontaneously under conditions of *thermodynamic equilibrium*.[1]

Therefore any first-principles explanation for the existence of elementary magnetic dipoles has to be sought elsewhere. There are only two plausible candidates: extrinsic, the electrons' quantized orbital motions and intrinsic, their quantized spins.[2] Let us start with the latter.

The once radical notion that each electron carries an *intrinsic* half unit of the quantum of angular momentum $\hbar$ was originally proposed by Goudsmit and Uhlenbeck[3] for the most practical of reasons. An extra quantum number was required to reconcile observed atomic spectra and in particular, the number of electrons within a closed shell, to the symmetry principles of

---

[1] Bohr-Van Leuween's theorem: originally formulated in the 1911 Ph.D. dissertation of Niels Bohr. See Sec. 1.6 for details.

[2] Spin is an *intrinsic* angular momentum; its magnitude is expressed in units of $\hbar = 1.054\,572\,66 \times 10^{-34}$ J sec.

[3] G. E. Uhlenbeck and S. Goudsmit, *Naturwiss.* **33**, 953 (1925) and *Nature* **117**, 264 (1926).

the nascent quantum theory. This quantum number could only assume two values, $\pm\hbar/2$. A short time later, Dirac[4] deduced this quantization from *spin*, a vector *operator* originating in the relativistic quantum mechanics of *fermions*.[5] Still, electrons need not move at relativistic speeds to exhibit this extra degree of freedom.

According to Dirac's theory, spin angular momentum and a point magnetic dipole associated with it are localized at the instantaneous position of the electron. Thus they are inherent properties of the electron, regardless of whether the electron is attached to a particular atom or ion or whether it is free to move within the entire solid. The *total* "spin" of an atom or ion composed of many electrons has magnitude $S\hbar$, with $S = \frac{1}{2}, 1, \frac{3}{2}, \ldots$, half-integer or integer values, fixed according to the number of participating electrons and the internal structure of the atom or ion, but rarely exceeding $\frac{7}{2}$, its value in some of the rare-earths.

Spin should be distinguished from *motional* angular momentum obtained by summing the orbital angular momenta of the individual particles (labeled i) within an atom or ion, $L\hbar = \sum_i r_i \times p_i$. The *orbital* quantum number L is also quantized but restricted to integer values $0, 1, 2, \ldots$. Actually, it vanishes in the most magnetic of atoms or ions, those with half-filled shells! Moreover, when atoms are embedded in a solid or are made part of certain molecules, orbital rotations of their electrons are easily hindered. This frequently results in an effective $L = 0$, regardless of the original value of L in the atom. By contrast, the spin magnitudes are sturdier and are generally (albeit, not always) insensitive to their nonmagnetic surroundings.

Therefore, to the extent that any such generalization is permitted, it is correct to think of magnetic materials as being composed of high-spin "magnetic" atoms or ions only, drawn typically from the middle columns of the d- or f-transition series of the periodic table. *Non*magnetic ions that may be present in a magnetic material serve various functions: mechanical (to stabilize the material), electronic (affecting its electrical conductivity) or in some instances, providing the "glue" connecting nearby spins with one another (as the oxygen ligand ions do in "superexchange"). Other than in these passive ways, nonmagnetic ions or atoms are irrelevant in determining the magnetic properties of the medium. This now brings us to the question,

---

[4] P. A. M. Dirac, *Proc. Roy. Soc.* **117A**, 610 (1928).
[5] Particles satisfying Fermi-Dirac statistics and the Pauli principle, as discussed elsewhere in this book.

what are the various plausible, or even just possible, mechanisms by which
the spins might interact and then cooperatively align themselves?

2.2.   The Magnetic Dipole

Just as a stationary electron couples to electric fields through its intrinsic
charge e, it couples to magnetic fields through the magnetic dipole associated with its spin. Such a spin S of electronic origin carries a magnetic mo-
ment of magnitude $\mu = g\mu_B S$, where $\mu_B = \frac{e\hbar}{2mc} = 9.274\,015\,4 \times 10^{-24}$ J/T
is the "Bohr magneton," a constant of nature. With $\hbar S$ the spin angular
momentum, a spin's gyromagnetic ratio as calculated from Dirac theory is
$ge/2mc$ — similar to the classical value derived previously — except that
$g = 2.002\,319\,304\ldots$ (The "*Landé g factor*" $g \equiv 2$ in the Dirac theory, but
there are small corrections due to quantum field-theoretic effects and possibly large corrections when the angular momentum $\neq 0$.) The factor 2 is in
excellent agreement with observation and indicates that spin is not just a
current loop shrunk to a point.[6]

Both the spin and its concomitant magnetic dipole transform like *three-
dimensional vector* quantities. Each of the three components of such a vector
is itself a quantum operator that fails to commute with the other two. (By
changing the order in which they occur in a product such as $\mu_x\mu_y$ one changes
the result to $\mu_y\mu_x \neq \mu_x\mu_y$.) According to the tenets of quantum mechanics
only one component out of the three can be measured. It is often convenient
to choose the z-axis for this preferred component, μ_z. This has an important consequence — and this is basically all that we need to know — that
the magnetic dipole moment becomes quantized along this axis and is only
permitted discrete projections along it. For spins S these are $\mu_z = g\mu_B S_z$,
with S_z restricted to a set of $2S + 1$ "eigenvalues" $S, S - 1, \ldots, -S$.[7]

In principle there are many ways to formulate the interactions among
the spins. In Heisenberg-type models all three components of each vector
spin are included in the interactions. The dynamics of Heisenberg-type
models has proved rather difficult to solve and therefore a number of simpler

---

[6] The value quoted here is for a free electron or to atoms with $L = 0$. In general g depends
somewhat on the environment and may even have to be generalized to a tensor quantity.

[7] If S is sufficiently large ($S \gg 1$) this equispaced discrete assignment correctly reduces
to the continuous "correspondence limit" value $S\cos\theta$ with measure $d\theta\sin\theta$; however, in
the extreme quantum limits of $S = 1/2$ and 1, which are in fact the most interesting, the
discreteness *is* significant.

variants have been considered in the literature. In "XY models" (also known as "plane-rotator" models) just *two* components of the spin vectors appear in the Hamiltonian while the third is either discarded or arbitrarily set zero. The symmetry is profoundly modified by this truncation.

In the simplest model of all, the "Ising model" of magnetism[8] or its various generalizations, a z-axis is fixed such that the only degree of freedom available to any given spin is to point "up" or "down" along this common axis. (The "transverse" components along the x- and y- axes and their concomitant magnetic moments are either omitted from the Hamiltonian or set to zero.) Whether physically justified or not, the Ising model has proved to be a wonderful mathematical tool and extraordinarily fruitful in the study of cooperative phenomena. Surprisingly, its most successful applications have been to *nonmagnetic* problems, ranging from the practical — such as the study of forest fires and epidemics, vacancies in solids and the ordering of binary alloys — to the most esoteric, with important applications in "critical phenomena" (phase transitions) and in pure mathematics.

In summary, a magnetic insulator containing "permanent" atomic or ionic spins *can* be fruitfully viewed as an array of N permanent magnetic dipoles $\boldsymbol{\mu}$. These could be aligned to form a maximum total magnetic moment $\mathbf{M} = N\boldsymbol{\mu}$ if the internal forces favor their alignment, or as small as $\mathcal{M} = |\mathbf{M}| = 0$ if the internal forces favor mutual cancellation of the individual magnetic moments. In determining from first principles which tendency carries the day, one must first decide on the nature of the mutual interactions between spins. We start by examining the most obvious, the dipole-dipole interactions, and find it surprisingly inimical to ferromagnetism in three dimensions (3D)!

2.3. The Magnetic Dipole-Dipole Interactions

This section examines the commonsense magnetostatic interactions.[9] Each dipole generates a magnetic field able to exert a torque on the other dipoles. We expect that in the *stationary state* of *lowest energy* (i.e., the ground state), the *net* torque vanishes on each and every dipole. The magnetostatic

---

[8] E. Ising, *Z. Phys.* **31**, 253 (1925).

[9] We find in this instance that common sense comes a cropper. In a subsequent paragraph we shall examine the *exchange* interactions between spins — the name given to a category of mechanisms as deeply rooted in the quantum theory as are the spins themselves — and it is there we find a plausible explanation for the cooperative behavior inherent in ferromagnetism and other significant magnetic phenomena.

interaction energy V of two dipoles at a distance R is a function of 4 variables:

$$V(R, \cos\theta_1, \cos\theta_2, \cos\theta_{1,2}) = \frac{1}{R^3}[\boldsymbol{\mu}_1 \cdot \boldsymbol{\mu}_2 - 3(\hat{R} \cdot \boldsymbol{\mu}_1)(\hat{R} \cdot \boldsymbol{\mu}_2)], \qquad (2.1)$$

where $\mathbf{R} = (X, Y, Z)$, $\hat{R} = \mathbf{R}/R$ is a unit vector, and the dependence on direction cosines is obvious. According to (2.1) the interaction energy ranges from $-2|\boldsymbol{\mu}_1\|\boldsymbol{\mu}_2|/R^3$ to $+2|\boldsymbol{\mu}_1\|\boldsymbol{\mu}_2|/R^3$. To find the ferromagnetic ground state of N such dipoles, let us first orient them all parallel and adjust the orientation in such a manner that *each and every bond V is optimized*.

Consider a linear molecule of N dipoles, all oriented parallel to a common axis we shall define as the z-axis. The line along which they are strung lies at an arbitrary angle θ to the z-axis in the x-z plane, as shown below.

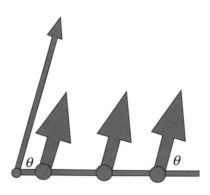

Then each 2-dipole bond is:

$$V(R_{ij}) = \frac{1}{R^3}[1 - 3\cos^2\theta]\mu_i\mu_j = J(R,\theta)S_iS_j. \qquad (2.2)$$

Here $J(R,\theta) = \frac{1}{R^3}[1 - 3\cos^2\theta](g\mu_B)^2$ increases with angle $|\theta|$ and decreases as $1/R^3$ with distance. In the state of least energy (the ground state), clearly $|\theta| = 0$ or π as shown below.

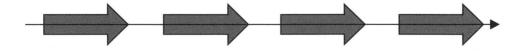

Fig. 2.1.   The ground state of magnetic dipoles constrained to lie along a common molecular axis exhibits an anisotropic form of *ferromagnetism*.

The nearest-neighbor distance between magnetic ions, a_0, the *magnetic lattice parameter*, can be as small as 2–3 Å , i.e. 2–3×10⁻¹⁰ m in a dense magnetic salt. (In dilute magnetic substances it could be far greater.) The interaction parameter $J(m)$ of two spins at a distance ma_0 on this chain $(m = 1, 2, \ldots)$ is $J(m) = -2(g\mu_B)^2/(ma_0)^3$. Thus the total ground state energy of this configuration, the sum of V's in (2.2),[10] is:

$$E_0 = -\frac{N}{2}\left\{2\frac{(g\mu_B S)^2}{a_0^3}\right\}\left\{\sum_{m=1}^{\infty}|m|^{-3} + \sum_{m=-\infty}^{-1}|m|^{-3}\right\}$$

$$= -N\left\{\frac{(g\mu_B S)^2}{a_0^3}\right\} \times 2\zeta(3), \tag{2.3}$$

in which $\zeta(3) = 1.20206\ldots$ is the Riemann zeta function. It is interesting to observe that the nearest-neighbor interactions at $m = \pm 1$ are responsible for $\frac{1}{1.202} = 83\%$ of the binding energy in this instance.[11] The *total* magnetic moment of N spins in this ground state has the maximum possible magnitude, μN. But to benefit from this (lowest) ground state energy (2.3), this ferromagnetic state has to be aligned at $\theta = 0$ or π, as in the illustration.

This is, in fact, what is observed in long thin bar magnets, such as magnetic needles or compass needles. So have we constructed a theory for magnetic needles of iron or steel? Let us examine this question more quantitatively. For a typical $a_0 = 2$ Å, Eq. (2.3) yields an energy *per* site $E_0/N = -(2S)^2 \times 2.5 \times 10^{-5}$ eV $\approx 0.3(2S)^2 k_B$ — too small by *any* measure![12] Let us investigate the effects of finite temperature T.

Excited states of ferromagnets serve to lower the total magnetization; with increasing temperature they ultimately destroy it. For example, suppose we twisted all dipoles past the nth by 180° so the first n have $\theta = 0$ and the remaining $N - n$ have $\theta = \pi$. The total energy exceeds the ground state (2.3) by a finite amount only, that we can estimate as $\Delta \approx 2|E_0/N| = 0.6(2S)^2 k_B$.

According to thermodynamics, the probability of one or more such breaks increases with temperature approximately as $N \exp -\Delta/kT$. Therefore once

---

[10] Note: we include a factor $1/2$ to avoid double-counting bonds. An alternative is to compute the total induced field energy $\frac{1}{8\pi}\int d^3 r B^2(t)$ and add to it the interaction energy of the magnetic field with the dipoles.

[11] This suggests a great simplification in two- or three-dimensional arrays: for purposes of *estimating* energies, the dipole-dipole interactions can be treated as being effectively short-ranged and the sums restricted to nearest-neighbors.

[12] With $k_B T \approx 2.5 \times 10^{-2}$ eV at room temperature (300 K), the interaction energy *per* site for spins $1/2$ thus corresponds to a mere 0.3 K.

T exceeds $\approx 0.6(2S)^2$ on the Kelvin scale, *there can be no long-range order.* In ordinary solids of iron, cobalt, chromium, manganese, and their oxides, $S = O(1)$, and this characteristic temperature is $O(1 \text{ or } 2 \text{ K})$ (one or two degrees Kelvin).

Hence we must look elsewhere if we are to understand the observed ferromagnetism of steel needles at room temperature (some 300 K). But supposing individual spins exceeded $S > 10$, could there be hope that dipolar room-temperature ferromagnetism exists? Indeed, such large spins *are* found in magnetic organic molecules. For example, the large, three-dimensional Fe8 molecule shown below (only the gridwork of Fe ions is shown: the organic framework is omitted) in which 6 iron atoms are parallel to an applied field and 2 are antiparallel, resulting in a total spin $S = 4 \times 5/2 = 10$. But here the physical *size* of such molecules ($a_0 \approx 10$ to 20 Å) decreases the relevant V by a factor 10^{-2} to 10^{-3} and the dipolar forces turn out to be even smaller than before!

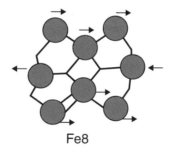

Fe8

Fig. 2.2. [Schematic adapted from S. J. Blundell and F. L. Pratt, *Organic and Molecular Magnets, J. Phys.: Conden. Matter* **16**, R771–828 (2004).]

Next, consider $\theta = \pi/2$, spins pointing transverse to the chain. The energy of the ferromagnetic state becomes a maximum! Instead, we now find an *entirely different ground state* illustrated below. The optimum configuration for this orientation requires nearest-neighbor dipoles to be *antiparallel*: all even-numbered spins are "up" and odd-numbered spins "down," or *vice versa*. In 3D such antiparallel configurations are known as *Néel states* and materials in which the Néel state is the ground state are "*antiferromagnets.*"[13]

---

[13] Louis Néel was awarded the 1970 Nobel prize for this work from the 1930's, in which he anticipated the existence of antiferromagnetism — found experimentally by H. Bizette, C. Squire and B. Tsai, *Comptes Rendus Acad. Sci.* **207**, 449 (1938) and ultimately confirmed using neutron scattering, by C. G. Schull and J. S. Smart, *Phys. Rev.* **76**, 1256 (1949).

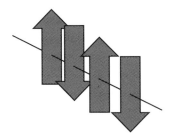

Fig. 2.3. Constrained $\perp$ to the axis, linearly arrayed spins exhibit *antiferro-magnetism*.

The total magnetic moment in the Néel state is zero. In the present example the energy is:

$$E_0 = +\frac{N}{2}\left\{\frac{(g\mu_\text{B}S)^2}{a_0^3}\right\} 2\sum_{m=1}^{\infty}(-1)^m|m|^{-3}$$

$$= -N\left\{\frac{(g\mu_\text{B}S)^2}{a_0^3}\right\} \times \frac{3}{4}\zeta(3)\,. \tag{2.4}$$

Although appropriately negative, the energy of this state (the Néel configuration) adds up to only 3/8 of the preceding value. Even though this is not the minimum energy, there is no net torque on any one dipole and this state remains *stationary*, i.e. it is a *local* energy minimum.[14] But whether this state is stable against *quantum* fluctuations remains to be seen. Generally, if there exists any channel for decay, the quantum states can and do "tunnel" into the more probable configurations of equal energy.

Next, extend the analysis to the simple cubic 3D lattice. Here the lattice sums are rather more difficult (but not impossible) to evaluate. However, we noted that dipolar interactions fall off quickly so that, judging from the preceding examples, if we just retained nearest-neighbor interactions the results should be accurate to within 20%.

Assume there is an ion on every point of a Cartesian grid $(n_x, n_y, n_z) \times a_0$ where the n_j are integers.

First consider the ferromagnetic state, in which the magnetic z-axis is parallel to one of the principal axes of the crystal and all spins are "up." If at first we restrict the interactions to nearest-neighbors, a spin at grid

---

[14] That is, if the spins were appropriately stimulated to surmount the energy barriers that keeps them antiparallel, they would ultimately re-align parallel to the direction of the chain and parallel to one another, as before, achieving a global minimum energy.

point (n_1, n_2, n_3) shares an attractive bond of strength $-2\{\frac{(g\mu_B S)^2}{a_0^3}\}$ with two neighbors at $(n_1, n_2, n_3 \pm 1)$, and a repulsive bond of strength $+\{\frac{(g\mu_B S)^2}{a_0^3}\}$ with each of four intraplanar neighbors, situated respectively at $(n_1 \pm 1, n_2, n_3)$ and $(n_1, n_2 \pm 1, n_3)$. *The sum total of nearest-neighbor interactions is zero.*

The same holds as well for the long-distance interactions in the ferromagnetic state. To see this, approximate lattice sums by three-dimensional integrals (as is justified at long distances). The angular average of $\cos^2 \theta$ over a unit sphere is $\frac{1}{3}$ and *each* interaction energy V in (2.1) vanishes on average. *Exact* lattice summation does not challenge this conclusion: the ferromagnetic state is indeed inherently unstable.

If, as we have seen, in an isotropic medium the dipole-dipole interaction *does not produce* a ferromagnetic ground state in 3D, then what does? And what *is* the ground state that results from purely dipole forces? One candidate for the lowest energy has each spin orientation "up" if $n_1 + n_2 =$ even integer, and "down" if $n_1 + n_2 =$ odd integer, independent of n_3. This describes a Néel state within each x-y plane but assumes all x-y planes to be parallel (identical). (In the true 3D Néel state, the spins in the vertical direction would *also* be staggered, but then the energy is zero once again!) Retaining only nearest-neighbor bonds we estimate the ground state energy in this *columnar* state. Noting that each bond is shared by 2 sites we divide the sum by 2:

$$E_0 = -\frac{1}{2} 8N \left\{ \frac{(g\mu_B S)^2}{a_0^3} \right\}, \tag{2.5}$$

with N the total number of sites. This configuration is clearly more favorable than either the Néel state or the ferromagnetic state. But is it sufficiently "sturdy" against small fluctuations or thermal excitations?

Choosing for the lattice parameter a typical value $a_0 \approx 2$ Å, E_0/N as calculated above yields $O(1k_B)$ *per* site.[15] For $S = \frac{1}{2}$, at all temperatures exceeding 1 K, thermal fluctuations allow the dipoles to re-orient, destroying any remnant ground state correlations. Such crude estimates rule out magnetostatics as a credible mechanism for *any* of the cooperative forms of magnetism at any temperature $T > O(1)$.

---

[15] For $S = \frac{1}{2}$. (For spins $S = \frac{7}{2}$, such configurations might be worth investigating at very low temperatures.) Here and throughout, $k_B = 1.380\,658 \times 10^{-23}$ Joules/K is Boltzmann's constant.

Still, magnetic dipolar forces do play a role in modifying the proper-ties of magnetic materials, in domain formation, and in the *shape-dependent anisotropy* of permanent magnets. We shall take this topic up again in the next Section.

2.4. The Exchange Interactions

Exchange "forces" are not forces in the usual sense. The energy splitting between quantized parallel and antiparallel spin configurations must be com-puted by some independent means and parametrized by a *scalar* "exchange constant" *J*. In the simplest instances where this splitting can be calculated explicitly, it is the result of a concatenation of two effects: the Pauli exclu-sion principle (affecting only electrons of parallel spin), and their two-body electrostatic repulsion. Although here we can anticipate the consequences of such a mechanism, details are more conveniently postponed to a later chapter.

There are two principal categories of exchange interactions: the *intra*-atomic interactions which favor spin alignment and are responsible for the operation of Hund's rules in unfilled atomic shells, and the *inter*-atomic interactions which govern relative spin alignment in neighboring ions or atoms. Both are canonically expressed through similar *scalar* two-body interaction Hamiltonians of the form

$$H_{ij} = J_{ij} S_i \cdot S_j , \tag{2.6}$$

where the $J_{ij}(R_{ij})$ are typically independent of the *orientation* of $\mathbf{R}_{ij}$ relative to any fixed axis. Both this isotropy and the strength of the J_{ij} (which typically range upward of $O(10^3 \text{ K})$) distinguish the exchange interactions from the dipolar. However, the sign of J_{ij} is not always negative, favoring parallel spins. In many instances it is positive and favors antiparallel spins, or it may even oscillate with distance R_{ij}, as is the case in metals.

For the sake of argument let us assume that the spins occupy fixed points on a regular lattice and that those $J(R_{ij})$ which are not zero (e.g., the nearest-neighbor interactions) are *all negative*. Under this assumption it is possible to prove rigorously that the ground state, *in any geometry and in any dimension, is* ferromagnetic.

The proof is immediate: if all the spins are parallel $\mathbf{S}_i \cdot \mathbf{S}_j = S^2$. If the J's are all nonpositive, the energy $\sum_{(ij)} J(R_{ij})S^2$ has mathematically the *lowest possible value*, i.e. it is the ground state. The total spin is then $S_\mathrm{T} = NS$

and it is macroscopic. Its $2S_T + 1$ quantized projections along any arbitrary z-axis range from $-NS$ to $+NS$; all projections have the same energy.

In his 1942 Ph.D. thesis at M.I.T., J. M. Luttinger pointed out the following amusing generalization. Suppose the discrete Fourier transform of $J(R_{ij})$, $\sum_{(ij)} J(R_{ij})e^{iq \cdot R_{ij}} = j(q)$, has a minimum at some point q_0 then among the ground state configurations one must include the spiral configuration $S_i = S(\cos q_0 \cdot R_i, \sin q_0 \cdot R_i, 0)$. The ferromagnetic state is the limiting case of this spiral at $q_0 = 0$. If there exist several values of q_0 that minimize $j(q)$, then more complicated phases may be manifest. For example, if $J(R)$ is positive and restricted to nearest-neighbors on a simple cubic lattice, the Néel state is the ground state (as the reader should be able to show). Ideally, the total magnetization $M = Ng\mu_B S$ in the ferromagnetic ground state (a vector quantity), is free to point in *any* direction in a "magnetically soft" material. Because the magnetostatic energy vanishes in the ferromagnetic

Fig. 2.4.   Domains on the Surface of Cobalt.
Different shadings reveal different axes of magnetization. Width of individual domains ranges from 10^4 Å to 10^5 Å.

configuration, commonsense dictates that magnetostatics should not enter into the ferromagnetic states nor into their dynamic excitation spectrum.

But it does! We recall that in a cubic medium the magnetostatic energy *can* be lowered for configurations in which the dipoles are antiparallel to one another. This small but not insignificant reduction in magnetostatic energy is achieved through the formation of *magnetic domains*. Spins within a given domain are aligned ferromagnetically along a given local *z*-axis, but as the *z*-axes of nearest-neighbor domains are not co-linear all flux lines close in the optimal state and the *net* magnetization of the ferromagnet actually vanishes. A number of domains are visible in the Fig. 2.4, a microphotograph of the surface of a generic cobalt crystal.

A notable exception to spontaneous domain formation is the long, thin, essentially one-dimensional, cylinder or needle. Preceding considerations have made it plausible that if the exchange interactions are negative, individual dipoles align along the long axis of the cylinder, thereby optimizing *both* exchange *and* magnetostatic energies in the ground state. In fact this shape for permanent bar magnets, with north and south poles lying at opposite ends of a long bar, "works" and has been traditional since ancient times. The picture below, purporting to show the lines of force around a cylindrical magnet, is taken from William Gilbert's *De Magnete*, a lengthy and learned monograph first published in London in 1600, arguably the first printed textbook in any branch of modern physics, discussed earlier in Sec. 1.2.

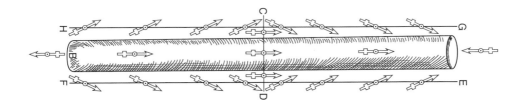

2.5.   Metals and Their Alloys

Magnetism in metals poses a number of special challenges. If the electrons transport their spins by appreciable distances a description based on localized spins is patently inappropriate.

In ordinary nonmagnetic metals the kinetic energy of motion is minimized when half the electrons have spin "up" and the other half spin "down." An external magnetic field pointing "up" promotes an excess of "up" over "down" electrons; a magnetization of magnitude $M = g\mu_B(N_\uparrow - N_\downarrow)$ ensues.

The gain (lowering) in energy $-B \cdot M$ occurs at a cost $+\frac{X}{2}(N_\uparrow - N_\downarrow)^2$ in total kinetic energy. X, a parameter characteristic of the metal and independent of T, has dimension of energy — i.e., $X \propto 1/\rho_F$, where ρ_F is defined as the "density of states" ($\rho_F de$ = number of distinct electronic energy levels between e_F and $e_F + de$). Minimizing the total energy *w.r.* to $(N_\uparrow - N_\downarrow)$ yields $M = (g\mu_B)^2 B/X$. We call this property, that the total energy is lowered in a magnetic field, a *paramagnetic* response. The *paramagnetic susceptibility* of the metal, defined as the ratio of M to B, is a positive quantity which, in metals, is proportional to ρ_F.

A ferromagnet such as iron builds up its *spontaneous* magnetization *via* internal atomic "exchange" forces that mimic the action of an external field. However, the precise mechanism that produces the exchange interaction is not entirely straightforward. Some magnetic metals and alloys have been dubbed "half-magnets," as it appears that the majority spin-up electrons are immobile while the minority of spin down electrons are itinerant! We discuss this paradoxical situation in due course.

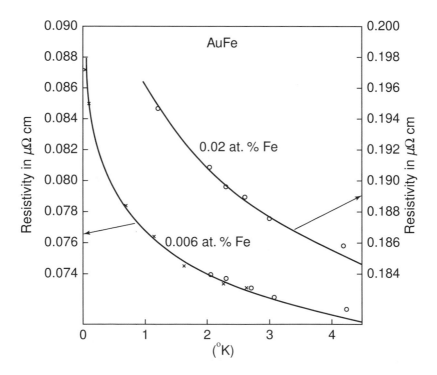

Fig. 2.5.   Resistivity of dilute Fe alloys, as function of T.

Surprisingly, even a *single* magnetic impurity affects the properties of a nonmagnetic metal, as was first observed in extremely dilute alloys of manganese (Mn) or iron (Fe) in the nonmagnetic metals copper (Cu) and gold (Au). In the preceding figure we show the experimentally observed[16] excess electrical resistance of nonmagnetic gold, after it is doped by iron impurities to concentrations of 0.02 and 0.006 atomic %. The excess resistivity that appears at low T is the so-called "Kondo effect," a condensation of conduction electrons' spins to screen the localized moment, *a phenomenon that becomes the more remarkable the greater the dilution* and *the lower the temperature.*

The Kondo effect could only be understood on the basis of many-body theory and therefore it challenged theorists for two decades starting from the mid-1960's. Together with the study of screening in Coulomb interactions among electrons in metals and of arcane phenomena associated with superconductivity, the study of the Kondo effect helped drive theoretical and experimental research along unprecedented directions during this period. In fact, the Kondo effect did not yield center stage of magnetic studies until the mid-1980's, when first the *quantum Hall effect*, followed by *high-temperature superconductivity*, then *giant-* (and even *colossal-*) *magnetoresistance*, entered the scene.

At the same time that the mysteries surrounding the dilute Kondo effect were being unraveled the study of more concentrated magnetic alloys revealed equally puzzling features.

At magnetic concentrations exceeding $O(1\%)$, the magnetic impurity atoms interact with one another, with nontrivial interaction energy. If they are disposed at random, a new phase — the "spin glass" — appears. The mechanism for their interaction is most interesting. Each individual magnetic impurity polarizes the metallic medium in its immediate vicinity. The response of these electrons is not monotonic, but exhibits "Friedel oscillations." That is, when measured or calculated as a function of distance R from a magnetic impurity, the polarization of the electron gas behaves asymptotically approximately as,

$$\sigma(R) \approx J_{\text{eff}} \frac{\cos(2k_{\text{F}} R + \varphi)}{R^3} \tag{2.7}$$

with J_{eff} a lumped constant, φ a phase shift, and k_{F} the characteristic wavevector at the Fermi surface of the metal (inversely proportional to the de

---

[16] Adapted from work by D. K. C. MacDonald, W. B. Pearson and I. M. Templeton.

Broglie wavelength of a conduction electron at the Fermi surface). The interaction of a spin S_j at a distance R_j from a spin S_0 at the origin takes the form
$\sigma(R_j)S_j \cdot S_0$, where σ can be of either sign: positive (antiferromagnetic) or
negative (ferromagnetic) depending on the phase $2k_F R_j + \varphi$.

The formula (2.7) is not applicable at small distances nor does it take into
account the Kondo effect mentioned previously. Nevertheless it signals that
in a dilute alloy in which the nearest-neighbor R is random, the spin-spin
interaction has a frozen-in *random* sign. There results a cooperative spin
"glass" in which the spins *can* become fixed at low temperatures, after
assuming one of possibly many metastable random configurations mandated
by the set of frozen-in random bonds. Observed properties shared by all
spin glasses include: electrical resistivity dependent on magnetic ordering,
hysteresis upon warming and cooling, and a specific heat linear in T at low
temperature. When a spin glass is cooled in an external magnetic field B
and the field is subsequently removed, the remnant magnetization depends
on the sign and magnitude of B. Spin glasses in their low-temperature phases
are in apparent violation of Nernst's theorem.[17]

2.6.   Superconductivity

Arguably the most remarkable discovery of the recent decades occurred
in 1987 when perovskites containing planes of the antiferromagnetic oxide
CuO_2 were found to exhibit *superconductivity* at temperatures $O(100 \text{ K})$
after appropriate doping. A race is now on to find new materials that super-
conduct at or above room temperature. Were this pursuit successful it would,
of course, extend the reign of quantum physics to room temperature while
enabling the implementation of dramatically new technologies — creating
an industrial "quantum" revolution with unimaginable consequences.

The search for a convincing theoretical explanation of high-temperature
superconductivity (HTS) has finally narrowed to some mechanisms virtually
identical to those that govern ordinary magnetic materials. There is agreement on general structural and electronic features, but not on what are the
precise mechanisms (superexchange?) nor even on the primordial role of the
oxide CuO_2 (is it unique in its properties?) in promoting the *pairing* of
electrons into Cooper pairs.

---

[17] Also known as the "third law of thermodynamics," it asserts that the ground state of
all substances is essentially unique, therefore entropy vanishes in the limit as $T \to 0$.

For example, suppose we postulated the emission of a magnon by a mobile electron and its subsequent re-absorption by a second, correlated electron of opposite spin. Is this similar to the "retarded phonon-exchange mechanism" that causes pairing the classic low-T_c superconductors, or is it more akin to quantum electrodynamics, in which the virtual exchange of photons by two equally charged objects results in an effective Coulomb *repulsion, not* attraction? At the present time there exist many conflicting opinions and theories for this most vital phenomenon but none has gained general acceptance nor confirmation.

2.7. The Need to Study Spin Angular Momentum

In many aspects of condensed matter physics and in some emerging fields such as "spintronics," magnetic phenomena play a prominent role — but one that cannot even be assessed without an accurate understanding of the underlying electron physics. In developing an adequate theoretical framework for what turns out to be a complex many-body problem we shall first turn to small systems that can be solved in closed form (including familiar examples taken from atomic and molecular physics), for clues. In so doing, it will be necessary to reëxamine what is meant by angular momentum, spin, etc., using the language of quantum mechanics. The following chapter deals with this topic. Chapter 4 surveys some uses to which spin polarized electrons may be put.

Chapter 3

Quantum Theory of Angular Momentum

This chapter affords a brief summary of the quantum theory of angular momentum. As the magnetization of a charged particle is proportional to its angular momentum, this subject is at the core of the theory of magnetism. We shall show that motional angular momentum is inadequate and introduce spin angular momentum. We develop operator techniques expressing angular momentum or spin operators in terms of more primitive fermion or boson operators. The topics of spin one-half and spin one are treated individually, for use in subsequent chapters on the theory of magnetism.

3.1. Kinetic Angular Momentum

In the absence of any external forces, the classical angular momentum of a point particle has a constant value

$$\mathbf{L} = \mathbf{r} \times \mathbf{p} \tag{3.1}$$

and therefore in quantum theory the analogous quantity, which shall be denoted "kinetic angular momentum" is an operator,

$$\mathbf{L} = \mathbf{r} \times \frac{\hbar}{i} \nabla \tag{3.2}$$

according to the usual rule for constructing quantum-mechanical momenta: $p_x = \hbar/i(\partial/\partial x)$, etc. *The angular momentum associated with a wavefunction ψ can be easily determined if this function is spherically symmetric* so that it does not depend on the angular coordinates φ and θ, but only on the magnitude of the radius vector $\mathbf{r}$. For in that case we can prove that $\mathbf{L} \equiv 0$ by noting that

$$\nabla\psi(r) = \hat{\mathbf{u}}_r \frac{d}{dr}\psi(r) \tag{3.3}$$

where $\hat{\mathbf{u}}_r$ = unit vector in the radial direction = $\mathbf{r}/r$, and therefore,

$$\mathbf{L}\psi(r) = \frac{\hbar}{i}(\mathbf{r} \times \hat{\mathbf{u}}_r)\frac{d}{dr}\psi(r) \equiv 0. \tag{3.4}$$

A function with *nontrivial* angular dependence has *nonvanishing* kinetic angular momentum, although it is not always so simple to discover its magnitude. Even if ψ is an eigenfunction of $\mathbf{L}$, this requires the solution of an eigenvalue problem; and moreover, the three components of $\mathbf{L} = (L_x, L_y, L_z)$ do *not* commute with one another and therefore cannot be simultaneously specified. Briefly, the commutator is:

$$\begin{aligned}
[L_x, L_y] &\equiv L_x L_y - L_y L_x \\
&= \frac{\hbar}{i}\frac{\hbar}{i}\left[\left(y\frac{\partial}{\partial z} - z\frac{\partial}{\partial y}\right)\left(z\frac{\partial}{\partial x} - x\frac{\partial}{\partial z}\right)\right. \\
&\quad \left. - \left(z\frac{\partial}{\partial x} - x\frac{\partial}{\partial z}\right)\left(y\frac{\partial}{\partial z} - z\frac{\partial}{\partial y}\right)\right] \\
&= \hbar^2\left(x\frac{\partial}{\partial y} - y\frac{\partial}{\partial x}\right) \\
&= i\hbar L_z.
\end{aligned} \tag{3.5}$$

With the other two components obtained by cyclic permutations of the above, we find a set of relations which can most conveniently be described by the vector cross product, as given by the usual determinantal rule

$$\mathbf{L} \times \mathbf{L} \equiv \begin{vmatrix} \hat{\mathbf{u}}_x & \hat{\mathbf{u}}_y & \hat{\mathbf{u}}_z \\ L_x & L_y & L_z \\ L_x & L_y & L_z \end{vmatrix} = i\hbar\mathbf{L}. \tag{3.6}$$

While it is true that this rule was obtained by the definition of angular momentum given in (3.2), it is in fact more general. There are operators (identified later) which obey (3.6) but not (3.2). It is therefore important that we define angular momentum as follows: *A vector operator is a genuine angular momentum operator (whether kinetic, spin, or generalized angular momentum) only if it satisfies (3.6) or the equivalent set of Eqs. (3.10–3.16).*

Example: If $\mathbf{L} = (L_x, L_y, L_z)$ is an angular momentum, then $\mathbf{L}' = (-L_x, -L_y, -L_z)$ is not. This is related to the pseudovector nature of the classical angular momentum $\mathbf{r} \times \mathbf{p}$, and is also directly a consequence of (3.6).

Returning to the motional angular momentum, express it in spherical polar coordinates defined by r, θ and ϕ:

$$\mathbf{r} = (r \sin \theta \cos \phi, \ r \sin \theta \sin \phi, \ r \cos \theta)$$

so that the cartesian components of $\mathbf{L}$ take the form

$$L_z = \frac{\hbar}{i} \frac{\partial}{\partial \phi} \tag{3.7}$$

$$L^+ = \hbar e^{i\phi} \left(\frac{\partial}{\partial \theta} + i \cot \theta \frac{\partial}{\partial \phi} \right) \tag{3.8}$$

$$L^- = \hbar e^{-i\phi} \left(-\frac{\partial}{\partial \theta} + i \cot \theta \frac{\partial}{\partial \phi} \right). \tag{3.9}$$

Here, for future convenience, we introduced the operators

$$L^\pm \equiv L_x \pm iL_y. \tag{3.10}$$

But if desired, the Cartesian components may be extracted by (what is equivalent to the last equation):

$$L_x = \frac{1}{2}(L^+ + L^-) \quad \text{and} \quad L_y = \frac{1}{2i}(L^+ - L^-). \tag{3.11}$$

The $\pm$ operators are the more useful, though, and have the names of angular momentum *raising* $(+)$ and *lowering* $(-)$ operators, respectively. The reason for this nomenclature is to be sought in the commutation laws for these operators. Because it is possible to derive these commutators using only (3.6), without making use of the actual differential forms for L_z, etc., given above, the following equations are generally valid for *any* angular momentum:

$$[L_z, L^+] = \hbar L^+ \tag{3.12}$$

and

$$[L_z, L^-] = -\hbar L^-. \tag{3.13}$$

Given an eigenfunction Ψ_m of L_z with eigenvalue m, that is,

$$L_z \Psi_m = \hbar m \Psi_m$$

let us apply both sides of (3.12) to this eigenfunction and after rearranging terms, obtain

$$L_z(L^+\Psi_m) = \hbar(m+1)(L^+\Psi_m) \,.$$

This shows that $L^+\Psi_m$ is an eigenfunction of L_z belonging to eigenvalue $m+1$. It is similarly shown that $L^-\Psi_m$ belongs to eigenvalue $m-1$, whence the terminology of "raising" and "lowering" operators. A final commutation law which can be obtained from (3.6) is

$$[L^+, L^-] = 2i[L_y, L_x] = 2\hbar L_z \,. \tag{3.14}$$

There is only one important operator which has not yet been introduced; it is the scalar associated with the vector angular momentum, its length.

$$L^2 = L_x^2 + L_y^2 + L_z^2 \tag{3.15a}$$

$$= \frac{1}{2}(L^+L^- + L^-L^+) + L_z^2 \tag{3.15b}$$

$$= L^+L^- + L_z(L_z - \hbar) \tag{3.15c}$$

$$= L^-L^+ + L_z(L_z + \hbar) \,. \tag{3.15d}$$

The various alternative forms for L^2 are obtained from each other by means of the commutation relations of (3.14), and the definitions (3.10, 3.11). Some will be more useful than others when dealing with eigenfunctions of L_z, for example. We intend to prove that L^2 commutes with all the components of **L**, that is,

$$[L^2, L_z] = [L^2, L_x] = [L^2, L_y] = [L^2, L^\pm] \equiv 0 \,. \tag{3.16}$$

The proof for L_z is by inspection, as L^2 does not involve the angle ϕ and therefore must commute with $\partial/\partial\phi$. If follows by rotational symmetry, that it also commutes with both other components of **L**.

Because L^2 and L_z commute, they share a complete set of eigenfunctions which can be labeled by the eigenvalue of each of these two operators. Because L_x and L_y (and also $L^\pm$) do not commute with L_z, they cannot be simultaneously diagonalized with the former two operators, and therefore their eigenvalues cannot *also* be specified. But there is no loss of generality in singling out the component L_z, since the z axis can be picked along any arbitrary direction. In the case of kinetic angular momentum, the desired eigenfunctions will be the well-known spherical harmonics, as we shall now discover.

3.2.   Spherical Harmonics

Because the radial dependence of the wavefunction plays no role in the determination of angular momentum, we may dispense with it altogether and consider functional dependence on the angles only. The wavefunctions are normalized on the unit sphere,

$$\int_0^{2\pi} d\phi \int_0^{\pi} d\theta \sin\theta \psi^*(\theta,\phi)\psi(\theta,\phi) = 1 \qquad (3.17)$$

and the two simultaneous eigenvalue equations are

$$L_z\psi \equiv \frac{\hbar}{i}\frac{\partial}{\partial\phi}\psi = \hbar m\psi \qquad (3.18)$$

and

$$L^2\psi = -\hbar^2\left[\frac{1}{\sin\theta}\frac{\partial}{\partial\theta}\left(\sin\theta\frac{\partial}{\partial\theta}\right) + \frac{1}{\sin^2\theta}\frac{\partial^2}{\partial\phi^2}\right]\psi = \hbar^2\lambda\psi \qquad (3.19)$$

with eigenvalues, respectively, of m (the "magnetic quantum number") and λ. We still show the dependence on the quantum-mechanical unit of angular momentum $\hbar$, but eventually it will be most convenient to use units in which $\hbar = 1$, and such units will be assumed throughout most of the book.

The L_z equation can be integrated directly, with the result

$$\psi(\theta,\phi) = e^{im\phi}\psi(\theta,0). \qquad (3.20)$$

A boundary condition must now be invoked to determine the quantum numbers m, one obvious choice being to require $\psi(\theta,\phi)$ to be single-valued on the unit sphere. Therefore we require

$$e^{im(\phi+2\pi)} = e^{im\phi} \qquad (3.21)$$

which is satisfied by the choice

$$m = \text{integer} = 0, \pm 1, \pm 2, \dots . \qquad (3.22)$$

Although other boundary conditions are possible, they would violate other requirements of quantum theory that have not yet been discussed, to be examined in the following section. The same boundary condition of (3.21) has also been used by *Dirac* (see *Note* at end of section) to prove that the electric charge is quantized, in units of a fundamental charge q.

The solution, (3.20), may now be introduced into the second differential equation, (3.19), which is the eigenvalue equation for L^2. One finds directly

$$\left(\frac{1}{\sin\theta} \frac{\partial}{\partial\theta} \sin\theta \frac{\partial}{\partial\theta} - \frac{m^2}{\sin^2\theta} + \lambda \right) \psi = 0, \tag{3.23}$$

recognizable as the equation obeyed by the *associated Legendre polynomials*, the properties of which are well established. But if they were not known, the following constructive procedure would solve this eigenvalue equation by elementary means:

Assume that the wavefunction $\psi(\theta,\phi) = e^{im\phi}\psi(\theta,0)$ is a solution of the following first-order partial differential equation,

$$L^+\psi(\theta,\phi) = 0. \tag{3.24}$$

Therefore, by (3.15d), ψ is an eigenfunction of L^2 with eigenvalue

$$\lambda = m(m+1). \tag{3.25}$$

This value of m is special by virtue of the previous equation, $L^+\psi = 0$. It *cannot be stepped up*. It is the maximum value of the azimuthal quantum number for the calculated value of λ. Therefore we shall denote it by symbol l, defined by

$$\lambda = l(l+1) \tag{3.26}$$

and let m continue to represent the azimuthal (or "magnetic") quantum number. For the function which obeys (3.24), $l = m$. By repeated application of L^-, we can generate functions always belonging to the same value of λ or l, but with decreasing magnetic quantum numbers $m = l-1$, $m = l-2, \ldots$. The series terminates with $m = -l$, by (3.15c).

The first eigenfunction, with $m = l$, is now obtained explicitly by integrating the linear first-order differential equation (3.24). Given

$$\left(\frac{\partial}{\partial\theta} - l\cot\theta \right) \psi = 0, \quad \text{i.e.} \quad \frac{d\psi}{\psi} = l\frac{d(\sin\theta)}{\sin\theta}, \tag{3.27}$$

integrate to obtain

$$\psi(\theta,\phi) = (\sin^l\theta)(e^{il\phi})\text{const}. \tag{3.28}$$

Except for the normalization constants, the remaining functions of the set having the common eigenvalue l are found by repeated applications of the operator L^- to this solution. These functions can be normalized and given standard phases, and are denoted the *spherical harmonics*, $Y_{l,m}$.

$$Y_{l,m}(\theta, \phi) = \frac{(-1)^{l+m}}{2^l l!} \sqrt{\frac{(2l+1)(l-m)!}{4\pi(l+m)!}}$$

$$\cdot (\sin \theta)^m \left(\frac{\partial}{\partial \cos \theta}\right)^{l+m} (\sin \theta)^{2l} e^{im\phi} . \tag{3.29}$$

These are, then, the associated Legendre polynomials of $(\cos \theta)$, multiplied by $\exp(im\phi)$ and suitably normalized over the unit sphere.[1] These functions form a complete, orthonormal set (see Table 3.1).

Note: Dirac assumed the existence, somewhere in the universe, of a magnetic *monopole*. Although such objects have never been found on earth or in astronomical observations, which have so far always indicated that the magnetic dipole is the fundamental source of magnetic fields, the search goes on for this elusive particle,[2] and we may follow Dirac in assuming the existence of at least one such source — of strength μ_0. The magnetic field of such a pole is

$$H = \frac{\mu_0}{r^2}$$

pointing in the radial direction, which implies a vector potential (in spherical polar coordinates)

$$\mathbf{A} = (A_r, A_\theta, A_\phi) \quad \text{with } A_r = A_\theta = 0 \quad \text{and} \quad A_\phi = \frac{\mu_0}{r} \tan\left(\frac{1}{2}\theta\right).$$

This may be verified by the relation $\nabla \times \mathbf{A} = \mathbf{H}$, defining $\mathbf{A}$. As we know, the Hamiltonian, as well as any other dynamical operator of a $\pm$ charged particle, is modified in a magnetic field because of a change in the momentum of the particle

$$\mathbf{p} \to \mathbf{p} \pm \frac{e}{c}\mathbf{A}$$

with $c =$ speed of light. But notice that $\mathbf{A} \to \infty$ along the line $\theta = \pi$, in just such a way that the integral along a path enclosing this line has a definite value, namely,

---

[1] The phase $(-1)^{l+m}$ is chosen in conformity with that found in A. R. Edmonds, *Angular Momentum in Quantum Mechanics*, Princeton Univ. Press, 1957 and E. U. Condon and G. Shortley, *The Theory of Atomic Spectra*, Cambridge Univ. Press, New York, 1935.
[2] See, e.g. E. Goto, H. Kolm and K. Ford, *Phys. Rev.* **132**, 387 (1963).

Table 3.1.

l	m	$Y_{l,m}$ (normalized on unit sphere)	
0	0	$\dfrac{1}{2\sqrt{\pi}}$	

$$1\begin{cases}\end{cases}$$

	1	$\dfrac{1}{r}\left[-\sqrt{\dfrac{3}{8\pi}}(x+iy)\right]$	$=-\sqrt{\dfrac{3}{8\pi}}\sin\theta e^{i\varphi}$
	0	$\dfrac{1}{r}\sqrt{\dfrac{3}{4\pi}}z$	$=\sqrt{\dfrac{3}{4\pi}}\cos\theta$

$$2\begin{cases}\end{cases}$$

	2	$\dfrac{1}{r^2}\left[\sqrt{\dfrac{15}{32\pi}}(x+iy)^2\right]$	$=\sqrt{\dfrac{15}{32\pi}}\sin^2\theta e^{2i\varphi}$
	1	$\dfrac{1}{r^2}\left[-\sqrt{\dfrac{15}{8\pi}}z(x+iy)\right]$	$=-\sqrt{\dfrac{15}{8\pi}}\cos\theta\sin\theta e^{i\varphi}$
	0	$\dfrac{1}{r^2}\left[\sqrt{\dfrac{5}{16\pi}}(3z^2-r^2)\right]$	$=\sqrt{\dfrac{5}{16\pi}}(2\cos^2\theta-\sin^2\theta)$

$$3\begin{cases}\end{cases}$$

	3	$\dfrac{1}{r^3}\left[-\sqrt{\dfrac{35}{64\pi}}(x+iy)^3\right]$	$=-\sqrt{\dfrac{35}{64\pi}}\sin^3\theta e^{3i\varphi}$
	2	$\dfrac{1}{r^3}\left[\sqrt{\dfrac{105}{32\pi}}z(x+iy)^2\right]$	$=\sqrt{\dfrac{105}{32\pi}}\cos\theta\sin^2\theta e^{2i\varphi}$
	1	$\dfrac{1}{r^3}\left[-\sqrt{\dfrac{21}{64\pi}}(5z^2-r^2)(x+iy)\right]$	$=-\sqrt{\dfrac{21}{64\pi}}(4\cos^2\theta\sin\theta-\sin^3\theta)e^{i\varphi}$
	0	$\dfrac{1}{r^3}\left[\sqrt{\dfrac{7}{16\pi}}(5z^2-3r^2)z\right]$	$=\sqrt{\dfrac{7}{16\pi}}(2\cos^3\theta-3\cos\theta\sin^2\theta)$

$$4\begin{cases}\end{cases}$$

	4	$\dfrac{1}{r^4}\left[\dfrac{3}{16}\sqrt{\dfrac{35}{2\pi}}(x+iy)^4\right]$	$=\dfrac{3}{16}\sqrt{\dfrac{35}{2\pi}}\sin^4\theta e^{4i\varphi}$
	3	$\dfrac{1}{r^4}\left[-\dfrac{3}{8}\sqrt{\dfrac{35}{\pi}}z(x+iy)^3\right]$	$=-\dfrac{3}{8}\sqrt{\dfrac{35}{\pi}}\sin^3\theta\cos\theta e^{3i\varphi}$
	2	$\dfrac{1}{r^4}\left[\dfrac{3}{8}\sqrt{\dfrac{5}{2\pi}}(7z^2-r^2)(x+iy)^2\right]$	$=\dfrac{3}{8}\sqrt{\dfrac{5}{2\pi}}(6\sin^2\theta\cos^2\theta-\sin^4\theta)e^{2i\varphi}$
	1	$\dfrac{1}{r^4}\left[-\sqrt{\dfrac{45}{64\pi}}(7z^2-3r^2)z(x+iy)\right]$	$=-\sqrt{\dfrac{45}{64\pi}}(4\sin\theta\cos^3\theta-3\cos\theta\sin^3\theta)e^{i\varphi}$
	0	$\dfrac{1}{r^4}\left[\sqrt{\dfrac{3}{16\sqrt{\pi}}}(35z^4-30r^2z^2+3r^4)\right]$	$=\dfrac{3}{16\sqrt{\pi}}(8\cos^4\theta+3\sin^4\theta-24\sin^2\theta\cos^2\theta)$

Note: $\dfrac{1}{r^4}\left(x^4+y^4+z^4-\dfrac{3}{5}r^4\right)=\dfrac{4\sqrt{\pi}}{15}\left[Y_{4,0}+\sqrt{\dfrac{5}{14}}(Y_{4,4}+Y_{4,-4})\right]$

$Y_{l,-m}=(-1)^m Y_{l,m}^*$

$$\int_{\theta=\pi} \mathbf{A} \cdot ds = \mu_0 4\pi \, .$$

A particle which, in the absence of the monopole, had magnetic quantum number m, now has magnetic quantum number

$$m' = m \pm \left(e \int \mathbf{A} \frac{ds}{2\pi\hbar c} \right) = m \pm \frac{e}{2\pi\hbar c} \mu_0 4\pi \, .$$

The wavefunction must still be single-valued, hence m' must also be an integer, from which it follows that $\pm e$ must be an integer multiple of the "fundamental charge"

$$q = \frac{\hbar c}{2\mu_0} \, .$$ Q.E.D.

By showing, moreover, that there is no fundamental unit of length associated with the magnetic monopole, Dirac proved that there could not be any bound states of a charged particle in the field $\mathbf{H}$, so that the main physical effect of the monopole is the quantization of charge.[3]

3.3.   Reason for Integer l and m

It is a straightforward matter to show that l and m are all integers, or all half-odd integers. The choice between these two sets is, however, not so simple.

Let us prove the first statement by recalling the boundary condition, (3.21), by which we required the wavefunctions to be single-valued on the unit sphere. This is too strict; in fact, the only physical requirement is that the probability density — a physical *observable* — be single-valued in an arbitrary state. That is, if we take a wave function which depends on ϕ as follows:

$$\psi(\phi) = \sum_m A_m e^{im\phi} \tag{3.30}$$

and require the probability $|\psi|^2$ to be single-valued,

$$|\psi(\phi + 2\pi)|^2 = |\psi(\phi)|^2$$

---

[3] P. A. M. Dirac, *Proc. Roy. Soc.* (*London*) **A123**, 60 (1931).

then the only way to satisfy this for arbitrary A_m's is to have all the m's integers, or all the m's half-odd integers. It is a simple exercise to verify that if $both$ are admitted, or other choices of m are made, then some probability densities can be constructed that are not single-valued.

The choice between the integers and half-integers is now made on the basis of the following reasoning. Physical quantities must be independent of the choice of coordinate system (whether Cartesian, spherical, or whatever) and of the origin of that coordinate system. For example, even for particles not undergoing circular motion, the complete set of eigenfunctions of angular momentum must be adequate to describe the angular part of the motion. And conversely, purely spherical motion must be describable in terms of other complete sets of wavefunctions, such as the plane-wave states. Let us take this for an example. The well-known formula for the expansion of plane waves in spherical harmonics is

$$e^{i\mathbf{k}\cdot\mathbf{r}} = 4\pi \sum_{l=0}^{\infty} \sum_{m=-l}^{+l} i^l j_l(kr) Y_{l,m}(\theta, \phi) Y_{l,m}^*(\Theta, \Phi) \qquad (3.31)$$

where the coefficients

$$j_l(kr) = \sqrt{\frac{\pi}{2kr}} J_{l+\frac{1}{2}}(kr) \qquad (3.32)$$

are spherical Bessel functions, here defined in terms of the ordinary Bessel functions $J_p(z)$. If Θ and Φ are the angles of the $\mathbf{k}$ vector and θ and ϕ are the angles of the coordinate vector $\mathbf{r}$, the expansion formula is further reduced by expressing it in terms of the relative angle ω,

$$\cos\omega = \frac{\mathbf{k}\cdot\mathbf{r}}{kr} = \cos\theta\cos\Theta + \sin\theta\sin\Theta\cos(\phi - \Phi) \qquad (3.33)$$

using the so-called $addition$ formula,

$$P_n(\cos\omega) = \frac{4\pi}{2n+1} \sum_{m=-n}^{+n} Y_{n,m}^*(\theta, \phi) Y_{n,m}(\Theta, \Phi). \qquad (3.34)$$

We make no attempt to derive (3.31–3.34) as they are well established in elementary texts in electromagnetic and quantum theory. The Legendre polynomials, for example,

$$P_0(x) = 1, \quad P_1(x) = x, \quad P_2(x) = \frac{1}{2}(3x^2 - 1) \qquad (3.35)$$

are particularly well known (almost elementary) functions, constituting a complete set of orthonormal polynomials on the interval $-1 \leq x \leq +1$.

The expansion of the plane waves given above (or of any other useful complete set of nonspherical wavefunctions) in a complete set of spherical wavefunctions requires only the *integer* values of l and m, and this is what finally fixes our choice.

Note that the plane waves are a complete set of states of the infinitesimal translation operator. The half-odd integer states must therefore be reserved for a space in which *only* rotations are permitted, so-called *spin space*.

3.4. Matrices of Angular Momentum

The spherical harmonics are the orthonormal eigenfunctions of L^2 and L_z, and by using them we can calculate the matrix structure of such operators as L_x and L_y, or $L^{\pm}$. For example, supplementing the relationship we derived earlier,

$$L^+ Y_{l,l} = 0,$$

we can obtain

$$L^+ Y_{l,m} = \hbar \sqrt{(l-m)(l+m+1)} Y_{l,m+1} \tag{3.36}$$

by operating directly on the spherical harmonics given in (3.29). The following relations are also very useful:

$$L^- Y_{l,m} = \hbar \sqrt{(l-m+1)(l+m)} Y_{l,m-1} \tag{3.37}$$

$$L_z Y_{l,m} = \hbar m Y_{l,m} \tag{3.38}$$

and

$$L^2 Y_{l,m} = \left[L_z^2 + \frac{1}{2}(L^- L^+ + L^+ L^-) \right] Y_{l,m}$$

$$= \hbar^2 \left[m^2 + \frac{1}{2}(l-m)(l+m+1) + \frac{1}{2}(l-m+1)(l+m) \right] Y_{l,m}$$

$$= \hbar^2 l(l+1) Y_{l,m}. \tag{3.39}$$

Note that in this last equation, no special property of the spherical harmonics was used. The eigenvalue of L^2 can be calculated using the previous three equations, that is, by knowing the matrix structure of $L^{\pm}$ and L_z. But this matrix structure, also, can be derived without explicit recourse to the spherical harmonics, by using the commutation relations, (3.12–3.14). Thus it would apply also to the half-odd-integer solutions in which we shall soon be interested. To distinguish these, we shall want to extend our notation.

The conventional notation of $L^{\pm}$, L_z, and L^2, l, and m, will continue to be used when we are dealing with kinetic angular momentum. For general or arbitrary angular momentum, when only the basic commutation relations are invoked but the arguments are not restricted to integer values of l and m, $\mathbf{J}$ will replace $\mathbf{L}$ and j will replace l, although the eigenvalue associated with $\mathbf{J}$ will still be denoted m. We shall also use $\mathbf{S}$ interchangeably with $\mathbf{J}$. The general matrix structure of the general angular momentum operators can be obtained using only (3.12–3.16) and the construction given in the first section. One obtains the following results:

$$(j'm'|J^+|jm) = \delta_{j,j'}\delta_{m+1,m'}\hbar\sqrt{(j-m)(j+m+1)} \tag{3.40}$$

$$(j'm'|J^-|jm) = \delta_{j,j'}\delta_{m-1,m'}\hbar\sqrt{(j-m+1)(j+m)} \tag{3.41}$$

$$(j'm'|J_z|jm) = \delta_{j,j'}\delta_{m,m'}\hbar m \tag{3.42}$$

$$(j'm'|J^2|jm) = \delta_{j,j'}\delta_{m,m'}\hbar^2 j(j+1). \tag{3.43}$$

Because of their common factor $\delta_{j,j'}$, angular momentum operators can be represented by $(2j+1)\times(2j+1)$ square matrices (*irreducible representations*) which act in a subspace of the $2j+1$ linearly independent eigenfunctions of J^2 belonging to a given j value. Such matrices satisfy the basic commutation relations (3.6) or (3.12–3.16), *by construction*, and are therefore genuine angular momenta in their own right.

Larger matrices can also be constructed that are angular momenta, but then they have sub-blocks with the above matrix structure, and are *reducible representations*. The most compact matrices having the properties of angular momentum are *irreducible representations* of the angular momentum operators. For example, the irreducible representation of angular momentum $l = 1$ is

$$\mathcal{L}^+ = \hbar \begin{bmatrix} 0 & \sqrt{2} & 0 \\ 0 & 0 & \sqrt{2} \\ 0 & 0 & 0 \end{bmatrix}$$

$$\mathcal{L}^- = \hbar \begin{bmatrix} 0 & 0 & 0 \\ \sqrt{2} & 0 & 0 \\ 0 & \sqrt{2} & 0 \end{bmatrix} \tag{3.44}$$

$$\mathscr{L}_z = \hbar \begin{bmatrix} 1 & 0 & 0 \\ 0 & 0 & 0 \\ 0 & 0 & -1 \end{bmatrix}$$

$$\mathscr{L}^2 = 2\hbar^2 \begin{bmatrix} 1 & 0 & 0 \\ 0 & 1 & 0 \\ 0 & 0 & 1 \end{bmatrix}.$$

For angular momentum $l = 2$ there are also four 5×5 matrices, for angular momentum $l = 3$ they are 7×7, etc. It is a striking fact, however, that all the *even*-dimensional representations are missing from the list if we insist on the integer values of angular momentum possessed by the spherical harmonics. The missing matrices, the even-dimensional ones, are the spin angular momentum operators for $j, m = $ half-odd integers. Upon including these, the matrix representation is made complete.

3.5.   Pauli Spin Matrices

Here we discuss the smallest of the even-dimensional irreducible representations of angular momentum, the matrices of *spin one-half*. They are very similar to the 2×2 matrices, which were invented by Pauli precisely for the purpose of describing the intrinsic spin angular momentum of the electron.

In this smallest subspace, the spin operators are, according to (3.40–3.43),

$$S^+ = \hbar \begin{bmatrix} 0 & 1 \\ 0 & 0 \end{bmatrix} \qquad S^- = \hbar \begin{bmatrix} 0 & 0 \\ 1 & 0 \end{bmatrix}$$

and

$$S_z = \hbar \begin{bmatrix} \dfrac{1}{2} & 0 \\ 0 & -\dfrac{1}{2} \end{bmatrix} \tag{3.45}$$

and

$$S^2 = \hbar^2 \dfrac{3}{4} \begin{bmatrix} 1 & 0 \\ 0 & 1 \end{bmatrix}. \tag{3.46}$$

The Pauli spin matrices are obtained from $S_z, S_x = \frac{1}{2}(S^+ + S^-)$, and $S_y = (S^+ - S^-)/2i$ by the relation

$$\mathbf{S} = \frac{1}{2}\hbar\boldsymbol{\sigma} = \frac{1}{2}\hbar(\sigma_x, \sigma_y, \sigma_z) \tag{3.47}$$

and so are explicitly

$$\sigma_x = \begin{bmatrix} 0 & 1 \\ 1 & 0 \end{bmatrix} \quad \sigma_y = i\begin{bmatrix} 0 & -1 \\ 1 & 0 \end{bmatrix} \quad \sigma_z = \begin{bmatrix} 1 & 0 \\ 0 & -1 \end{bmatrix}$$

and the unit operator:

$$\mathbf{1} = \begin{bmatrix} 1 & 0 \\ 0 & 1 \end{bmatrix}. \tag{3.48}$$

The eigenvectors of S_z, and σ_z are the two-component *spinors*

$$\chi_+ = \begin{bmatrix} 1 \\ 0 \end{bmatrix} \quad \text{and} \quad \chi_- = \begin{bmatrix} 0 \\ 1 \end{bmatrix} \tag{3.49}$$

corresponding to $m = \pm\frac{1}{2}$, respectively. Conventionally, $\sigma^\pm$ are defined by $\sigma^\pm = S^\pm/\hbar$.

The higher-dimensional even matrices, 4×4, 6×6, etc., can be similarly constructed using the general set of rules for the matrix elements given in (3.40–3.43) and will represent the higher spins $3/2, 5/2$, etc. There is no need to give these explicitly, considering that these matrices become unwieldy with increasing size and given that we shall find far more convenient *operator* representations.

3.6. Compounding Angular Momentum

It is frequently necessary to compound constituent angular momenta into a total angular momentum operator and, conversely, to decompose and simplify operators with complex structure. Let us go into this matter briefly and directly, even at the expense of omitting formal proofs.

Given two angular momenta $\mathbf{J}_1$ and $\mathbf{J}_2$, it seems certain (intuitively) that $\mathbf{J} = \mathbf{J}_1 + \mathbf{J}_2$ will be an allowable angular momentum, obeying the commutation laws of Eqs. (3.6) and those following. However, no other linear combination of $\mathbf{J}_1$ and $\mathbf{J}_2$ will do (cf. Problem 3.1).

Problem 3.1.   Given angular momenta $\mathbf{J}_1$ and $\mathbf{J}_2$, show that $a\mathbf{J}_1 + b\mathbf{J}_2$ is itself an angular momentum only if $a = b = 1$; or $a = 1$, $b = 0$; or $a = 0$, $b = 1$.

How does one express the eigenfunctions of $\mathbf{J}_1$ and $\mathbf{J}_2$ in terms of the eigenfunctions of $\mathbf{J}$ and vice versa?

Supposing we start by knowing a complete set of angular momenta eigenfunctions of $\mathbf{J}_1$ and $\mathbf{J}_2$, whether spinors or spherical harmonics, or whatever, which we label by the two sets of quantum numbers

$$|j_1 m_1 j_2 m_2\rangle . \tag{3.50}$$

For fixed j_1 and j_2 there are $(2j_1 + 1) \cdot (2j_2 + 1)$ orthonormal eigenfunctions of this type corresponding to the various choices of m_1 and m_2. Each one is, therefore, also an eigenfunction of J^z, with eigenvalue $m = m_1 + m_2$. (Henceforth, subscripts will label individual angular momenta, and superscripts their components; for example, J_i^x.) This has a maximum value $j_1 + j_2$, therefore it is the maximum value of m, which is by definition j, the quantum number of J^2 [recall (3.24–3.26)]. By repeated applications of the operator $J^- = J_1^- + J_2^-$ to this state of maximal m, we generate a total of $2(j_1+j_2)+1$ orthogonal states all belonging to the same j value, but with $m = j_1 + j_2$, $j_1 + j_2 - 1, \ldots, -j_1 - j_2$. (Note that j is not changed, because J^- commutes with J^2.) Only *one* of these states has $m = j_1 + j_2 - 1$, for indeed, every m value occurs only once in this list. But there are *two* of the original states that belong to this eigenvalue of J^z : $|j_1 j_1 - 1 j_2 j_2\rangle$ and $|j_1 j_1 j_2 j_2 - 1\rangle$. The function which has been included is in fact the *sum* of these two functions,

$$|j_1 j_1 - 1 j_2 j_2\rangle + |j_1 j_1 j_2 j_2 - 1\rangle \tag{3.51}$$

because $J^- = J_1^- + J_2^-$. This leaves us free to consider the difference of the two

$$|j_1 j_1 - 1 j_2 j_2\rangle - |j_1 j_1 j_2 j_2 - 1\rangle . \tag{3.52}$$

This function is orthogonal to the previous one, it belongs to the same m value, and when $J^+ = J_1^+ + J_2^+$ is applied to it the result vanishes. Therefore the m value to which it belongs must be maximal, and (3.52) must be an eigenfunction belonging to $j = j_1 + j_2 - 1$. Repeated applications of $\mathbf{J}^-$ to this function, (3.52), results indeed in $2(j_1+j_2-1)+1$ orthogonal functions, corresponding to all the attainable m values.

Two of the three states belonging to $m = j_1 + j_2 - 2$ are thus accounted for, and the third can now be used to construct the set of functions belonging to $j = j_1 + j_2 - 2$. This procedure may be continued until all $(2j_1+1)\cdot(2j_2+1)$ initial states are exhausted. Such a stage is reached when $j = |j_1 - j_0|$, which is therefore the *minimum* magnitude of the compound angular momentum, just as $j_1 + j_2$ is the *maximum* magnitude. The proof of this so-called "triangle inequality" is given in Problem 3.2 and by an operator method in (3.75, 3.76).

Problem 3.2.   Show that the procedure indicated in the text does in fact account for every one of the $(2j_1 + 1) \times (2j_2 + 1)$ initial functions, by proving the identity

$$\sum_{j=|j_1-j_2|}^{j_1+j_2} (2j + 1) = (2j_1 + 1) \cdot (2j_2 + 1) .$$

Having constructed the complete set of eigenfunctions of J^2 and J^z from the set of eigenfunctions of J_1^z and J_2^z, one may formalize this procedure somewhat. The two sets of orthonormal states are related by a canonical (unitary) transformation, that is, we can express the new wavefunctions $|j_1, j_2 jm)$ in terms of the old by

$$|j_1 j_2 jm) = \sum_{m_1 m_2} |j_1 m_1 j_2 m_2)(j_1 m_1 j_2 m_2 | j_1 j_2 jm)$$

where the coefficients are the matrix elements of the transformation operator

$$(j_1 m_1 j_2 m_2 | j_1 j_2 jm) \tag{3.53}$$

are known as the vector-coupling, or *Clebsch-Gordan coefficients*. A somewhat symmetric form for these coefficients was derived first by Wigner using group-theoretical methods, and later by Racah and by Schwinger.[4] (Later in this chapter there is an introduction to Schwinger's operator technique in angular momentum.) We quote the result:

---

[4] For references to Wigner's and other work, and for detailed formulas see A. R. Edmonds, *op. cit.*[1]

$(j_1 m_1 j_2 m_2 | j_1 j_2 jm)$

$$= \delta_{m_1 + m_2, m} \cdot \left[\frac{(2j + 1)(j_1 + j_2 - j)!(j_1 - j_2 + j)!(-j_1 + j_2 + j)!}{(j_1 + j_2 + j + 1)!} \right.$$

$$\times (j_1 + m_1)!(j_1 - m_1)!(j_2 + m_2)!(j_2 - m_2)!(j + m)!(j - m)! \left. \right]^{1/2}$$

$$\times \sum_z (-1)^z [z!(j_1 + j_2 - j - z)!(j_1 - m_1 - z)!(j_2 + m_2 - z)!$$

$$\times (j - j_2 + m_1 + z)!(j - j_1 - m_2 + z)!]^{-1} . \tag{3.54}$$

This unwieldy formula is evaluated in a form useful in problems involving angular momenta $\frac{1}{2}$ or 1 in Table 3.2.

For the purpose of nuclear and atomic physics, a symmetrized form of these coefficients, the *Wigner 3-j symbols*, is simpler to manipulate. However, it does not serve any purpose to include them here, and we refer the reader to Edmond's book.[1]

Table 3.2.   Nonvanishing Clebsch-Gordan coefficients (3.53) for $j_2 = \frac{1}{2}$ and 1.

$$\left(j_1 m_1 \frac{1}{2} m_2 \Big| j_1 \frac{1}{2} jm \right)$$

	$m_2 = \dfrac{1}{2}$	$m_2 = -\dfrac{1}{2}$
$j = j_1 + \dfrac{1}{2}$	$+\sqrt{\dfrac{j_1 + m + \frac{1}{2}}{2j_1 + 1}}$	$+\sqrt{\dfrac{j_1 - m + \frac{1}{2}}{2j_1 + 1}}$
$j = j_1 - \dfrac{1}{2}$	$-\sqrt{\dfrac{j_1 - m + \frac{1}{2}}{2j_1 + 1}}$	$+\sqrt{\dfrac{j_1 + m + \frac{1}{2}}{2j_1 + 1}}$

$$(j_1 m_1 1 m_2 | j_1 1 jm)$$

j	$m_2 = 1$	$m_2 = 0$	$m_2 = -1$
$j_1 + 1$	$+\sqrt{\dfrac{(j_1 + m)(j_1 + m + 1)}{(2j_1 + 1)(2j_1 + 1)}}$	$+\sqrt{\dfrac{(j_1 - m + 1)(j_1 + m + 1)}{(2j_1 + 1)(j_1 + 1)}}$	$+\sqrt{\dfrac{(j_1 - m)(j_1 - m + 1)}{(2j_1 + 1)(2j_1 + 1)}}$
j_1	$-\sqrt{\dfrac{(j_1 + m)(j_1 - m + 1)}{2j_1(j_1 + 1)}}$	$+\dfrac{m}{\sqrt{j_1(j_1 + 1)}}$	$+\sqrt{\dfrac{(j_1 - m)(j_1 + m + 1)}{2j_1(j_1 + 1)}}$
$j_1 - 1$	$+\sqrt{\dfrac{(j_1 - m)(j_1 - m + 1)}{2j_1(2j_1 + 1)}}$	$-\sqrt{\dfrac{(j_1 - m)(j_1 + m)}{j_1(2j_1 + 1)}}$	$+\sqrt{\dfrac{(j_1 + m + 1)(j_1 + m)}{2j_1(2j_1 + 1)}}$

We shall return to the subject of compound angular momentum, after introducing an operator technique for expressing the angular momenta.

3.7. Equations of Motion of Interacting Angular Momenta

An interaction Hamiltonian involving two or more angular momenta (e.g., spin-orbit coupling of a valence electron in an atom) must be free of unwarranted dependence on an arbitrary coordinate system, so must be a scalar. For two angular momenta, the Hamiltonian must therefore be a function only of $\mathbf{J}_1 \cdot \mathbf{J}_2$. The Heisenberg exchange operator, $-J(\mathbf{S}_1 \cdot \mathbf{S}_2)$, is the simplest example of this. The time development of all other operators are determined by the Hamiltonian, through commutator bracket equations of motion (Poisson bracket equations of motion classically). Therefore we are interested in the commutation relations of the components $J_{1x}, J_{2y}, \ldots, J_{2z}$ with the bilinear scalar. They are, compactly,

$$\frac{1}{i\hbar}[\mathbf{J}_1 \cdot \mathbf{J}_2, \mathbf{J}_1] = \mathbf{J}_1 \times \mathbf{J}_2 \quad \text{and} \quad \frac{1}{i\hbar}[\mathbf{J}_1 \cdot \mathbf{J}_2, \mathbf{J}_2] = \mathbf{J}_2 \times \mathbf{J}_1 . \tag{3.55}$$

Antisymmetry of the vector product then establishes that the total angular momentum operator, $\mathbf{J}_1 + \mathbf{J}_2$, commutes with $\mathbf{J}_1 \cdot \mathbf{J}_2$. This motivates representations such as the one introduced in the preceding section.

3.8. Coupled Boson Representation

In semiclassical theories of magnetism it is common to approximate the unwieldy spin operators or matrices by harmonic-oscillator operators, for the general matrix structure of the spins, (3.40–3.43), resembles in many respects the matrix structure of harmonic oscillator operators. That this is no coincidence has been proved by Schwinger in his theory of angular momentum[5] based on coupled harmonic-oscillator fields.

As is well known, one may take linear combinations of momentum and coordinate operators, to obtain harmonic-oscillator "raising" and "lowering" operators $\mathbf{a}^*$ and $\mathbf{a}$; for example,

$$\mathbf{a}^* = \frac{1}{\sqrt{2\hbar}}\left(\frac{\hbar}{i}\frac{\partial}{\partial x} + ix\right), \quad \mathbf{a} = \frac{1}{\sqrt{2\hbar}}\left(\frac{\hbar}{i}\frac{\partial}{\partial x} - ix\right).$$

---

[5] J. Schwinger: "On Angular Momentum," U.S. Atomic Energy Commission Report NYO-3071 (1952), reprinted in L. Biedenharn and H. Van Dam (eds.), *Quantum Theory of Angular Momentum*, Academic Press, New York, 1965.

Let us make clear the reason for this terminology. The commutation relations which may be derived for the operators $\mathbf{a}$, $\mathbf{a}^*$ are those of *Bosons*

$$a_i a_j - a_j a_i \equiv [a_i, a_j] = [a_i^*, a_j^*] = 0 \quad \text{and} \quad [a_i, a_j^*] = \delta_{ij} \tag{3.56}$$

with i, j referring to different particles, from which we deduce also

$$[\mathbf{n}_i, a_j^*] = \delta_{ij} a_j^* \quad \text{and} \quad [\mathbf{n}_i, a_j] = -\delta_{ij} a_i, \tag{3.57}$$

introducing $\mathbf{n}_i = a_i^* a_i = $ *occupation-number operator*, referring to the degree of excitation (occupation) of the ith harmonic oscillator. One may construct a complete set of states, labeled by occupation numbers, first introducing a "vacuum." This is the ground state of all the harmonic oscillators, the no-particle state, denoted $|0)$, which is annihilated by every lowering operator

$$a_i |0) \equiv 0 \quad \text{for all } i. \tag{3.58}$$

The one-particle states are clearly

$$a_i^* |0)$$

and the (normalized) two-particle states are

$$a_1^* a_2^* |0) \quad \text{or} \quad \frac{(a_1^*)^2}{\sqrt{2}} |0),$$

etc. The general formula for the many-particle, normalized state is

$$\frac{(a_1^*)^{n_1} (a_2^*)^{n_2} \cdots}{\sqrt{n_1! n_2! \cdots}} |0) \tag{3.59}$$

as proved in Problem 3.3. In this so-called "occupation-number" representation, the $\mathbf{n}_i$ are diagonal, with non-negative integer eigenvalues $n_i = 0, 1, 2, \ldots$. The total occupation of a state is given by the sum of the eigenvalues,

$$\sum n_i.$$

Problem 3.3. Prove that the state in Eq. (3.59) is normalized by evaluating $(0| \cdots (a_2)^{n_2} (a_1)^{n_1} (a_1^*)^{n_1} (a_2^*)^{n_2} \cdots |0)$, using only the definition of the vacuum, (3.58), and the commutation relations of (3.56, 3.57). Show that n_i is the eigenvalue of $\mathbf{n}_i$.

Schwinger showed that with the aid of only two harmonic oscillators the entire matrix structure of a single angular momentum, as summarized in (3.40–3.43), could be exactly reproduced. The advantages are great, even if no new results were obtained, most especially for large values of j for which the angular-momentum matrices are large and unwieldy. New features can also be studied with Schwinger's approach, due to the great understanding that has been achieved concerning the harmonic-oscillator fields, as compared to the more obscure angular momentum-operators. Labeling the two oscillators by subscripts 1 and 2, let us introduce the *spinor operators*

$$\mathbf{a}^+ = (a_1^*, a_2^*) \quad \text{and} \quad \mathbf{a} = \begin{pmatrix} a_1 \\ a_2 \end{pmatrix}, \tag{3.60}$$

which are merely two-component vectors with operator components. If, moreover, we contract these operators with the Pauli spin matrices, we obtain the desired representation. That is, let

$$J^z = \frac{\hbar}{2}\mathbf{a}^+ \cdot \sigma_z \cdot \mathbf{a} = \frac{\hbar}{2}(a_1^* a_1 - a_2^* a_2) = \frac{\hbar}{2}(n_1 + n_2)$$

and similarly for the other components, with the following compact result:

$$\mathbf{J} = \frac{\hbar}{2}\mathbf{a}^+ \cdot \sigma \cdot \mathbf{a} \quad \text{(cf. Eq. (3.47))}. \tag{3.61}$$

One-half times the contraction with the unit operator will be denoted the **j** operator

$$\mathbf{j} = \frac{1}{2}\mathbf{a}^+ \cdot \mathbf{a} = \frac{1}{2}(a_1^* a_1 + a_2^* a_2) = \frac{1}{2}n_1 + \frac{1}{2}n_2. \tag{3.62}$$

It may be verified that the eigenvalues of this operator are indeed $j = 0, \frac{1}{2}, 1, \frac{3}{2}, \ldots$, and that $\mathbf{J}^2$ has eigenvalue $\hbar^2 j(j+1)$.

Problem 3.4.   Show that $J^+ = \hbar a_1^* a_2$ and $J^- = \hbar a_2^* a_1$. (These formulas and the ones in the text may be remembered by associating a change $\Delta m = +\frac{1}{2}$ with each type-1 particle, and $\Delta m = -\frac{1}{2}$ with each type-2 particle.)

Problem 3.5.   Prove the operator identity $\mathbf{J}^2 = \hbar^2 \mathbf{j}(\mathbf{j}+1)$.

Problem 3.6. Quantization of the electric and magnetic fields of light waves is the first step in quantum electrodynamic theory. Assuming these are harmonic-oscillator fields, let subscript 1 refer to the electric field and subscript 2 to the magnetic field in operators a_1, a_1^* and a_2, a_2^* (e.g., for a particular plane-wave mode). Use the fact that half the total energy must be contained in each field to prove that the eigenvalues j can only take on integer values, thus showing that the photon has a quantized spin of unity. Compare with (3.103, 3.104).

Normalized eigenfunctions of a single angular momentum, possessing definite m and j eigenvalues shall be denoted $|jm\rangle$, and are simply

$$|jm\rangle = \frac{(a_1^*)^{j+m}(a_2^*)^{j-m}}{\sqrt{(j+m)!(j-m)!}}|0\rangle. \tag{3.63}$$

The coupled-Boson operators have more flexibility than the original angular-momentum operators. For example, we can make use of the extra degrees of freedom to construct the so-called *hyperbolic operators* which conserve m, but *raise* or *lower* j

$$K^+ = \hbar a_1^* a_2^*, \quad K^- = \hbar a_2 a_1 \quad \text{and} \quad K^z = \frac{\hbar}{2}(n_1 + n_2 + 1). \tag{3.64}$$

They obey the commutation relations,

$$[K^z, K^+] = \hbar K^+ \quad [K^z, K^-] = -\hbar K^- \quad \text{and} \quad [K^+, K^-] = -2\hbar K^z. \tag{3.65}$$

Only the last of these differs, by a sign, from the commutation relations of the angular momentum operators. The following equations may also be verified:

$$(J^z)^2 - \frac{1}{4}\hbar^2 = (K^z)^2 - \frac{1}{2}(K^+K^- + K^-K^+) \tag{3.66a}$$

$$= K^z(K^z - \hbar) - K^+K^- \tag{3.66b}$$

$$= K^z(K^z + \hbar) - K^-K^+. \tag{3.66c}$$

3.9.   Rotations

The study of rotations of coordinate systems and of particles is intimately tied in with the theory of angular momentum. For example, given an arbitrary function $f(\phi)$ of the azimuthal angle ϕ, its argument may be rotated through an angle α by applying a differential operator

$$e^{\alpha(d/d\phi)} f(\phi) = f(\phi + \alpha).$$

This formula may be checked by a Taylor series expansion of both sides, in powers of α. The operator which is exponentiated will be recognized as proportional to L^z, and is just one of the three differential operators, or linear combinations thereof, that may be exponentiated to effect desired rotations in the coordinate system. The most general rotation is expressible in terms of the three Euler angles α, β, and γ, and takes the form of the unitary operator

$$D(\alpha\beta\gamma) = e^{i(\alpha/\hbar)J^x} e^{i(\beta/\hbar)J^y} e^{i(\gamma/\hbar)J^z} = (D^\dagger)^{-1}. \tag{3.67}$$

One writes $\mathbf{J}$ instead of $\mathbf{L}$ because the rotations are not limited to the integer angular momenta. Of course, the three-Euler-angle formula is equivalent to a single rotation about a suitably chosen axis, with its direction along a unit vector $\hat{\mathbf{u}}$, and so

$$D(\alpha\beta\gamma) = e^{i\alpha'\hat{\mathbf{u}}\cdot\mathbf{J}} \tag{3.68}$$

for a suitable angle α'. Thus, two or more successive rotations can be expressed by a single one. One of the practical consequences of this equality, after the rotations are expressed in terms of a complete set of spherical harmonics (for integer angular momentum), is a very useful formula for expressing *products* of spherical harmonics as a *linear* combination of spherical harmonics, viz.,

$$Y_{l_1 m_1}(\theta, \phi) Y_{l_2 m_2}(\theta, \phi) = \sum_{l,m} Y_{l,m}(\theta, \phi) \sqrt{\frac{(2l_1 + 1)(2l_2 + 1)}{4\pi(2l + 1)}}$$

$$\cdot (l_1 m_1 l_2 m_2 | l_1 l_2 l m)(l_1 0 l_2 0 | l_1 l_2 l 0). \tag{3.69}$$

We shall not prove this formula here, but proceed instead to show in what way the rotations are related to the unitary transformations of Boson operators. The transformation that takes a operators into a linear combination of each other is a unitary canonical transformation if it preserves the Boson commutation relations and the Hermitean nature of Hermitean operators. If it also preserves the eigenvalue j [that is, commutes with $\mathbf{j}$, the operator of Eq. (3.62)] it corresponds precisely to a rotation. In particular,

$$a_1^* \rightarrow \left[e^{+(i/2)(\alpha+\gamma)} \cos \frac{1}{2}\beta \right] a_1^* + \left[e^{(i/2)(\gamma-\alpha)} \sin \frac{1}{2}\beta \right] a_2^*$$

$$a_2^* \rightarrow - \left[e^{(i/2)(\alpha-\gamma)} \sin \frac{1}{2}\beta \right] a_1^* + \left[e^{(i/2)(\alpha+\gamma)} \cos \frac{1}{2}\beta \right] a_2^*$$

$$a_1 \rightarrow \left[e^{-(i/2)(\alpha+\gamma)} \cos \frac{1}{2}\beta \right] a_1 + \left[e^{-(i/2)(\gamma-\alpha)} \sin \frac{1}{2}\beta \right] a_2$$

$$a_2 \rightarrow - \left[e^{-(i/2)(\alpha-\gamma)} \sin \frac{1}{2}\beta \right] a_1 + \left[e^{-(i/2)(\alpha+\gamma)} \cos \frac{1}{2}\beta \right] a_2 \qquad (3.70)$$

is the transformation which results, if in (3.67) we express the cartesian components of $\mathbf{J}$ in the coupled-Boson representation, and calculate

$$a_i \rightarrow D(\alpha\beta\gamma)a_i D^{-1}(\alpha\beta\gamma) \,.$$

For complex angles, this is no longer a unitary but a *similarity* transformation, preserving the commutation relations but not Hermiticity. One may also imagine unitary or similarity transformations that do not conserve j, mixing the **a** operators with **a*** operators. Such transformations mix the angular-momentum operators with the hyperbolic ones and are generated by including the components of $\mathbf{K}$ as well as of $\mathbf{J}$ in D.

3.10.   More on Compound Angular Momentum

The addition of the spin of an electron to its mechanical angular momentum, the addition of the angular momenta or spins of two distinct electrons, and in general the coupling of two or more angular momenta of various origins, require mathematical techniques beyond the simple formulation given so far. Here we shall show some of the machinery established by Schwinger for handling these problems, without, however, entering into the details. We shall study the coupling of two angular momenta, and it might appear that by induction one may use this theory to couple an arbitrary number of angular momenta. However, in practice this is not quite so, as even for three angular momenta certain additional simplifications must be sought if the problem is to remain manageable.

Two angular momenta require four Bose particles for their description. But with four Boson operators, far more than the components of $\mathbf{J}_1$ and $\mathbf{J}_2$ can be constructed: we can obtain $\mathbf{J} = \mathbf{J}_1 + \mathbf{J}_2$, and also all the relevant hyperbolic operators. The Clebsch-Gordan coefficients, rotations in one or the other angular-momentum space, and all the other possible transformations of interest also may be investigated by studying the Bose operators and their eigenfunctions.

Let the Bose operators referring to angular momentum 1 be labeled a_1 and a_2, and those referring to angular momentum 2 be labeled b_1 and b_2. Any **a** operator commutes with any **b** operator. The notation and terminology will be that introduced in (3.56–3.63). This takes care of the usual operators, the components of J_i, and J_j^2. In addition, the following new ones are useful:

$$I^+ = \hbar \mathbf{a}^+ \mathbf{b} \quad I^- = \hbar \mathbf{b}^+ \mathbf{a} \quad I^z = \hbar(\mathbf{j}_1 - \mathbf{j}_2), \quad \text{where} \tag{3.71}$$

$$\mathbf{a}^+ \mathbf{b} = a_1^* b_1 + a_2^* b_2, \tag{3.72}$$

etc., in the spinor notation, and

$$\cdot\, K^+ = \hbar(a_1^* b_2^* - a_2^* b_1^*) \quad K^- = \hbar(a_1 b_2 - a_2 b_1) \quad K^z = \hbar(\mathbf{j}_1 + \mathbf{j}_2 + 1). \tag{3.73}$$

The following identities may be verified by substituting the definitions of I, $J = J_1 + J_2$ and $K = K_1 + K_2$,

$$J^2 = J_x^2 + J_y^2 + J_z^2$$

$$= I^z(I^z - \hbar) + I^+ I^- \tag{3.74a}$$

$$= I^z(I^z + \hbar) + I^- I^+ \tag{3.74b}$$

$$= K^z(K^z - \hbar) - K^+ K^- \tag{3.74c}$$

$$= K^z(K^z + \hbar) - K^- K^+ . \tag{3.74d}$$

Using the definition of I^z (3.71), and the fact that $I^+ I^-$ in (3.74a) is a positive semidefinite operator (so is $I^- I^+$), one proves directly that $j(j+1) \leq (j_1 - j_2)(j_1 - j_2 - 1)$. The choice $j_1 \leq j_2$ [if $j_2 < j_1$, then use (3.74b)] establishes

$$j \geq |j_2 - j_1| . \tag{3.75}$$

Similarly, the two equations involving K can be used to prove

$$j \leq (j_1 + j_2) \tag{3.76}$$

and thus to establish the "triangle inequality" first encountered in the discussion following (3.52) and in Problem 3.2.

A state of definite eigenvalues $j_1 m_1$ and $j_2 m_2$ is

$$|j_1 m_1 j_2 m_2\rangle = \frac{(a_1^*)^{j_1+m_1} (a_2^*)^{j_1-m_1} (b_1^*)^{j_2+m_2} (b_2^*)^{j_2-m_2}}{\sqrt{(j_1+m_1)!(j-m_1)!(j_2+m_2)!(j_2-m_2)!}}|0\rangle. \qquad (3.77)$$

A state of definite j, m, j_1 and j_2 may equally well be labeled j, m, μ, and v, where

$$\mu = j_1 - j_2 \quad \text{and} \quad v = j_1 + j_2 + 1 \qquad (3.78)$$

and denoted

$$|jm\mu v\rangle \qquad (3.79)$$

where $\mu \leq j$ and $v \geq j+1$. For $m = j$ and $v = j+1$, the appropriate state is

$$|jj\mu j + 1\rangle = \frac{(a_1^*)^{j+\mu}(b_1^*)^{j-\mu}}{\sqrt{(j+\mu)!(j-\mu)!}}|0\rangle. \qquad (3.80)$$

Application of the spin-lowering operator yields the states $|jm\mu j + 1\rangle$, and finally, arbitrary states are given by

$$|jm\mu v\rangle = \sqrt{\frac{(2j+1)!}{(v+j)!(v-j-1)!}}(K^+)^{v-j-1}|jm\mu j + 1\rangle. \qquad (3.81)$$

The inner product of these wavefunctions with the $|j_1 m_1 j_2 m_2\rangle$ of (3.77) yields the Clebsch-Gordan coefficients by Schwinger's method.

The Bose operator method may be extended in various directions to recover well-known results (e.g., for the spherical harmonics) or to discover new ones. A connection with the continuous representations may be made using the operator identity

$$a_i F(a_i^*)|0\rangle = \frac{\partial}{\partial a_i^*} F(a_i^*)|0\rangle \qquad (3.82)$$

which may be used to construct generating functions for various sets of orthogonal polynomials. But this subject is properly outside the scope of the present elementary treatment.

3.11.   Other Representations

The first formal justification of the Bloch theory of spin waves may be found in the work of Holstein and Primakoff.[6] Bloch had naturally assumed that spin waves obey Bose-Einstein statistics, but these authors showed how spin operators could be expressed in terms of true Bose fields, much as we shall show it in the chapter on spinwave theory. The Holstein-Primakoff represen- tation is best understood as a special case of the Schwinger coupled-Boson representation, that is, as an irreducible representation of the latter in a subspace of fixed j. Recall (3.62), which we rewrite as

$$a_2^* a_2 = \sqrt{2\mathbf{j} - \mathbf{n}_1} \cdot \sqrt{2\mathbf{j} - \mathbf{n}_1} \,. \tag{3.83}$$

In a subspace of fixed eigenvalue j and fixed $\mathbf{n}$, this equation is solved by treating a_2 and its conjugate as two diagonal operators,

$$a_2 = a_2^* = (2j)^{1/2} \sqrt{1 - \frac{\mathbf{n}}{2j}} \quad \text{hence}, \tag{3.84}$$

$$J^+ = \hbar a^* (2j)^{1/2} \sqrt{1 - \frac{\mathbf{n}}{2j}} \quad J^- = \hbar (2j)^{1/2} \sqrt{1 - \frac{\mathbf{n}}{2j}} a \quad \text{and}$$

$$J^z = \hbar (\mathbf{n} - j) \tag{3.85}$$

omitting the subscript $(_1)$ on a and $\mathbf{n}$. The formalism is incorrect whenever n exceeds $2j$, but within the allowed range, it may be verified that the basic commutation relations (3.12–3.16) are correctly obeyed. (The rationalization of the square root is discussed, following (3.101) in the section on "spins one.")

If all that one requires is the satisfaction of these commutation laws, and relaxes the requirement that J^+ and J^- be Hermitean conjugate operators, then one may perform the so-called Maléev similarity transformation to a new set of operators

$$J^+ = a^* (2j)^{1/2} \left(1 - \frac{\mathbf{n}}{2j} \right) \hbar \quad J^- = (2j)^{1/2} a \hbar \quad \text{and} \quad J^z = (\mathbf{n} - j) \hbar \,. \tag{3.86}$$

The obvious advantage, the rationalization of the square root, is somewhat offset by the complications introduced by the nonunitary transformation.

---

[6] T. Holstein and H. Primakoff, *Phys. Rev.* **58**, 1048 (1940).

A variant of the above is due to Villain.[7] Assuming $\mathbf{J} \leq |J_z|$, he writes $J^{\pm}$ as

$$J^+ = e^{i\varphi}\sqrt{\left(J + \frac{1}{2}\hbar\right)^2 - \left(J_z - \frac{1}{2}\hbar\right)^2}, \quad \text{and} \quad J^- = (J^+)^\dagger \qquad (3.87)$$

from which (3.15d, 3.40–3.42) flow easily. The correct commutation relations are ensured by requiring, as in (3.7), or (3.42),

$$[\varphi, J_z] = i\hbar, \quad \text{i.e.,} \quad \varphi = \hbar i\, \partial/\partial J_z \quad \text{or} \quad \mathbf{J}_z = -i\hbar\partial/\partial\varphi. \qquad (3.88)$$

Thus, $\exp(-ix\varphi)$ is a translation operator for J_z and

$$[J_z, e^{\pm im\varphi}] = \pm\hbar m\, e^{\pm im\varphi}. \qquad (3.89)$$

In the semi-classical limit, one just drops the terms $\pm\frac{1}{2}\hbar$ in (3.87). The extreme quantum limit $j = \frac{1}{2}$ or 1 deserves separate investigation and will be discussed in the following sections.

3.12.   Spins One-Half: Special Tricks

For spins one-half, $n = 0$, 1 in the preceding and the Holstein-Primakoff and Maléev representations are identical within the physically allowed subspace.

Problems involving N interacting spins one-half may be more easily formulated with the aid of one of several Fermion representations. This takes us back eighty years, to the historic paper of Jordan and Wigner on second quantization[8] in which the Fermion anticommuting operators were explicitly constructed out of Pauli spin matrices. It is, however, the converse which is of present interest.

Fermion operators, which we denote by the letter c, are a set of *anticommuting* operators; for example, for any state $|\Phi\rangle$,

$$c_1^* c_2^* |\Phi\rangle = -c_2^* c_1^* |\Phi\rangle \qquad (3.90)$$

and therefore, care must be taken of the order in which the operators are written, see Chapter 4 for further discussion. The vacuum is defined, just as for spins, as the state annihilated by all c's,

$$c_i|0\rangle = 0 \quad \text{and} \quad \mathbf{n}_i|0\rangle = 0 \quad (i = 1, \ldots, N) \qquad (3.91)$$

---

[7] T. Villain, *J. de Phys.* **35**, 27 (1974).
[8] P. Jordan and E. Wigner, *Z. Phys.* **47**, 631 (1928).

and is the state in which the particle number operators

$$\mathbf{n}_i = c_i^* c_i \tag{3.92}$$

all have zero eigenvalue.

The anticommutation relations are indicated by curly brackets and are

$$c_i c_j + c_j c_i \equiv \{c_i, c_j\} = 0, \quad \{c_i^*, c_j^*\} = 0, \quad \{c_i, c_j^*\} = \delta_{ij}. \tag{3.93}$$

Setting $i = j$ in the above equations yields the relations

$$c_i^2 = (c_i^*)^2 = 0, \quad c_i^* c_i + c_i c_i^* = 1 \tag{3.94}$$

which are identically the equations obeyed by the Pauli spin matrices $\sigma^{\pm}$. It is only the fact that Pauli spin matrices referring to different particles $(i \neq j)$ *commute* with one another, that differentiates them from the anticommuting Fermion operators above. This may be remedied by introducing a set of drone operators d_i and d_i^*, equal in number to the original set of c's, that anticommute with the latter and obey amongst themselves anticommutation relations entirely analogous to (3.93). As a consequence,

$$(d_i + d_i^*)^2 \equiv 1 \quad \text{and} \quad \{d_i + d_i^*, d_j + d_j^*\} = 0 \quad \text{for } i \neq j \tag{3.95}$$

so that finally

$$S_i^+ = \hbar c_i^* (d_i + d_i^*) \quad S_i^- = \hbar (d_i + d_i^*) c_i \quad \text{and} \quad S_i^z = \hbar \left(c_i^* c_i - \frac{1}{2} \right) \tag{3.96}$$

is the desired set of spin one-half operators, which *commute* when $i \neq j$.

Problem 3.7. Prove that the various representations in this section obey (3.12–3.16), and also that different spins *commute*:

$$[S_i^k, S_j^{k'}] = 0 \quad \text{for } i \neq j$$

for all components $k, k' = x, y, z$ of the spin vectors.

A second useful representation is the exact analogue of the coupled-Boson picture, but replaces the Bosons by Fermions. The paired Fermions, instead of being labeled 1, 2, will be labeled by $\uparrow$ and $\downarrow$ to make more explicit the role of each operator. Thus,

$$S_i^+ = \hbar c_{i\uparrow}^* c_{i\downarrow}, \quad S_i^- = \hbar c_{i\downarrow}^* c_{i\uparrow}, \quad \text{and} \quad S_i^z = \frac{\hbar}{2} (c_{i\uparrow}^* c_{i\uparrow} - c_{i\downarrow}^* c_{i\downarrow}), \tag{3.97}$$

where the anticommutation relations are

$$\{c_{i,m}, c_{j,m'}\} = \{c^*_{i,m}, c^*_{j,m'}\} = 0,$$

$$\{c_{i,m}, c^*_{j,m'}\} = \delta_{i,j}\delta_{m,m'} \quad (m, m' = \uparrow \text{ or } \downarrow). \tag{3.98}$$

This is a very important representation, corresponding to the second quantization of electrons *cum* spin and we shall return to it in Chapter 6.

A final representation of the spins one-half brings us closest to the work of Jordan and Wigner. The technique we use has been useful in the solution of one-dimensional problems, and also in the two-dimensional Ising model, an important subject in the statistical mechanics of magnetism. Therefore, last but not least, set

$$S^+_i = \hbar c^*_i Q_i \quad S^-_i = \hbar Q_i c_i \quad \text{and} \quad S^z_i = \frac{\hbar}{2}(2c^*_i c_i - 1), \tag{3.99}$$

where

$$Q_i = Q^*_i = Q^{-1}_i = \exp\left(i\pi \sum_{j<i} c^*_j c_j\right) \tag{3.100a}$$

$$= \exp\left(i\pi \sum_{j<i} s^+_j s^-_j\right) \tag{3.100b}$$

$$= \Pi_{j<i}(c^*_j + c_j)(c^*_j - c_j), \tag{3.100c}$$

etc. In actual problems, the choice of ordering $i = 1, \ldots, N$ is crucial, because the phase factors Q_i introduce great complexity into a problem unless means are found to eliminate them. When this is possible, however, then the representation above is the simplest of all, because it is the only one to establish a one-to-one correspondence between spins and Fermions, and their respective eigenstates, i.e. it is *irreducible*.

3.13.   Spins One

There is no special representation for spins one, except for that which deals specifically with vector fields, such as the electromagnetic field. However, the Holstein-Primakoff representation may be used after rationalizing the square root. For, in the physically admissible range $n = 0, 1, 2$, the equation

$$\sqrt{1 - n/2} \equiv 1 - \left(\frac{3}{2} - \sqrt{2}\right)n - \frac{1}{2}(\sqrt{2} - 1)n^2 \tag{3.101}$$

is exact, and only fails for $n \geq 3$, which is outside the domain of validity of this particular representation. In fact for any j, a polynomial of order $2j$ can be found which has the same structure as the Holstein-Primakoff root (but is quite tedious to construct for large j). Because the Taylor series expansion of the root,

$$\sqrt{1 - \frac{\mathbf{n}}{2j}} \doteq 1 - \frac{\mathbf{n}}{4j} + \cdots ,$$

becomes asymptotically correct in the limit $j = \infty$, it is commonly used in approximate theories.

We have not dealt with vector fields, except in Problem 3.6, where it was suggested that the "intrinsic spin" of the photon was unity. There are good reasons why this should be so. When angular momentum generates infinitesimal rotations of the coordinates, it changes not only the arguments of a vector field, as it does a scalar field (e.g., ordinary wavefunctions), but also mixes the various components of the field amongst themselves. The total effect might be described by an operator

$$J^z = \frac{\hbar}{i} \left(\frac{\partial}{\partial \phi} - \hat{\mathbf{u}}_z \times \right) = L^z + S^z \tag{3.102}$$

of which the first term takes care of arguments, and the second of the rotations of the vector field components. Similar expression for the two other Cartesian components results in an operator $\mathbf{J} = \mathbf{L} + \mathbf{S}$. Recall that $\mathbf{J}$ can be a true angular momentum only if $\mathbf{L}$ and $\mathbf{S}$ each are, in their own right. Therefore, let us investigate this new angular-momentum operator,

$$\mathbf{S} = i\hbar(\hat{\mathbf{u}}_x \times , \ \hat{\mathbf{u}}_y \times , \ \hat{\mathbf{u}}_z \times) \tag{3.103}$$

specifically constructed to operate on vector fields, such as the vector potential $\mathbf{A}(\mathbf{r})$. Calculating $S^2 = S_x^2 + S_y^2 + S_z^2$, we readily find

$$S^2 = \hbar^2(2) \tag{3.104}$$

and therefore the spin magnitude $s = 1$.

Field amplitudes may then be expanded in the eigenvectors of S^z. There are 3 of them,

$$\mathbf{e}_{\pm 1} = \frac{-1}{2}(i\hat{\mathbf{u}}_y \pm \hat{\mathbf{u}}_x) \quad \text{and} \quad \mathbf{e}_0 = \hat{\mathbf{u}}_z \tag{3.105}$$

with eigenvalues,

$$S^z \mathbf{e}_r = \hbar r \mathbf{e}_r , \quad r = -1, 0, 1 . \tag{3.106}$$

These are particularly useful in problems with some spherical symmetry, e.g., the field of a radiating atom. Linear combinations of functions having definite J and total azimuthal quantum number M are called *vector* spherical harmonics

$$\mathscr{Y}_{JlM}(\theta, \phi) = \sum_{mr} Y_{lm}(\theta, \phi) \mathbf{e}_r (lm1r|l1JM). \tag{3.107}$$

In addition to J and M, only the l of the spherical harmonic need be indicated, for the spherical unit vectors invariably have unit angular momentum, as we have shown.

The vector spherical harmonics form a complete, orthonormal set for the description of vector fields. The normalization integral, including scalar product, is

$$\int_0^{2\pi} d\phi \int_0^{\pi} d\theta \sin\theta \mathscr{Y}^*_{Jlm}(\theta, \phi) \cdot \mathscr{Y}_{J'l'M'}(\theta, \phi) = \delta_{JJ'} \delta_{ll'} \delta_{MM'}. \tag{3.108}$$

3.14.   Quadratic Forms

We have already discussed the scalar product of two angular momentum operators, say $\mathbf{S}_1 \cdot \mathbf{S}_2$. We now obtain the eigenvalues of this operator and find them to be parameterized only by invariants.

We start by completing the square

$$\mathbf{S}_1 \cdot \mathbf{S}_2 = \frac{1}{2}(\mathbf{S}_1 + \mathbf{S}_2)^2 - \frac{1}{2}(\mathbf{S}_1^2 + \mathbf{S}_2^2) \tag{3.109}$$

then replace $\mathbf{S}_1^2$ by its magnitude $s_1(s_1 + 1)$, $\mathbf{S}_2^2$ by $s_2(s_2 + 1)$, and $\mathbf{S}_1 + \mathbf{S}_2$ by the compound operator $\mathbf{S}_{\text{tot}}$, obtaining

$$\mathbf{S}_1 \cdot \mathbf{S}_2 = \frac{1}{2}[s_{\text{tot}}(s_{\text{tot}} + 1) - s_1(s_1 + 1) - s_2(s_2 + 1)] \tag{3.110}$$

with s_{tot} taking on any possible integer value ranging from $|s_1 - s_2|$ to a maximum $(s_1 + s_2)$. Substituting in the above we find for the maximum and minimum values of $\mathbf{S}_1 \cdot \mathbf{S}_2$

$$-s_2(s_1 + 1) \leq \mathbf{S}_1 \cdot \mathbf{S}_2 \leq s_1 s_2 \tag{3.111}$$

assuming $s_2 \leq s_1$. The maximum, occurring when both spins are parallel, could have been predicted classically. The minimum, corresponding to antiparallel orientation, is lower by a fraction $1/s_1$ than the classical

estimate; this foreshadows the complications of antiferromagnetism compared to ferromagnetism.

For the special case of spins one-half only, the quadratic form can be
diagonalized by a nonlinear transformation.[9] We introduce two new spin
one-half operators denoted $\mathbf{J}$ and $\mathbf{P}$, constructed out of the components of
$\mathbf{S}_1$ and $\mathbf{S}_2$ in the following manner ($\hbar = 1$):

$$(J_x, J_y, J_z) = (S_{1x}, 2S_{1y}S_{2z}, 2S_{1z}S_{2z}), \quad \text{and}$$

$$(P_x, P_y, P_z) = (S_{2z}, 2S_{1x}S_{2y}, -2S_{1x}S_{2x}). \tag{3.112}$$

The inverse of these relations is

$$(S_{1x}, S_{1y}, S_{1z}) = (J_x, 2J_yP_x, 2J_zP_x), \quad \text{and}$$

$$(S_{2x}, S_{2y}, S_{2z}) = (-2P_zJ_x, 2P_yJ_x, P_x). \tag{3.113}$$

In the new representation, the quadratic form is diagonal

$$\mathbf{S}_1 \cdot \mathbf{S}_2 = \frac{1}{2}(J_z - P_z) + J_zP_z \tag{3.114}$$

where $J_z = \pm\frac{1}{2}$ and $P_z = \pm\frac{1}{2}$ in the units $\hbar = 1$. For some problems this
formulation may be superior to (3.110).

3.15. Projection Operators

Supposing we are given a state function of many spins and wish to know
which of them are entangled, whether in singlet pairs or otherwise. There is
then some benefit to having recourse to projection operators. Let us illustrate
with two-spin operators for spins 1/2 and also for spins 1, setting $\hbar = 1$ in
all applications below.

Spins 1/2. Two such spins (label them 1 and 2) can be either in a singlet or in a triplet configuration, or in a linear combination within a state
vector of all the spins, $|\Phi\rangle$. In the first instance, $S_1 \cdot S_2|\Phi\rangle = -3/4|\Phi\rangle$ and
in the second $S_1 \cdot S_2|\Phi\rangle = +1/4|\Phi\rangle$. Thus when the rotationally invariant
"orthogonal singlet projection operator" $P_{12} \equiv (S_1 \cdot S_2 + 3/4)$ is applied to
an arbitrary state vector of many spins, such as $|\Phi\rangle$), it yields zero if the
two spins are in a singlet configuration and 1 if they are in a triplet state.
Similarly $Q_{12} \equiv (1/4 - S_1 \cdot S_2)$ is a complementary operator that yields 1 in

---

[9] D. Mattis and S. Nam, *J. Math. Phys.* **13**, 1185 (1972).

the singlet configuration and vanishes when applied to any triplet. Naturally, $P \cdot Q \equiv 0$.

Problem 3.8.   With the aid of Pauli operators construct the joint operator $P \cdot Q = (S_1 \cdot S_2 + 3/4) \cdot (1/4 - S_1 \cdot S_2)$ and show explicitly it is identically zero when operating on an *arbitrary* configuration of the two spins one-half.

Spins **1.** Let us construct orthogonal-singlet, orthogonal-triplet and orthogonal-quintet projection operators. Using the identity $S^2 = 2$ for spins $s = 1$, the rotationally invariant operator of two such spins, that vanishes on joint singlet states, is proportional to

$$P^{(0)} = (S_1 + S_2)^2 \quad \text{(which simplifies to } 2(2 + S_1 \cdot S_2)). \tag{3.115}$$

Similarly, any operator that vanishes on triplet states must be proportional to

$$P^{(1)} = (S_1 + S_2)^2 - 2 \quad \text{(simplifies to } 2(1 + S_1 \cdot S_2)). \tag{3.116}$$

Finally, an orthogonal projection operator that vanishes when applied to quintuplet states, where the total angular momentum of the two spins adds up to 2, is

$$P^{(2)} = 6 - (S_1 + S_2)^2 \quad \text{(simplifies to } 2(1 - S_1 \cdot S_2)). \tag{3.117}$$

With these one is able to construct an operator that yields a nonvanishing result *only on joint singlet states*. It has to be proportional to the product of $P^{(1)} P^{(2)}$. When normalized such that it yields 1 on singlets and zero otherwise it is denoted by Q:

$$Q^{(0)} = ((S_1 \cdot S_2)^2 - 1)/3. \tag{3.118}$$

Similarly, the normalized operator that yields a nonvanishing result (i.e., 1) on joint triplet states only, $P^0 P^2$, is,

$$Q^{(1)} = (2 + S_1 \cdot S_2)(1 - S_1 \cdot S_2)/2$$
$$= (2 - S_1 \cdot S_2 - (S_1 \cdot S_2)^2)/2. \tag{3.119}$$

Finally, the operator that yields 1 only on the joint quintuplet states and zero otherwise, *viz.*, $P^{(0)} P^{(1)}$, is

$$Q^{(2)} = (2 + 3 S_1 \cdot S_2 + (S_1 \cdot S_2)^2)/6. \tag{3.120}$$

This last has been used to construct a model spin-one linear chain antiferromagnet, the ground state and elementary excitations of which are exactly calculable.[10] Higher spins and higher dimensions are also amenable to this particular methodology. This remarkably interesting topic is examined in Chapter 5.

If desired, three- or four-spin correlation functions may be constructed by similar means. The need for such correlations arises when one wishes to calculate, say, the specific heat of a magnetic substance for which the Heisenberg model of magnetism is applicable. The magnetic specific heat is calculated using the fluctuation-dissipation theorem, and is expressible as a sum over bond-bond correlation terms such as

$$\langle S_i \cdot S_j S_k \cdot S_l \rangle_{\text{TA}}$$

("TA" = thermal average). We revisit this topic in a later chapter on thermodynamics.

---

[10] I. Affleck, T. Kennedy, E. H. Lieb and H. Tasaki, *Phys. Rev. Lett.* **59**, 799 (1987) and *Commun. Math. Phys.* **115**, 477 (1988).

Chapter 4

Magnetism and the Many-Body Problem

We study one-, two- and many-electron systems for a clue to their collective behavior and the emergence of nonzero spins. The examples of collective magnetic behavior are taken from atomic, molecular and solid-state physics and include the simplest theories that explain them.

4.1. Hamiltonian Physics and Degeneracy

Here, as in much of this book, we make use of the Hamiltonian formalism of quantum mechanics in conjunction with the time-independent Schrödinger equation $\mathcal{H}\psi = E\psi$. Generally speaking, the "wave function"[1] ψ (or Ψ) contains complete information concerning the state of the particle(s)[2] while $\mathcal{H}$ contains the dynamics, or at least that the part of the dynamics in which energy is *conserved* (not converted into "heat").

In the case of a single particle with spatial coordinates $\mathbf{r} = (x, y, z)$ and spin coordinate $\sigma_z = \uparrow$ or $\downarrow$ (i.e., $\pm\hbar/2$), ψ is explicitly a function of the independent variables: $\psi(x, y, z; \sigma_z)$. In non-relativistic physics and *in the absence of an external magnetic field*, $\mathcal{H}(\mathbf{p}, \mathbf{r})$ is just the sum of kinetic energy $\mathbf{p}^2/2m$ and potential energy $V(\mathbf{r})$ and is independent of spin.

In the Hamiltonian formulation of classical mechanics it is the potential energy $V(\mathbf{r})$, not the external forces $\mathbf{F}$, that enter into $\mathcal{H}$. If this V is constructed with "conservative potentials," the Hamiltonian is a "constant

---

[1] We use ψ in describing a single particle, Ψ to indicate $N \geq 2$ particles.
[2] Where $|\psi|^2$ is the related probability, the integral of which is conventionally normalized to 1.

of the motion," i.e., $\mathcal{H}(\mathbf{p}, \mathbf{r}) = E$, a constant. This constant is the particle's energy, determined by initial conditions. However, in the context of classical mechanics, there is no such thing as "spin."

Wave mechanics, which reflects Nature more accurately, requires $\mathbf{p}$ to be a differential operator in $\mathbf{r}$, *viz.*

$$\mathbf{p} = (p_x, p_y, p_z) \quad \text{where} \quad p_x = \frac{\hbar}{i}\frac{\partial}{\partial x}, \quad p_y = \frac{\hbar}{i}\frac{\partial}{\partial y} \quad \text{and} \quad p_z = \frac{\hbar}{i}\frac{\partial}{\partial z}.$$

$$(4.1)$$

Then $\mathcal{H} = E$ where E is a continuous variable has to be replaced by an eigenvalue problem known as the time-independent one-particle Schrödinger equation $\mathcal{H}\psi = E\psi$. It is a second-order differential equation in the variables x, y, and z. The spectrum of allowed energy eigenvalues E is determined partly by spatial boundary conditions and partly by the strength and symmetries of the potential $V(\mathbf{r})$. Let us now recapitulate some truisms from elementary quantum theory (actually from linear algebra), starting with the Schrödinger equation:

$$\mathcal{H}\psi = -\frac{\hbar^2}{2m}\nabla^2\psi + V\psi = E\psi \, ; \qquad (4.2)$$

it is subject to a specified condition on ψ on some bounding surface. (Generally, both ψ and $\nabla\psi$ are continuous and $\psi(\infty) = 0$.)

Each solution ψ of (4.2) and its corresponding E is labeled by a set of spatial quantum numbers α. If a particle carries spin this adds an extra label. Because a particle has to be *somewhere*, the integral over its probability is $\int |\psi|^2 d^3r = 1$. This condition is called *normalization* of ψ. Any two distinct solutions of (4.2) are, or can be made, *orthogonal*, i.e., $\int \psi_{\alpha'}^* \psi_\alpha d^3r = 0$ if $\alpha' \neq \alpha$. These last two statements can be combined as,

$$\int \psi_{\alpha'}^* \psi_\alpha d^3r = \delta_{\alpha,\alpha'} \, , \quad \text{the "orthonormality condition."} \qquad (4.2a)$$

The obverse of the orthonormality condition is the "completeness relation":

$$\sum_\alpha \psi_\alpha^*(r)\psi_\alpha(r') = \delta(r - r') \, , \qquad (4.2b)[3]$$

that allows expanding an arbitrary function in a generalized Fourier series in the ψ_α's. Additionally, if the particle whose wave function satisfies (4.2) is

---

[3] Here enters the Dirac delta function $\delta(\mathbf{r})$, zero everywhere except at $\mathbf{r} = 0$ where it is infinite, such that its integral is 1.

an electron with spin, each state requires, in addition to the quantum label α, a spin label $\sigma_z = \pm 1/2$. In the absence of magnetic fields, two states that differ only in the values of the spin quantum number, have the same energy. This additional two-fold degeneracy is called "Kramers' degeneracy." Because it is discrete, one can use σ_z either as a label (quantum number) or as a coordinate, with equal justification.

Example: The Harmonic Oscillator. The isotropic harmonic oscillator $V = \frac{K}{2}r^2$ provides an example that is easy to work out. Because H is separable in Cartesian coordinates (also in many other coordinate systems: spherical, etc.) we can write $\psi(\mathbf{r}) = f_1(x)f_2(y)f_3(z)$, thereby reducing (4.2) to three equations of identical form in a single variable each:

$$-\frac{\hbar^2}{2m}\frac{\partial^2}{\partial x^2}f_\alpha(x) + \frac{K}{2}x^2 f_\alpha(x) = e_\alpha f_\alpha(x),$$

$$-\frac{\hbar^2}{2m}\frac{\partial^2}{\partial y^2}f_\beta(y) + \frac{K}{2}y^2 f_\beta(y) = e_\beta f_\beta(y),$$

$$-\frac{\hbar^2}{2m}\frac{\partial^2}{\partial z^2}f_\gamma(z) + \frac{K}{2}z^2 f_\gamma(z) = e_\gamma f_\gamma(z),$$

$$\text{with energy } E \equiv e_\alpha + e_\beta + e_\gamma. \tag{4.3}$$

The particle's quantum number is (α, β, γ), actually a set of 3 quantum numbers. The energies are the sum of partial energies along each axis. In the x-direction the partial energy eigenvalue is $e_\alpha = \hbar\omega(n_\alpha + \frac{1}{2})$, and similarly in y and z directions. Here ω is the classical radial frequency $(K/m)^{1/2}$ and each n_κ is a non-negative integer. The functions f, each of them *normalized*, come from solving the identical harmonic oscillator equation and differ only in their subscript indicating the number of nodes n_κ and the partial energy e_κ. We know f_α has precisely n_α nodes, f_β has precisely n_β nodes, etc.[4]

Except for the ground state, all eigenstates of this Hamiltonian are degenerate, increasingly so at higher quantum numbers. There is only one way to achieve energy $E_0 = \hbar\omega(3/2)$. Setting $(n_\alpha, n_\beta, n_\gamma) = (1,0,0),(0,1,0)$ and $(0,0,1)$ yields three distinct ways to attain $E_1 =$

---

[4] A "node" in f is any root ($f = 0$), except one dictated by the boundary condition. Thus, a *nodeless* function is positive (or negative) everywhere (except possibly at the boundaries). In 1D, and for *any* V (including the harmonic potential), the energy e increases monotonically with the number of nodes.

$\hbar\omega(5/2)$. There are six ways to attain $E = \hbar\omega(7/2)$, etc. Briefly, the number of ways to have $E_k = \hbar\omega(k + 3/2)$ increases monotonically as the integer $k = (n_\alpha + n_\beta + n_\gamma)$ is increased.

Problem 4.1. Find an expression for the "degeneracy" $g(k)$, defined as the number of distinct ways to obtain an energy $E_k = \hbar\omega(k + 3/2)$ in the isotropic three-dimensional harmonic oscillator. Relate it to the number of nodal surfaces in $\psi(\mathbf{r})$.

With these features taken into account, *any and all one-body eigenfunctions* of this isotropic oscillator must take the form,

$$\psi_a = f_n(x)f_{n'}(y)f_{n''}(z)\chi(\sigma_z), \tag{4.4}$$

labeled by a set of 4 quantum numbers, $a = (\alpha, \sigma_z)$ where $\alpha = (n_\alpha, n_\beta, n_\gamma)$ are three integers corresponding to the number of nodes in x, y, z directions respectively and σ_z assumes one of 2 allowed values. *All a's that are of this form are admissible.*[5]

4.2. The Pauli Principle

The situation changes dramatically when two or more electrons occupy the same space and are subject to the same potential, *regardless* of whether they have a physical interaction (i.e., exert forces on each other) or not. The Pauli principle states that under the transposition of (all) the coordinates of *any* two fermions or, more generally, under *any odd* permutation of the particles, $\Psi(r_1, \sigma_{z,1}|r_2, \sigma_{z,2}|r_3, \sigma_{z,3}|r_4, \sigma_{z,4}|\cdots)$ is *antisymmetric*,[6] i.e., $\Psi(r_2, \sigma_{z,2}|r_1, \sigma_{z,1}|r_3, \sigma_{z,3}|r_4, \sigma_{z,4}|\cdots) = -\Psi(r_1, \sigma_{z,1}|r_2, \sigma_{z,2}|r_3, \sigma_{z,3}|r_4, \sigma_{z,4}|\cdots)$. This poses a severe constraint on the wave functions of many-fermion systems and limits the number of them that are physically admissible. A similar symmetry principle: Ψ is *symmetric* under permutations, can be formulated for bosons, where the consequences are less dramatic (if you do not count Bose-Einstein condensation as a dramatic consequence).

---

[5] Hence the number of eigenstates of $\mathcal{H}$ is a countable infinity.
[6] Amusingly, in order to render the particles indistinguishable we must first label, i.e., distinguish them!

In the case of *several* identical electrons the global wave function Ψ depends on all the coordinates of each and every particle. For ease of notation let $(x_1, y_1, z_1; \sigma_1) \equiv 1$ be short-hand for all the coordinates of the first particle, $(x_2, y_2, z_2; \sigma_2) \equiv 2$ for all of the second, etc. (In this notation the Hamiltonian in Eq. (4.3) is labeled $\mathcal{H}_1$.) The Pauli principle requires anti-symmetry under any odd permutation:

$$\Psi(1, 2, 3, \ldots, j, \ldots) = -\Psi(2, 1, 3, \ldots, j, \ldots)$$

$$= +\Psi(2, 1, j, \ldots, 3, \ldots), \text{ etc.} \tag{4.5}$$

It ensures that the probability density $|\Psi|^2$ is appropriately *symmetric* under permutations of the particles. Given that the particles are indistinguishable their permutations cannot be observed. This is a physical requirement; for unlike Ψ, the probability density $|\Psi|^2$ *is* experimentally observable. But what is meant by indistinguishable particles?

First, the many-particle Hamiltonian of indistinguishable particles has to be invariant (i.e., totally symmetric) under the interchange of the individual particles themselves. To take advantage of this symmetry, any eigenstate Ψ of the many-particle Hamiltonian has to be either totally antisymmetric (for fermions) or totally symmetric (for bosons). Given that electrons are fermions, we are interested in the former. But it is not just the Hamiltonian that is invariant in the case of indistinguishable particles: the total momentum operator is the sum of the individual momenta, hence is totally symmetric, ditto for the total angular momentum operator, the total spin operators, etc.

Example: Several (non-interacting) fermions in the same harmonic well. Each has its own energy eigenvalue $E(k_a)$ (where $k_a = (n_\alpha + n_\beta + n_\gamma)$) and wave function $\psi_a(j)$. One might suppose the many-particle eigenfunction $\Psi(1, 2, \ldots, j, \ldots)$ to be a *product state*:

$$\psi_a(1)\psi_{a'}(2) \cdots = \Psi_{a,a',a'',\ldots}(1, 2, \ldots) \text{ and the energy to be additive:}$$

$$E_{\text{total}} = \sum_a E(k_a).$$

When used as a variational approximation to the energy of a collection of *interacting* particles, this product state yields what is known as the *Hartree approximation*. However, in order to eliminate unphysical states, it *must* be supplemented by the following artificial subsidiary condition: *no two sets of quantum numbers can be identical*, i.e., $a \neq a' \neq a'' \cdots$ This condition was discovered experimentally when quantum theory was first applied to atomic physics. It is at the heart of the "building up" principle used to explain the atomic periodic table and also nuclear structure; this condition is sometimes referred to, somewhat imprecisely, as the "exclusion principle."

The chosen Hartree function $\Psi_{a,a',a'',...}(1,2,...)$ has no particular symmetry under permutations and thus it *fails to satisfy the Pauli principle*, Eq. (4.5). Nor is it in any way unique: any permutation $\Psi_{a,a',a'',...}(2,1,...)$ has precisely the same energy E_{total}, the same total momentum, etc. Given the subsidiary condition mentioned above, there are $N!$ such permutations, any one of which is essentially equivalent to the others. Note that none of them satisfies the Pauli principle. So, which to choose?

Now, any linear combination of such *degenerate eigenstates* is itself an eigenstate. Noting this, J. C. Slater concocted his eponymous determinantal function that *does* satisfy the Pauli principle:

$$\Psi = \frac{1}{\sqrt{N!}} \begin{vmatrix} \psi_a(1) & \psi_{a'}(1), \cdots \\ \psi_a(2) & \psi_{a'}(2), \cdots \\ \psi_a(3) & \psi_{a'}(3), \cdots \\ \cdot & \cdot, \cdots \end{vmatrix} \tag{4.6}$$

The main diagonal is just the Hartree function, $\Psi_{a,a',a'',...}(1,2,...)$. More importantly, if two or more sets of quantum numbers a, a' coincide the determinant has two identical columns and vanishes identically; the subsidiary condition is no longer artificially tacked on but arises quite naturally from the form of (4.6). Because of its antisymmetry under transposition of any two *rows*, Slater's determinant perfectly expresses the Pauli principle (4.5). When used as a variational expression with which to extract the energy of a system of interacting particles, this determinantal function gives rise to what is known as the *Hartree-Fock (H-F) approximation*.[7]

Stated in other words, the Slater determinant (4.6) is the appropriately normalized *totally antisymmetric* component of the Hartree product wave function. But how does spin enter into this? Given Kramers' degeneracy,

---

[7] The accuracy of the best H-F approximation in determining the ground state energy of filled (or half-filled) shells in atomic physics is excellent, typically better than 1%; however its accuracy decreases in the cases of partly-filled shells and for all excited states, as it does not deal well with *correlations*. This limits its applicability to metal physics, where levels are closely spaced and correlations are important. For more insight into atomic many-electron physics, the book by I. Lindgren and J. Morrison, *Atomic Many-Body Theory*, Springer-Verlag, Berlin, 1982 and 2nd Edition, 1986.

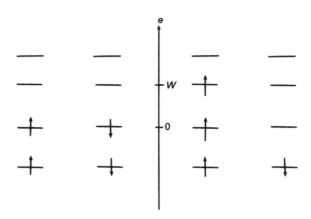

Fig. 4.1. The ground state of 4 noninteracting fermions (left-hand side) has total spin $S_{tot} = 0$. In this example the excited state (right-hand side) of total spin 1 *costs* a *minimum* energy W to create.

two electrons of opposite spins ($\sigma_1 = \uparrow$ and $\sigma_2 = \downarrow$ and/or vice versa) *can* share the subset of *spatial* quantum numbers n. The net (i.e., composite) spin of such a pair is then identically zero. Often this pairing leads to a state of minimal energy. One good example in the theory of metals is the "Cooper pair," consisting of an electron of spin $\uparrow$ and momentum $\mathbf{p}$ paired with a second electron of spin $\downarrow$ and momentum $-\mathbf{p}$. When combined with a state in which the electron of spin $\downarrow$ and momentum $\mathbf{p}$ **is** paired with a second electron of spin $\uparrow$ and momentum $-\mathbf{p}$, the result is a Cooper pair of zero total momentum and zero spin. Such a pair is at the heart of the BCS theory of superconductors, and similar pairings have been shown to lower the energy of even-even nuclei.

    An important consequence of the determinantal form of the wave functions is that the ground state for N *noninteracting* fermions (e.g., an Hamiltonian in which the two-body forces are replaced by suitably averaged, spin-independent, one-body potentials) has a total spin closer to zero than to $N/2$. The proof is trivial: in the ground state we place two particles in the lowest level, 2 in the next lowest level, etc. If the number of particles is even (or odd), the *total spin* will be 0 (or 1/2) in the ground state. To polarize the N particles' spins in a magnetic field, particles have to be de-paired (see Fig. 4.1).

    Recapitulating: of all $N!$ possible Hartree product states that one can construct using a given set of quantum numbers $\{a\}$ of space and spin, *only the totally antisymmetric linear combination, i.e., the Slater determinant, satisfies the Pauli principle.* The other $N! - 1$ states are inadmissible.

In the following section we study just two electrons. In this special case, just as in general for free (noninteracting) fermions, we find it *always* "costs" energy to create a net, nonzero total spin just as it costs energy to have a nonzero total angular momentum. This feature of electron physics has to be understood and overcome before it becomes possible to formulate *any* credible theory of magnetism in solids.

4.3.   The Two-Fermion Problem

Among well-known examples of two-electron systems we note that the helium (^{4}He) atom and the hydrogen molecule (H_2) are devoid of net spin and also devoid of total angular momentum in the ground state. We shall see that this is not a coincidence, but a general feature of two-electron systems subject to the Pauli principle. For three or more electrons in 2 or 3 spatial dimensions, the situation changes dramatically.

Consider two identical, non-relativistic fermions subject to any one- and/or two-body potential $V(r_1, r_2)$. If the two particles are to be *indistin-guishable*, (a) their masses must be equal and (b) $V(r_1, r_2) = V(r_2, r_1)$ has to be invariant under their interchange.[8] (There are no other requirements of the potential energy V except that it be real.)

Generalizing (4.2), the two-body Schrödinger equation reads,

$$\mathcal{H}\Psi(1,2) = -\frac{\hbar^2}{2m}(\nabla_1^2 + \nabla_2^2)\Psi(1,2) + V(r_1, r_2)\Psi(1,2) = E\Psi(1,2). \quad (4.7)$$

This Hamiltonian $\mathcal{H}$ is not separable, as it cannot be written in the form $\mathcal{H}_1 + \mathcal{H}_2$. Consequently the eigenfunctions of space and spin, the Ψ's, are not products of one-body eigenfunctions and the energy eigenvalues E cannot be written as the sum of the individual particles' energies. But because spin does not explicitly enter into $\mathcal{H}$ we may consider a simpler eigenvalue problem, $\mathcal{H}\Phi(r_1, r_2) = E\Phi(r_1, r_2)$, in which $\mathcal{H}$ and E are the same as in (4.7) but the spin labels are missing. Given that $\mathcal{H}$ is invariant under the interchange of the spatial coordinates and that it is real, its eigenstates $\Phi(r_1, r_2)$ can be

---

[8] If the two fermions were distinguishable, as in the case of an electron (r) and a proton (R) that have different masses, then $\mathcal{H}$ is not invariant under their permutation even though $V(r - R)$ is. In this case, Ψ is not, and need not be, antisymmetric.

chosen real and *either* symmetric or antisymmetric under permutation of the two arguments. We now prove the following.

Theorem: The nodeless function has the lowest energy ("lies lowest"), hence the *ground state* Φ of 2 electrons is *symmetric* under transposition of the two electrons' coordinates.[9] It is also unique (non-degenerate). By the Pauli principle, the two particle's spin function must then be *antisymmetric*, i.e., a "*spin singlet*" $\chi_0 = \frac{1}{\sqrt{2}}(\uparrow_1\downarrow_2 - \uparrow_2\downarrow_1)$, where $\uparrow_1$ stands for $\sigma_{z,1} = +1/2$, etc. The function χ_0 is the prototype spin zero state.

The simplest proof is *by contradiction*. One starts by *assuming an anti-symmetric ground state* and finds an inconsistency. Suppose then that the *antisymmetric* eigenstate $A(r_1, r_2) = -A(r_2, r_1)$ is unique and belongs to the (lowest) energy eigenvalue E_0. Its energy is:

$$E_0 = \frac{\int\{-\frac{\hbar^2}{2m}(|\nabla_1 A(r_1, r_2)|^2 + |\nabla_2 A(r_1, r_2)|^2) + V(r_1, r_2)A^2(r_1, r_2)\}d^3r_1 d^3r_2}{\int A^2(r_1, r_2)d^3r_1 d^3r_2}.$$

(4.8a)

Next we construct the *symmetric* state $S(r_1, r_2) \equiv |A(r_1, r_2)|$. This last is not an eigenstate of $\mathscr{H}$ because, perpendicular to the nodal surfaces (defined by $A(r_1, r_2) = S(r_1, r_2) = 0$) its normal derivatives are discontinuous ($S(r_1, r_2)$ has a cusp). Using $S(r_1, r_2)$ to calculate the expectation value of $\mathscr{H}$ we obtain a "variational" energy, E_{Sym} that *has to be* higher than the ground state because $S(r_1, r_2)$ does not satisfy the Schrödinger equation (cusps are not allowed if the potential is nonsingular).

$$E_{\text{Sym}} = \frac{\int\{-\frac{\hbar^2}{2m}(|\nabla_1 S(r_1, r_2)|^2 + |\nabla_2 S(r_1, r_2)|^2) + V(r_1, r_2)S^2(r_1, r_2)\}d^3r_1 d^3r_2}{\int S^2(r_1, r_2)d^3r_1 d^3r_2}$$

$$= E_0.$$

(4.8b)

According to the variational principle, $E_{\text{Sym}} > E_0$. Yet, by construction, $E_{\text{Sym}} \equiv E_0$.

This conclusion is untenable. It follows that an antisymmetric state cannot be the ground state. By similar argumentation one shows that the symmetric state of lowest energy, i.e., the true ground state, *is nodeless*. Because two nodeless states cannot be orthogonal the ground state of the two particles is *nondegenerate* (unique).   Q.E.D.

---

[9] An antisymmetric function has, by definition, at least one node.

4.4.   Electrons in One Dimension: A Theorem

> ... no one could move to the right or left to make way for passers-by, it
> followed that no Linelander could ever pass another. Once neighbors,
> always neighbors. Neighborhood with them was like marriage with
> us. Neighbors remained neighbors till death did them part.[10]

Because electrons in one dimension can be ordered like beads on a string,
the topology of their nodal surfaces turns out relatively simply. One can
generalize the result found for two electrons, and prove the following:

*In one dimension, not even interacting and overlapping electrons can be
ferromagnetic.*[11] A somewhat milder extension of this theorem, in which
the type of interactions is restricted, can also be proved for N electrons in
three dimensions (as we shall see in a subsequent section).

Before stating the theorem and proceeding to the proof, it is appropriate
to comment on its relevance. This theorem occupies a special place when
we speculate concerning the origins of ferromagnetism. Any fundamental
explanation or theory, that conceivably would lead to ferromagnetism in
one dimension, must *ipso facto* be false! *That* is the reason we must discuss
angular momentum, atomic theory, and the origins of Hund's rules, all of
which are intimately connected with the three-dimensionality of space and,
by virtue of the present theorem, are an essential part of the phenomena.
Deferring further such considerations, we now go on with the

Theorem: For interacting electrons in one dimension, the ground state in
any M subspace is nondegenerate, and belongs to $S_{tot} = |M|$. From this it
follows that $E_0(S) < E_0(S + 1)$, that the current vanishes in the ground
state of any M subspace, and that electrons in one dimension are entirely
nonferromagnetic.

Proof: The properties of noninteracting particles will now be adapted to
the present case of interacting electrons, for which the Hamiltonian is

$$\mathcal{H} = -\frac{\hbar^2}{2m} \sum_{j=1}^{N} \frac{\partial^2}{\partial x_j^2} + V(x_1, x_2, \ldots, x_N). \tag{4.9}$$

The potential V includes all external forces, periodic or other, and the interactions among electrons which may be taken to be arbitrary. The only

---

[10] E. A. Abbott, *Flatland*, Grant Dahlstrom, Pasadena, California, 1884.
[11] E. Lieb and D. Mattis, *Phys. Rev.* **125**, 164 (1962).

exclusions are as follows: no velocity or spin-dependent forces or nonlocal potentials will be admitted, and V must be integrable, for reasons which are or will become clear. We shall adopt a standard boundary condition: If the size of Lineland is D, the boundary condition will be that the wave functions vanish whenever any x_n reaches $\pm\frac{1}{2}D$. Admissible Pauli eigenfunctions are $^M\Psi^S(1,2,\ldots,N)$, where S labels total spin and M the eigenvalue of total S_Z along the z-axis.

For compactness, $n = 1,2,\ldots$ stands for the couple of space and spin coordinates (x_n, σ_n). First, we write the eigenvalue equation,

$$\mathscr{H}\,^M\Psi^S = E(S)\,^M\Psi^S \tag{4.10}$$

noting that the energy cannot depend on M. (Because $\mathscr{H}$ commutes with the spin operators, we may apply spin raising and lowering operators to both sides of Schrödinger's equation, changing M without affecting the energy eigenvalue.)

Let us now eliminate the spins, and consider the space functions $^Mf^S$, which themselves obey Schrödinger's equation (4.10) with precisely the eigenvalue $E(S)$. All of them can be brought into the $M = 0$ subspace, if the total number of electrons is even, by application of spin raising or lowering operators if necessary, and therefore we start by examining this particular subspace in which the ground-state eigenfunction is surely to be found. Consider an arbitrary eigenfunction, of unknown S_{tot}, in which particles $1\cdots N/2$ are associated with spin up, the rest spin down:

$$^0f(x_1,\ldots,x_{N/2}|x_{N/2+1},\ldots,x_N) \tag{4.11}$$

in the region R defined as follows:

$$R \equiv \begin{cases} -\dfrac{1}{2}D < x_1 < x_2 < \cdots < \dfrac{1}{2}D \\[2mm] -\dfrac{1}{2}D < x_{N/2+1} < \cdots < x_N < \dfrac{1}{2}D. \end{cases} \tag{4.12}$$

The natural boundaries of R occur where the inequalities are just barely violated, e.g.,

$$x_1 = -\frac{1}{2}D \quad \text{or } x_n = x_{n\pm1} \quad \text{or } x_{N/2} = \frac{1}{2}D, \tag{4.13}$$

etc., and it should be noted that ^0f vanishes identically on these boundaries.

According to the conditions (4.13), the wave function vanishes whenever any two coordinates to the left of the bar coincide or whenever two coordinates to the right of the bar coincide. That is because the Pauli principle requires the wave function to be antisymmetric in particles of spin up and, separately, antisymmetric in the particles of spin down. There is no particular symmetry for permutations across the bar that separates the ups from the downs.

Although the total configuration space consists of the variables $x_1, \ldots,$ x_N in any ordering, *when ^0f is known in R it is known everywhere*, because the other regions can be found by trivial permutations $\mathscr{P}$ of the two sets of coordinates. For example, if $(x_1, \ldots)$ is a point in R and $(x'_1, \ldots)$ is a point in R' and $\mathscr{P}$ is a permutation that connects R to R', $\mathscr{P}R = R'$, then we obtain ^0f in R' by the trivial permutation, $^0f(x'_1, \ldots) \equiv (-1)^P \mathscr{P}[^0f(x_1, \ldots)]$. And in this self-evident manner, we can find the function everywhere. Indeed, with the sign convention used above, it is guaranteed that ^0f will have the proper antisymmetries under its trivial permutations $\mathscr{P}$.

Now, we state and prove a *lemma* which is at the core of the theorem: The eigenfunction of $\mathscr{H}$ that belongs to the lowest energy eigenvalue (and is subject only to the boundary condition that it vanish on the boundaries of R) is *nodeless* in R.

The proof proceeds exactly as in the case of two electrons. If the ground-state function had nodes, then its absolute value would be a nodeless function, which would have identically the same variational energy but which would not itself be an eigenfunction. Since the variational energy of any but a ground-state eigenfunction must exceed the ground-state energy, there is a contradiction unless the absolute value of the ground-state eigenfunction is itself a ground-state eigenfunction; which is only possible if the ground state is nodeless. The nodeless state is, moreover, nondegenerate. For two nodeless functions cannot be orthogonal, but all the eigenfunctions of $\mathscr{H}$ can be made mutually orthogonal, and other than nodeless functions have been excluded by the preceding arguments.

The "nodeless" property is independent of the magnitude of V, and so is the shape of R. We may therefore compare in R the ground state of a Hamiltonian $\mathscr{H}$ with arbitrary V, to that of the noninteracting Hamiltonian $\mathscr{H}_0$ with $V = 0$. In each case, the ground state is nodeless; the two corresponding functions cannot be orthogonal, therefore they share the same quantum numbers.

What these quantum numbers are for noninteracting electrons ($V = 0$) is something which was already discussed: viz., *zero*. And by the above it follows that both these quantum numbers vanish for arbitrary V as well, in the ground state. Next, one introduces the fundamental region suitable for $M = 1$, simply by moving the bar to the right by one notch:

$$R \equiv \begin{cases} -\frac{1}{2}D < x_1 < \cdots < x_{N/2+1} < \frac{1}{2}D \\ -\frac{1}{2}D < x_{N/2+2} < \cdots < x_N < \frac{1}{2}D. \end{cases} \tag{4.14}$$

By analogy, one proves that the ground-state function in this region is nodeless and belongs to the same quantum numbers regardless of V. It follows by comparison with the results for $V = 0$ that this function has $S_{\text{tot}} = 1$; the proof for higher M proceeding in identically the same manner, until maximum S is attained and the theorem for the interacting particles is proved.

As for the current operator, it will be the same for the interacting electrons as for the noninteracting ones, namely zero. That is because the ground state functions are real and unique, and a finite current can only flow if the state is degenerate.

4.5.   The Wronskian[12]

A somewhat different approach may serve to clarify the results obtained so far. Suppose we have a set of real space eigenfunctions of the Schrödinger equation

$$\mathscr{H} f_j(x_1, \ldots) = E_j f_j(x_1, \ldots)$$

with

$$\mathscr{H} = -\frac{\hbar^2}{2m} \sum \frac{\partial^2}{\partial x_n^2} + V(x_1, \ldots)$$

and it is desired to order the eigenvalues E_j in a sequence of increasing energy.

The lowest energy belongs to the nodeless eigenfunction; since the permutation operators commute with the Hamiltonian, it can also be established that this function is totally symmetric under the interchange of any coordinates and therefore is not of any use for more than two Fermions. But there

---

[12] E. Lieb and D. Mattis, unpublished work, 1962.

can only be one nodeless function and the others can be examined as to the topology of their nodal surfaces (defined as the locus of $f = 0$).

We shall not go into any detail beyond proving the simple but interesting *lemma*:

If there are two functions f_i and f_j such that a nodal surface S_i of f_i, encloses a region R_i which neither contains, nor is intersected by any nodal surface of f_j, then $E_i > E_j$. For example, consider any eigenfunction f_i together with the nodeless eigenfunction f_0; the lemma implies $E_i > E_0$, which is obviously true for all i.

To prove the lemma, it is necessary to multiply the equation on the left by f_i; to multiply the analogous equation for f_i on the left by f_j, substract the first from the second, and obtain

$$-\frac{\hbar^2}{2m}\nabla \cdot (f_j\nabla f_i - f_i\nabla f_j) = (E_i - E_j)f_if_j \,.$$

The potential has been eliminated from this expression, and the left-hand side involves only the kinetic energies or the gradient of the Wronskian, for which we have used the generalized Laplacian notation

$$\nabla \cdot \nabla = \nabla^2 \equiv \sum_{n=1}^{N} \frac{\partial^2}{\partial x_n^2} \,.$$

By assumption, neither function changes sign in R_i so we may, in considering this region, assume both functions are positive, and if not, make them so. Next, integrate over R_i, which by Green's theorem gives an integral of the Wronskian over S_i on the left-hand side, and a positive integral times the energy difference on the right.

$$-\frac{\hbar^2}{2m}\int_{S_i} dS \cdot (f_j\nabla f_i - f_i\nabla f_j) = (E_i - E_j)\int_{R_i} d\tau f_if_j \,.$$

But $f_i = 0$ on the nodal surface S_i; and the normal derivative of f_i is negative on S_i because the function is positive inside the volume bounded by the surface and negative outside. So the left-hand side is positive and the equation can only be satisfied if $E_i > E_j$, which proves the lemma.

Evidently, we have used nothing about dimensionality in this demonstration, therefore it is indeed equally valid for electrons in three dimensions.

But the difficulty of finding nodal surfaces has so far frustrated attempts to apply these results to realistic problems for $N > 2$ in three dimensions.

4.6. States of Three Electrons

Let us describe the states of three electrons by means of Slater determinants. The most general single determinant is

$$
\begin{vmatrix}
\varphi_a(r_1)\chi_r(1) & \varphi_a(r_2)\chi_r(2) & \varphi_a(r_3)\chi_r(3) \\
\varphi_b(r_1)\chi_s(1) & \varphi_b(r_2)\chi_s(2) & \varphi_b(r_3)\chi_s(3) \\
\varphi_c(r_1)\chi_t(1) & \varphi_c(r_2)\chi_t(2) & \varphi_c(r_3)\chi_t(3)
\end{vmatrix}
\qquad (4.15)
$$

aside from a normalization constant. For fixed $\varphi_a, \varphi_b, \varphi_c$, the two possibilities for *each* spin function result in eight distinct determinantal functions in total. The spin functions are defined as follows: $\chi_r(1)$ is a spinor of particle 1, which can be up or down. Similarly for particles 2 and 3. $\chi_s(2)$ refers to particle 2 and can be the same as or different from $\chi_r(2)$, and similarly for χ_t. If some of the space functions are not linearly independent, then the number of independent determinantal functions could be less than eight, e.g., if $a = b = c$ and $r = s = t$, the number of possible determinantal functions is zero. But in no case does it exceed eight for three particles. Note that if there were no exclusion principle, and no requirement of antisymmetrization, the number of allowed product configurations for $a \neq b \neq c$ would be $6 \times 6 \times 6 = 216$ in the present case. So, a considerable number of states have been pruned.

Next let us look at this function in more detail. There are two instances of all spins "up" or "down" simultaneously,

$$
\begin{vmatrix}
\varphi_a(r_1)\chi_r(1) & \varphi_a(r_2)\chi_r(2) & \varphi_a(r_3)\chi_r(3) \\
\varphi_b(r_1)\chi_r(1) & \varphi_b(r_2)\chi_r(2) & \varphi_b(r_3)\chi_r(3) \\
\varphi_c(r_1)\chi_r(1) & \varphi_c(r_2)\chi_r(2) & \varphi_c(r_3)\chi_r(3)
\end{vmatrix}
\qquad (4.16a)
$$

in which the spin coordinate $r = +$ or $-$. The remaining six determinants are, in turn,

$$
\begin{vmatrix}
\varphi_a(r_1)\chi_r(1) & \varphi_a(r_2)\chi_r(2) & \varphi_a(r_3)\chi_r(3) \\
\varphi_b(r_1)\chi_s(1) & \varphi_b(r_2)\chi_s(2) & \varphi_b(r_3)\chi_s(3) \\
\varphi_c(r_1)\chi_s(1) & \varphi_c(r_2)\chi_s(2) & \varphi_c(r_3)\chi_s(3)
\end{vmatrix}
\qquad (4.16b)
$$

with $r = +$ and $s = -$, or the converse; and

$$\begin{vmatrix} \varphi_a(r_1)\chi_s(1) & \varphi_a(r_2)\chi_s(2) & \varphi_a(r_3)\chi_s(3) \\ \varphi_b(r_1)\chi_r(1) & \varphi_b(r_2)\chi_r(2) & \varphi_b(r_3)\chi_r(3) \\ \varphi_c(r_1)\chi_s(1) & \varphi_c(r_2)\chi_s(2) & \varphi_c(r_3)\chi_s(3) \end{vmatrix} \qquad (4.16c)$$

and finally

$$\begin{vmatrix} \varphi_a(r_1)\chi_s(1) & \varphi_a(r_2)\chi_s(2) & \varphi_a(r_3)\chi_s(3) \\ \varphi_b(r_1)\chi_s(1) & \varphi_b(r_2)\chi_s(2) & \varphi_b(r_3)\chi_s(3) \\ \varphi_c(r_1)\chi_r(1) & \varphi_c(r_2)\chi_r(2) & \varphi_c(r_3)\chi_r(3) \end{vmatrix} \qquad (4.16d)$$

all with $r = +$ and $s = -$, or the converse, corresponding to one spin up and two down, or the converse. Of all these determinants, the spin functions factor out of only the two quartet states of (4.16a). Two more quartet states must therefore be intimately mixed in with the two sets of doublet states in the determinants of (4.16b–4.16d).

We see that the determinantal method does not necessarily generate eigenfunctions of S_{tot}^2, although by construction all determinants are eigenfunctions of total S_z.

The doublet states can be projected from the above by means of the obviously constructed projection operator that acts only on the χ's,

$$\mathbf{D} \equiv \frac{1}{3}\left(\frac{15}{4} - \mathbf{S}_{\text{tot}}^2\right) = \frac{1}{2} - \frac{2}{3}(\mathbf{S}_1 \cdot \mathbf{S}_2 + \mathbf{S}_2 \cdot \mathbf{S}_3 + \mathbf{S}_3 \cdot \mathbf{S}_1) \qquad (4.17)$$

which has zero eigenvalue on the quartet states, and eigenvalue unity on the doublet states. The resulting states are not single Slater determinants, but linear combinations of them.

Problem 4.2. Find eigenfunctions of the projection operator of (4.17) using the set of determinantal functions (a–d) as a basis set.

Trial functions to calculate the low-lying eigenvalues of a Hamiltonian for three electrons could be made of linear combinations of the eight determinantal functions. Moreover, if $\mathcal{H}$ does not involve the spins explicitly, we can always decompose it into noninteracting blocks by using the projection operators $\mathbf{D}$ and $1-\mathbf{D}$. The "quartet," i.e., fourfold degenerate states, $S_{\text{tot}} = \frac{3}{2}$

belong to the totally antisymmetric space function. The two sets of doublet states, $S_{\text{tot}} = \frac{1}{2}$, give rise to two distinct doubly degenerate doublet levels.

Problem 4.3.   Construct the projection operator **R** which has zero eigenvalue for triplet states and eigenvalue unity for singlet states of two electrons. Operate with **R** on the most general Slater determinant of two electrons, and show that the singlet state is always symmetric under the interchange of the spatial coordinates of the two particles, and the triplet antisymmetric.

4.7.   Eigenfunctions of Total S^2 and S_z

If a function of space and spin coordinates is totally antisymmetric, we shall denote it a "Pauli function": the Slater determinants are a special case. If in addition it is an eigenfunction of S_{tot}^2 and S_{tot}^z, then there are severe limitations on it.

Let us assume first that the Pauli function is an eigenfunction of S_{tot}^z with eigenvalue M — just like the Slater determinants of Eqs. (4.15–4.16).

$$S_{\text{tot}}^z \, {}^M\Psi = M \, {}^M\Psi \tag{4.18}$$

indicating the quantum number explicitly as a superscript. Let us now expand (4.18) in a complete set of product spin functions,

$$\chi_{r_1}(1)\chi_{r_2}(2)\cdots\chi_{r_N}(N). \tag{4.19}$$

Normally 2^N such functions would be needed, but the restriction to a definite M value means that

$$M + \frac{1}{2}N \text{ have spin up } (+) \text{ and} - M + \frac{1}{2}N \text{ have spin down } (-). \tag{4.20}$$

We can take a typical such function, with each ξ a spinor (up or down),

$$^M\chi(\xi_1, \xi_2, \ldots, \xi_{M+N/2} | \xi_{M+N/2+1}, \ldots, \xi_N)$$

$$\equiv \chi_+(1)\cdots\chi_+(N/2)\cdots\chi_-(N) \tag{4.21}$$

and generate all the others by permutations. The permutations of the spins *up* among themselves do not generate a new function, and neither do permutations of the spins *down* among themselves. We therefore call these the *trivial permutations* of the spin function $^M\chi(\xi_1, \ldots | \ldots, \xi_N)$. This function is totally symmetric under the trivial permutations, those of spin coordinates to the left of the vertical bar amongst themselves, or of the spin coordinates to the right of the bar amongst themselves.

There are, however, $[N!/(\frac{1}{2}N + M)!(\frac{1}{2}N - M)!] = n_M$ *nontrivial* permutations generating n_M orthogonal spin functions including the original function in (4.21) and these comprise the complete set for the expansion of $^M\psi$. Note that

$$\sum_{M=-N/2}^{N/2} n_M = \sum_{M=-N/2}^{N/2} \frac{N!}{(\frac{1}{2}N + M)!(\frac{1}{2}N - M)!} = 2^N \qquad (4.22)$$

the dimensionality of all the M subspaces adding up to the total correct number 2^N of orthogonal spin functions.

When $^M\psi$ is expanded in these functions, the coefficient of $^M\chi$ in the expansion must be a *space* function of the type,

$$^Mf(r_1, r_2, \ldots, r_{M+N/2}|r_{M+N/2+1}, \ldots, r_N) \qquad (4.23)$$

which is totally antisymmetric under the trivial permutations of coordinates to the left of the bar amongst themselves, or of coordinates to the right of the bar amongst themselves. In general, however, it has complicated transformation properties under the n_M nontrivial permutations.

If in addition, the Pauli function is an eigenfunction of S_{tot}^2, with eigenvalue S,

$$S_{\text{tot}}^2 \, {}^M\Psi^S = S(S+1)\,{}^M\Psi^S, \qquad S_{\text{tot}}^z \, {}^M\Psi^S = M\,{}^M\Psi^S \qquad (4.24)$$

then we can actually figure out the nontrivial symmetries of the space functions $^Mf^S$ using only simple notions about angular momentum. We do this systematically, starting with $^{-S}f^S$, which is the coefficient of the spin function of (4.21) when $M = -S$. Recall that a fundamental property of angular momentum is that $|M| \leq S$, and therefore it should be impossible to further decrease M. In more formal terms,

$$S_{\text{tot}}^-(^{-S}\Psi^S) \equiv 0. \qquad (4.25)$$

This requires that there be some linear relationship among the f functions and their nontrivial permutations, and indeed, by using (4.21, 4.23, 4.25), we find

$$^{-S}f^S(r_1,\ldots,r_{N/2-S}|r_{N/2-S+1},\ldots,r_N)$$

$$-\sum_{j=N/2-S+1}^{N}\mathscr{P}_{ij}{}^{-S}f^S(r_1,\ldots,r_{-S}|r_{N/2-S+1},\ldots,r_N)=0 \qquad (4.26)$$

with

$$i=1,2,\ldots,\frac{1}{2}N-S$$

and $\mathscr{P}_{ij}$ being the transposition operator for the coordinates of the particles i and j.

We can construct the f functions belonging to higher M values, using $^{-S}f^S$'s that properly obey the equation above. This is done by repeated applications of the total-spin raising operator,

$$S_{\text{tot}}^+ \, ^{-S}\Psi^S = \, ^{-S+1}\Psi^S \quad \text{unnormalized} \qquad (4.27)$$

and leads directly to

$$^{-S+1}f^S(r_1,\ldots,r_{N/2-S+1}|r_{N/2-S+2},\ldots)$$

$$= \, ^{-S}f^S - \sum_{j=1}^{N/2-S}\mathscr{P}_{j,N/2-S+1}{}^{-S}f^S \qquad (4.28)$$

with

$$^{-S}f^S \equiv \, ^{-S}f^S(r_1,\ldots,r_{N/2-S}|r_{N/2-S+1},\ldots)$$

a function that obeys (4.26).

Thus, the attempt to move the bar to the right does *not* fail. By further application of the spin-raising operator we progressively generate $^{-S+2}f^S,\ldots$, up to $^{-S}f^S$, all expressed as *linear combinations of the original function* $^{-S}f^S$ *and its nontrivial permutations*. Finally, the angular-momentum rule

$$S_{\text{tot}}^+({}^{+S}\Psi^S) \equiv 0 \qquad (4.29)$$

gives us an additional set of constraints on the f functions, similar to the first ones we found *supra*. But now, these constraints are that the bar cannot be moved farther to the *right*, i.e.,

$$
{}^{S}\!f^{S}(r_1,\ldots,r_{N/2+S}|r_{N/2+S+1},\ldots,r_N)
$$

$$
-\sum_{j=1}^{N/2+S}\mathscr{P}_{j,q}{}^{S}\!f^{S}(r_1,\ldots,r_{N/2+S}|r_{N/2+S+1},\ldots,r_N)=0 \qquad (4.30)
$$

with

$$
q=\frac{1}{2}N+S+1,\ldots,N\,.
$$

Given an arbitrary space function, it is always a simple matter to generate a function of the type ${}^{M}\!f$, that is, the space partner of the spin eigenfunction of S_{tot}^{z}. We merely antisymmetrize in the first set of $p=\frac{1}{2}N+M$ coordinates, then antisymmetrize in the set of the remaining coordinates. Only those permutations which we have denoted as "trivial" are used in this process, the others being used to raise or lower M, as we have already seen. If for some reason f is unsuitable, the process of antisymmetrization will yield identically *zero* instead of the desired function. For example: if f is of the form ${}^{0}\!f^{0}$ and we wish to construct a function ${}^{M}\!f(M\neq 0)$ with it, we get zero instead (i.e., like trying to antisymmetrize a symmetric function).

There is no correspondingly simple rule for constructing functions of the type ${}^{M}\!f^{S}$ from an arbitrary function. It is necessary to project out the unwanted components belonging to values of the total spin other than S by repeated use of the spin raising and lowering operators $S_{\mathrm{tot}}^{\pm}$, after first constructing ${}^{M}\!f$, with $M=\pm S$. But in some special cases this can be done by inspection, as in the following example:

Example: A product wave function describes noninteracting particles, and also has its uses in variational approximations, as we have seen. Let us use such a product space function as the raw material in an example where we construct the proper space functions corresponding to definite M and S eigenvalues.

Starting with such a product as

$$
\varphi_1(r_1)\varphi_2(r_1)\cdots\varphi_N(r_N) \qquad (4.31)
$$

we can immediately write the ^Mf function as a product of two determinants

$$
^Mf = \det \begin{vmatrix} \varphi_1(r_1) & \varphi_1(r_2) & \cdots & \varphi_1(r_{N/2+M}) \\ \varphi_2(r_1) & & & \\ & & & \\ \varphi_{N/2+M}(r_1) & & \cdots & \varphi_{N/2+M}(r_{N/2+M}) \end{vmatrix}
$$

$$
\times \det \begin{vmatrix} \varphi_{N/2+M+1}(r_{N/2+M+1}) & \cdots & \varphi_{N/2+M+1}(r_N) \\ & & \\ & & \\ \varphi_N(r_{N/2+M+1}) & \cdots & \varphi_N(r_N) \end{vmatrix} \tag{4.32}
$$

which is explicitly antisymmetric in the two sets of variables. Now assume $M \leq 0$. (The case $M > 0$ can be handled by simply interchanging the two determinants.) Thus the first determinant is smaller than the second. Now, if all the functions $\varphi_1, \varphi_2, \dots, \varphi_{N/2+M}$ that appear in the smaller determinant are also present in the bigger one, then the antisymmetrization procedure of (4.26) will obviously yield *zero*, and therefore the above function belongs to the definite S value,

$$
S = |M|. \tag{4.33}
$$

Of course this requires that the various φ_j not be all distinct functions. When the φ_j are all distinct, the construction of a function of definite S is possible but more tedious.

The various φ_j in *each* determinant must be distinct or else ^Mf will vanish identically. But each can appear in both determinants (this is merely another expression of Kramers' degeneracy) for the appearance of a φ_j in the first determinant means there is an electron spin "up" in orbital φ_j, whereas its appearance in the second determinant means there is an electron spin "down" in that orbital.

This completes the demonstration, of how the Pauli wave function could be totally antisymmetric while the space functions, presumably eigenfunctions of a space Hamiltonian, could have complex transformation properties under the permutation group. We have also shown that these transformation properties are a consequence of the rules concerning total spin angular momentum. We now proceed to specific applications.

4.8. Hund's Rules

The Hartree-Fock procedure allows one to solve the many-electron Schrö-dinger equation approximately by supposing each electron to move in a self-consistent, averaged potential. Numerical solution of the effective one-body Schrödinger equations is still required, as the self-consistent potentials are not strictly Coulombic so that the eigenfunctions are not obtainable in terms of elementary functions. Still, the one-electron angular momentum is useful to organize the electrons into shells, and the periodic structure of the atomic table is understood to be the consequence of the filling up of successive shells.

The first closed shell, that of helium, being devoid of net spin or angular momentum, is magnetically inert. Let us add two more electrons, to make beryllium, Be. It may be thought the outer shell obeys an effective two-electron Hamiltonian, the solutions of which must be orthogonalized to the core (helium) states. If the theorem of the preceding section holds for the two new electrons then the total spin and angular momentum of Be must also vanish, which is the case.

Next, we add two more electrons to form carbon, C, and we find the first surprise. Carbon has $S_{\text{tot}} = 1$, and $L_{\text{tot}} = 1$, and the theorem fails completely for the third pair of electrons! Indeed, if we look further, of the first eighteen elements to have an even number of electrons (which takes us up to krypton, $Z = 36$), only eight have no spin or angular momentum in the ground state; and of these, half are inert gases. Why does the argument go awry? One might argue that the theorem breaks down for the following mathematical reason: that orthogonality to core states is equivalent to nonlocal potential forces, for which the theorem is inapplicable. But there are more convincing physical explanations. The question is, how can one predict the magnetic properties of an atom in the ground state?

The answers are given by Hund's rules. These laws, originally based on the abundant spectroscopic evidence, were of course confirmed by complete atomic calculations. Fortunately, there exists a less exhaustive explanation of Hund's rules in terms of exchange, or the vector model, of which we shall give some examples.

First Rule: Electrons occupy the $2 \times (2l + 1)$ states of a shell in such a manner as to maximize the total spin. This may determine the configuration uniquely. When it does not, then one appeals to the second rule.

Second Rule: Any ambiguities in the first rule are resolved in favor of the highest value of L_{tot}. But the second rule is not always a unique prescription either, although these two rules are adequate to describe *grosso modo* the magnetic properties of the atom. Other rules determine whether the spin and angular momenta are parallel or not, but it is not necessary for us to become too involved with atomic structure. For with the exception of the rare earths, angular momentum is almost always quenched in the process of chemical binding or in the solid state, and only the spin and Hund's first rule survive.

Qualitatively, both Hund's rules follow from degenerate-state perturbation theory. Electrons in the outer shells move in the spherical potential of the nucleus and of the filled inert-gas configurations. To a first approximation, all the configurations within the outer shell are degenerate. It is the Coulomb repulsion between these electrons which lifts the degeneracy. Now if any two electrons are in the same orbital state, the Coulomb repulsion is maximal. These configurations should be avoided:

Therefore, in a first crude attempt to "prove" Hund's rules, or rather to make them more plausible without too much work, only those configurations of the outer shell electrons need be considered in which every electron occupies a different orbital state. This would imply that whenever a shell contains $2l + 1$ electrons, all the degenerate orbital states in that shell are occupied: for each ml there is a $-ml$; and therefore presumably $L_{\text{tot}} = 0$.

The validity of this simple guess is striking: there appear to be no exceptions in the periodic table. For example, elements with a half-filled p shell N, P, As, Sb, Bi, all have vanishing orbital angular momentum; elements with half-filled d shell, Cr, Mn, Mo, etc., also obey this rule.

For shells less than half-filled, one speaks of electrons, whereas for shells more than half-filled, one commonly speaks of "holes." There is indeed a great similarity between the ground terms of elements on either side of half-filled shells in the periodic table: with a general exception that the total J value (spin *plus* orbital angular momenta) equals $|S - L|$ for electrons and $S + L$ for holes. This difference is caused by spin-orbit coupling, a relatively small energy responsible for the phenomenon of "magnetic anisotropy" in the solid state. It changes sign with the "charge" of the particle.

To understand Hund's rules, the vector model, and particularly their physical origins, it is important to work out some examples. This is done in the following sections, but only for illustrative purposes, and the reader who wishes to become proficient in the true art of atomic calculation and quantum chemistry is referred to specialized treatises and review articles.

Hund's rules give only the ground state of the atom. The methods of atomic calculation which we shall illustrate give the (very important) low-lying excited states as well, together with their energy, spin, and angular-momentum quantum numbers. These results are central to the theory of magnetism, for it is impossible to understand magnetic solids without understanding the magnetic atoms and the magnetic molecules first.

Now we can return to our earlier question: why doesn't the two-particle theorem hold for the valence electrons of carbon? Are Hund's rules contra-dictory to our theorem? The answer, briefly, is this. The orbital states, in atoms heavier than helium, must be orthogonal to the (inert) core, and there-fore *cannot* be nodeless. The theorem, particularly the method of proof we used, just does not apply to these "excited state" orbitals. On the contrary; if the functions are already required to suffer a number of nodes, and to pay the ensuing price in kinetic energy, it is then advantageous for the potential energy to be minimized *in exactly the manner predicted by Hund's rules*, as we shall now see by actual calculation.

4.9. p^3 **Configuration**

First, recall that in hydrogenic atoms four atomic quantum numbers specify each electron: the principal quantum number $n = 1, 2, \ldots$, the orbital angular momentum $l \leq n - 1$, the magnetic quantum number $m_l = -l, \ldots, +l$, and the spin eigenvalue $m_s = \pm\frac{1}{2}$. But in the description of an electron, instead of giving $l = 0, 1, 2, 3, \ldots$, it is conventional to use the letters $s(l = 0), p(l = 1), d(l = 2), f(l = 3), g(l = 4)$, and thence in alpha-betical sequence. Therefore, $n = 2$ and $l = 0$ *is a 2s electron*. One does not indicate m_l or m_s, and the number of particles with given n and l is shown as a superscript. For example, two electrons with $n = 3$, $l = 2$ are denoted $3d^2$; three such electrons by $3d^3$. The m values not being specified, there are many possible wave functions for a given configuration. With the total angular momentum of the configuration indicated by a capital letter, for example, $S, P, D, F, \ldots$ for $L = 0, 1, 2, 3$, and the total spin S_{tot} *via* an exponent: $2S_{\text{tot}} + 1 =$ multiplicity, the various configurations are indicated in Table 4.3. The ground state has *maximum* multiplicity.

Consider N, nitrogen: outside the helium core, two electrons are in the $2s$ shell, and three in the $2p$ shell. The s electrons have no angular momentum

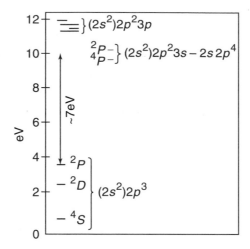

Fig. 4.2.   Energy-level diagram of nitrogen (N). Vector model applies to lowest three levels only; but note that they are well below the other levels. The ground state belongs to maximum $S_{\text{tot}} = 3/2$.

or spin, and we ignore them in zeroth order. The important particles are the three $2p$ electrons (the $2p^3$ configuration, in atomic notation). If such is the case, the ground-state and low-lying terms of any atom or ion with a p^3 configuration outside of closed shells must be similar. Figure 4.2 and Table 4.1 confirm this, and show the good agreement between theory and experiment.

We first tackle this problem by first-order perturbation theory.

First we write out the spatial basis functions, the three p states specified by the quantum numbers $m_l = -1, 0, +1$. We allow each to be occupied just once: this reduces both the Coulomb repulsion, and the number of spatial configurations necessary to consider.

The $3! = 6$ spatial configurations are

$$\psi_1 = \varphi_1(r_1)\varphi_0(r_2)\varphi_{-1}(r_3)$$
$$\psi_2 = \varphi_1(r_2)\varphi_0(r_1)\varphi_{-1}(r_3)$$
$$\psi_3 = \varphi_1(r_3)\varphi_0(r_2)\varphi_{-1}(r_1)$$
$$\psi_4 = \varphi_1(r_1)\varphi_0(r_3)\varphi_{-1}(r_2)$$
$$\psi_5 = \varphi_1(r_2)\varphi_0(r_3)\varphi_{-1}(r_1)$$
$$\psi_6 = \varphi_1(r_3)\varphi_0(r_1)\varphi_{-1}(r_2). \tag{4.34}$$

The subscripts on the one-particle functions refer to m_l, the eigenvalue of the one-electron operators L_i^z. These functions are all part of an orthonormal set and are by definition eigenstates of the best possible self-consistent one-particle Hamiltonian, $\mathcal{H}_0$. As $\mathcal{H}_0$ usually turns out to be spherically symmetric, we shall assume this to be the case here and factor the functions as follows:

$$\varphi_{m_l}(r) = Y_{l,m}(\theta, \phi) R(r). \tag{4.35}$$

Fortunately, everything is known about the angular part of the wave function, the spherical harmonics, $Y_{l,m}$, and as for the common radial function (which is unknown), it will be eliminated from the problem by calculating *ratios* of the energy splittings, such that the radial integrals appear in the numerator and denominator.

The perturbation is the sum of 2-body interactions:

$$\mathcal{H}' = e^2 \left(\frac{1}{r_{12}} + \frac{1}{r_{23}} + \frac{1}{r_{31}} \right). \tag{4.36}$$

An eigenvalue equation is constructed for the energies in terms of the matrix elements of $\mathcal{H}'$ between the six basis states.

Define the diagonal term as

$$A \equiv \mathcal{H}'_{1,1} = \cdots = \mathcal{H}'_{6,6} > 0 \tag{4.37}$$

and nondiagonal elements $B_1 A$ by

$$B_1 A \equiv \mathcal{H}'_{1,2} = \mathcal{H}'_{1,4} = \cdots = \mathcal{H}'_{4,6}$$

$$= \int \frac{e^2}{r_{12}} [\varphi_1^*(r_1)\varphi_0(r_1)\varphi_0^*(r_2)\varphi_1(r_2)] d^3 r_1 d^3 r_2. \tag{4.38}$$

Other nondiagonal elements $B_2 A$ are defined as,

$$B_2 A \equiv \mathcal{H}'_{1,3} = \cdots = \mathcal{H}'_{6,2}$$

$$= \int \frac{e^2}{r_{13}} [\varphi_1^*(r_1)\varphi_{-1}(r_1)\varphi_{-1}^*(r_3)\varphi_1(r_3)] d^3 r_1 d^3 r_3 = 2 B_1 A. \tag{4.39}$$

All matrix elements are proportional to the same constant A, also all turn out to be real or can be made real. We have made use of the orthonormality of the one-electron functions to eliminate all the irrelevant terms in $\mathcal{H}'$ from these expressions, and have indicated the difference between the two types of exchange integrals by the subscript B_1 or B_2. A third type of matrix element, for example, $\mathcal{H}'_{1,5}$, automatically vanishes by virtue of the orthogonality of the φ's.

Finally, the eigenvalue equation is,

$$
\mathcal{H}' \cdot \mathbf{v} = A
\begin{bmatrix}
1 & B_1 & B_2 & B_1 & 0 & 0 \\
B_1 & 1 & 0 & 0 & B_1 & B_2 \\
B_2 & 0 & 1 & 0 & B_1 & B_1 \\
B_1 & 0 & 0 & 1 & B_2 & B_1 \\
0 & B_1 & B_1 & B_2 & 1 & 0 \\
0 & B_2 & B_1 & B_1 & 0 & 1
\end{bmatrix}
\cdot \mathbf{v} = E \mathbf{v} .
\tag{4.40}
$$

The eigenfunctions are obtained from the basis vectors of this matrix, (a, b, c, d, f, g) as $\Phi = a\psi_1 + b\psi_2 + c\psi_3 + d\psi_4 + f\psi_5 + g\psi_6$. We observe that there is one eigenvector totally symmetric under permutations of the particles, one that is totally antisymmetric, and that the rest have a mixed symmetry. There is an additional operator of which we should like to know the eigenvalues, that is L^2_{tot}:

$$
L^2_{\text{tot}} = L_1^2 + L_2^2 + L_3^2 + 2(L_1^z L_2^z + L_2^z L_3^z + L_3^z L_1^z)
$$

$$
+ (L_1^+ L_2^- + L_2^+ L_3^- + L_3^+ L_1^- + \text{H.c.}) .
\tag{4.41}
$$

All that is required to calculate the matrix elements of this operator on our set of states is such information as

$$
L_i^2 = l(l+1) = 2 \quad \text{and}
\tag{4.42a}
$$

$$
(m_i + 1|L_i^+|m_i) = \sqrt{2}(\delta_{m_i,-1} + \delta_{m_i,0})
\tag{4.42b}
$$

for angular momentum l. Thus in the diagonal matrix element $(L^2_{\text{tot}})_{1,1}$ there enter three contributions of the form of (4.42a), making 6, minus 2 from $2L_1^z L_3^z$, for a total of 4. The other matrix elements are just as easily found, and finally,

$$
L^2_{\text{tot}} =
\begin{bmatrix}
4 & 2 & 0 & 2 & 0 & 0 \\
2 & 4 & 0 & 0 & 2 & 0 \\
0 & 0 & 4 & 0 & 2 & 2 \\
2 & 0 & 0 & 4 & 0 & 2 \\
0 & 2 & 2 & 0 & 4 & 0 \\
0 & 0 & 2 & 2 & 0 & 4
\end{bmatrix} .
\tag{4.43}
$$

It is now a simple exercise (for the reader) to write the eigenvalues E as functions of A, B_1, and B_2 and calculate the value of the angular momentum and spin for these same states:

With $B_1 \equiv b$ and $B_2 = 2B_1$ in Eq. (4.40) [cf. Eqs. (4.38 and 4.39)], the set of six eigenvalues of the Hamiltonian matrix comes out simply[13] as,

$$A\{1 - 4b, 1 - b, 1 - b, 1 + b, 1 + b, 1 + 4b\}, \tag{4.44}$$

each attached to a distinct eigenvector forming an othonormal set:

$$\{\{1, -1, -1, -1, 1, 1\}/\sqrt{6}, \ \{-1, -1, 1, 0, 0, 1\}/\sqrt{4},$$

$$\{-1, 0, 1, -1, 1, 0\}/\sqrt{4}, \ \{-1, 1, -1, 0, 0, 1\}/\sqrt{4},$$

$$\{-1, 0, -1, 1, 1, 0\}/\sqrt{4}, \ \{1, 1, 1, 1, 1, 1\}/\sqrt{6}\}. \tag{4.45}$$

The L_{tot}^2 matrix in Eq. (4.43) commutes with H and shares this set of eigenvectors. Operating on each of the 6 eigenvectors it yields,

$$\{\{0, 0, 0, 0, 0, 0\}/\sqrt{6}, \ \{-6, -6, 6, 0, 0, 6\}/\sqrt{4},$$

$$\{-6, 0, 6, -6, 6, 0\}/\sqrt{4}, \ \{-2, 2, -2, 0, 0, 2\}/\sqrt{4},$$

$$\{-2, 0, -2, 2, 2, 0\}/\sqrt{4}, \ \{8, 8, 8, 8, 8, 8\}/\sqrt{6}\}. \tag{4.46}$$

Reading off this result, we infer that $L = 0$ is associated with the first eigenvalue in (4.44), $1 - 4b$, that $L = 2$ (*note*: $2(2 + 1) = 6$) is the angular momentum for a state with energy $1 - b$ and finally, that $L = 1$ is the angular momentum of the state with $1 + b$. That takes care of 5 states.

A sixth — totally symmetric — state is inadmissible and therefore it is not paradoxical that its value of $L^2 = 8$ cannot be expressed as $L(L + 1)$.

In the usual spectroscopic notation, we write the angular momentum of the atom as a capital letter $(S, P, D, F, G\ldots)$ with superscript equal to the multiplicity $(2S_{\text{tot}} + 1)$. Now for three electrons, the total spin $S_{\text{tot}} = 1/2$ or $3/2$, hence the multiplicity can only be 2 (*doublet*) or 4 (*quartet*). That is, the quartet state ^4S has energy $A(1 - 4b)$, etc.

---

[13] For example, with the aid of software from Mathematica or Matlab or Maple.

Problem 4.4.   In this subspace of $S_{z,\text{tot}} = +1/2$, only 3 linear combinations of the 6 basis states are admissible Pauli functions. We already disposed of the totally symmetric function. Show which are the 3 acceptable functions and assign to them the correct value of $S_{\text{tot}} = 1/2$ or $3/2$. Construct the corresponding Pauli functions of space and spin. Explain how the apparent degeneracy of states with $L = 1$ and 2 is resolved.

The ratio of energy-level splittings is independent of A. Thus we calculate ratios:

$$\frac{E(^2P) - E(^2D)}{E(^2D) - E(^4S)} = \frac{2}{3} \tag{4.47}$$

a ratio independent of A and B_1, or of the integrals involving the radial parts of the wave functions.

The agreement with experiment is pleasant. Not only are all p^3 configurations found to possess low-lying terms of the type discussed, and ordered in energy as we have found them, but the ratio (4.47) is experimentally well obeyed within our self-imposed limits of accuracy. Figure 4.2 reproduces the term level scheme of nitrogen, together with the electronic configurations associated with the various terms. In Table 4.1 is given the experimentally observed ratio of (4.47) for various atoms and ions:

Table 4.1.   Configuration and ratio of term splittings in various atoms and ions versus theory.[14]

	Theory	Experimental			
		N	O^+	S^+	As
Configuration	np^3	$2p^3$	$2p^3$	$3p^3$	$4p^3$
Ratio[a]	0.667	0.500	0.509	0.651	0.751

[a]See (4.47).

Because each one-electron state is occupied, there is no difficulty in principle to the construction of an equivalent Heisenberg, or *vector model*, Hamiltonian for spins one-half. This is illustrated in Fig. 4.3.

---

[14] E. U. Condon and G. Shortley, *The Theory of Atomic Spectra*, Cambridge Univ. Press, New York, 1935, pp. 174–179.

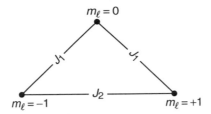

Fig. 4.3.   Labels (m_l) and bonds in vector model of nitrogen atom.

Assign a spin to each 1-particle state $-1, 0, +1$:

$$\mathcal{H}_{\text{Heis}} = -J_1(\mathbf{S}_{-1} \cdot \mathbf{S}_0 + \mathbf{S}_0 \cdot \mathbf{S}_{+1}) - J_2(\mathbf{S}_{-1} \cdot \mathbf{S}_{+1}). \qquad (4.48)$$

The proper choice of exchange constants is easily determined to be,

$$J_1 = 2B_1 A \quad \text{and} \quad J_2 = 2B_2 A \qquad (4.49)$$

both ferromagnetic. This is shown in the following Problem.

Problem 4.5.   Find the eigenvalues of the Heisenberg Hamiltonian (4.48) for three spins one-half. Use the Pauli principle, and space inversion symmetry considerations (on the true Pauli wave functions of space and spin) to classify the solutions of this exchange Hamiltonian according to L_{tot} as well as S_{tot}. Find the ratio $[E(^2P) - E(^2D)]/[E(^2D) - E(^4S)]$ and verify that it equals $\frac{2}{3}$ when $J_2 = 2J_1$. Hint: define $\mathbf{T} = \mathbf{S}_{-1} + \mathbf{S}_1$, express $\mathcal{H}_{\text{Heis}}$ in terms of $\mathbf{T}$ and $\mathbf{T} + \mathbf{S}_0$.

4.10. p^2 and p^4 Configurations

Carbon, silicon, germanium, tin all have two electrons in the p shell. Oxygen, sulphur, selenium, and others have two holes in the p shell. All have ^3P ground states. (The term level scheme for carbon, which is representative, is shown in Fig. 4.4.) Can we show this theoretically, and find the low-lying terms as well?

This is an exercise in Hund's second rule, which has not come into play before. Also, here is a problem in which the Heisenberg Hamiltonian is of dubious validity since there are only *two* particles, with *three* orbital states in which to put them. This configuration is studied in lieu of the complex transition series atoms, or rare-earth series, which are of greatest interest in

magnetism. For it is not intended to take the reader into involved computations, regardless of their relevancy, but only to provide her or him with a
conceptual introduction to the theory of atomic structure.

It is essential to reduce the number of configurations that have to be
considered to a manageable few. In the L–S coupling scheme (so-called),
states which can be obtained from each other by repeated applications of
$S_{\text{tot}}^{\pm}$, or $L_{\text{tot}}^{\pm}$, are equivalent except for the splitting due to spin-orbit coupling,
(which is usually treated separately). So, it is only necessary to consider the
largest attainable value of M_S and M_L for each term, viz.,

$$M_S = S_{\text{tot}} \quad \text{and} \quad M_L = L_{\text{tot}}, \tag{4.50}$$

and in order to determine S_{tot} and L_{tot} and the associated energy eigenvalues, to remain in the subspace of $M_S = 0, 1$ and $M_L = 0, 1, 2$.

$M_L = 2$: The space function must be

$$\varphi_1(\mathbf{r}_1)\varphi(\mathbf{r}_2) \tag{4.51}$$

in which $M_L = m_{l1} + m_{l2} = 1 + 1 = 2$. As this is symmetric under interchange
of the spatial coordinates, the Pauli principle requires an antisymmetric spin
function, the singlet state

$$\frac{1}{\sqrt{2}}[\chi_+(\xi_1)\chi_-(\xi_2) - \chi_+(\xi_2)\chi_-(\xi_1)]. \tag{4.52}$$

Thus, $L_{\text{tot}} = 2$ and $S_{\text{tot}} = 0$, and this is the ^1D state. The energy will be
computed subsequently, but now we continue the classification.

$M_L = 1$: The space function of interest must be

$$\frac{1}{\sqrt{2}}[\varphi_1(\mathbf{r}_1)\varphi_0(\mathbf{r}_2) - \varphi_1(\mathbf{r}_2)\varphi_0(\mathbf{r}_1)] \tag{4.53}$$

since the other possibility, the linear combination with $+$ sign, is symmetric
under the interchange of the spatial coordinates and is therefore the $M_L = 1$
projection of the $L_{\text{tot}} = 2$ state derived above. If the space part is antisymmetric, the spin part must be symmetric [the spin-triplet $\chi_+(\xi_1)\chi_+(\xi_2)$, or
$\chi_-(\xi_1)\chi_-(\xi_2)$, or (4.52) with $+$ replacing $-$]; thus $L_{\text{tot}} = 1$ and $S_{\text{tot}} = 1$,
and this is the ^3P configuration.

$M_L = 0$: The three functions with $M_L = 0$ have $m_{l1} = m_{l2} = 0$, or
$m_{l1} = -m_{l2} = \pm 1$. Two linear combinations of these are merely the $M_L = 0$
projections of the two states found above, (4.51, 4.53). The linear combina-

tion orthogonal to both of these is the new function of interest,

$$\frac{1}{\sqrt{3}}[\varphi_0(\mathbf{r}_1)\varphi_0(\mathbf{r}_2) - \varphi_{-1}(\mathbf{r}_1)\varphi_1(\mathbf{r}_2) - \varphi_{-1}(\mathbf{r}_2)\varphi_1(\mathbf{r}_1)]. \tag{4.54}$$

$L_{\text{tot}}^{\pm}$ applied to this state yields identically zero, and it follows that this is the ^1S configuration. The list is now complete (see Table 4.3).

The perturbation is

$$\mathscr{H}' = \frac{e^2}{r_{12}} \tag{4.55}$$

which we expand in Legendre polynomials,

$$\frac{1}{r_{12}} = \sum_{n=0}^{\infty} \frac{r_<^n}{r_>^{n+1}} P_n(\cos\omega) \quad \text{where} \quad \cos\omega = \frac{\mathbf{r}_1 \cdot \mathbf{r}_2}{r_1 r_2} \tag{4.56}$$

in which $r_<$ is the lesser, and $r_>$ is the greater of $\mathbf{r}_1$ and $\mathbf{r}_2$, and ω is the angle between the two vectors. Because one takes expectation values using functions of definite parity, only terms with $n = 0$ and $n = 2$ can survive the integration.

The atomic functions have normalized angular factors (cf. Table 3.1):

$$Y_{1,0} = (3/4\pi)^{1/2}\cos\theta \tag{4.57}$$

and

$$Y_{1,\pm1} = \mp(3/8\pi)^{1/2}\sin\theta e^{\pm i\varphi} \tag{4.58}$$

as well as a common (unknown) radial factor $R(r)$.

The term energies are the expectation values of the perturbation in the various states listed previously. The leading term in the expansion, corresponding to $n = 0$, is

$$\frac{1}{r_>} \tag{4.59}$$

which does not depend on angles and is therefore the same in all three states. This can be eliminated by a simple shift in the zero of energy.

The next term in the expansion, $n = 1$,

$$\frac{r_<}{r_>^2} P_1(\cos\omega)$$

vanishes by inversion symmetry, but finally $n = 2$,

$$\frac{r_<^2}{r_>^3} P_2(\cos\omega) \tag{4.60}$$

contributes differently in all three states. The radial factor is common to all three. It is readily seen that all succeeding terms in the expansion in Legendre polynomials will not contribute. So that apart from an additive constant arising from (4.59), all the energies will be proportional to the same, albeit unknown, radial integral. And it remains an amusing exercise in manipulating trigonometric identities to show that

$$\frac{E(^1S) - E(^1D)}{E(^1D) - E(^3P)} = \frac{3}{2}. \qquad (4.61)$$

This relationship is of course independent of the additive energy shift from (4.59) or of the common radial integral. Its validity depends only on the validity of the Hartree-Fock procedure and on the validity of our neglect of states outside the p shell. In Table 4.2 we compare the theory with experiment.

Problem 4.6.   Using the procedure indicated in the text, derive (4.61). Hint: no integrals need be evaluated to establish this result, but various integrals need to be compared. This can be done solely by use of the mathematical identity,

$$\int d\varphi_1 \int d\theta_1 \sin\theta_1 P_2(\cos\omega) \cos^2\theta_1$$

$$= -\int d\varphi_1 \int d\theta_1 \sin\theta_1 P_2(\cos\omega) \sin^2\theta_1.$$

The agreement is *not* always as perfect as for the atoms in Table 4.2. For example, the failure of one or another of the hypotheses gives for lanthanum ion, La^+, the ratio 18.43, which differs from theory by an order of magnitude! In such instances more terms must be taken into account, and while ultimately the calculations show good agreement with experiment they may

Table 4.2.   Configurations and ratio of term splittings in various atoms versus theory.[14]

	Theory	Experimental				
		C	Si	Ge	Sn	O
Configuration	$np^{3\pm1}$	$2p^2$	$3p^2$	$4p^2$	$5p^2$	$2p^4$
Ratio	1.50	1.13	1.48	1.50	1.39	1.14

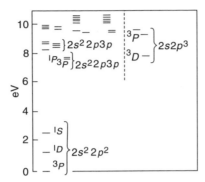

Fig. 4.4. Energy-level diagram of carbon (C), as example of Hund's two rules.

become very involved. But by now the ultimate validity of the many body Schrödinger equation for the atoms is not in dispute. Even the remarkably simple organization into atomic shells gives reasonably accurate results with minimum calculation.

Thus, atomic magnetism which is the consequence of Hund's rule, is documented by experiment, explained by elementary theory, and confirmed by elaborate calculations.

In Fig. 4.4, we show the spectrum of carbon. It is to be noted that as in nitrogen, the terms *not* described by the simple theory lie several volts above the ^1S term. As a consequence, theories of neutral carbon in solids probably need to take only the lowest three terms into account, which in the atom are adequately explained by the sort of first-order perturbation theory just expounded.

Finally, in Table 4.3, there is given a list of possible terms (configurations) that can be constructed out of a specified number of equivalent electrons. The lowest, according to Hund's rule, is given in boldface.

Practically all known magnetic substances involve d- or f-shell electrons, the study of which would take us well beyond the scope of an introductory work. The interested reader will readily find specialized treatises on this subject.[15] In pursuing this line of inquiry, we shall find it advantageous to review some fairly general and useful techniques in the study of large numbers of identical particles, namely second quantization.

[15] J. S. Griffith, *The Theory of Transition-Metal Ions*, Cambridge Univ. Press, New York, 1961; H. F. Schaefer (ed.), *Methods of Electronic Structure Theory*, Plenum, New York, 1977.

Table 4.3.   Possible terms for a number of equivalent electrons.[16]

²S (i.e., multiplicity = 2, total angular momentum = 0)

Configuration	Singlet	Doublet	Triplet	Quartet	Quintet	Sextet	Septet	Octet
s^2	^{1}S							
s		**^{2}S**						
p or p^5		**^{2}P**						
p^2 or p^4	^{1}S D		**^{3}P**					
p^3		^{2}P D		**^{4}S**				
d or d^9		**^{2}D**						
d^2 or d^8	^{1}S D G		**^{3}P F**					
d^3 or d^7		^{2}P D F G H (2)		**^{4}P F**				
d^4 or d^6	^{1}S D F G I (2 2 2)		^{3}P D F G H (2 2)		**^{5}D**			
d^5		^{2}S P D F G H I (3 2 2)		^{4}P D F G		**^{6}S**		
f or f^{13}		**^{2}F**						
f^2 or f^{12}	^{1}S D G I		**^{3}P F H**					
f^3 or f^{11}		^{2}P D F G H I K L (2 2 2 2)		**^{4}S D F G I**				
f^4 or f^{10}	^{1}S D F G H I K L N (2 4 4 2 3 2)		^{3}P D F G H I K L M (3 2 4 3 4 2 2)		**^{5}S D F G I**			
f^5 or f^9		^{2}P D F G H I K L M N O (4 5 7 6 7 5 5 3 2)		^{4}S D F G H I K L M (2 3 4 4 3 3 2)		**^{6}P F H**		
f^6 or f^8	^{1}S P D F G H I K L M N Q (4 6 4 8 4 7 3 4 2 2)		^{3}P D F G H I K L M N O (6 5 9 7 9 6 6 3 3)		^{5}S P D F G H I K L (3 2 3 2 2)		**^{7}F**	
f^7		^{2}S P D F G H I K L M N O Q (2 5 7 10 10 9 9 7 5 4 2)		^{4}S P D F G H I K L M N (2 2 6 5 7 5 5 3 3)		^{6}P D F G H I		**^{8}S**

Note: A number under a term symbol indicates the number of distinct levels of this type. The lowest level is normally that with the highest L of the highest multiplicity (Hund's rule). It is indicated in bold type.

16 *American Institute of Physics Handbook*, 2nd ed., McGraw-Hill, New York, 1963, pp. 7–21.

4.11.   Second Quantization

The essential information contained in the determinantal function (4.15), is that 3 particles satisfying the Pauli principle occupy (j, r) [using short-hand notation for the spatial state $f_j(r)$ and spin state $r = \pm\frac{1}{2}$], (k, s) and (n, t). The set of 3 pairs of quantum numbers summarizes this information, but does not explicitly reveal the antisymmetry of the wave function. If, however, we introduced 3 operators $c_{jr}^\dagger$, $c_{ks}^\dagger$ and $c_{nt}^\dagger$ with the following property

$$c_a^\dagger c_b^\dagger = -c_b^\dagger c_a^\dagger$$

where b, a are any of the three sets of quantum numbers above, then $c_{jr}^\dagger c_{ks}^\dagger c_{nt}^\dagger$ is an operator carrying the requisite information and which changes sign upon interchange of any pair of particles. It is on this observation that the formalism of second quantization is based.

Second quantization allows one to deal with variable numbers of particles as conveniently as with a fixed number. The "particles" can be electrons as stated above, or any other variety of interest such as lattice vibrations (phonons), and spin excitations (magnons) with differing symmetry properties. We shall first analyze the case of electrons, then discuss examples not subject to the Pauli principle. It is convenient to deal with non-interacting particles at first, and then to express the interactions in the new language.

The unique state of zero particles is denoted the "vacuum" and written,

$$|0\rangle \quad \text{(with } \langle 0| \text{ the conjugate state)} . \tag{4.62}$$

To introduce an electron of spin s at the spatial point $\mathbf{r}$ one uses the wave-operator $\Psi_s^*(\mathbf{r})$ as follows:

$$\Psi_s^*(\mathbf{r})|0\rangle , \quad \text{with } \langle 0|\Psi_s(\mathbf{r}) \text{ the conjugate state} . \tag{4.63}$$

The requirement that this state be normalized and orthogonal to that of a particle having a different spin index or at a different point of space is the following:

$$\langle 0|\Psi_{s'}(\mathbf{r}')\Psi_s^*(\mathbf{r})|0\rangle = \delta_{s,s'}\delta(\mathbf{r} - \mathbf{r}') \tag{4.64}$$

using the Kronecker delta for the discrete index and the Dirac delta for the continuous one. Equation (4.64) has two interpretations, the first as the inner product of a state (4.63) with a similar (but conjugate, such as column to row vectors) state; and the second as the inner product of the (conjugate) vacuum $\langle 0|$ with the state

$$\Psi_{s'}(\mathbf{r}')\Psi_s^*(\mathbf{r})|0\rangle . \tag{4.65}$$

For $s = s'$ and $\mathbf{r} = \mathbf{r}'$ this must be proportional to the vacuum, thus (4.65) must not contain any particles. Hence the interpretation: successive applications of Ψ^* raise the number of particles in a state by 1, successive applications of Ψ lower it by 1. (For the conjugate states, the opposite is true.) As the vacuum has the lowest possible occupancy, we find

$$\Psi_s(\mathbf{r})|0\rangle \equiv 0 \quad (\text{or } \langle 0|\Psi_s^*(\mathbf{r}) = 0). \tag{4.66}$$

The Pauli principle comes into play with 2 or more particles

$$\Psi_s^*(\mathbf{r})\Psi_{s'}^*(\mathbf{r}')|0\rangle . \tag{4.67}$$

It requires a change of sign under interchange of two particles, i.e.,

$$\Psi_{s'}^*(\mathbf{r}')\Psi_s^*(\mathbf{r})|0\rangle = -\Psi_s^*(\mathbf{r})\Psi_{s'}^*(\mathbf{r}')|0\rangle \tag{4.68}$$

which has to be true even if $|0\rangle$ is replaced by an arbitrary state. This requirement, and the normalization (4.64) can be conveniently combined into a compact set of *anti*commutation relations as follows:

$$\{\Psi_s(\mathbf{r}), \Psi_{s'}(\mathbf{r}')\} = 0 = \{\Psi_s^*(\mathbf{r}), \Psi_{s'}^*(\mathbf{r}')\}$$
$$\{\Psi_s(\mathbf{r}), \Psi_{s'}^*(\mathbf{r}')\} = \delta_{ss'}\delta(\mathbf{r} - \mathbf{r}') \tag{4.69}$$

in which the anticommutator bracket $\{A, B\}$ is short-hand for $AB + BA$.

The linear combination of $\Psi^*(\mathbf{r})$'s that corresponds to a given spatial wave function, e.g., to one of the states $\varphi_j(\mathbf{r})$ used in the Slater determinantal wave function early in the chapter, is simply

$$c_{j,s}^\dagger \equiv \int \varphi_j(\mathbf{r})\Psi_s^*(\mathbf{r})d^3r \tag{4.70}$$

and thus the second-quantized equivalent of the three-particle wave function (4.15) is precisely

$$c_{jr}^\dagger c_{ks}^\dagger c_{nt}^\dagger|0\rangle . \tag{4.71}$$

Problem 4.7. Prove that (4.15) is given by:

$$\sum_{\substack{s', s'' \\ s'''}} \langle 0| \Psi_{s'}(\mathbf{r}_3) \Psi_{s''}(\mathbf{r}_2) \Psi_{s'''}(\mathbf{r}_1) c_{jr}^{\dagger} c_{ks}^{\dagger} c_{nt}^{\dagger} |0\rangle$$

as function of coordinates r_1, r_2, r_3 and spins r, s, t.

Problem 4.8. Prove the anticommutation relations

$$\{c_a, c_b\} = \{c_a^{\dagger}, c_b^{\dagger}\} = 0 \quad \text{and} \quad \{c_a, c_b^{\dagger}\} = \delta_{a,b}$$

by use of (4.69), (4.70).

Given a state such as (4.67) or (4.71) with a certain number of particles, we can define the particle number operator n as a useful probe of this number:

$$\mathsf{n} = \sum_{s=\pm 1/2} \int \Psi_s^*(\mathbf{r}) \Psi_s(\mathbf{r}) d^3 r \,. \tag{4.72}$$

Given an Hamiltonian $\mathscr{H}$ for noninteracting particles with a complete set of one-particle eigenstates $f_j(\mathbf{r})$, we can always invert the type of relation (4.70) to obtain

$$\Psi_s(\mathbf{r}) = \sum_j f_j(\mathbf{r}) c_{j,s} \,. \tag{4.73}$$

Inserting this into (4.72) yields for the number operator

$$\mathsf{n} = \sum_s \sum_j c_{js}^{\dagger} c_{js} = \sum_s \sum_j n_{js} \,. \tag{4.74}$$

The occupation of the state (j, s) is given by the operator $n_{js} \equiv c_{js}^{\dagger} c_{js}$.

Like the number operator, the Hamiltonian of noninteracting particles is a quadratic form in field operators

$$\mathbf{H}_{\mathrm{op}} \equiv \sum_s \int \Psi_s^*(\mathbf{r}) \mathscr{H}(\mathbf{r}) \Psi_s(\mathbf{r}) d^3 r = \sum_s \sum_j \epsilon_j c_{js}^{\dagger} c_{js} = \sum_s \sum_j \epsilon_j n_{js} \,, \tag{4.75}$$

making use of (4.73) and of $\mathscr{H} f_j(\mathbf{r}) = \epsilon_j f_j(\mathbf{r})$, the time-independent Schrödinger equation for the jth eigenfunction. The interpretation is fairly direct: the energy of the jth state is the product of the energy (ϵ_j) and the occupancy $(n_{j+} + n_{j-})$.

The simplicity ceases when two-body forces are introduced. Where in first-quantization they might be written

$$\mathscr{H}_2 = \sum_{(i,j)} V(\mathbf{r}_i, \mathbf{r}_j) , \tag{4.76}$$

they now appear as

$$\mathbf{H}_{2\mathrm{op}} = \frac{1}{2} \sum_s \sum_{s'} \int\!\!\int \Psi^*_{s'}(\mathbf{r}')\Psi^*_s(\mathbf{r})V(\mathbf{r},\mathbf{r}')\Psi_s(\mathbf{r})\Psi_{s'}(\mathbf{r}')d^3r\,d^3r' . \tag{4.77}$$

In terms of the raising and lowering operators of the unperturbed Hamiltonian (4.75), $c^\dagger_{js}$ and c_{js}, the interaction Hamiltonian takes the form

$$\mathbf{H}_{2\mathrm{op}} = \frac{1}{2} \sum_{\substack{s,s' \\ j_1,j_2 \\ j_3,j_4}} K(j_1 j_2 j_3 j_4) c^\dagger_{j_3 s'} c^\dagger_{j_1 s} c_{j_2 s} c_{j_4 s'} \tag{4.78}$$

with

$$K(j_1, \ldots, j_4) = \int f^*_{j_1}(\mathbf{r}) f_{j_2}(\mathbf{r}) V(\mathbf{r}, \mathbf{r}') f^*_{j_3}(\mathbf{r}') f_{j_4}(\mathbf{r}') d^3r\, d^3r' . \tag{4.79}$$

The factor $\frac{1}{2}$ in (4.77), (4.78) corrects for double-counting the bonds. First-quantized operators and wave functions refer to an explicit number of particles [e.g., the 3×3 determinant (4.15) for 3 particles], whereas the second-quantized operators are valid for all possible occupation numbers. The variational and perturbation-theoretic formalisms are equally valid in the new language as in the old; even better, pictorial Feynman diagrams can be used to describe the various terms in the perturbation-theoretic studies of $\mathbf{H}_{2\mathrm{op}}$. For example, (4.78) is represented pictorially as a vertex with four arrows: two in, two out. For further details, we refer to any specialized text.[17]

A very important result, known as Wick's theorem,[17] allows the computation of the matrix elements of $\mathbf{H}_{2\mathrm{op}}$ and higher powers of this operator, in terms of simple contractions such as $\langle 0|c_{j_1 s}c^\dagger_{j_2 s}|0\rangle$, their products and sums, e.g., determinants. This theorem is available for fermion (anticommuting) operators of the type we have been discussing, for boson operators as we shall now discuss, *but not for spin operators* which, as previously studied in Chapter 3, have fairly complex commutation relations.

---

[17] E.g., R. D. Mattuck, *A Guide to Feynman Diagram in the Many-Body Problem*, 2nd ed., McGraw-Hill, New York, 1976, also C. P. Enz, *A Course on Many-Body Theory Applied to Solid-State Physics*, World Scientific, 1992.

Particles subject to Bose-Einstein statistics include ^{4}He atoms (even numbers of fermions) as well as any system of particles for which the wave function remains symmetric under permutations of two or more of the particles. The main modification one makes to the relations developed above for fermions is the substitution of commutation relations for the anticommutators of (4.69). Thus,

$$\Phi(\mathbf{r})\Phi(\mathbf{r}') - \Phi(\mathbf{r}')\Phi(\mathbf{r}) \equiv [\Phi(\mathbf{r}), \Phi(\mathbf{r}')] = 0$$

and similarly,

$$[\Phi^*(\mathbf{r}), \Phi^*(\mathbf{r}')] = 0$$

together with

$$[\Phi(\mathbf{r}), \Phi^*(\mathbf{r}')] = \delta(\mathbf{r} - \mathbf{r}') \tag{4.80}$$

writing the wave-operator as Φ for the bosons, and omitting the spin index, if any. Among fields satisfying these commutation relations we can also count quantized lattice vibrations (phonons), photons, and quantized spin-waves (magnons). Concerning the latter we shall have much more to say in later sections of this text. Particle-conservation laws are far less stringent for bosons than for fermions, and it is not uncommon to have odd powers of Φ or Φ^* in the Hamiltonian in addition to the terms discussed above for the fermions, whereas odd powers of the fermions Ψ or Ψ^* are inconceivable, physically or mathematically.

4.12.   Double Exchange

In this and the following sections, we examine some well-known mechanisms for creating magnetic order in molecules and in the solid state. Let us consider two "magnetic" ions S_1 and S_2, and one mobile electron that connects them. Typically this electron will align its spin *antiparallel* to whichever fixed spin it is closest to; so normally the sign of the coupling to either fixed spin, is positive. (But if the tendency is to parallelism we can choose it negative. Encompassing both possibilities, the Hamiltonian is,

$$\mathcal{H} = -t \sum_{m=\uparrow,\downarrow} (c^\dagger_{1,m} c_{2,m} + \text{H.C.}) + J\{\sigma_1 \cdot S_1 + \sigma_2 \cdot S_2\}. \tag{4.81}$$

The 2nd quantized notation is self explanatory; σ_j is the electron's spin operator when it is on the jth site and zero otherwise. For simplicity first assume the stationary spins S each carry spin $1/2$. The eigenstates and eigenvalues are obtained by diagonalizing a matrix representing the Hamiltonian. The 16 basis states are: $c_{i,m}^\dagger |0\rangle \otimes \gamma_1^+ \gamma_2^+ |\downarrow\downarrow\rangle$, where $i = 1$ or 2 labels the position of the electron and $m = \uparrow$ or $\downarrow$ its spin; the operators γ_i^+ are 1 if the fixed spin at i is $\downarrow$ and S_i^+ if it is $\uparrow$. Number these states $|1\rangle \cdots |16\rangle$ in some arbitrary sequence and evaluate the resulting Hamiltonian matrix:

$$M_{i,j} = (i|\mathscr{H}|j). \qquad (4.82)$$

(For example, $M_{1,1}$ might represent $(\downarrow\downarrow| \otimes \langle 0|c_{1,\uparrow}\mathscr{H}c_{1,\uparrow}^\dagger|0\rangle \otimes |\downarrow\downarrow\rangle = -J/4$, while $M_{1,2} = (\downarrow\downarrow| \otimes \langle 0|c_{2,\uparrow}\mathscr{H}c_{1,\uparrow}^\dagger|0\rangle \otimes |\downarrow\downarrow\rangle = -t$, etc.)

Basically, we now need to solve the eigenvalue equation $\sum_{j=1}^{16} M_{i,j} f_j |j) = E f_i |i)$ that represents the Schrödinger equation in this basis set. The 16 equations have no common solution unless a *secular determinant* vanishes. This last takes the form,

$$\det \begin{bmatrix} M_{1,1} - E, & M_{1,2},\ldots, & M_{1,16} \\ M_{2,1}, & M_{2,2} - E, & M_{2,3},\ldots \\ & \cdots & \\ M_{16,1}, & \cdots, & M_{16,16} - E \end{bmatrix} = 0, \qquad (4.83)$$

which, when expanded, is a 16th order polynomial equation in E. By the fundamental theorem of algebra, there are 16 real roots (because $\mathscr{H}$ and by extension M are Hermitean). To each there is a set of f_j's that defines the corresponding *eigenvector*.

Setting $t = 1$ (or, rather, expressing and E in units of t), we find the 16 roots to be:

$$\Big\{ \frac{1}{4}(-4 + j), \ \frac{1}{4}(-4 + j), \ \frac{1}{4}(-4 + j), \ \frac{1}{4}(-4 + j), \ \frac{4+j}{4}, \ \frac{4+j}{4}, \ \frac{4+j}{4}, \ \frac{4+j}{4},$$

$$\frac{1}{4}(-j - 2\sqrt{4 - 2j + j^2}), \ \frac{1}{4}(-j - 2\sqrt{4 - 2j + j^2}),$$

$$\frac{1}{4}(-j + 2\sqrt{4 - 2j + j^2}), \ \frac{1}{4}(-j + 2\sqrt{4 - 2j + j^2}),$$

$$\frac{1}{4}(-j - 2\sqrt{4 + 2j + j^2}), \ \frac{1}{4}(-j - 2\sqrt{4 + 2j + j^2}),$$

$$\frac{1}{4}(-j + 2\sqrt{4 + 2j + j^2}), \ \frac{1}{4}(-j + 2\sqrt{4 + 2j + j^2}) \Big\}. \qquad (4.84)$$

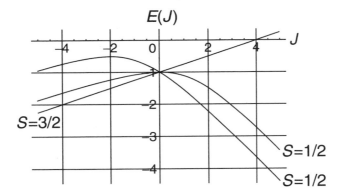

Fig. 4.5. $E(J)$ **versus** J **in "double exchange" of two localized spins** $1/2$ *via* **1 electron,** $t = 1$.

For $J < 0$, ferromagnetic coupling, the lowest state has $S_{\text{tot}} = 3/2$. The other low-lying states belongs to $S_{\text{tot}} = 1/2$. The two localized spins are parallel in the ground state for either sign of J.

In the above Fig. 4.5, we plot the energies E/t of the lowest-lying states (for $J > 0$ as well as $J < 0$) as function of J/t. The lowest levels cross at $J = 0$. This figure exhibits *asymmetry* between ferro- and antiferromagnetic couplings, attributable to quantum fluctuations (the asymmetry disappears in the classical limit treated next).

If the spins are sufficiently large, i.e., $s \gg 1/2$, they may be treated as classical vectors. Then the model is easily solved once again. Given that the angle between the two spins in *fixed* (it will be a parameter in the Hamiltonian along with J and the magnitude of the spins, s) there are only 4 states we need to label: the position of the wandering electron (1 or 2) and its spin relative to some fixed axis ($\uparrow$ or $\downarrow$). Without loss of generality we take the z-axis as the direction of S_1, and set $S_2 \cdot S_1 = s^2 \cos x$. It is a good exercise for the reader to set up this problem and reproduce the four energy eigenvalues explicity given below:

$$\left\{ -\sqrt{g^2 + t^2 - 2\sqrt{g^2t^2 \cos^2(x/2)}}, \ \sqrt{g^2 + t^2 - 2\sqrt{g^2t^2 \cos^2(x/2)}}, \right.$$

$$\left. -\sqrt{g^2 + t^2 + 2\sqrt{g^2t^2 \cos^2(x/2)}}, \ \sqrt{g^2 + t^2 + 2\sqrt{g^2t^2 \cos^2(x/2)}} \right\}.$$

$$(4.85)$$

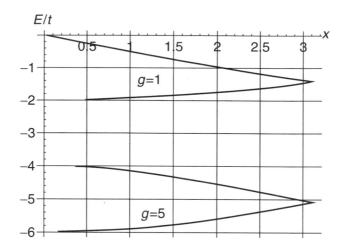

Fig. 4.6. $E(g,x)/t$ versus angle $x \equiv \cos^{-1} S_1 \cdot S_2 / s^2$, in double exchange for classical spins.

This figure shows the two lowest energies in "double exchange" for classical spins, for $g = Js/t = 5$ and 1, as functions of the angle illustrating the definite *bias for ferromagnetic alignment* of the immobile spins. (For a given g, $x = 0$ always has the lowest energy.) Unlike the preceding example, the present results are symmetric in J, therefore the energies for $g = -1$ and -5 are identical to those shown for 1 and 5.

For each value of $g = Js/t$ (we choose $g = 1$ and 5) we plot the two lowest eigenvalues as a function of x. We find ferromagnetic alignment of the localized spins is always preferred.

   In summary, the double exchange mechanism, whether quantum or classical, has as its general outcome that *a lone itinerant electron has lower energy if the localized spins are all aligned.* The distortion this effect may cause in antiferromagnets has been discussed by de Gennes.[18]

4.13.   Superexchange

Here we deal with a mechanism that is ubiquitous in magnetic salts and oxides, one that aligns neighboring spins *antiparallel.* The mathematics is a slight variant of double-exchange (the subject of preceding section), but it generally involves *two* electrons, not necessarily itinerant. The situation is illustrated in Fig. 4.7.

---

[18] P.-G. de Gennes, *Phys. Rev.* **118**, 141 (1960).

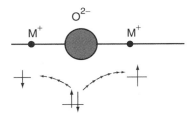

Fig. 4.7. Superexchange.
Showing a neutral metal-oxide molecule in which the metal ions carry spin $1/2$ each, and the *ligand* oxygen spin zero (but charge $-2e$). When a *second* electron "jumps" (makes a transition with matrix element $-t$) onto either metal site (as shown by dotted arrows) it acquires a high energy U. Because the oxygen is in a single state, the total energy of this complex is slightly lowered if the metal spins are *also* in a total *singlet* configuration as opposed to being in a joint *triplet* configuration. Thus, superexchange favors low spin or *anti*ferromagnetism.

Let us label the states on sites 1, 2 (magnetic ions, each assumed to carry spin $1/2$) and on the oxygen ion that separates them. The hopping is from a central oxygen ion to either magnetic neighbor (hopping matrix element $-t$). The neighbors are "magnetic" because the cost of double-occupancy of either one of them (changing its spin from $1/2$ to zero) is a large 2-body repulsion energy U; double occupancy on *both* therefore costs $2U$.

There are 9 normalized, orthogonal states belonging to $S^z_{\text{total}} = 0$ satisfying the Pauli principle for this system, although not all participate. Two normalized candidate states enter into a *triplet* configuration, call them A and B:

$$c^\dagger_{1\uparrow} c^\dagger_{0\uparrow} c^\dagger_{0\downarrow} c^\dagger_{2\uparrow} |0\rangle = |A\rangle$$

$$c^\dagger_{1\uparrow} c^\dagger_{0\uparrow} \frac{c^\dagger_{1\downarrow} + c^\dagger_{2\downarrow}}{\sqrt{2}} c^\dagger_{2\uparrow} |0\rangle = |B\rangle\,.$$

The matrix element connecting (A) to (B) is $-\sqrt{2}t$. The energy of (B) exceeds that of (A) by U. We set $t = 1$ as usual. Diagonalizing the 2×2 secular determinant relevant to these 2 configurations we obtain the 2 triplet eigenvalues $(\pm)$,

$$E = \frac{1}{2}(U \pm \sqrt{8 + U^2}) \quad \text{(total spin 1)}, \tag{4.86}$$

for a total of 6 states (2 with $S^z_{\text{total}} = 0$).

Three *singlet* configurations participate in the singlet ground state:

$$\sqrt{\frac{1}{2}}(c_{1\uparrow}^{\dagger}c_{0\uparrow}^{\dagger}c_{0\downarrow}^{\dagger}c_{2\downarrow}^{\dagger} - c_{1\downarrow}^{\dagger}c_{0\uparrow}^{\dagger}c_{0\downarrow}^{\dagger}c_{2\uparrow}^{\dagger})|0\rangle = |C\rangle\,.$$

$((C|\mathcal{H}|C) = 0$, as it is chosen to have the same reference energy as A.) It "connects" to the following normalized linear combination of states

$$\sqrt{\frac{1}{4}}(c_{1\uparrow}^{\dagger}c_{2\uparrow}^{\dagger}c_{0\downarrow}^{\dagger}c_{2\downarrow}^{\dagger} - c_{1\downarrow}^{\dagger}c_{1\uparrow}^{\dagger}c_{0\downarrow}^{\dagger}c_{2\uparrow}^{\dagger} + c_{1\uparrow}^{\dagger}c_{0\uparrow}^{\dagger}c_{1\downarrow}^{\dagger}c_{2\downarrow}^{\dagger} - c_{1\downarrow}^{\dagger}c_{0\uparrow}^{\dagger}c_{2\downarrow}^{\dagger}c_{2\uparrow}^{\dagger})|0\rangle = |D\rangle$$

$$\text{(where } (D|\mathcal{H}|D) = U)$$

which in turn connects to,

$$c_{1\uparrow}^{\dagger}c_{2\uparrow}^{\dagger}c_{1\downarrow}^{\dagger}c_{2\downarrow}^{\dagger}|0\rangle = |F\rangle\,. \quad \text{(where } (F|\mathcal{H}|F) = 2U)$$

In addition to the diagonal elements given above, non-vanishing matrix elements (and their Hermitean conjugates) are $(C|\mathcal{H}|D) = -\sqrt{2}t$ and $(D|\mathcal{H}|F) = -2t$. The singlet eigenvalues are calculated numerically, and the lowest of them is plotted in Fig. 4.8 alongside the lower of the two triplet state energies. The singlet-triplet splitting vanishes at large U as $4t^4/U^3$, in agreement with the *fourth-order* terms in a perturbation expansion in powers of t.

Problem 4.9.   Calculate the superexchange splitting as function of the angle ϑ subtended by two *classical* spins (length s) on M_1 and M_2. Does it favor $\vartheta = 0$ as in double-exchange, or π? In addition to an extra U when an electron wanders onto an M site, the interaction includes the atomic exchange coupling $J\Sigma\sigma_j \cdot S_j$ (summing j from 1 to 2). E/t depends on *two* parameters: U/t and $g = Js/t$. For definiteness, consider only the representative case $U/t = 5g = 10$.

[Hint: use a formulation similar to the double-exchange model for classical spins in the preceding section, taking the orientation of one spin to define the z-axis, and the other to be at an angle ϑ to the first. The present case differs somewhat, in that there are now 2 electrons on 3 possible sites, two of them with "built-in" spins S_j. As two electrons initially on the oxygen have opposite spins, there are 9 configurations and H is a 9×9 matrix. For parallel spins, this reduces to 3 configurations.]

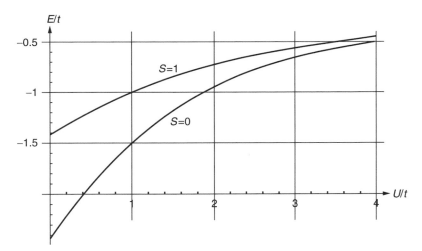

Fig. 4.8. Superexchange.

Energy eigenvalues (in units t) of lowest singlet and triplet states of two local-ized spins 1/2 connected through a singlet "ligand" ion, as functions of two-body repulsion parameter U. The difference in energy between the two curves yields *super-exchange splitting* as a function of U. At large U/t, this splitting vanishes as $4t^4/U^3$.

In summary, in *superexchange* the various valencies are enforced by a "charge transfer gap" U and ionic valency fluctuations are minimal. *Double-exchange* is representative of a more general metallic situation in which a gas of itinerant fermions (electrons or holes) finds it advantageous to align immobile spins present in the lattice. We pursue this theme in Chapter 6, esp. Sec. 6.6. But how does a low-density gas of fermions acquire a net magnetic moment? We turn to this now.

4.14. Jellium

In any attempt to understand *any* metallic substance, whether the density of electrons is high or low, one is faced with the difficulty of the many-body problem compounded with that of periodicity on an atomic scale.[19]

---

[19] The history of jellium can be traced back to the pioneering work on the binding energy of metals by E. P. Wigner and F. Seitz, *Phys. Rev.* **43**, 804 (1933), thence to Bohm and Pines (see D. Pines and P. Nozières, *The Theory of Quantum Liquids*, Addison-Wesley, New York, 1989, vol. I) who introduced the useful "random-phase approximation" (RPA) into this field. Although almost devoid of parameters, this subject has proved extremely difficult and a good guide is useful. Advanced students will benefit from the book by A. L. Fetter and J. D. Walecka, *Quantum Theory of Many-Particle Systems*, McGraw-Hill, New York, 1971.

"Jellium" is an amusing name for a simplified quantum electron fluid in which this periodicity is absent and the closed atomic shells are ignored. Simply put, it describes an N-body electron fluid in a box L^3 with Coulomb repulsions among the particles. However, a box of volume 1 cm^3 containing 10^{22} negatively charged electrons would literally explode unless a compensating positive charge of $+10^{22}e$ were also present. In jellium, the positive ionic cores *are* presumed present but are "smeared out." This eliminates the lattice periodicity, restores charge neutrality and frees one up to study the effects of the sole remaining variable, the electron density $\rho = N/L^3$, where $N = N_\uparrow + N_\downarrow$.

The 2-body Coulomb repulsion $\frac{e^2}{|r_{ij}|}$ is a positive definite operator, meaning that all its eigenvalues are nonnegative. Despite being divergent it can be used directly in expressions such as (4.77), although its Fourier transform is even more useful and takes the form $V(q) = 4\pi \frac{e^2}{q^2}$. Here q is the momentum transferred when 2 plane waves scatter: $(k, k') \to (k+q, k'-q)$. *Why plane waves?* In a medium that is homogeneous and has no periodicity on an atomic scale, the *net* charge density is zero, point by point. The self-consistent solution to Poisson's equation in jellium is a constant or zero potential. Electronic wave functions in a constant potential (when subject to periodic boundary conditions), are plane waves. Thus, plane wave electron states and a zero electrostatic potential are *both* consequences of a self-consistent solution of this highly nonlinear model.

As for the 2-body interactions, one must realize that electric fields are always screened in an electronic medium. Wherever there is an electric field $\mathscr{E}$, a current $j \propto \mathscr{E}$ is sure to flow. But no currents can flow in thermodynamic equilibrium, hence bare charges *must* disappear at any macroscopic distance. *With* screening, the two-body repulsion reduces (in some crude approximation) to $V(q) = 4\pi \frac{e^2}{\varepsilon(q)q^2} \approx 4\pi \frac{e^2}{q^2+q_0^2}$, a short-ranged 2-body potential (its spatial Fourier transform vanishes exponentially). This very notion may be what inspired Hubbard to construct a model with total screening — in which *only* zero-range potentials are retained. His simplification allows to keep both two-body repulsion *and* the atomic periodicity of the lattice and it has proved popular in the theory of magnetism. (We return to it later on.)

Now, as the electron density is decreased by one or two orders of magnitude, transitions out of the homogeneous jellium into condensed phases are suspected to occur. In the very lowest range of density, Eugene Wigner predicted there would occur a spontaneous symmetry breaking from an homogeneous fluid → a close-packed lattice. In an electronic lattice there would be just 1 electron in each unit cell. Hence Wigner's solid is an antiferromagnet, as localized electrons *always* have an energetic preference to form a singlet state with their nearest-neighbors. The total spin S_{total} would remain constant (zero for an even number of electrons) through the transition from the fluid phase into a Wigner solid, even while the spatial long-range order (LRO) changes from zero in the fluid, to a finite value in the solid. Such a scenario describes a typical *first-order* thermodynamic phase transition. *However, it might not describe the reality.*

Recall that in Chapter 1 we mentioned F. Bloch's proposal, ca. 1930, that exchange corrections to the Coulomb interaction favored ferromagnetism at low density. This primitive theory of metallic ferromagnetism, based on the Hartree-Fock approximation to the unscreened electron gas, was quickly dismissed by Pauli and others, as being too crude. But what if the idea itself was correct? Then the first instability of jellium, as the density is decreased, would be into a *ferromagnetic fluid phase* and *not* into an antiferromagnetic Wigner crystal.

Before discussing these two phase transitions let us see where Bloch's original theory came from. Expressed in a plane wave representation and including a factor $\frac{1}{2}$ for double counting, the two-body interaction Hamiltonian of Eq. (4.78) for the bare Coulomb interactions is:

$$H_2 = \frac{2\pi e^2}{\text{Vol.}} \sum_{\sigma,\sigma'} \sum_k \sum_{k'} \sum_q \frac{1}{q^2} c^\dagger_{k+q,\sigma} c^\dagger_{k'-q,\sigma'} c_{k',\sigma'} c_{k,\sigma}. \qquad (4.87)$$

For parallel spins ($\sigma = \sigma'$) there are *two* ways a pair (k, k') can scatter into $(k+q, k'-q)$: with direct matrix element $4\pi e^2/q^2$, and with *exchange* matrix element $-4\pi e^2/|k - k'|^2$. These have opposite signs and interfere destructively. Conventionally one expresses the relevant energies in terms of r_s, the characteristic separation of 2 electrons.[20] The kinetic energy in the Fermi sea *per* particle $\propto k_F^2 \propto 1/r_s^2$. The energy $E(k)$ of an electron near the Fermi surface (which is at k_F), relative to the Fermi energy, is then,[21]

$$E(k) - E_F = \frac{A}{r_s^2}((k/k_F)^2 - 1)$$
$$+ \frac{B}{r_s}\left\{ \frac{k_F}{k}((k/k_F)^2 - 1) \log\left|\frac{k_F + k}{k_F - k}\right| \right\} \qquad (4.88)$$

---

[20] With ρ the electron density, the volume *per* electron in $1/\rho \equiv 4\pi r_s^3/3$, or $r_s = (3/4\pi\rho)^{1/3}$ in 3D.
[21] P. A. M. Dirac, *Proc. Camb. Phil. Soc.* **26**, 376 (1930); J. Bardeen, *Phys. Rev.* **49**, 653 (1936).

in which A, B are suitable constants. The constant A expresses the kinetic energy of the individual particle after mass renormalization and other effects are taken into account. The term in B reflects the *exchange correction*. Because this exchange energy has a logarithmic singularity at the Fermi surface (tantamount to an energy gap of zero width at the Fermi surface, at which the density of states actually vanishes) the low-temperature specific heat calculated using Eq. (4.88) has logarithmic terms. The dispersion also predicts the electrical resistance of a pure metal should increase logarithmically at low temperatures (instead of decreasing), etc. As such anomalies have never been observed, whether in pure metals or in degenerate semiconductors where r_s can be arbitrarily large, modern treatments favor the use of the screened Coulomb $2\pi e^2/\varepsilon(q)q^2$ and exchange $-2\pi e^2/\varepsilon(|k-k'|)|k-k'|^2$ potentials, expressions devoid of singularities as $q \to 0$ or $k \to k'$, because $\varepsilon(q)q^2 \to$ constant. The screening is calculated self-consistently from the dynamics of density fluctuations, the paramagnetism from spin fluctuations.

Returning now to Bloch's ferromagnetism: when the unscreened exchange correction in (4.88) is averaged over the Fermi sphere, it becomes substantial $\approx -0.458\ e^2/r_s$.

Then the exchange correction soon comes to dominate the kinetic energy ($\propto 1/r_s{}^2$) as r_s is increased. It is then possible to lower the total energy by having more electrons with spin up than with spin down, either partial or total ferromagnetism. By direct calculation, Hartree-Fock predicts a paramagnetic-to-ferromagnetic phase transition at $r_s = 5.45$.

We now know the prediction was correct even though the derivation is flawed. By taking advantage of cutting-edge Quantum Monte Carlo methodology, Ortiz, Harris and Ballone[22] have been able to follow 2000 electrons dynamically at various densities. Their non-perturbational numerical computer experiments reveal that the paramagnetic jellium-to-ferromagnetic phase transition occurs as a continuous, second-order phase transition, with partial ferromagnetism commencing at $r_s = 20$ (with a statistical error estimate of ± 5). The degree of spin polarization $\zeta = (N_\uparrow - N_\downarrow)/(N_\uparrow + N_\downarrow)$ increases with increasing r_s, approaching 1 in a fully polarized fluid at $r_s = 40 \pm 5$. This fluid remains stable until $r_s = 65 \pm 10$, when it condenses into a hitherto-undiscovered *ferromagnetic body-centered cubic (bcc) lattice*. At some $r_s > 100$ the "correlation energy" finally beats out the exchange

[22] G. Ortiz, M. Harris and P. Ballone, *Phys. Rev. Lett.* **82**, 5317 (1999). See also, R. Palanichamy and K. Iyakutti, *Int. J. Mod. Phys. B* **16**, 1353 (2002).

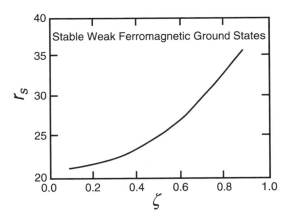

Fig. 4.9. Rise of magnetization parameter ζ with increasing r_s (decreasing density). This figure shows ζ increasing smoothly from zero at $r_s = 20$ to fully polarized at $r_s = 40$.

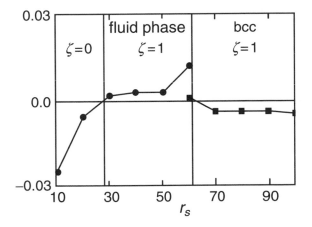

Fig. 4.10. Phase diagram. Energy $\times r_s$ versus r_s. We see paramagnetic (jellium) phase at small r_s, magnetic fluid in range $25 < r_s < 60$, and ferromagnetic solid bcc phase up to $r_s \approx 100$. Transition into Wigner antiferromagnetic lattice should occur at $r_s > 100$.

energy and Wigner's antiferromagnetic lattice becomes the stabler solid-state phase. We show the results in Figs. 4.8–4.10 adapted (by permission) from a paper by Ortiz et al.[22]

The theoretical results appear to be in good agreement with experiments[23] on $Ca_{1-x}La_xB_6$, a material that appears to show weak ferromag-

---

[23] D. P. Young et al., Nature **397**, 412 (1999).

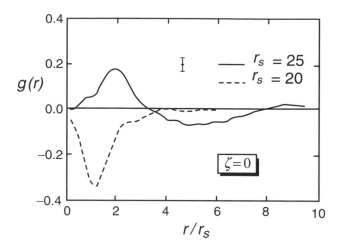

Fig. 4.11. Spin-spin correlation function $g(r)$ as a function of r/r_s.
Curves shown are calculated for just below and just above the ferromagnetic transition, with ζ kept 0 in both. The magnitude of the statistical error bar is indicated. These curves show how anticorrelation (the familiar "exchange hole" of Hartree-Fock theory at $r \approx r_s$) changes over into a positive correlation at $\approx 2r_s$ when r_s is increased only slightly, from 20 to 25.

netism, $\zeta \approx 0.1$, at $r_s = 28$, with a relatively high Curie T_c of 600 K. Unfortunately, these experiments and the conclusions drawn from them have been controversial and unambiguous confirmation remains for the future.

4.15.   Hubbard Model: An Introduction

Since its introduction by John Hubbard,[24] Martin Gutzwiller[25] and others in the early 1960's, this model of an interacting electron gas has garnered a lot of attention — so much so that several books dedicated to the topic have since been published and well received.[26] Yet we have a lot to learn on the topic and this section will serve only as an introduction to what is known and what is not. Although review books may help, the interested reader is urged to go to the most recent ongoing literature.

---

[24] J. Hubbard, *Proc. Roy. Soc. A* **276**, 238 (1963).
[25] M. C. Gutzwiller, *Phys. Rev. Lett.* **10**, 159 (1963).
[26] *The Hubbard Model*, a reprint volume, A. Montorsi, ed., World Scientific, 1992; *The Hubbard Model, Recent Results*, a reprint volume, M. Rasetti, ed., World Scientific, 1991; *Exactly Solvable Models of Strongly Correlated Electrons*, a reprint volume, V. E. Korepin and F. H. L. Essler, eds., World Scientific, 1994, *inter alia.*

In its simplest form, the model is described by an Hamiltonian for electrons "hopping" from points R_i on a given lattice to R_j (and back), repelling when on the same site:

$$\mathcal{H} = -\sum_{i,j,\sigma}(t_{ij}c_{i,\sigma}^{\dagger}c_{j,\sigma} + \text{H.c.}) + U\sum_{i}c_{i,\uparrow}^{\dagger}c_{i,\uparrow}c_{i,\downarrow}^{\dagger}c_{i,\downarrow}, \qquad (4.89)$$

where a zero-range potential $U > 0$ is what remains of the Coulomb interaction when screening has weakened or made irrelevant the n.-n. and more distant interactions. Notice that "exchange corrections" have been eliminated *ab initio*, by means of the fermion identity $n_{i,\sigma}{}^2 = n_{i,\sigma}$; so the interaction is $n_{i\uparrow}n_{i\downarrow}$ only. Now, for fermions or bosons, we can always add 1-body terms to an Hamiltonian without it affecting the physics in any way. For example, one frequently subtracts $\mu\sum_{i,\sigma}n_{i,\sigma}$ to measure one-particle energies from the reference Fermi surface at μ instead of from a less convenient zero. In practice, (4.89) is functionally equivalent to the "symmetric" Hubbard model,

$$\mathcal{H}_{\text{sym}} = -\sum_{i,j,\sigma}(t_{ij}c_{i,\sigma}^{\dagger}c_{j,\sigma} + \text{H.c.})$$

$$+U\sum_{i}\left(n_{i,\uparrow} - \frac{1}{2}\right)\left(n_{i,\downarrow} - \frac{1}{2}\right) \qquad (4.90)$$

which is invariant under the transformation from holes to particles *on a bipartite lattice* (i.e., a lattice that can be decomposed into 2 interpenetrating sublattices, such that $t_{i,j}$ connects a site i on the A sublattice to a site j on the B sublattice, but *not* an A to an A or a B to a B). A particle-hole transformation is useful when dealing with a more-than-half-filled band, and consists simply of letting $c_{i,\sigma} \Leftrightarrow c_{i,-\sigma}^{\dagger}$ and vice versa. It must be succeeded by a simple gauge transformation: $c_{j,\sigma}$ and $c_{j,\sigma}^{\dagger} \Leftrightarrow -c_{j,\sigma}$ and $-c_{j,\sigma}^{\dagger}$ for all j on the B sublattice only, in order that $\mathcal{H}$ maps onto itself. Hence, in symmetric models on bipartite lattices all the eigenvalues are evenly distributed about half-occupancy.

Problem 4.10. Show that under these transformations, $(n_{i,\sigma} - 1/2) \Leftrightarrow -(n_{i,-\sigma} - 1/2)$ and therefore $\mathcal{H}_{\text{sym}} \to \mathcal{H}_{\text{sym}}$.

For a bipartite lattice, then, we need only consider $0 < \rho < 0.5$, where $\rho = N_{\text{ele}}/2N_{\text{sites}}$. In the range $0.5 < n < 1$ one substitutes $\rho_h = 1 - \rho$ for ρ

and obtains the same results. Thus $\rho = 1/2$ (half-occupancy) is a symmetry point, as is seen in the following phase diagram (Fig. 4.12):

It is instructive to solve the Hubbard model for 2 particles, then extrapolate to low density. The one-body energies are $E(k) = \sum_j t_{ij} e^{ik \cdot R_{ij}} = E(-k)$, at any fixed R_i. (For example, in the *sc* lattice with $t_{ij} \neq 0$ only on n.-n. sites, $E(k) = -2t(\cos k_x + \cos k_y + \cos k_z)$ and the bandwidth is $24t$.)

Two-body triplet states,

$$ c_{k,\uparrow}^\dagger c_{-k',\uparrow}^\dagger |0\rangle , \quad \frac{1}{2}(c_{k,\uparrow}^\dagger c_{-k',\downarrow}^\dagger + c_{k,\downarrow}^\dagger c_{-k',\uparrow}^\dagger)|0\rangle , \quad \text{and} \quad c_{k,\downarrow}^\dagger c_{-k',\downarrow}^\dagger |0\rangle $$

$$(4.91)$$

are not scattered by the interaction in U hence they have energies $E_{k,k'} = E(k) + E(k')$. On the other hand, the *singlet* state $\frac{1}{\sqrt{2}}(c_{k,\uparrow}^\dagger c_{-k',\downarrow}^\dagger - c_{k,\downarrow}^\dagger c_{-k',\uparrow}^\dagger)|0\rangle \equiv b_{k,k'}$ is scattered by the zero-range potential and it is instructive to figure out exactly how.

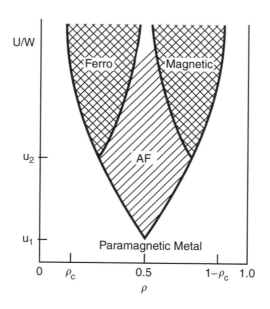

Fig. 4.12.   Phase diagram for One-Band Hubbard model in 3D at $T = 0$. The unit of energy in the figure is $W = $ bandwidth, ($W = 24t$ for n.-n. hoppings on a *sc* lattice). Near half-filling the antiferromagnetic phase (*AF*) is stable provided $U/W > u_1$. (N.B., $u_1 = 0$ in a bipartite lattice). Ferromagnetism is never stable for $\rho < \rho_c$ or $\rho > 1 - \rho_c$ (regardless of how large is U), or $U/W < u_2$ (regardless of ρ). Derivation is in the text.

Its Bloch energy is also $E_{kk'}$. Then let the eigenstate have energy $W_{kk'}$ and be

$$B_{kk'} \equiv b_{kk'} + \frac{1}{N} \sum_{q \neq 0} L_q b_{k+q,k'+q} \tag{4.92}$$

the usual form, an incoming wave and a scattered part. The Schrödinger equation consists of three parts:

$$\mathcal{H}_0 B_{kk'} = E_{kk'} b_{kk'} + \frac{1}{N} \sum_q L_q E_{k+q,k'+q} b_{k+q,k'+q} \tag{4.93a}$$

$$\mathcal{H}' B_{kk'} = \frac{U}{N} B_{kk'} + \frac{U}{N} \left(\sum_{q \neq 0} b_{k+q,k'+q} + \frac{1}{N} \sum_{q', \, q \neq q'} L_{q'} b_{k+q,k'+q} \right) \tag{4.93b}$$

$$W_{kk'} B_{kk'} = W_{kk'} b_{kk'} + \frac{1}{N} \sum_q L_q W_{kk'} b_{k+q,k'+q} . \tag{4.93c}$$

Equating the coefficients of $b_{kk'}$ yields an expression for the energy eigenvalue

$$W_{kk'} = E_{kk'} + \frac{U}{N} \left\{ 1 + \frac{1}{N} \sum_{q \neq 0} L_q \right\} . \tag{4.94}$$

The scattering amplitudes L_q are obtained from the coefficients of the $b_{k+q,k'+q}$. After minor algebra, the result is

$$L_q = \frac{1}{W_{kk'} - E_{k+q,k'+q}} \times \frac{U}{1 + U G_0} \tag{4.95}$$

where

$$G_0 \equiv \frac{1}{N} \sum_q \frac{1}{E_{k+q,k'+q} - W_{kk'}} . \tag{4.96}$$

Consequently, the energy is

$$W_{kk'} = E_{kk'} + \frac{1}{N} \left(\frac{U}{1 + U G_0} \right) \equiv E_{kk'} + \frac{1}{N} t(W_{kk'}) \tag{4.97}$$

which also serves to define the scattering t matrix, the expression of the interaction energy between the two particles with all multiple scattering taken into account. For all practical purposes the $W_{kk'}$ can be replaced by the unperturbed $E_{kk'}$ from which it differs by only $O(1/N)$. (The exception is for bound states, i.e., the zeros of $1 + UG_0 = 0$ which, if they exist, lie above the highest $E_{kk'}$ and thus will not be relevant to the subsequent discussion.) (In passing, one notes that for *negative U*, no matter how weak, such a pole in the t matrix could develop, *below* the lowest $E_{kk'}$; the lowest of these bound states belongs to $q = 0$, and is the zero-momentum Cooper pair, famous in the BCS theory of superconductivity. Our interest, however, is in fairly large, *positive U*.)

The modification of 2-body scattering required for the low density electron gas is straightforward: the range of $\mathbf{q}$'s in (4.92) et seq., especially G_0 (4.96), must be restricted to the unoccupied states: $\mathbf{k} + \mathbf{q}, -\mathbf{k}' - \mathbf{q}$ must lie outside the Fermi volume, i.e., $E(\mathbf{k} + \mathbf{q})$ and $E(-\mathbf{k}' - \mathbf{q})$ must exceed E_F. Denote the suitably modified G_0 by $\widetilde{G}_0$ and similarly for $\widetilde{t}(E_{kk'})$ to obtain the effective Hamiltonian for the singlet pairs

$$\mathscr{H}_\mathrm{eff} = \sum_{k,m} E(\mathbf{k})\mathfrak{n}_{km} + \frac{1}{N} \sum_{k,k'} \widetilde{t}(E_{kk'})\mathfrak{n}_{k\uparrow}\mathfrak{n}_{-k'\downarrow}, \quad \widetilde{t} \equiv \frac{U}{1 + U\widetilde{G}_0}. \tag{4.98}$$

To proceed we need to evaluate $\widetilde{t}(E)$. Our approach is greatly simplified, yet retains essential features. In a dilute electron gas, the Fermi level is near enough to the bottom of the conduction band(s) (or, for holes, the top) that the effective mass approximation $E(k) = \hbar^2 k^2 / 2m^*$ is appropriate. With this, all the integrals that go into the calculation of the ground state energy become simple.

First, we introduce the cutoff k_0 to retain, properly, the volume of the first BZ

$$\frac{1}{N} \sum_{k \leq \mathrm{BZ}} 1 = 1 = \frac{4\pi}{3} \left(\frac{k_0 a_0}{2\pi} \right)^3. \tag{4.99}$$

The density parameter ρ is defined: *the number of electrons per atom per band in a given spin direction* $\uparrow$ *or* $\downarrow$. With a spherical Fermi surface at k_F, ρ is given as

$$\frac{1}{N} \sum_{k < k_\mathrm{F}} 1 \equiv \rho = (k_\mathrm{F}/k_0)^3 \quad \text{(by comparison with (4.99))}.$$

Similarly the kinetic energy (KE) per band per spin component is

$$KE = \sum_{k<k_F} E(k)$$

$$= \frac{N \sum_{k<k_F} E(k)}{\sum_{all} 1} = N \left[\frac{\int_0^{k_F} dk\, k^4 \hbar^2/2m^*}{\int_0^{k_0} dk\, k^2} \right]$$

$$= \frac{9}{10} N \frac{\hbar^2 k_0^2}{3m^*} \rho^{5/3}. \tag{4.100}$$

We start with the exact expression for $\widetilde{G}_0$:

$$\widetilde{G}_0 = (a/2\pi)^3 \int_{\substack{|k+q|>k_F \\ |k'-q|>k_F \\ q<k_0}} d^3q \frac{1}{(\hbar^2/2m^*)[(k+q)^2 + (k'-q)^2 - k^2 - k'^2]} \tag{4.101}$$

and evaluate it, neglecting $k, k' \ll k_0$, in the dilute limit. An approximate but very convenient expression, independent of k and k', is what results:

$$\widetilde{G}_0 \approx (a/2\pi)^3 \int_{k_0>q>k_F} d^3q \frac{1}{(\hbar^2/2m^*)2q^2} = \frac{3m^*}{\hbar^2 k_0^2}(1 - \rho^{1/3}) \tag{4.102}$$

using (4.99) to eliminate k_F. With this, we obtain a $\tilde{t}$ matrix which is independent of k, k' and depends explicitly on ρ.

As the combination $\hbar^2 k_0^2/3m^*$ occurs in the interaction and in the KE, we eliminate it. It is recognized as a measure of the bandwidth, in units appropriate to the effective mass approximation. We therefore denote it W as usual

$$W \equiv \hbar^2 k_0^2/3m^*. \tag{4.103}$$

With $\widetilde{G}_0$ and $\tilde{t}$ suitably re-expressed in terms of W, and with the KE also simplified, we now have for the ground state energy of (4.98) the perspicuous expression

$$E_0 = NW \left[2 \cdot \frac{9}{10}\rho^{5/3} + \frac{U\rho^2}{W + U(1 - \rho^{1/3})} \right]$$

$$= NW\rho^{5/3} \left[\frac{9}{5} + \frac{U\rho^{1/3}}{W + U(1 - \rho^{1/3})} \right]. \tag{4.104}$$

Both KE and interaction energies have been expressed in terms of the sole physical parameters of interest: the bandwidth W, the coupling constant U, and the electron density ρ. Writing a similar expression for the *totally* ferromagnetic state at this stage (we do not bother with the added complications of partially magnetized states), so $\rho_\uparrow = 2\rho$, and $\rho_\downarrow = 0$. Now $\tilde{t}$ vanishes and we find

$$E_{\text{ferro}} = NW(2\rho)^{5/3}\left(\frac{9}{5}\right). \qquad (4.105)$$

A criterion for the occurrence of ferromagnetism can be obtained by comparing the two expressions. If (4.105) is lower, then the ground state has spin $\mathcal{N}\frac{1}{2}\hbar$. After a few most elementary algebraic manipulations, this criterion is seen to be

$$\frac{U\rho^{1/3}}{U + W} > 0.5139. \qquad (4.106a)$$

There is no solution for $\rho < (0.5139)^3 = 0.136$. Otherwise, we find ferromagnetism when

$$\frac{U}{W} > \frac{1}{\rho^{1/3} - (0.136)^{1/3}} \quad \text{for } \rho > 0.136. \qquad (4.106b)$$

A more sophisticated treatement by Kotliar and Ruckenstein,[27] believed valid at all densities, results in a similar phase diagram with $\rho_c = 0.19$ instead of 0.136. Comparing the energy in the ferromagnetic phase with that in the AF phase, these authors find a minimum U *below which* there is no ferromagnetism at *at any density*, viz. $u_2 \approx 15$. These two numbers serve to delimit the ferromagnetic phase.

However we have yet to analyze the AF phase that is dominant near half-filling. This may be simpler than it seems at first sight. Suppose we look for a Néel-type phase at first: all the A sites would contain electrons of spin $\uparrow$ and B sites electrons of spin $\downarrow$. If the number in each sublattice is the same, this is a Néel antiferromagnet. If the numbers differ, it is a *ferrimagnet*. Its motional energy is exactly the same as is in the saturated ferromagnetic phase, because in either phase each electron is initially localized — one to each site. But whereas the ferromagnetic state is an eigenstate the Néel state connects to states in which electrons have "hopped" onto nearest-neighbor

---

[27] G. Kotliar and A. E. Ruckenstein, *Phys. Rev. Lett.* **57**, 1362 (1986).

sites. The cost is U for any such hop, and the matrix element is t. Thus, in second-order perturbation theory, the energy of this modified Néel state lies lower than the ferromagnetic state in the amount, $N_{sites} z \, t^2/U$ (here z is the coordination number of the lattice, i.e., the number of n.-n. sites onto which each electron can hop). But is this perturbation expansion valid in the first place?

Fortunately, some properties of the one-band Hubbard model are known exactly without recourse to approximation or perturbation theory. Here is one: *there can be no ferromagnetism in the half-filled band.* Consider the ferromagnetic phase of $N_{elec} = N_{sites} = N$ (all spins up, no spins down), at arbitrary U. This state belongs to total spin $S_{tot} = N/2$. We show that a state belonging to $S_{tot} = (N/2) - 1$ has a lower energy than the initial ferromagnetic state. Continuing by induction to flip spins, one finds that a state belonging to $S_{tot} = (N/2) - 2$ has lower energy still, etc. That is, in the ground-state of the half-filled Hubbard model a small — or even zero — magnetic moment is what should be expected, *not* ferromagnetism.

The proof, using $\mathcal{H}$ in (4.89), now follows. The maximally ferromagnetic state $|\Psi_F\rangle$, (belonging to $S_{tot} = N/2$), has $E_{total} = 0$: no site is doubly occupied and every site is filled (so by Pauli principle, no hopping takes place). The total spin-lowering operator $S_{tot}^- = \sum_i c_{i,\downarrow}^\dagger c_{i,\uparrow}$ commutes with $\mathcal{H}$, hence when it is applied to any eigenstate of $\mathcal{H}$ it cannot change the energy eigenvalue. (Reader: proof?) Thus, the (un-normalized state) $S_{tot}^- |\Psi_F\rangle$ *also* has energy eigenvalue 0 even though it belongs to $S_{tot}^z = (N/2) - 1$. *All other states* in the subspace $S_{tot}^z = (N/2) - 1$ belong to $S_{tot} = (N/2) - 1$. (Reader: why? Can there be a second state belonging to $S_{tot} = N/2$ in this subspace?) Now we show that at least one such state has energy *lower than* 0, thus establishing that each spin-reversal lowers the energy as well as total spin. The calculation involves the spin-reversed ($\downarrow$) fermion, the energy of which is $U + E(k)$, [cf. $E(k)$ given just above Eq. (4.91)], the "hole" in the spin up sea, and their interactions. Measuring the wavevector of the hole from the point, (π, π, π), its energy is $E(k')$. The two-body potential energy that connects the spin-reversed fermion and the hole is a zero-range *attractive* potential of strength U. Now we just have to adapt the preceding two-body calculations (4.91 ff. for the repulsive potential U) to the present case. Let us guess total momentum $= 0$ for the lowest energy, which is certainly true although there is a wide range of momenta $\neq 0$ for which binding also occurs.

For a bound state B_0 there is no incoming wave, hence we write

$$B_0 = \frac{1}{\sqrt{N}} \sum_k F_k b_{kk} ,$$

$$H_0 B_0 = \frac{1}{\sqrt{N}} \sum_k F_k (2E(k) + U) b_{kk} ,$$

$$H_2 B_0 = \frac{-U}{N} \sum_{k'} F_{k'} \frac{1}{\sqrt{N}} \sum_k b_{kk}$$

$$\text{and} \quad W_0 B_0 = \frac{W_0}{\sqrt{N}} \sum_k F_k b_{kk} .$$

(4.107)

The Schrödinger equation is, in effect, $(\mathcal{H}_0 + \mathcal{H}_2) B_0 = W B_0$. Equating coefficients of b_{kk} on both sides yields F_k, i.e.,

$$F_k = \left(\frac{1}{N} \sum_{k'} F_{k'} \right) \frac{U}{2E(k) + U - W} . \tag{4.108}$$

Now multiply both sides by $1/N$ and sum over k. Either (a) the sum in parentheses in (4.108) vanishes, or (b) we have to satisfy the following transcendental equation for W:

$$\frac{1}{U} = \frac{1}{N} \sum_k \frac{1}{2E(k) + U - W} . \tag{4.109}$$

One assumption we shall make is that $E(k + (\pi, \pi, \pi)) = -E(k)$ (for symmetric bipartite lattices there is no essential loss of generality because $\cos(k_x + \pi) = -\cos k_x$). The lhs of this equation can be rewritten $\frac{1}{U} = \frac{1}{N} \sum_k \frac{1}{U}$ and combined with the rhs. After a few more obvious manipulations one obtains what is equivalent to (4.109),

$$1 = \frac{U(U - W)}{N} \sum_k \frac{1}{(U - W)^2 - (2E(k))^2} . \tag{4.110}$$

When $W = 0$ the rhs of this equation is clearly > 1. As W is made negative, the rhs becomes monotonically smaller, all the way to zero; hence there exists a solution at some finite, negative, W, for any magnitude of U.   Q.E.D.

An amusing and more difficult problem arises if $N_{\text{elec}} = N_{\text{sites}} - 1$ originally. Then after a spin flip, there are 2 fermionic holes in the up spin fluid and 1 electron with spin down; the 2 holes have to be in a state that is antisymmetric under their permutation, i.e., in angular momentum 1, whereas the electron has to be in the smoothest, nodeless, function relative to the two hole "particles." Does the ground state with these symmetries have energy lower than the 2-body bound state in (4.110) plus one hole far away, with its energy near the bottom of the band? That is, is there a 3-body bound state? The answer is unknown, although a moment's analysis indicates that the result is no longer universal but depends on the magnitude of U/W. In one limit ($U = 0$), the more spin flips the lower the energy, while in $\lim U = \infty$, the ground state *is* ferromagnetic according to Nagaoka.[28]

As one traverses Fig. 4.12 on p. 156, at constant U, a knowledge of the phase diagram would inform just where the transitions from AF to ferromagnetism occur and similarly for transitions from ferromagnetism to paramagnetism. Approximate equations for these curves are obtained in Sec. 6.11 where this problem is reëxamined from a more general point of view.

4.16. Nagaoka's Ferromagnetism

Because there are so few *rigorous* results concerning the Hubbard model we should first mention that Brant and Giesekus[29] have found some exact states on cleverly *decorated* hypercubic lattices in $\lim U \to \infty$, that is ordinary lattices with extra sites and bonds, at special occupation numbers.

What is more, the Hubbard model is trivially solvable in the $U \to \infty$ limit in 1D, as the particles are prevented from crossing either by the potential or by the Pauli principle. Hence Ψ is a determinantal function of plane waves in space $\times$ an *arbitrary* spin function. This is tantamount to a huge magnetic degeneracy akin to that of free spins. At large but not infinite values of U, it is legitimate to assume that this charge-spin decoupling persists, to the extent that the spins satisfy their own Heisenberg Hamiltonian (with nearest-neighbor antiferromagnetic couplings $J \propto t^2/U$), while the remaining spinless degrees of freedom continue to be expressed by a single determinantal function. Of course such guesswork is unnecessary, given that the one-dimensional Hubbard model is solvable in all limits for all occupation numbers in $\lim N \to \infty$ and is *never* found to be *ferromagnetic*. (We return to this topic later.)

However, small systems behave differently! When one calculates the energies of three electrons on a loop of 4 sites (i.e., 1 hole) there is found a "quantum" phase transition at $U/t \approx 18.5$ ($U/W \approx 4.6$). Below this value

---

[28] Y. Nagaoka, *Phys. Rev.* **147**, 392 (1966); D. C. Mattis and R. E. Peña, *Effect of Band Structure on Ferromagnetism*, *Phys. Rev.* **B10**, 1006 (1974); M. W. Long and X. Zotos, *Phys. Rev.* **B48**, 317 (1993) studied correlations among the *holes* in Nagaoka's model.
[29] U. Brandt and A. Giesekus, *Phys. Rev. Lett.* **68**, 2648 (1992).

of U the ground state belongs to (minimal) spin 1/2, above it, to (maximal) spin 3/2.[30] Because a single plaquette such as this is closer to 2D than it is to the one-dimensional chain, the reason for this strange behavior may be related to Nagaoka's "theorem."

Nagaoka studied the Hubbard model in dimensions $d \geq 2$ in $\lim U \to \infty$. In this limit, Eq. (4.110) has no solution other than $W = 0$. However, Nagaoka found that if $N_{\text{elec}} = N_{\text{sites}} - 1$, the ground state 2^N-fold degeneracy is lifted in favor of the fully ferromagnetic state by a very small amount. Let us examine this further.

Suppose we expanded the Néel state or any arbitrary state, whether random or not, (call it $|\Psi\rangle$), in eigenstates of total spin (S) and total $S^z(M)$. A single hole has ground state energy $\varepsilon_0(S)$, hence the energy in any rotationally invariant state is just, $\langle\varepsilon\rangle = \langle\Psi|\varepsilon(S)|\Psi\rangle = \frac{\int_0^{N/2} dS(2S+1)|A_S|^2\varepsilon(S)}{\int_0^{N/2} dS(2S+1)|A_S|^2}$. We shall see that $\varepsilon(N/2) = -2tz$ is the minimum. (Z = coordination number = # of n.-n. to any given site.) If the amplitudes A are random this integral could be evaluated with few additional assumptions, in any dimension $d > 1$.

In two or higher dimensions the configurations that result from moving the hole to some specified point depend on the path taken, except in the ferromagnetic case. Supposing for the ferromagnetic case a configuration ϕ_a connects to a configuration ϕ_b through two different paths: the matrix element will be -2 (in some units). If in the AF state the two final configurations are different, say $\phi_{b'}$ and $\phi_{b''}$, then the matrix element connecting ϕ_a to $2^{-1/2}(\phi_{b'} + \phi_{b''})$ in the same units will be $-\sqrt{2}$. This difference favors the ferromagnetic state over all other configurations, as we now see with a specific example.

We illustrate with a square lattice, starting with the hole at some initial site

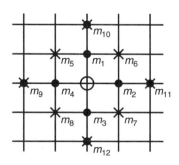

(I)

[30] D. C. Mattis, *Int. J. Nanoscience* **2**, 165 (2003).

with spin indices noted, and distinguishing n.-n.-n.'s according to whether they are neighbors to a single n.-n. (✸) or to two (×).

The state to which (I) connects is the sum over 4 configurations: (II), which, in turn, connect to a linear combination of configurations (III), in which the hole has moved to the n.-n. spot. It is the motion into the spots marked **X** that creates a new situation. These spots are accessible two ways. For example, the two configurations shown explicitly in (II) connect to the two (of 12 configurations) making up (III). But *they are not necessarily distinct*; if $m_1 = m_4 = m_5$, they are identical. If the m's are random, there is 1/4 probability that

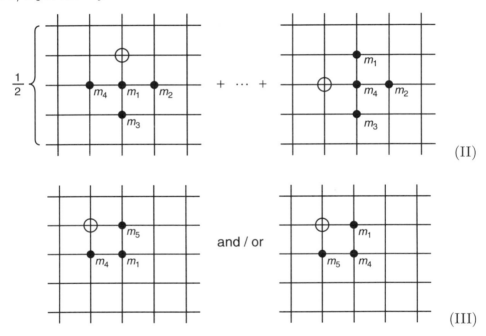

(II)

(III)

they are, indeed, identical. If the state is ferromagnetic, they are *necessarily* identical and, finally, if the state is AF all odd subscripted m's are + and even subscripted —, so that they are *never* identical. It is easily seen that the fewer the distinct configurations, the larger the matrix elements. In all cases, $\mathscr{H}_{I,II}$ are $-B\sqrt{4}$, regardless of the m's. But at the second step

$$\mathscr{H}_{II,III} = -B\sqrt{5} \quad \text{(ferro)}$$
$$\mathscr{H}_{II,III} = -B\sqrt{3} \quad \text{(AF)}.$$

(4.111)

Asymptotically, the matrix elements must be

$$\mathscr{H}_{M,M+1} = -2B \quad \text{(ferro)}$$

and, approximately,

$$\mathscr{H}_{M,M+1} = -2(3/5)^{1/2}B \quad (\text{AF}) \tag{4.112}$$

to yield a ground state eigenvalue

$$\begin{array}{ll} -4B & (\text{ferro}) \\ -4(3/5)^{1/2}B & (\text{AF}) \end{array} \tag{4.113}$$

and something intermediate for the random case. The smoothness of the ferromagnetic state results in the lowest energy. The AF state splinters into many extra nonidentical configurations such as shown in (III) above, and results in smaller matrix elements, hence a ground state which is not nearly so favorable. In 1D where only a single path connects any two states regardless of spins, this tendency is lacking. In 3D, where the number of paths connecting various configurations is greater than in 2D, the tendency to ferromagnetism is substantially increased.

For a small density n_h of holes, the energy favoring ferromagnetic order is thus $\propto n_h W$, taking the bandwidth W to be proportional to B. On the other hand, at finite U the energy favoring AF near a half-filled band is $\propto W^2/U$. With $n_h = \frac{1}{2} - \rho$, we set the opposing energies equal to obtain the phase boundary in Fig. 4.12 that separates AF from ferromagnetism. By symmetry the equation for a more-than-half-filled band is similar, with the final result for the bounding curves being,

$$\left| \rho - \frac{1}{2} \right| = \alpha W/U , \tag{4.114}$$

α being some lumped constant.

Richmond and Rikayzen[31] extended Nagaoka's work formulating a trial function in which all but one of the electron spins are aligned. The odd electron is localized about a particular site, whilst the remaining electrons adjust their motion to the presence of the repulsive potential that it represents. This is exactly soluble by the scattering theory that we have used in many connections throughout this book, and the bound state (if any!) energy is a variational upper bound to the true energy. If a bound state is found, the ferromagnetic state is unstable; thus their calculation also provides an equation for the ferro-antiferro phase boundary. Their findings are quite similar to (4.114).

---

[31] P. Richmond and G. Rikayzen, *J. Phys.* **C2**, 528 (1969).

Many other analyses and extensions of Nagaoka's work have appeared over the years, some of them containing new information and insights. For example, Zhang *et al.*, and especially Long and Zotos[32] looked at the 2-hole correlations in $\lim U \to \infty$ in the presence of a finite concentration of holes. At low density, they are found to be just like fermions: holes in the fermi sea of all spins up. At higher densities the correlations apparently morph into those of hard-core bosons. For his part, Hirsch[33] has concentrated on the n.-n. exchange integral, as the first major correction to Hubbard's model in a 1-band approximation and has emphasized the *asymmetry* between electrons and holes in the band.

4.17. One Dimension: Exact Solutions

Part of the attraction of the Hubbard model is its deceptive simplicity in 1D and its ultimate intractability in 2D and 3D. Although the first exact or quasi-exact solutions were found in 1D by Lieb and Wu[34] using a compound Bethe ansatz, and subsequently reproduced and improved by a slew of investigators testing this or that hypothesis, none of the studies could be generalized to higher dimensions where they could be useful.

Building on the original Lieb and Wu paper the interested reader will find 24 related works reprinted in the first chapter of Korepin and Essler.[26] Most deal with the excitation spectrum in the model, issues of whether the k's are real or come in complex pairs, connection to the 8-vertex model,[35] thermal properties, etc. Among other notable works one should study, let me single out Essler, Korepin and Schoutens' use of SO(4) symmetry to classify and count all the individual eigenstates[36] and Schulz' approach to obtaining the power law decays of correlation functions.[37]

The Lieb-Wu work starts with all spin-up particles ordered ($x_1 < x_2 < \cdots < x_M$) and all spin-down particles also ordered ($x_{M+1} < x_{M+2} < \cdots < x_N$). The order cannot change except if periodic boundary conditions are

[32] X. Y. Zhang, E. Abrahams, and G. Kotliar, *Phys. Rev. Lett.* **66**, 1236 (1991); M. W. Long and X. Zotos, *Phys. Rev.* **B48**, 317 (1993).

[33] J. Hirsch, *Phys. Rev.* **B65**, 184502 (2002) and references therein.

[34] Elliott H. Lieb and F. Y. Wu, *Phys. Rev. Lett.* **20**, 1445 (1968). Also, Elliott H. Lieb and F. Y. Wu, The one-dimensional Hubbard model: A reminiscence, *Physica* **A321**, 1 (2003). Their remarkable solution builds on the work of M. Gaudin, *Phys. Lett.* **24A**, 55 (1967) and C. N. Yang, *Phys. Rev. Lett.* **19**, 1312 (1967).

[35] B. S. Shastry, *Phys. Rev. Lett.* **56**, 1529 and 2453 (1986).

[36] F. H. L. Essler, V. E. Korepin and K. Schoutens, *Nucl. Phys.* **B384**, 431 (1982).

[37] H. J. Schulz, *Int. J. Mod. Phys.* **B5**, 57 (1991). A recent and complete review is given by D. Schlottmann, *Int. J. Mod. Phys.* **B11**, 355 (1997).

applied at the ends. Except when they collide at internal boundaries (say where $x_n = x_{n+1}$) the particles propagate freely. Hence one takes a set of N unequal real wavevectors $k_1, k_2, \ldots$ and expands the spatial eigenfunction $f(x_1, x_2, \ldots, x_M | x_{M+1}, \ldots, x_N)$ in plane waves and their permutations: $f = \sum A(P, Q) e^{i \sum_j k_{P(j)} x_{Q(j)}}$. The energy is just the sum $E = -2t \sum_j \cos k_j$. The k_j's $=$ unperturbed $k_j^{(0)} +$ phase-shift. The k's and their phase shifts satisfy some transcendental equations.

The complexity of these equations puts the theory squarely outside the scope of the present text. The diligent reader is referred to the literature on the topic. Interestingly, the solution (by a generalized Bethe ansatz) proves that the linear chain (like the Hubbard model on other symmetric bipartite lattices in higher dimensions) is an antiferromagnetic insulator at half-filling where it exhibits a massive excitation spectrum above an energy gap that, like the ground state energy, depends nonanalytically on U.

Away from half-filling, many of the properties of the one-dimensional Hubbard model are similar to those of a far simpler model, the so-called Luttinger-Tomonaga liquid, much studied and better understood than its Hubbard counterpart. It is the solid-state version of certain quantum field theories in $1 + 1$ dimensions that were introduced over the years in high-energy physics, starting in the 1930's. As in the large U Hubbard model, spin and charge degrees of freedom are naturally decoupled in the Luttinger model and propagate at different speeds. The energy gap is normally absent (this is a "weak-coupling" theory) but can be introduced *via* back-scattering matrix elements. The Luttinger model too has a vast literature, some of which is repertoried in a book by the present author.[38] Alas, it has no more applicability in 2D and 3D than does the Bethe ansatz solution of Hubbard model. We must leave this topic to the specialists and turn to the next item in our survey of the many-body theory of magnetism.

4.18.   Ferrimagnetism

In ferrites (e.g., lodestone, Fe_3O_4) different ions of the same metal ion are nearest-neighbors, e.g., Fe^{2+} and $2 \times Fe^{3+}$. While their spins are normally antiparallel the magnitudes are different, thus they cannot quite cancel and there is a net moment in each unit cell. A similar sort of ferrimagnetism can

---

[38] D. C. Mattis, *The Many-Body Problem*: An encyclopedia of exactly solved models in 1D, World Scientific, 1992, Chapter 4.

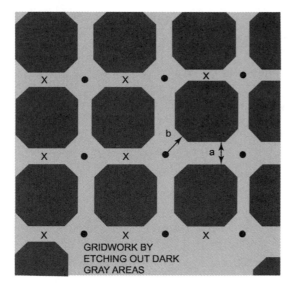

Fig. 4.13. **2D "antidot" lattice made of intrinsic silicon (light gray areas; insulator in dark gray areas).**[40]
This network of intrinsic silicon semiconductor is a decorated antidot square lattice that emulates the CuO_2 gridwork in high T_c superconductors. In CuO_2 the copper atoms are at the corners (black dots) and the oxygen ions are in-between, as indicated by X's.

also be created by purely geometrical means.[39] For purposes of illustration, let us consider the "decorated sq lattice" of a single CuO_2 plane, such as those found in the high-T_c superconductors. Assuming electrons can hop to and from n.-n. sites only we shall now construct the energy band structure for this geometry.

In a single plane of copper oxide, found in any of the high-temperature superconducting pervoskites, the copper sites have 1 active electron (spin 1/2) and are connected by oxygen ions with 2 electrons each (spin 0). In the artificial network made out of a film of intrinsic silicon, whose architecture (see in Fig. 4.13), emulates that of copper oxide, the intersections take the place of copper sites and the X's take the place of the mid-point, ligand oxygens. There are 3 viable sites in each unit cell: 2 X's and 1 corner site, hence a practical maximum of 6 electrons *per cell*. With a 2-electron singlet on each X and 1 electron in each corner, the picture illustrates an archetype "superexchange antiferromagnet."

---

[39] E. H. Lieb, *Phys. Rev. Lett.* **62**, 1201 and 1927 (1989).

But unlike chemical compounds in which valencies are fixed, in nanocircuits it is possible to vary the electron concentration from zero to a fairly large number and, at 3 electrons *per cell*, create an artificial ferrimagnet for this geometry.[40] The basis for this hypothetical magnet (it has not yet been made!) is the 1989 theorem by Lieb.[39,41] But it is easier to see why this is so than it is to calculate it. If one electron is at each corner site and there is one at each of the interstitial sites marked X, then with the known tendency of 2 n.-n. electrons to have antiparallel spins, there will be 2 electrons of spin up on each X site and 1 of spin down in each corner sites — for a total of 1 Bohr magneton *per cell*.

A tight-binding calculation confirms this should be so. Assume each site connects to its neighbors with a matrix element t. The starting point is,

$$\mathscr{H}_0 = -t \sum_{j,\sigma} (c_{j,\sigma}^\dagger (a_{j+(1/2,0),\sigma} + a_{j+(0,1/2),\sigma}) + \text{H.c.}). \tag{4.115}$$

We now transform to plane waves. Let the F.T. of the a's on the horizontal be $a(k)$ and on the verticals $b(k)$. Then,

$$\mathscr{H}_0 \Rightarrow -2t \sum_{k,\sigma} (c_\sigma^\dagger(k)(a_\sigma(k)\cos k_x/2 + b_\sigma(k)\cos k_y/2) + \text{H.c.}). \tag{4.116}$$

We can adopt $\alpha_\sigma(k) \equiv \frac{a_\sigma(k)\cos k_x/2 + b_\sigma(k)\cos k_y/2}{\sqrt{\cos^2 k_x/2 + \cos^2 k_y/2}}$ as a new operator replacing $a(k)$ and $b(k)$, together with an orthogonal operator, $\beta_\sigma(k) \equiv \frac{a_\sigma(k)\cos k_y/2 - b_\sigma(k)\cos k_x/2}{\sqrt{\cos^2 k_x/2 + \cos^2 k_y/2}}$. The latter is absent from $\mathscr{H}_0$ but the former is not:

$$\mathscr{H}_0 \Rightarrow -2t \sum_{k,\sigma} \left(c_\sigma^\dagger(k)\alpha_\sigma(k)\sqrt{\cos^2 k_x/2 + \cos^2 k_y/2} + \text{H.c.} \right). \tag{4.117}$$

It is now trivially diagonalized and yields 2 energy bands:

$$\mathscr{H}_0 \Rightarrow -2t \sum_{k,\sigma} \sqrt{\cos^2 k_x/2 + \cos^2 k_y/2} (c_{k,\sigma}^\dagger c_{k,\sigma} - \alpha_{k,\sigma}^\dagger \alpha_{k,\sigma}) \tag{4.118}$$

with dispersion $\varepsilon_\pm(k) = \pm 2t\sqrt{\cos^2 k_x/2 + \cos^2 k_y/2}$. A third band, the β's, is dispersionless (the energy of each and every $\beta_\sigma(k)$ being 0 as it is absent from $\mathscr{H}$).

---

[40] D. C. Mattis, *J. Stat. Phys.* **116**, 773 (2004).

[41] The proof of ferrimagnetic long-range order is by S.-Q. Shen, Z.-M. Qiu and S.-S. Tian, *Phys. Rev. Lett.* **72**, 1280 (1994).

Now let us populate these states. The first $2N$ fill the lower of the two bands (4.118). The next $2N$ will occupy the β band at energy 0. The last $2N$ populate the higher of the 2 bands in (4.118).

As one increases N_{elec} in the interval $2N_{\text{sites}} < N_{\text{el}} < 4N_{\text{sites}}$, their spins are arbitrary (for the first N). So each $\beta(k)$ state has spin σ that is up or down, at no cost in energy. Then, because of the Pauli principle, the next N must be antiparallel to those that came before. The maximum spin when we add the first electron into this band is $1/2$, just as it is when we are left with just 1 hole. The symmetry point $N_{\text{elec}} = N$ when each k can belong to either spin up or down, making for a degeneracy 2^N. Lieb's theorem[39] states that for *interacting electrons* the ground state that emerges from this degenerate bunch is the state of maximal total spin, $S_{\text{tot}} = N/2$. This Hund's rule for N electrons (!) can best be understood in a Hubbard model with large, but not infinite, U.

In that case, the band structure is not useful. Electrons are not itinerant but remain on their individual sites, with the occasional excursion to the neighboring site. Two n.-n. electrons with parallel spin cannot perform this excursion. But if their spins are antiparallel, the energy of this pair is the lowest eigenvalue of:

$$
\begin{bmatrix}
0 & 0 & -t & -t \\
0 & 0 & -t & -t \\
-t & -t & U & 0 \\
-t & -t & 0 & U
\end{bmatrix}. \tag{4.119}
$$

Problem 4.11.   Write the 4 basis states for 2 neighboring electrons with hopping matrix element t and energy U when either site is doubly occupied. Show the Hamiltonian can be written in the form (4.119) and find its 4 eigenvalues. Show the lowest one (for the singlet state) reduces to $E_0 = -4t^2/U$ at large U while triplet states have energy zero. Show this yields a coupling constant $J = 4t^2/U$ that favors n.-n. antiparallelism.

Because there are twice as many interstitial sites X as there are corner sites the antiparallelism produces a net spin. The situation has similarities with double-exchange, Sec. 4.12, or ferrimagnetism, a topic for Secs. 5.9 and 5.10.

4.19.   Spin-Dependent Tunneling

Magnetoresistive tunneling junctions have been recently developed that test *macroscopic* parallelism (or antiparallelism) of two magnetic substances separated by a tunneling barrier. The purpose is to create a sort of *spin valve*, that is, a device that allows a small "tunneling" current to flow between the two magnetic metals when their magnetizations are parallel but utterly prevents such flow when the magnetizations are antiparallel.[42] This phenomenon has also been dubbed "Giant Magneto-Resistance," i.e., GMR for short.

Consider the magnetic metal to be a "half-magnet," that is a substance in which the majority spin band is fully occupied while the minority spin band is partly empty. The small tunneling current is proportional to the product of the density of states (at the Fermi level) of both metals: $j \propto \sum_\sigma \rho_{1,\sigma}(\mu)\rho_{2,\sigma}(\mu)$. Electrons do not change their spin orientations when tunneling through a nonmagnetic medium. Thus, when the magnetizations are parallel, the minority spins can tunnel back and forth while the majority spins are prevented from transferring by the Pauli principle.

Conversely, if the magnetizations are antiparallel, the minority electrons at or near the Fermi level in either metal have no corresponding state in the other, so that the current is totally blocked. In this instance, a picture is worth a thousand words. The situation described above is illustrated in Fig. 4.14.

The reader will want to adapt the theory of double-exchange in the case of partial parallelism, as in the following Problem.

Problem 4.12.   If the angle between the magnetizations is ϑ, in the range $0 < \vartheta < \pi$, derive an expression for the tunneling current $j(\vartheta)$ in units of j_0, the current at $\vartheta = 0$.

Practical uses of this phenomenology have been widespread in recording devices. Recently, S.S. Parkin of IBM has designed non-volatile magnetic

---

[42] Applications to magnetic drives are obvious: if the tunneling barrier consists of air or vacuum or an insulating layer just above a rotating magnetic disk, a magnetic reading head will read the magnetically stored information by the amount of current that flows through it.

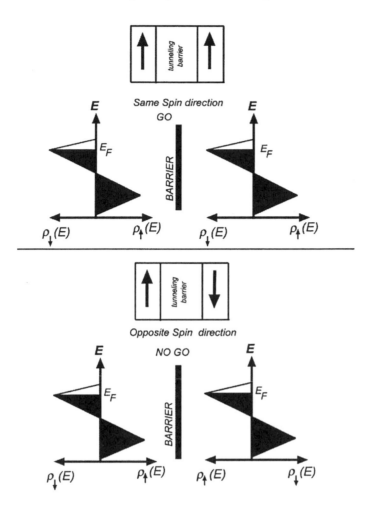

Fig. 4.14. Spin-dependent tunneling junction.
If a small voltage is applied to the two metals separated by a tunneling barrier, a tunneling current flows in the upper situation (parallel magnetizations) but not if the magnetizations are antiparallel, as in the lower picture, in the absence of a strong spin-flipping mechanism (derived from spin-orbit coupling).

memory elements (MagRAM) that are outlined in Figs. 4.15 and 4.16 immediately following.[43] Finally, Su and Suzuki[44] have predicted that Josephson-like fluctuations should be seen in such spin valves.

---

[43] Reproduced by permission of IBM research, Almaden Research Center. Unauthorized use not permitted.
[44] G. Su and M. Suzuki, "Josephson-like effect in a spin valve," *Mod. Phys. Lett.* **B16**, 711 (2002).

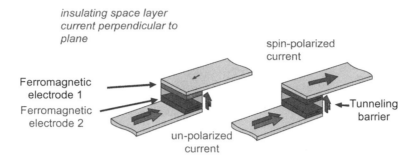

Fig. 4.15.   Elements of a magnetic spin valve.[43]

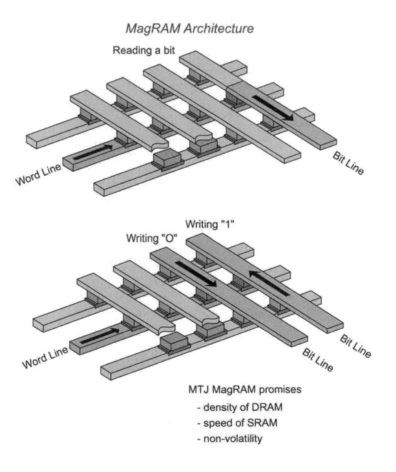

Fig. 4.16.   Application of IBM spin valve as memory element.[43]

Chapter 5

From Magnons to Solitons: Spin Dynamics

Ground State of a Ferromagnet, of an Antiferromagnet, and Their Spectra of "Elementary Excitations"

There exist striking similarities between the elementary excitations of a ferromagnet and those in an elastic solid. In the latter, we know that an atom displaced from equilibrium position will oscillate with the motion and frequencies of the *normal modes* of the crystal. The effect of quantum mechanics on this motion is to quantize the amplitudes of the individual normal modes, into units known as *phonons*. The analogous normal modes in the magnetic system are the spin waves; when the quantum mechanical nature of spins is taken into account, these also are quantized, with the basic unit being the *magnon*.

In the ferromagnet, for which the ground state has all spins parallel, the small amplitude oscillations are accurately described by harmonic-oscillator type dynamical variables. The interaction between neighboring oscillators is responsible for the dispersion, the dependence of frequency ω of a normal mode, on the wave vector $\mathbf{k}$ (the wavelength $\lambda = 2\pi/k$). Indeed, as the ground state of a ferromagnet is known exactly, we are able to extend the analysis to obtain the exact one- and two-magnon eigenstates. The latter number both bound states and scattering states, the interactions being caused by an inherent nonlinearity in the spin dynamics due to the coupling of the motion in various directions by the commutation relations elaborated in Chapter 3. The analysis of 3- or more-magnon states requires the solution of a full many-body problem.

In the *anti*ferromagnet (a solid in which neighboring spins tend to be antiparallel), the inherent nonlinearities even prevent us from obtaining the exact ground state or elementary excitations, with the exception of one dimension (and there, only for spins $1/2$ or spins $s \to \infty$ with the coupling constant $J \propto 1/s^2$, the classical limit). Nevertheless, spin-wave theory is as useful for an approximate analysis, as are the phonons in the study of the *an*harmonic solid. The ground state and elementary excited states, and the ground state correlation functions, can all be computed with spin-wave theory. A check on the accuracy exists in a variety of numerical experiments on representative systems.

There are instances where magnons yield only part of the answers, perhaps not even the qualitatively significant part. In two-dimensional magnets, the *vortex* has been found to be the most significant topological excitation, and in one-dimensional magnets, magnons and *solitons* are both required in the dynamics. Such relatively new developments are treated in this chapter which, with a few exceptions, is consecrated to the study of the Heisenberg Hamiltonian with nearest-neighbor interactions, i.e., to insulators.

5.1.   Spin Waves as Harmonic Oscillators

The approximation of spin waves by harmonic-oscillator functions, as opposed to a description "based on creation and annihilation operators, is to a considerable extent only a semantic one, but nevertheless is probably of use to those readers to whom harmonic oscillators are more intuitive than the techniques of quantum-mechanical field theory."[1]

For concreteness, let us assume (for the present) a ferromagnetic Heisenberg Hamiltonian with nearest-neighbor interactions, in an external homogeneous magnetic field H

$$\mathcal{H} = -J \underset{\substack{i>j= \\ \text{nearest neighb.}}}{\sum\sum} \mathbf{S}_i \cdot \mathbf{S}_j - H g \mu_B \sum_i S_j^z, \quad J > 0 \qquad (5.1)$$

where

$$\mathbf{S}_i \cdot \mathbf{S}_j \equiv S_i^x S_j^x + S_i^y S_j^y + S_i^z S_j^z$$

$$= \frac{1}{2}(S_i^+ S_j^- + S_j^+ S_i^-) + S_i^z S_j^z.$$

---

[1] The present section has been modeled on "Spin Waves," by J. Van Kranendonk and Van Vleck (*Rev. Mod. Phys.* **30**, 1 (1958)) from which this quotation is taken.

Beside the isotropic interactions which are included therein, we may also consider anisotropic interactions of various origins. Those of lowest order are of "dipolar" structure,

$$\frac{1}{2}\sum_i\sum_j D_{ij}\frac{\mathbf{S}_i\cdot\mathbf{S}_j r_{ij}^2 - 3\mathbf{S}_i\cdot\mathbf{r}_{ij}\mathbf{S}_j\cdot\mathbf{r}_{ij}}{r_{ij}^2} \tag{5.2}$$

which represents the magnetic dipole-dipole interaction (the magnetostatic potential of each spin in the field of the others) with

$$D_{ij} = g^2\mu_{\mathrm{B}}^2 r_{ij}^{-3}. \tag{5.3}$$

But there may be additional contributions to D_{ij}, principally those arising from the spin-orbit coupling mechanism.[2] These, like the exchange force itself, tend to be very short-ranged. For concreteness, let us limit the present discussion to cubic materials and the anisotropic interactions to (5.3) in lowest order.

In Chapter 3 on angular momentum, we have shown rigorous representations of spins constructed with the aid of harmonic-oscillator operators; in particular, the Holstein-Primakoff representation, and approximations to it, were introduced. Let us review the latter from a new point of view. First, the spins are defined by the three nonvanishing matrix elements:

$$\langle n_i|S_j^x|n_i+1\rangle = \langle n_i+1|S_i^x|n_i\rangle^* = \frac{1}{2}\sqrt{(n_i+1)(2s-n_i)}$$

$$\langle n_i|S_i^y|n_i+1\rangle = \langle n_i+1|S_i^y|n_i\rangle^* = \frac{1}{2}i\sqrt{(n_i+1)(2s-n_i)} \tag{5.4}$$

$$\langle n_i|S_i^z|n_i\rangle = s-n_i, \quad 0\leqslant n_i\leqslant 2s, \quad \hbar=1.$$

The quantum numbers of other spins (n_j) have been suppressed, because the components of $\mathbf{S}_i$ cannot change these; but all occupation numbers $n_1, n_2, \ldots, n_N$ must be specified in a definite state of the N spins, and therefore there are exactly $(2s+1)^N$ such states if all spins have the same magnitude s.

Matrix elements of harmonic-oscillator operators have a structure very similar to the above. For example, if we symbolize the integrals

$$\int \psi_n^*(x_i)(\mathrm{op})\psi_m(x_i)\,dx_i$$

---

[2] F. Keffer and T. Oguchi, *Phys. Rev.* **117**, 718 (1960).

where $\psi_n(x)$ is the nth harmonic-oscillator function, in the Dirac bracket notation, $\langle n|(\text{op})|m\rangle$, then for the very important operators x_i coordinate and p_i momentum, we have

$$\langle n_i|x_i|n_i+1\rangle = \langle n_i+1|x_i|n\rangle^* = \sqrt{\frac{\hbar}{2m\omega}(n_i+1)}$$

$$\langle n_i|p_i|n_i+1\rangle = \langle n_i+1|p_i|n\rangle^* = -i\sqrt{\frac{\hbar m\omega}{2}(n_i+1)}$$

(5.5)

and the quantum number n_i may also be defined as the eigenvalue of an operator, now the energy operator

$$\left\{\frac{p_i^2}{2m} + \frac{1}{2}m\omega^2 x_i^2 - \frac{1}{2}\hbar\omega\right\}|n_i\rangle = n_i\hbar\omega|n_i\rangle .$$

(5.6)

The connection between harmonic oscillator and spin is next established, by use of the dimensionless canonical variables P, Q, defined by

$$Q_i = x_i\sqrt{\frac{m\omega}{\hbar}} , \quad P_i = p_i\sqrt{\frac{1}{\hbar m\omega}} , \quad [P_i, Q_j] = \delta_{ij}\frac{1}{i} ,$$

(5.7)

by comparison of (5.4) and (5.5).

In terms of P, Q the matrix elements are

$$\langle n_i|Q_i\sqrt{s}|n_i+1\rangle = \langle n_i+1|Q_i\sqrt{s}|n_i\rangle^* = \frac{1}{2}\sqrt{(n_i+1)2s}$$

$$\langle n_i|P_i\sqrt{s}|n_i+1\rangle = \langle n_i+1|P_i\sqrt{s}|n_i\rangle^* = -\frac{1}{2}i\sqrt{(n_i+1)2s}$$

(5.8)

$$\langle n_i|s - \frac{1}{2}(P_i^2 + Q_i^2 - 1)|n_i\rangle = s - n_i ,$$

quite similar to (5.4) at *small values of the* n_i. Note that for values of $n_i \approx s$, the harmonic-oscillator approximation becomes quantitatively incorrect, although the matrix structure is still qualitatively similar to the correct one. The error however becomes catastrophic only when some n_i equals or exceeds the value $2s$, for although (5.8) continues to define matrix elements for the harmonic oscillators, there is no corresponding structure in angular-momentum space. So it must be understood that the whole theory which will be developed on the base of similarity between spins and harmonic oscillators will only be valid for low occupation numbers of every harmonic oscillator.

In the Hamiltonian of (5.1) we now make the substitutions,

$$S_i^x = Q_i\sqrt{s}\,,$$

$$S_i^y = P_i\sqrt{s}\,,$$

$$S_i^z = s - \frac{1}{2}(P_i^2 + Q_i^2 - 1)$$

(5.9)

and in accord with the above instructions, systematically discard cubic and quartic terms. One obtains the "linearized" Hamiltonian (i.e., the equations of motion are linearized, the Hamiltonian itself is of course quadratic)

$$\mathcal{H}_{\text{lin}} = E_0 + g\mu_{\text{B}}H \sum_i \frac{1}{2}(P_i^2 + Q_i^2 - 1) - Js \sum_{(ij)}(P_iP_j + Q_iQ_j) \qquad (5.10)$$

where (i,j) means the sum is over nearest-neighbor pairs. The constant energy term

$$E_0 = -NHg\mu_{\text{B}}s - \frac{1}{2}NzJs^2 \qquad (5.11)$$

is the energy of the completely saturated state in which all spins are parallel to the applied field.

The linearized Hamiltonian bears an obvious resemblance to the Hamiltonian of lattice vibrations, which also incorporates interactions among neighboring harmonic oscillators. An important difference arises from the presence of "velocity-dependent forces" P_iP_j, to which may be attributed the difference between the spectra of spin waves and sound.

We now make the plane-wave transformation to the new set of generalized coordinates and momenta

$$Q_i = \frac{1}{\sqrt{N}} \sum_k e^{i\mathbf{k}\cdot\mathbf{R}_i}Q_k\,,$$

$$P_i = \frac{1}{\sqrt{N}} \sum_k e^{i\mathbf{k}\cdot\mathbf{R}_i}P_k$$

(5.12)

note

$$Q_k^* = Q_{-k} \ \text{etc.,} \quad \text{and} \quad [P_k^*, Q_{k'}] = \frac{\delta_{kk'}}{i} \qquad (5.13)$$

where the components k_x, k_y, k_z are integer multiples of $2\pi/L$. This transformation diagonalizes (5.10) for which we find

$$\mathcal{H}_{\text{lin}} = \sum_k \frac{1}{2}(P_k^* P_k + Q_k^* Q_k - 1)\hbar\omega(\mathbf{k}) + E_0$$

$$= \sum_\mathbf{k} n_\mathbf{k} \hbar\omega(\mathbf{k}) + E_0 . \qquad (5.14)$$

It may be verified that the operators, P_k, $Q_{k'}^*$, etc., which refer to different momenta, commute, and so we have N new oscillators, which are uncoupled in the above approximation. For each of these, the quadratic "number operator"

$$n_\mathbf{k} \equiv \frac{1}{2}(P_k^* P_k + Q_k^* Q_k - 1) \qquad (5.15)$$

has integer eigenvalues $= 0, 1, 2, \ldots$. Fortunately, it is not at all necessary that each of these eigenvalues be restricted to the range $\ll s$, for it is sufficient that $\sum n_k \ll Ns$ for the linearization to be accurate overall.

As for the energies of the plane-wave "magnons,"

$$\hbar\omega(\mathbf{k}) = Hg\mu_B + Js \sum_\delta (1 - \cos \mathbf{k} \cdot \boldsymbol{\delta}) \qquad (5.16)$$

$$\simeq Hg\mu_B + Jsa^2\mathbf{k}^2 + O(\mathbf{k}^4)$$

the sum runs over the vectors $\boldsymbol{\delta}$ which connect a typical spin to the z nearest neighbors with which it interacts; $a = |\boldsymbol{\delta}|$ for the simple-cubic (sc) lattice, and the wave vectors $\mathbf{k}$ are restricted to the range $|k_x|, |k_y|, |k_z| < \pi/a$.

Let us now include the dipolar interactions, to see what is their effect on the magnon energy spectrum. Defining the direction cosines, α, β, and γ of the unit vector joining any two sites,

$$\frac{\mathbf{r}_{ij}}{r_{ij}} = (\alpha_{ij}, \beta_{ij}, \gamma_{ij}) \qquad (5.17)$$

we find three contributions to the dipolar Hamiltonian of (5.2), after omitting cubic and higher terms outside the scope of the linearized theory:

$$\mathcal{H}_{d,0} = \frac{1}{2}s^2 \sum_i \sum_j D_{ij}(1 - 3\gamma_{ij}^2) \qquad (5.18)$$

$$\mathcal{H}_{d,1} = -3s^{3/2} \sum_i \sum_j D_{ij}(\alpha_{ij}\gamma_{ij}Q_j + \beta_{ij}\gamma_{ij}P_j) \qquad (5.19)$$

$$\mathcal{H}_{d,2} = \frac{1}{2}s^2 \sum_i \sum_j D_{ij}[(1 - 3\alpha_{ij}^2)Q_iQ_j - 3\alpha_{ij}\beta_{ij}(Q_iP_j + P_iQ_j)$$

$$+ (1 - 3\beta_{ij}^2)P_iP_j - (1 - 3\gamma_{ij}^2)(P_i^2 + Q_i^2 - 1)]. \tag{5.20}$$

For the present, it is convenient to imagine that our (cubic) ferromagnet has the shape of an ellipsoid with one of the principal axes in the z direction. The zeroth order term above is then readily interpreted in terms of the classical demagnetizing factor N_z,

$$N_z = a^3 \sum_j \frac{1}{r_{ij}^3}(1 - 3\gamma_i^2) + \frac{4\pi}{3} \tag{5.21}$$

with α and β replacing γ above in the similar definitions of N_x and N_y. The sum in (5.21) being independent of $\mathbf{r}_i$, provided this point is within the crystal, we may express the zeroth order dipolar contribution in terms of the demagnetizing factor, introducing

$$M_0 = \frac{g\mu_B s}{a^3} = \text{saturation magnetization per unit volume} \tag{5.22}$$

where a^3 = volume per spin, and we find

$$\mathcal{H}_{d,0} = -\frac{1}{2}VM_0\left(\frac{4\pi}{3}M_0 - N_zM_0\right). \tag{5.23}$$

The factor in brackets is the sum of the Lorentz field $(4\pi/3)M_0$ and the demagnetizing field (which is uniform in an ellipsoidally shaped material) — N_zM_0, and is therefore the "effective field" acting at a given lattice site; $V = Na^3$ is the total volume.

Aside from this constant energy term and from the constant E_0 in the exchange Hamiltonian, (5.10), the dynamic terms include: the nontrivial terms in (5.10) plus $\mathcal{H}_{d,1}$ and $\mathcal{H}_{d,2}$, which all together are of the form

$$\mathcal{H} = \sum_{i,j}(V_{ij}P_iP_j + W_{ij}Q_iQ_j + U_{ij}P_iQ_j)$$

$$- 3s^{3/2}\sum_j\left[\left(\sum_i D_{ij}\alpha_{ij}\gamma_{ij}\right)Q_j + \left(\sum_i D_{ij}\beta_{ij}\gamma_{ij}\right)P_j\right] \tag{5.24}$$

allowing V_{ij}, W_{ij}, and U_{ij} to indicate schematically the coefficients in (5.10, 5.20). The linear term may be eliminated altogether by means of the canonical transformation,

$$P_i \rightarrow P_i + f_i \quad \text{and} \quad Q_i \rightarrow Q_i + g_i \tag{5.25}$$

where f_i, g_i are appropriately determined constants. (As it happens, perturbation theory also converges for these linear terms, and the shift in
ground-state energy which can be calculated by treating $\mathcal{H}_{d,1}$ by second-
order perturbation theory agrees identically with the result of the exact
canonical transformation above.) Clearly the result is small, for it is second
order in such quantities as

$$\sum_j D_{ij}\alpha_{ij}\gamma_{ij} \tag{5.26}$$

which vanish at any point $\mathbf{r}_i$ which is a point of symmetry. This would be at
every point not on the surface of the cubic material if D_{ij} were not of such
long range. Calculation of this term is left as an exercise for the reader, a
tedious one.

The complete elimination of $\mathcal{H}_{d,1}$ still leaves the first line in (5.24), that
is $\mathcal{H}_{\text{lin}} + \mathcal{H}_{d,2}$ unaffected; and this is now to be diagonalized by a transformation, first to running waves, (5.13),

$$\mathcal{H} \rightarrow \frac{1}{2}\sum_k [A(\mathbf{k})Q_k^*Q_k + B(\mathbf{k})P_k^*P_k + 2C(\mathbf{k})Q_k^*P_k] + \text{const.} \tag{5.27}$$

where, with $\hbar\omega(\mathbf{k})$ given in (5.16),

$$A(\mathbf{k}) = \hbar\omega(\mathbf{k}) + A_{xx}(\mathbf{k}) - A_{zz}(\mathbf{0})$$
$$B(\mathbf{k}) = \hbar\omega(\mathbf{k}) + A_{yy}(\mathbf{k}) - A_{zz}(\mathbf{0}) \tag{5.28}$$
$$C(\mathbf{k}) = A_{xy}(\mathbf{k})\,.$$

The various A's are seen by comparison with the original equations, (5.10,
5.20) to be dipolar lattice sums,

$$A_{xx}(\mathbf{k}) = \frac{s}{N}\sum_{i,j} D_{ij}(1 - 3\alpha_{ij}^2)e^{i\mathbf{k}\cdot\mathbf{R}_{ij}} \tag{5.29}$$

and A_{yy}, A_{zz} are given similarly by replacing α_{ij} by β_{ij} or γ_{ij}. Also

$$A_{xy}(\mathbf{k}) = \frac{-3s}{N}\sum_{i,j} D_{ij}\alpha_{ij}\beta_{ij}e^{i\mathbf{k}\cdot\mathbf{R}_{ij}} \tag{5.30}$$

and A_{xz}, etc., can also be obtained by obvious permutations. These dipolar sums occur in several other problems of interest in solid-state physics,
and have been investigated numerically, quite thoroughly, by Cohen and

Keffer.[3,4] As an example of the size effect, these authors give $A_{xx}(\mathbf{k})$ calculated for a spherical sample of radius R, and $\mathbf{r}_i$ near the center of the crystal

$$A_{xx}(\mathbf{k}) \sim \left(1 - 3\frac{k_x^2}{k^2}\right)\left[1 - \frac{3j_1(kR)}{kR}\right] \tag{5.31}$$

(omitting constant multiplicative factors). We see a not entirely unexpected phenomenon: as k is decreased to a value $\approx 10/R$, the dipolar sums in a finite crystal begin to differ significantly from their value in an infinite material, and the shape of the surface, and the position of the origin become of importance. However, precisely at $\mathbf{k} = \mathbf{0}$, this lattice sum approaches a well-defined limit related to the demagnetizing factor defined in (5.21). We shall make use of this fact.

Assuming, then, that the coefficients $A(\mathbf{k})$, $B(\mathbf{k})$, and $C(\mathbf{k})$ are known, it is a straightforward matter to perform a canonical transformation to a new set of normal modes

$$P'_{\mathbf{k}} = a_{\mathbf{k}}P_{\mathbf{k}} + b_{\mathbf{k}}Q_{\mathbf{k}}, \quad Q'_{\mathbf{k}} = c_{\mathbf{k}}Q_{\mathbf{k}} + d_{\mathbf{k}}P_{\mathbf{k}} \tag{5.32}$$

choosing the numerical coefficients $a_{\mathbf{k}}, \ldots, d_{\mathbf{k}}$ so as to diagonalize the Hamiltonian yet preserve the canonical commutation relations,

$$[P'^*_{\mathbf{k}_1}, Q'_{\mathbf{k}_2}] = \frac{\delta_{\mathbf{k}_1,\mathbf{k}_2}}{i} \tag{5.33}$$

all other commutators $= 0$. The final result is,

$$\mathscr{H} = \sum \frac{1}{2}(P'^*_{\mathbf{k}}P'_{\mathbf{k}} + Q'^*_{\mathbf{k}}Q'_{\mathbf{k}} - 1)\hbar\omega(\mathbf{k}) + \text{const.} \tag{5.34}$$

with

$$\hbar\omega(\mathbf{k}) = \sqrt{A(\mathbf{k})B(\mathbf{k}) - C^2(\mathbf{k})} \tag{5.35}$$

an anisotropic function of the direction of $\mathbf{k}$, as is seen from the definition of $A(\mathbf{k})$, $B(\mathbf{k})$, and $C(\mathbf{k})$ given in (5.28), and as plotted in Fig. 5.1.

In *ferromagnetic resonance* an oscillating electromagnetic field is applied to the sample. Usually, the electromagnetic wavelengths are sufficiently long that only the $\mathbf{k} = \mathbf{0}$ mode need be considered, and we therefore obtain a resonance phenomenon at applied frequencies in the neighborhood of $\omega'(\mathbf{0})$.

---

[3] M. H. Cohen and F. Keffer, *Phys. Rev.* **99**, 1128, 1135 (1955).
[4] Is the energy properly extensive ($\propto N$)? This is answered in the affirmative by R. Griffiths (*Phys. Rev.* **176**, 655 (1968)).

The integrals (5.74) making up the 3×3 matrix $\mathbb{M}$ are difficult to express in terms of elementary functions (although they simplify in the long wavelength limit ka, $k'a \to 0$, precisely the limit studied by Dyson[8] because of its applicability to low-temperature and low-energy behavior). In leading semiclassical approximation, we just ignore $\mathbb{M}$ which is $O(1/s)$, and evaluate

$$\delta E(\mathbf{kk'}) \approx 4J\mathbf{C}(\mathbf{q_0}) \cdot [\mathbf{C}(1/2\mathbf{K}) - \mathbf{C}(\mathbf{q_0})]$$
$$= -(1/s)[\hbar\omega(\mathbf{k}) + \hbar\omega(\mathbf{k'}) - \hbar\omega(\mathbf{k} - \mathbf{k'}) - \hbar\omega(\mathbf{0})] \qquad (5.78)$$

with $\hbar\omega(\mathbf{k})$ given in (5.16), as usual.

In $1D$ $\mathbb{M}$ is just a number.

Equation (5.74) yields

$$M = \frac{2J}{2\pi} \int_{-\pi}^{+\pi} dq \frac{(\cos 1/2K - \cos q)\cos q}{4sJ \cos 1/2K(\cos q - \cos q_0) + i\varepsilon}$$
$$= \frac{1}{2s\cos 1/2K} \left(\cos \frac{1}{2}K - \cos q_0\right)(1 - i\cot q_0) \qquad (5.79)$$

or, in terms of k and k'

$$M = \frac{-1}{s\cos 1/2(k + k')} \sin\frac{1}{2}k \sin\frac{1}{2}k' \left[1 - i\cot\frac{1}{2}(k - k')\right]. \qquad (5.80)$$

Insertion into (5.76) yields the scattering amplitudes characteristic of the continuum of scattering states, expressed in terms of the unperturbed k values of (5.57). Later [see (5.150)] we shall re-examine this point; energy and momentum can be conserved with just two values of $\mathbf{k}$, provided the k values are shifted somewhat. This is part of the so-called *Bethe ansatz*; its applications to date have been limited to 1D, its relation to the above, somewhat obscure.

We now turn our attention to the two-magnon *bound* states, and look for E lying outside the range of values given in (5.67). As there is no corresponding (real) q_0 there is no special amplitude (5.68). The eigenvalue equation (5.63) is now a homogeneous equation, and $f(\mathbf{q})$ is given by

$$f(\mathbf{q}) = \frac{2J\mathbf{C}(\mathbf{q}) \cdot \mathbf{V}}{E - \delta_{\mathbf{K}}(\mathbf{q})} \qquad (5.81)$$

with

$$\mathbf{V} = \mathbb{M} \cdot \mathbf{V}, \quad \text{the matrix } M_{\alpha\beta}(E) \text{ given in (5.73).} \qquad (5.82)$$

The bound state energy E is adjusted until $\mathbf{V}$ is an eigenvector of $\mathbb{M}$. This determines $\mathbf{V}$ to within a multiplicative constant, required for normalization.

5.4. Bound States in One Dimension

The easiest application of this theory is to the linear chain. The eigenvalue equation reduces to

$$1 = (J/\pi) \int_{-\pi}^{+\pi} dq \frac{(\cos 1/2K - \cos q) \cos q}{E - E_0 - 2g\mu_{\mathrm{B}}H - 4sJ + \cos 1/2K \cos q}. \tag{5.83}$$

For greater clarity, one introduces dimensionless parameters. Let

$$A = (E - E_0 - 2g\mu_{\mathrm{B}}H - 4sJ)/4sJ \cos \frac{1}{2}K$$

where K lies in the interval $-\pi < K < +\pi$, and therefore,

$$1 = \frac{1}{2s\pi} \int_0^\pi dx \frac{1}{A + \cos x} \left(\cos x - \frac{\cos^2 x}{\cos 1/2K} \right)$$

$$= \frac{1}{2s} \left(1 - \frac{|A|}{\sqrt{A^2 - 1}} \right) \left(1 + \frac{A}{\cos 1/2K} \right). \tag{5.84}$$

There are solutions only for $A < -1$, proving that the bound state exists only below the continuum, and that there is no bound state above the continuum (which would correspond to positive values of A). This is not surprising in view of the fact that the effective interaction is attractive: two spin deviations have lower energy when they are nearest neighbors than when they are farther apart.

On the other hand, one might have thought of the effective interaction when two spin deviations occur on the same site, as a repulsive (or "hard core") potential which could lead to bound states above the continuum. The actual calculation shows this latter point of view to be false.

The equation is easiest to solve analytically when $s = \frac{1}{2}$, in which case we find

$$A \cos \frac{1}{2}K = -\frac{1}{2} \left(1 + \cos^2 \frac{1}{2}K \right)$$

$$= -1 + \frac{1}{2} \sin^2 \frac{1}{2}K$$

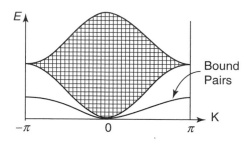

Fig. 5.2. Energy of 2-magnon states in linear chain. Shading indicates continuum, single line represents bound state.

and therefore,

$$E = E_0 + 2g\mu_B H + J\sin^2\frac{1}{2}K \quad \text{for the bound state.} \tag{5.85}$$

A similar formula will be obtained when we study solitons, (5.211). Equation (5.85) may be compared to the lowest energy of a scattering state, i.e., the bottom of the continuum, which occurs at $\mathbf{q} = \mathbf{0}$ [cf. (5.67)]

$$E_{\min}(\mathbf{q} = \mathbf{0}) = E_0 + 2g\mu_B H + 4J\sin^2\frac{1}{4}K. \tag{5.86}$$

The bound state lies lower than this by a "binding energy" δE in the amount of

$$\delta E = 4J\sin^2\frac{1}{4}K - J\sin^2\frac{1}{2}K, \tag{5.87}$$

which ranges from

$$\frac{J}{4}\left(\frac{K}{2}\right)^4 \quad \text{for } K \ll 1 \quad \text{to } J \quad \text{for } K = \pi.$$

A schematic plot is given in Fig. 5.2 for arbitrary value of the spin s.

5.5. Bound States in Two and Three Dimensions

The results obtained by Wortis[5-7] show that in two dimensions as in one, there is at least one bound state below the continuum *for all* $\mathbf{K}$, and that at some $\mathbf{K}$ there may be *two* bound states. In three dimensions, he found that there are no bound states whatever for small $\mathbf{K}$, and it is only at sufficiently large values of $\mathbf{K}$ that one, two, or up to a maximum of *three* bound states make their appearance. This is a most important result, for the absence of any bound-state solution over a significant range of $\mathbf{K}$ confers validity to the linear spin-wave theory in three dimensions even though the linearization be invalid in one or two dimensions.

Before considering the region of small $\mathbf{K}$, which because of the low spin-wave and bound-state energies will be of the greatest importance in thermo-dynamics, let us first examine the special case when all components of $\mathbf{K}$ are equal to π, their maximum possible value, in which case

$$\cos \frac{K_x}{2} = \cos \frac{K_y}{2} = \cos \frac{K_z}{2} = \cos \frac{\pi}{2} = 0$$

and all the integrals in the eigenvalue equation become trivial. Because the denominators are now all constant, the nondiagonal matrix elements $M_{x,y}$ defined in (5.73) all vanish, while the diagonal ones are all equal. The bound state will therefore be D-fold degenerate. Note that the continuum has collapsed to a single point at

$$\gamma_\pi \equiv \gamma_\pi(\mathbf{q}) \equiv E_0 + 2g\mu_B H + 2sJz \tag{5.88}$$

and therefore the integral equation reduces to an algebraic one. The bound-state solutions of (5.82) are easily found to have energy

$$E = \gamma_\pi - J, \tag{5.89}$$

by solving this algebraic equation. The bound state is identified as the complex formed by keeping the two spin deviates nearest neighbors; and the degeneracy can be understood as the D distinct directions along which two spins can be nearest neighbors.

The existence of at least one bound state for every value of $\mathbf{K}$ in *two dimension* is guaranteed by the behavior of the various integrals near $\mathbf{q} = \mathbf{0}$. Let us denote the energy of the lowest state in the continuum by $E_{\min}$, then the integrands in the vicinity of $\mathbf{q} = \mathbf{0}$ all contribute on the order of

$$\approx \int \frac{\text{numerator}}{E - E_{\min} - D'q^2} q\,dq$$

for suitable positive constant D', and appropriate nonvanishing numerator. This contribution, as a function of E, ranges from zero to $-\infty$ as E varies from $-\infty$ to $E_{\min}$, and in the limit, $E_{\min} - E = \delta E \to 0$, varies as $\approx \log \delta E$. Thus one expects, even without the benefit of explicit calculation, that the "binding energy" δE will depend exponentially on the various parameters. Indeed, for $K_x = K_y \approx 0$, Wortis found an essential singularity,

$$\delta E \sim e^{+2\pi sC/(1-C)}$$

where $C = \cos(K_x/2)$; for more general $\mathbf{K}$ he obtained the binding energy in terms of various elliptic integrals.

Turning finally to the more realistic case of a three-dimensional structure, for which the integrals cannot be evaluated analytically, one may find it advantageous to represent them by Laplace transform methods. That is, the substitution

$$\frac{1}{D} = \int_0^\infty dt\, e^{-Dt}$$

combined with the definition of the Bessel functions of imaginary argument

$$I_n(z) = \frac{1}{\pi} \int_0^\pi e^{z\cos x} \cos nx\, dx$$

permits the replacement of three-dimensional integrals by a rapidly converging one-dimensional integral. But these substitutions are not required to determine the threshold for the existence of a bound-state solution.

For this, we may presume $K_x = K_y = K_z$, which yields the lowest edge of the continuum and is more likely to produce the bound state than some less isotropic direction. In this case the three diagonal matrix elements $M_{i,i}$ are all equal, and the off-diagonal elements are also equal to each other. This results in $V_x = V_y = V_z$ and in an eigenvalue equation

$$1 = M_{x,x} + 2M_{x,y}$$

$$= \frac{2J}{N} \sum_q \frac{\cos q_x [3\cos(K_x/2) - \cos q_x - 2\cos q_y]}{E - E_0 - 2g\mu_{\rm B}H - 12sJ + 4sJ(\cos q_x + \cos q_y + \cos q_z)\cos(K_x/2)}$$

$$= \frac{E - E_0 - 2g\mu_{\rm B}H - 12sJ\sin^2(K_x/2)}{24s^2J\cos^2(K_x/2)}$$

$$\times \left[1 - \frac{E - E_0 - 2g\mu_{\rm B}H - 12sJ}{N} \sum_q \frac{1}{E - \gamma_{\mathbf{K}}(\mathbf{q})} \right]. \tag{5.90}$$

Use has been made only of the cubic symmetry, and of the equality of the three components of $\mathbf{K}$ to arrive at the last, simplified expression. The bound-state threshold occurs for that $\mathbf{K}$ at which E just drops below the bottom of the continuum at $\gamma_{\mathbf{K}}(\mathbf{0})$. This occurs for K_{x0}

$$E = E_0 + 2g\mu_{\rm B}H + 24sJ\sin^2\frac{K_{x0}}{4}. \tag{5.91}$$

Replacing the sum in (5.90) by an integral in the usual manner, we use the first line to determine the threshold K_{x0}

$$1 = \frac{-1 + \cos 1/2K_{x0}}{2s \cos 1/2K_{x0}}(1 - W), \quad \text{i.e.,} \quad \sin^2 \frac{1}{4}K_{x0} = \frac{s}{2s + W - 1}, \quad (5.92)$$

where W is Watson's integral (see Note at end of section)

$$W = \frac{1}{(2\pi)^3} \iiint_{-\pi}^{\pi} \frac{dq_x\, dq_y\, dq_z}{1 - 1/3(\cos q_x + \cos q_y + \cos q_z)} = 1.516386. \quad (5.93)$$

Thus, the bound state exists in the range

$$0 \leq \cos \frac{K_x}{2} \leq \frac{0.516386}{2s + 0.516386} \quad (5.94)$$

provided $K_x = K_y = K_z$. This range is $140° \leq K_x \leq 180°$ for $s = \frac{1}{2}$, $157° \leq K_x \leq 180°$ for $s = 1$, $167° \leq K_x \leq 180°$ for $s = 2$, etc. As K is increased beyond the minimum, other solutions make their appearance. For example, we have already found that there are three solutions at $K_x = K_y = K_z = 180°$, which is the largest total momentum wavevector in the positive direction. This small range of momentum space over which bound-state solutions are available, even in the most favorable case of spins one-half, and their *high* energies relative to the bottom of the continuum, is in sharp contrast with the results of the one- and two-dimensional calculations; and *in the correspondence limit $s \to \infty$, the bound states simply disappear.*

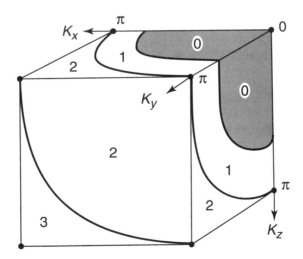

Fig. 5.3.   Octant of Brillouin zone showing regions of 0, 1, 2, and 3 bound-state solutions for two spin waves, $s = \frac{1}{2}$. For $s > \frac{1}{2}$, the region of 0 solutions grows, and for $s \to \infty$ occupies the entire Brillouin zone. (Sketch after Wortis[5].)

$$E_{\mathbf{k}} = \cos ka, \quad \text{with } |ka| < \pi. \tag{5.98}$$

Periodic boundary condition $\exp(ikNA) = 1$ results in the discrete set,

$$k = \frac{2\pi}{Na} \times \text{integer} = \frac{\pi(2p)}{Na} \quad p = 0, \pm 1, \ldots. \tag{5.99}$$

Next, a product of two plane waves does not vanish in the configurations $S_i^+ S_i^+$, as it should, but a determinant does. However, a determinant is antisymmetric under the interchange of the coordinates, whereas spins on different sites commute and therefore have a wavefunction symmetric under interchange. The following choice thus imposes itself:

$$\psi_{\mathbf{k},\mathbf{k}'} = \begin{cases} \dfrac{1}{\sqrt{N(N-1)}} \displaystyle\sum_{i,j>i} [e^{i(\mathbf{k}\cdot\mathbf{R}_i + \mathbf{k}'\cdot\mathbf{R}_j)} - e^{i(\mathbf{k}\cdot\mathbf{R}_j + \mathbf{k}'\cdot\mathbf{R}_i)}] S_i^+ S_j^+ |0\rangle \\[2ex] \dfrac{1}{\sqrt{N(N-1)}} \displaystyle\sum_{i,j<i} [e^{i(\mathbf{k}\cdot\mathbf{R}_i + \mathbf{k}'\cdot\mathbf{R}_j)} - e^{i(\mathbf{k}\cdot\mathbf{R}_j + \mathbf{k}'\cdot\mathbf{R}_i)}] S_i^+ S_j^+ |0\rangle. \end{cases} \tag{5.100}$$

When j is increased to $N-1$, then to N, and finally to $N+1$, the second spin becomes the first and the wavefunction changes discontinuously from the upper form to the lower unless the proper boundary condition is imposed. Because the position of the origin of the numbering system is completely arbitrary for the cyclic problem we are solving, such discontinuities at a particular site are inadmissible. The resolution of this difficulty is to take, instead of (5.99),

$$e^{ikNa} = -1, \quad \text{or} \quad k = \frac{\pi(2p+1)}{Na} \tag{5.101}$$

(and similarly for k'), an *antiperiodic* boundary condition. The generalization to any number of spins up is straightforward. Let

$$\psi_{\mathbf{k}_1 \mathbf{k}_2, \ldots} = C \sum_{i_1, i_2, \ldots} F_{\mathbf{k}_1, \mathbf{k}_2, \ldots}^{i_1, i_2, \ldots} S_{j_1}^+ S_{j_2}^+ \ldots |0\rangle \tag{5.102}$$

where C is the normalization constant, and F includes a determinant,

$$F_{k_1, \ldots}^{i_1, \ldots} = \varepsilon_P \begin{vmatrix} e^{ik_1 i_1} & e^{ik_2 i_1} & \cdots \\ e^{ik_1 i_2} & e^{ik_2 i_2} & \cdots \\ \vdots & \vdots & \end{vmatrix} \tag{5.103}$$

$\varepsilon_P = +1$ when the spins are in a natural order, $i_1 < i_2 < i_3 < \ldots$, or an even permutation of this order, and $\varepsilon_P = -1$ when the spins are arranged in an odd permutation of the natural order. If i_n is the farthest spin, then the translation of i_n from N to $N+1$ involves a reordering equivalent to an odd permutation for $n+1$ odd, and an even permutation for $n+1$ even. Thus

$$n + 1 = \text{odd} \rightarrow k = \frac{\pi}{Na}(2p + 1) \quad \text{and}$$

$$n + 1 = \text{even} \rightarrow k = \frac{\pi}{Na}(2p).$$

(5.104)

This is a very interesting situation. For although the many-spin wave-functions at first appear to be essentially independent particle wavefunctions, yet when one "particle" is added (that is, when one more spin is turned up) *all* the other plane-wave states are modified. This is in the nature of a cooperative effect, due to the effective hard core repulsion [from $(S_i^+)^2 = 0$] of two nearby spin deviations. Whether the k's are members of the even or of the odd set, the energy corresponding to the wavefunction above is

$$E_{k_1, k_2, \ldots} = \sum_{i=1}^{n} \cos k_i a = \sum_i \cos \frac{\pi}{N} \times \begin{cases} 2p_i \\ 2p_i + 1 \end{cases}.$$

(5.105)

Note that all k's must be distinct, or else $F \equiv 0$. The ground state energy is achieved by allowing all the states of negative energy to be occupied, that is, all k's in the range

$$\frac{\pi}{2} \le ka \le \frac{3\pi}{2} \quad \text{i.e.,} \quad \frac{N}{4} \le p \le \frac{3N}{4}.$$

(5.106)

However, we must consider *two* ground states: the ground state for the even k's and the ground state for the odd ones. Denote these by E_0' and E_0'', respectively, with values

$$E_0' = \sum_p \cos \frac{\pi}{N}(2p + 1) \quad \text{and} \quad E_0'' = \sum_p \cos \frac{\pi}{N}(2p).$$

(5.107)

One finds $\lim_{N \to \infty} E_0' = E_0'' = -N/\pi$ with the *sums* in either set *restricted to the range*, (5.106). The separation of energies E_0' and E_0'' depend on whether N is divisible by 4, or merely by 2, or whether it is odd. In any event, it is $O(1/N)$, not $O(1)$.

Either ground state may be represented by an energy-level diagram shown in Fig. 5.4. The Fermi level intersects the cosine curve at $ka = \pi/2$ and $3\pi/2$, and all allowed states below this are filled and all those above it are empty in either ground state. Excited states correspond to occupying states above the Fermi level, or emptying states below it. If k is a wave vector of either

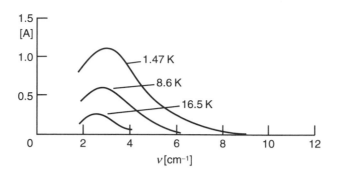

Fig. 5.5.   Absorption coefficient A for PrCl₃ at various temperatures.[16,17]

between the ground state and the low-lying excitations, viz., $(T_k)_{\alpha,0}$ and in principle obtain the absorption line shape. The experimental line is shown in Fig. 5.5 and is seen to peak at 3 cm⁻¹. Instead of computing this line shape, one can try to calculate this peak (which corresponds to the average absorption energy as well). The result is

$$\bar{E}_k = J\frac{(0| - S_{y,m-1}S_{y,m} + S_{z,m-1}S_{z,m} \cos ka|0)}{1/4 + 2\sum_n \cos nka(0|S_{x,m-n}S_{x,m}|0)} \qquad (5.113)$$

where m is any fixed site. The numerator is evaluated from the fermion representation, and yields $1/2\pi - \cos(ka/\pi^2)$. The denominator was evaluated numerically[16,17] to yield 0.0386, at $k = 0$. The result is $\bar{E}_0 = 1.50J$. With $J = 2.0$ cm⁻¹, this leads to $\bar{E}_0 = 3$ cm⁻¹, in excellent agreement with the location of the experimental absorption peak. Of course, the evaluation of the exact matrix element and fitting of the absorption line shape, Fig. 5.5 would be even more compelling evidence for this theory, but the onerous calculations have not been performed.

5.7.   Bethe's Solution of One-Dimensional Heisenberg Antiferromagnet

Progress in solving the Heisenberg antiferromagnet, even in one dimension, has been slow. Although Bethe[18] gave the wavefunctions in 1931, the ground-state energy was not obtained until 1938, by Hulthén.[19] It was

---

[18] H. Bethe, *Z. Phys.* **71**, 205 (1931). English language translation in D. Mattis, *The Many-Body Problem*, World Scientific, 1993, pp. 689ff.
[19] L. Hulthén, *Ark. Met. Astron. Fysik* **26A**, Na. 11 (1938).

subsequently generalized by Orbach,[20] who discussed an anisotropic Hamiltonian, and Walker[21] who studied the analytic properties of the Orbach solutions. Finally, fully 32 years after the initial progress, Des Cloizeaux and Pearson[22] calculated the one-magnon spectrum and Griffiths the magnetic susceptibility.[23]

It is desired to calculate the eigenstates of

$$\mathcal{H} = \sum_{i=1}^{N}\left[\frac{1}{2}(S_i^+ S_{i+1}^- + \text{H.c.}) + g S_i^z S_{i+1}^z\right] \quad \text{for } s = \frac{1}{2}, \, g = 1. \quad (5.114)$$

Other limits are:

$g = \pm\infty$ (Ising model), $g = 0$ (XY model), $g = -1$ (transformable into ferromagnet by trivial rotation of every second spin). For *all* $g > -1$ the ground state belongs to $M = 0$, i.e., contains Ns "particles."

We start with 2 "particles," using a procedure valid only for spins one-half and assign to the unphysical amplitudes f_{ii} the value determined by

$$f_{i,i} + f_{i+1,i+1} = f_{i,i+1} + f_{i+1,i} = 2f_{i,i+1} \quad (5.115)$$

where the physical amplitudes f_{ij} ($= f_{ji}$ for $j \neq i$) obey the equations

$$(E - E_f + 2)f_{ij} - \frac{1}{2}(f_{ij+1'} + f_{ij-1} + f_{i+1j} + f_{i-1j}) = 0. \quad (5.116)$$

This is the left-hand side of (5.55), with $s = \frac{1}{2}$, $J = -1$, $H = 0$, and $z = 2$. (5.115), playing the role of a boundary condition, ensures that the right-hand side of that equation always vanishes. The homogeneous equation is solved by plane waves, via Bethe's ansatz:

$$f_{ij} = e^{i(ki+k'j+1/2\psi)} + e^{i(kj+k'i-1/2\psi)} \quad (j \geq i) \quad (5.117)$$

with f_{ij} in the range $j < i$ given by symmetry, $f_{ij} = f_{ji}$. The phase factor ψ ranges over the interval $-\pi$, $+\pi$. (In the XY model, (5.100), the effective ψ is a constant $\pm\pi$.) We make no attempt to normalize solutions here, as it is not required in the calculation of energy eigenvalues and requires some fairly involved analysis.

[20] R. Orbach, *Phys. Rev.* **112**, 309 (1958).
[21] L. R. Walker, *Phys. Rev.* **116**, 1289 (1959).
[22] J. Des Cloizeaux and J. J. Pearson, *Phys. Rev.* **128**, 2131 (1962); M. Fowler, *Phys. Rev.* **B17**, 2989 (1978); *J. Phys.* **C11**, L977 (1978); M. Fowler and M. Puga, *Phys. Rev.* **B18**, 421 (1978); A. Ovchinnikov, *Sov. Phys. JETP* **29**, 727 (1969).
[23] R. B. Griffiths, *Phys. Rev.* **133**, A768 (1964).

Inserting the above form of f_{ij} into the boundary condition and performing some algebra, we obtain

$$2 \cot \frac{1}{2} \psi = \cot \frac{1}{2} k - \cot \frac{1}{2} k' \tag{5.118}$$

with a second, periodic, boundary condition,

$$f_{iN} = f_{0i} \tag{5.119}$$

determining the modified k, k'

$$k = \frac{\pi(2p) + \psi}{N} \quad \text{and} \quad k' = \frac{\pi(2p') - \psi}{N} \tag{5.120}$$

with p, p' = integers. Thus, a two-magnon state is described by two wave vectors and a phase shift; there is refraction but no scattering. The sum $k + k'$ is independent of ψ and is the center of mass wave vector $\mathbf{K}$ which parametrized the solution in an earlier section of this chapter. The difference k–k' corresponds to the variable $\mathbf{q}$. The bound state that appeared below the continuum threshold at $\mathbf{q} = 0$ is now *above* the continuum, due to the change in sign of the interaction. Now, as k approaches k', (5.118) shows that ψ approaches $\pm\pi$, the upper sign applying if $k < k'$. From (5.120), if $p = p' - 1$, $k = k'$ and $\psi = \pi$ so that $f_{ij} = \exp ik(i + j) \times 2 \cos \frac{1}{2}\pi \equiv 0$. Therefore, $|p - p'| \geq 2$.

The energy, measured from the ferromagnetic level $E_f = \frac{1}{4}N$, is given by the above eigenvalue equation as

$$E - E_f = -(1 - \cos k) - (1 - \cos k') \tag{5.121}$$

just the energy of two noninteracting magnons; the interactions are included implicitly, *via* the phase shifts in k, (5.118, 5.120).

Bethe's ansatz consists partly in the statement that in the many-particle state, the amplitudes are subject to phase shifts that are simply given, as the sum of the two-particle phase shifts. Thus, for a given ordering $i \leq j \ldots \leq n$,

$$f_{ij \cdots mn} = \left\{ \exp\left[i\left(k_1 i + k_2 j + \cdots + k_n n + \frac{1}{2}\sum_{r<t}^{n}\sum^{n} \psi_{k_r k_t}\right)\right] \right.$$

$$\left. + \left(\begin{array}{c} \text{all } (n! - 1) \\ \text{remaining permutations} \\ \text{of the } k'\text{s} \end{array}\right)\right\}. \tag{5.122}$$

If we need $f_{i,j,\dots}$ for any other ordering of the particles, we obtain it from the symmetry under permutations: $f_{i,j,\dots,m,\dots} = f_{i,m,\dots,j,\dots}$ so we need to know it only in the given interval. In this region, the amplitude f consists of a product over plane-wave factors, summed over all permutations of the wave vectors with the ψ's antisymmetric in their subscripts and satisfying the equations

$$2 \cot \frac{1}{2}\psi_{kk'} = \cot \frac{1}{2}k - \cot \frac{1}{2}k' \tag{5.123}$$

$$k = \frac{\pi(2p) + \sum_1^n \psi_{kk'}}{N} \qquad p = 0, \pm 1, \dots . \tag{5.124}$$

The sum in this equation, as well as in the next, is over the set of k's present in (5.122). The energy, measured relative to the ferromagnetic reference energy $E_f = N/4$, is

$$E - E_f = -\sum_{t=1}^n (1 - \cos k_t) . \tag{5.125}$$

Note that if we translate the entire chain by one site, that is, let $i, j, \dots \rightarrow i+1, j+1, \dots$, the wavefunction, (5.122), is multiplied by a phase factor

$$\exp\left(i \sum_i^n k \right) = \exp\left(\pi i \sum_1^n 2p/N \right). \tag{5.126}$$

The exponent is identified as the total momentum, modulo 2π, which is again independent of the phase shifts $\psi_{kk'}$. This is important, because the phase shifts themselves are now rather large. Each $\psi_{kk'}$ is $O(1)$; there are a number of them contributing to each k, and the total shift in k is $O(n/N)$ or a substantial fraction of k.

In the two-particle problem there was no solution of (5.118, 5.120) for $\mathbf{k} - \mathbf{k'} \approx 0$. Similarly, in the many-particle state, one must choose the interval between k's such that no $p = 0$, and such that for all p and p',

$$|p - p'| \geq 2 \tag{5.127}$$

for a non-vanishing solution to exist. The ground state, for $N/2$ particles subject to this restriction, has the set of $\{p\}$:

$$\{p\} = 1, 3, \dots, N - 1 . \tag{5.128}$$

Notice an interesting effect of the $S_i^z S_{i+1}^z$ "interaction," which is to spread the set of $\frac{1}{2}N$ integers p (restricted over the range $N/4 < p < 3N/4$ in the ground state of the XY model) to cover the entire range of phase space at present.

The regular spacing permits us to replace sums by integrals in the limit $N \to \infty$, and thus, with $x = p/N$ and $\Delta p = \frac{1}{2} N \, dx$

$$E - E_{\mathrm{f}} = -\frac{N}{2} \int_0^1 [1 - \cos k(x)] \, dx$$

$$= -N \int_0^1 \sin^2 \frac{k(x)}{2} \, dx \tag{5.129}$$

where

$$2 \cot \frac{\psi(x,y)}{2} = \cot \frac{k(x)}{2} - \cot \frac{k(y)}{2} \tag{5.130}$$

and

$$k(x) = 2\pi x + \frac{1}{2} \int_0^1 \psi(x,y) \, dy \tag{5.131}$$

as $|\psi|$ cannot exceed π, (5.130) indicates that ψ has a jump discontinuity of -2π at $x = y$. More precisely, the derivative is

$$\frac{\partial \psi(x,y)}{\partial x} = -2\pi\delta(x-y) + \frac{1}{4}(\xi^2 + 1)\frac{dk(x)/dx}{1 + 1/4(\xi - \eta)^2} \tag{5.132}$$

where we use the short-hand notation

$$\xi \equiv \cot \frac{1}{2}k(x) \quad \text{and} \quad \eta \equiv \cot \frac{1}{2}k(y). \tag{5.133}$$

We also find it convenient to introduce the functions

$$f(\xi) = -(d\xi/dx)^{-1} \quad \text{and} \quad f(\eta) = -(d\eta/dy)^{-1} \tag{5.134}$$

and to differentiate (5.131) w.r.t. x to obtain

$$dk(x)/dx = \pi + \frac{1}{2} \int_{-\infty}^{+\infty} \frac{f(\eta)/f(\xi)}{1 + 1/4(\xi - \eta)^2} \, d\eta. \tag{5.135}$$

With the identity $f(\xi) \, dk/dx = 2(1 + \xi^2)^{-1}$ this becomes

$$\frac{2}{1 + \xi^2} = \pi f(\xi) + 2 \int_{-\infty}^{+\infty} \frac{f(\eta)}{4 + (\xi - \eta)^2} \, d\eta. \tag{5.136}$$

This integral equation is exactly soluble because the kernel is a function only of the difference $(\xi - \eta)$. One takes Fourier transforms,

$$F_k \equiv \int_{-\infty}^{\infty} d\theta\, f(\theta) e^{ik\theta} \quad \text{and} \quad f(\theta) = \int_{-\infty}^{\infty} \frac{dk}{2\pi} F_k e^{-ik\theta} \tag{5.137}$$

to obtain a simple solution of (5.136),

$$F_k = (2 \cosh k)^{-1}. \tag{5.138}$$

The ground state energy (5.129) has the form

$$E - E_f = -N \int_{-\infty}^{\infty} d\xi \frac{f(\xi)}{1 + \xi^2} = -N \int_{-\infty}^{\infty} \frac{dk\, F_k}{2\pi} \int_{-\infty}^{\infty} \frac{e^{-ik\xi}}{1 + \xi^2}$$

$$= -2N \int_0^{\infty} dk\, F_k e^{-|k|} = -2N \int_0^{\infty} \frac{dk\, e^{-2|k|}}{1 + e^{-2|k|}} = -N \ln 2. \tag{5.139}$$

This is a very famous result, being one of the first exact solutions ever obtained of a nontrivial many-body problem in quantum mechanics.

In the usual terms of energy *per* spin, the Bethe-Hulthén result for E/N represents $-\ln 2 + \frac{1}{4} = -0.443$. We can compare this to the ground state energy *per* spin in the XY model which is $-\pi^{-1} = -0.318$. While the XY model energy is a variational *upper* bound to the Heisenberg model ground state energy, it is evidently not a very close one. On the other hand, $\frac{3}{2} \times$ ground state energy of the XY model is an exact *lower* bound on the Heisenberg model ground state energy which yields $-3\pi/2 = -0.477$ reasonably close to the exact answer. By way of contrast, we note that the Ising model $(S_i^z S_{i+1}^z)$ has energy -0.250 *per* spin in these same units, which is a variational upper bound (although a poor one) to both XY and Heisenberg models. Moreover, $2 \times (-0.25) = -0.500$ is a poor lower bound to the XY model and $3 \times (-0.25) = -0.750$ is an even worse lower bound to the Heisenberg energy *per* spin. (The reasons for these larger discrepancies may be attributed to the lack of low-energy spinwave dynamics in the Ising model.)

Des Cloizeaux and Pearson,[22] Fig. 5.6, extended Bethe's analysis to obtain the energy of some of the lowest excited states, known to be triplet states.[10] $(N/2 - 1)$ values of k are chosen by them, to obtain the lowest-lying elementary excitations for a given total momentum $\sum k$. This procedure is very delicate, however, and somewhat too complicated to reproduce or to justify here. There is an unanswered question of whether bound states participate among the low-lying excitations. If they do, then Bethe's *ansatz* formalism is incapable of describing them since the validity of the various

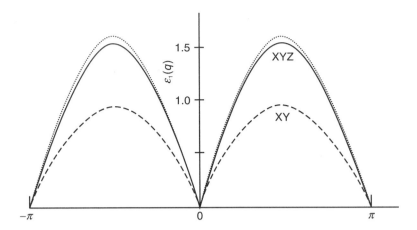

Fig. 5.6. Quasiparticle spectrum $\varepsilon_1(ka)$ obtained by Des Cloizeaux and Pearson for the infinite Heisenberg antiferromagnetic chain, $\pi/2 \sin ka$ (*upper curve*), compared with their numerical results for chain of 48 spins (*dotted curve above the exact curve*). Also, spectrum for infinite XY antiferromagnetic chain (*lower, dashed curve*). (Anderson's linearized antiferromagnons coincide precisely with the XY model.)

equations above seems predicated on having real k's. Nevertheless, it is entirely possible that further mathematical study may allow the extension of the known results to complex k plane and that a complete classification scheme will result. For example, following Orbach one might consider the $S_i^z S_{i+1}^z$ coupling terms with variable parameter g, and study the energy levels as g is adiabatically increased. At $g = 0$, we recover the XY model (of which the solutions are completely determined) and then increase it to $g = 1$ which is the present Heisenberg model. As g is further increased, the Ising model is approached. In fact, $g = \pm 1$ are two "critical points": for $g > 1$ the properties approach those of an Ising antiferromagnet (e.g., the ground state energy approaches $-\frac{1}{4}g$ *per* spin at large g, the excitation spectrum has a gap), and for $g < -1$ they resemble those of an Ising ferromagnet. It is only in the range $-1 \leq g \leq +1$ that results are obtained similar to the gapless spectra we have seen here. For more details, the papers of Orbach[20] or Walker[21] of C. N. Yang and C. P. Yang[24] or the generalization by Baxter[25] to the completely anisotropic linear chain by a procedure that side-steps

[24] C. N. Yang and C. P. Yang, *Phys. Rev.* **150**, 321 (1966).
[25] R. J. Baxter, *Ann. Phys. (N.Y.)* **70**, 323 (1972).

the Bethe ansatz, can be consulted. The case $g > 1$ has been analyzed in Johnson et al.[26],[27]

Des Cloizeaux and Pearson's results for the lowest-lying one-magnon triplet ($S = 1$) excitation spectrum is shown in Fig. 5.6, together with some numerical results on chains of 48 spins and compared there to the spectrum of the XY model ($g = 0$). This comparison will have further significance when we study approximate theories of antiferromagnetism in the following sections. If we denote their exact lower bounds on the triplet excitation spectrum ε_1, it is given by

$$\varepsilon_1 = \frac{1}{2}\pi |\sin q| . \tag{5.140}$$

Yamada,[28] and Bonner et al.[29] noted the existence of a continuum of triplet excitations above ε_1, with a curve ε_2 defining the upper edge given by

$$\varepsilon_2 = \pi \left| \sin \frac{1}{2}q \right| . \tag{5.141}$$

This continuum is illustrated in Fig. 5.7. Undoubtedly, excitations belonging to $S = 0, 2, 3, \ldots$ lie in other continua that partly intersect this. However, in an experimental situation where one unit of angular momentum is transferred to the magnetic system, only the triplet continuum can contribute. For such processes, Hohenberg and Brinkman[30] computed that the matrix elements peak near the lower, des Cloizeaux-Pearson, threshold. This topic, and its nontrivial extension to the case of an applied magnetic field, has been studied theoretically[31] and experimentally.[32]

One important result follows immediately from Fig. 5.7, viz., *the size of the Brillouin Zone (BZ) remains 2π for the linear chain antiferromagnet.* With only the spectrum of Fig. 5.6 one might suspect that the BZ could be folded in half, which is equivalent to the choice of 2 spins as the primitive unit cell. Physically, this is a very attractive idea: the couple of spin "up" and "down" nearest-neighbors make an ideal unit cell, and this is indeed the

[26] J. D. Johnson and B. McCoy, *Phys. Rev.* **A6**, 1613 (1972).

[27] J. D. Johnson, *Phys. Rev.* **A9**, 1743 (1974).

[28] T. Yamada, *Prog. Theor. Phys. Jpn.* **41**, 880 (1969).

[29] J. C. Bonner, B. Sutherland and P. Richards, *AIP Conf. Proc.* **24**, 335 (1975).

[30] P. Hohenberg and W. Brinkman, *Phys. Rev.* **B10**, 128 (1974).

[31] M. Steiner, J. Villain and C. Windsor, *Adv. Phys.* **25**, 87 (1976); G. Müller, H. Beck and J. Bonner, *Phys. Rev. Lett.* **43**, 75 (1979); M. Puga, *Phys. Rev. Lett.* **42**, 405 (1979).

[32] R. Birgeneau and G. Shirane, *Phys. Today*, Dec. (1978), p. 32; H. Mikeska and W. Pesch, *J. Phys.* **C12**, L37 (1979); S. Satija, G. Shirane, Y. Yoshizawa and K. Hirakawa, *Phys. Rev. Lett.* **44**, 1548 (1980).

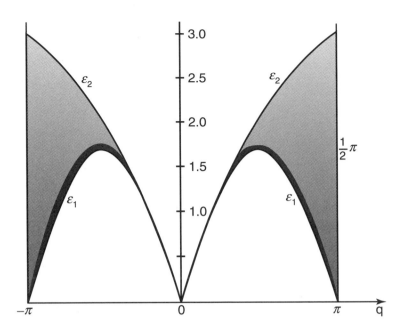

Fig. 5.7.   The continuum of magnon ($S = 1$) excitations for the Heisenberg XYZ linear chain antiferromagnet, indicating the two bounding curves ε_1 and ε_2 given in the text, Eqs. (5.140) and (5.141).

starting point of many approximate analyses.[32] With the *complete* excitation spectrum, however, one sees that the translational invariance of the ground state is preserved *and the larger BZ must be retained*.[33]

In recent years attention has turned to the correlation functions, of which the following is a typical example:

$$C_{\alpha,\beta}(i - j, t_1 - t_2) = \langle S_{i,\alpha}(t_1) S_{j,\beta}(t_2) \rangle_{\text{TA}}$$

where α, β refer to the Cartesian coordinates (x, y or z), i and j to the position on the lattice and t_1, t_2 to two different times. The thermal averaging is indicated by the subscript, TA. Fourier transforms are sometimes preferable, in which i–j is replaced by q and t_1–t_2 by ω. Such type of correlation

---

[33] The general topic of correlation functions for the ground state (and also, at finite T) for this model has been much studied in the recent past. The interested reader may wish to study the papers in the following footnote. Note also that the ground state must be nodeless, which excludes a Néel-like, degenerate, ground state!

[34] J. Johnson, S. Krinsky and B. McCoy, *Phys. Rev.* **A8**, 2526 (1973); A. Luther and I. Peschel, *Phys. Rev.* **B9**, 2911 (1974); **12**, 3908 (1975); H. Fogedby, *J. Phys.* **C11**, 4767 (1978), J.-S. Caux and J.-M. Maillet, cond-mat/0502365 v1, 15 Feb. 2005 calculate *dynamical* correlations.

functions can be measured directly by spin-polarized neutron scattering and are helpful in the analysis of the dynamics of the model. Later in this book we investigate a Green's function approach to the calculation of correlations. In the instance of the Heisenberg model, for which all the eigenstates are known and understood, this should have been child's play. However, the Bethe's ansatz merely simplifies the *eigenvalue* problem, merely allowing the ground state and elementary excited states and properties at finite temperatures (free energy and the like) to be extracted.

But when it comes to formulas for the normalization of the Bethe ansatz states, or the analytic calculation of matrix elements of various operators between normalized states, or the calculation of all correlation functions of the type shown above, the mathematics has proved rather difficult. This issue has been satisfactorily resolved only in recent years. We refer the interested reader to a book[35] and to a series of articles for worked-out examples.[36]

In higher dimensions $d > 1$ there is little hope of analytic solutions and all known results are the fruit of intuition and numerical work.

5.8. Linearized Antiferromagnetic Magnons

The quantization of antiferromagnons in the linear approximation is performed by analogy with the procedure in the theory of ferromagnetism. Such an approximate quantum theory, in which only terms quadratic in the P's and Q's are retained in the Hamiltonian, was originally proposed by Anderson,[37] and was very successful in the early interpretation of the dynamics of antiferromagnets.[14] As can be expected, the frequencies are identically the same as found in the classical linearized equations of motion of the antiferromagnet (discussed earlier); only the amplitudes are quantized as bosons. Below, we shall indicate in what vital manner procedures that were satisfactory in ferromagnetism have to be changed for the present problem.

The paramount difficulty is that the ground state of the Heisenberg antiferromagnet is not known, with the exception of the linear chain. Alas, it

---

[35] M. Jimbo and T. Miwa, *Algebraic Analysis of Solvable Lattice Problems*, AMS, Providence, 1995.

[36] N. Kitanine, J.-M. Maillet and V. Terras, *Nucl. Phys.* **B554**, 647 (1999); *ibid.* **B567**, 554 (2000); V. E. Korepin, *Commun. Math. Phys.* **86**, 391 (1982); D. Biegel, M. Karbach and G. Müller, *Europhys. Lett.* **59**, 882 (2002); *J. Phys.* **A36**, 5361 (2003) and *J. Phys. Soc. Jpn.* **73**, 3008 (2004). A very readable review: D. Schlottmann, *Int. J. Mod. Phys.* **B11**, 355 (1997).

[37] P. W. Anderson, *Phys. Rev.* **86**, 694 (1952).

is quite apparent that the procedure which worked in one dimension is not generalizable to three because of the heavy reliance on ordering the spin deviates along a line. The somewhat simpler XY model is similar to the quantum-mechanical hard-sphere Boson problem, for which also no adequate intermediate- or high-density theory exists in two or three dimensions.

Fortunately, there exists a perturbation theory of sorts that gives us a "handle" on an approximate, variational ground state. It is a series in powers of $1/s$. The off-diagonal parts of the Hamiltonian are linearized, and diagonalized as best can be, in the standard way. The starting, or zeroth, point of this perturbation procedure is the classical ground state: the configuration of lowest energy of the Hamiltonian with spins replaced by classical vectors of fixed length and position but variable orientations. The choice of this classical ground state is usually trivial, especially if we stick to "bipartite lattices," defined as those in which two spins that are nearest neighbors of a third are not nearest neighbors of each other. These requirements are met, e.g., in the linear chain, simple square, simple cubic, and body-centered cubic lattices (but not in the face-centered cubic lattice). The classical ground state is then the *Néel state*: every spin *up* is surrounded by nearest neighbors which are *down*, and vice versa.

To give them a name, denote the spins down the *A sublattice*, and the spins up the *B sublattice*. Perform a canonical transformation on the *B* (but not on the *A*) spins: rotate them by 180° about the S^x axis, to obtain the so-called Néel state

$$S_j^\pm \to +S_j^\mp, \quad S_j^z \to -S_j^z \quad (j \text{ in } B). \tag{5.142}$$

At first, assume that i is in A and j is a nearest neighbor of i in B to write the Hamiltonian in the form

$$\mathcal{H} = \sum_{i \leq A} \sum_{\substack{j \leq B \\ (i,j)}} \left[\left(\frac{1}{2} S_i^+ S_j^+ + \text{H.c.} \right) - S_i^z S_j^z \right]. \tag{5.143}$$

It is convenient to take the exchange constant J to be the unit of energy, so that it need not be written explicitly. Now, because this latest form is explicitly symmetric in the two sublattices, one allows i to range over the entire crystal and j over all nearest neighbors of i, denoted by $j(i)$, and divides by 2 so as not to double-count bonds.

$$\mathcal{H} = \frac{1}{2} \sum_{\text{all } i} \sum_{j(i)} \left[\frac{1}{2} (S_i^+ S_j^+ + \text{H.c.}) - S_i^z S_j^z \right]. \tag{5.144}$$

In the "linear" approximation, the Hamiltonian becomes

$$\mathcal{H}_{\text{lin}} = \frac{1}{2} \sum_i \sum_{j(i)} [s(a_i^* a_j^* + \text{H.c.}) - s^2 + s(a_i^* a_i + a_j^* a_j)]$$

$$\rightarrow -\frac{1}{2} N z s^2 + z s \sum_{\mathbf{k}} a_{\mathbf{k}}^* a_{\mathbf{k}} + \frac{s}{2} \sum_{\mathbf{k}} \left(a_{\mathbf{k}}^* a_{-\mathbf{k}}^* \sum_{\boldsymbol{\delta}} e^{i\mathbf{k} \cdot \boldsymbol{\delta}} + \text{H.c.} \right). \quad (5.145)$$

After the transformation to running-wave Boson operators, it is not yet diagonal. We eliminate the pair-creation terms by a method first due to Holstein and Primakoff.[38] Subsequently rediscovered by Bogolubov in his theory of Bose condensation, and now generally known as the *Bogolubov transformation*, it is:

$$a_{\mathbf{k}} \rightarrow (\cosh u_{\mathbf{k}}) a_{\mathbf{k}} + (\sinh u_{\mathbf{k}}) a_{-\mathbf{k}}^*$$

$$a_{\mathbf{k}}^* \rightarrow (\cosh u_{\mathbf{k}}) a_{\mathbf{k}}^* + (\sinh u_{\mathbf{k}}) a_{-\mathbf{k}} \quad \text{with } u_{\mathbf{k}} = u_{-\mathbf{k}} = u_{\mathbf{k}}^*. \quad (5.146)$$

One must choose the function $u_{\mathbf{k}}$ so that the terms $a_{\mathbf{k}}^* a_{-\mathbf{k}}^*$ are eliminated. It is a matter of some simple algebra to determine that what is required is

$$\tanh 2u_{\mathbf{k}} = \frac{-1}{z} \sum_{\boldsymbol{\delta}} \cos \mathbf{k} \cdot \boldsymbol{\delta} \quad (5.147)$$

with $\boldsymbol{\delta}$, a vector connecting any spin with any of its z nearest neighbors. This reduces the linearized Hamiltonian to the diagonal form,

$$\mathcal{H}_{\text{lin}} = -\frac{1}{2} N z s (s+1) + \sum_{\mathbf{k}} \left(n_{\mathbf{k}} + \frac{1}{2} \right) (zs) \sqrt{1 - \tanh^2 2u_{\mathbf{k}}}. \quad (5.148)$$

Before discussing the magnon spectrum it is not unwise to digress somewhat and examine the ground-state energy. In (5.148), it is the energy eigenvalue when all $n_{\mathbf{k}} = 0$, that is, the vacuum energy. Conventionally, one writes it in the following form:

$$E_0 = -\frac{1}{2} N s^2 z \left(1 + \frac{\gamma}{zs} \right). \quad (5.149)$$

The parameter γ may be shown to be bounded between 0 and 1. Indeed, Hulthen's exact result for the linear chain ($z = 2$, $s = \frac{1}{2}$) yields $\gamma = 0.7726$, which is perhaps the largest attainable value. The results in Table 5.1 were computed by the linear theory[37] (5.148), for nearest-neighbor interactions.

---

[38] T. Holstein and H. Primakoff, *Phys. Rev.* **58**, 1048 (1940). Also, see Chapter 3 on this topic.

Table 5.1. γ for various lattices.

Lattice	z	γ
Linear chain	2	0.726 (exact: 0.7726 ... for spins 1/2)
Square	4	0.632
Simple cubic	6	0.58
Body-centered cubic	8	0.58

But it is interesting to see the good agreement (within ≈ 6 percent) with the exact result for the linear chain, where the spin-wave theory might have been expected to fail badly. Although for the one-dimensional magnon spectrum the agreement will not appear so favorable, it is not far-fetched to conceive that the ground state and excited states of the simple theory may be in good agreement with the exact result for lattices with higher coordination numbers.

Again for one dimension, using (5.147, 5.148), we find for the magnon energy,

$$\varepsilon(k) = (2s) \sin ka \tag{5.150}$$

with $a =$ lattice spacing. For spins $\frac{1}{2}$ the result is similar to the exact result for the XY model, but is too low by a factor $2/\pi$ for spins $\frac{1}{2}$ in the Heisenberg model, which the spin-wave theory is supposed to approximate. The discrepancy is seen in Fig. 5.6.

In any number of dimensions, the magnon spectrum

$$\varepsilon_{\mathbf{k}} = s\sqrt{z^2 - \left(\sum_{\delta} \cos \mathbf{k} \cdot \boldsymbol{\delta}\right)^2} \tag{5.151}$$

is *doubly* degenerate, and is linear in k for small k [it should be noted that within finite length systems, magnons are *triplet* ($S = 1$), not *doublet* excitations].

Problem 5.2.   Derive the magnon spectrum (5.150) from (5.147, 5.148) evaluated for nearest-neighbors. Generalize (5.151) to arbitrary range interactions.

The existence of two distinct modes of zero energy, one at $\mathbf{k} = \mathbf{0}$ and the other at the edge of the Brillouin zone, means that magnons belonging to these wave vectors can be emitted at no cost of energy, and indicates degeneracy of the approximate ground state and of elementary excitations. This is a manifestation of the rotational invariance of the spin Hamiltonian. By the emission of a sufficiently large number of zero energy magnons, a state can also be reached in which the A and B sublattices have been interchanged. However, the *true* ground state is *non*degenerate.[10]

The diagonalization of $\mathscr{H}_{\text{lin}}$ leads to a reduction of the S^z spin components from their saturation magnitude in the Néel state. Calculating this reduction we find the reduction in sublattice magnetization due to this:

$$\langle \delta S_i^z \rangle \equiv \left\langle \frac{1}{N} \sum_i (s - |S_i^z|) \right\rangle = \frac{1}{2N} \sum_k \left(\frac{1}{\sqrt{1 - \tanh^2 2u_k}} - 1 \right) \approx 0.078 \text{ sc}.$$

(5.152)

The numerical value of 0.078 is calculated for the simple cubic (sc) structure. Note that for the linear chain this very same formula yields

$$\langle \delta S_i^z \rangle = \infty \quad \text{lin. chain} \tag{5.153}$$

due to the divergence of the integral at *long* wavelengths. This invalidates the linearization procedure in 1D and nicely also explain why the linearized magnons have energies that are some 50% too low, and we shall see later that quantum effects can affect the results profoundly for quantum spins $s = 1, 2, \ldots$.

Table 5.2. Comparison of various calculations of the ground state energy parameter $e_0 \equiv 1 + \frac{1}{2}\gamma$ and of $\langle \delta S_i^z \rangle$ for the $s = \frac{1}{2}$ Heisenberg antiferromagnet on a square lattice $(z = 4)$.

Method	Reference	e_0	$\langle \delta S_i^z \rangle$	γ
LSW	[37]	1.32	0.20	0.64
Perturbation theory	[39]	1.33	0.12	0.66
Variational	[41]	1.29	0.07	0.58
Variational	[42]	1.32	0.10	0.64
Cellular	[40]	1.31	0.25	0.62

[39] H. L. Davis, *Phys. Rev.* **120**, 789 (1960).
[40] J. Oitmaa and D. D. Betts, *Canad. J. Phys.* **56**, 897 (1978); R. Jullien *et al.*, *Phys. Rev. Lett.* **44**, 1551 (1980).
[41] T. Oguchi, *J. Phys. Chem. Sol.* **24**, 1649 (1963).
[42] R. Bartkowski, *Phys. Rev.* **B5**, 4536 (1972).

We now compare the results of linear spinwave theory (LSW) with other calculations, including the most accurate cellular method of Betts and Oitmaa, in Table 5.2 adapted from one of their works.[40] LSW is in good agreement with the last, and presumably best, result.

The method by which Betts and Oitmaa have been obtaining the ground state parameters in 2 and 3 dimensions deserves comment. Let us illustrate with the square lattice, then quote their results for the XY antiferromagnet and for both Heisenberg and XY models in 3D.

Figure 5.8 illustrates finite cells of 4, 8, 10, 16 and 18 spins on the square lattice. In each case the infinite lattice can be filled by periodic extensions of the cell; thus, the wavefunction in the cell is subjected to periodic boundary conditions. It consists of linear combinations of states belonging to the symmetry appropriate to the ground state. For example, starting with the Néel state, one overturns a pair of n.n. spins in all possible ways. One then overturns a pair in any of these states, etc., until all the states which can mix in the ground state are found. For the 16 spin diamond illustrated in Fig. 5.8 there are 6 generations of spin flips before the total of 153 basis states is achieved (a great deal less than the total number of 2^{16} basis states)! For $N = 18$, 398 states are generated, and the resulting matrices are easily diagonalized to yield the lowest eigenvalue, the ground state energy. When plotted versus $1/N$, as in Fig. 5.9, these cellular ground state energies typically lie on a straight line that extrapolates to $1/N = 0$.

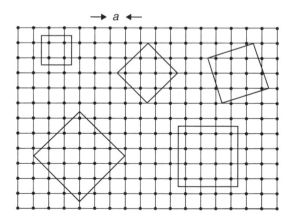

Fig. 5.8. Finite cells of 4, 8, 10, 16, 18 spins on the square lattice. In each case the infinite lattice is filled by periodic repetition of the cells.

Table 5.3. Ground state correlations for the 16 spin cell on the square lattice, $s = 1/2$.

R_{ij}/a	XY		Heis. A.-F.
	$4\langle S_i^z S_j^z \rangle$	$4\langle S_i^x S_j^x \rangle$	$4\langle S_i^z S_j^z \rangle$
1	0.182	−0.562	−0.468
$2^{1/2}$	0.027	0.487	0.285
2	0.027	0.487	0.285
$5^{1/2}$	0.023	−0.468	−0.270
$2 \times 2^{1/2}$	0.018	0.458	0.240

From the ground state energy one can readily extract the nearest neighbor correlation function $(\phi_0|S_i^x S_j^x|\phi_0) = (\phi_0|S_i^y S_j^y|\phi_0) = (\phi_0|S_i^z S_j^z \phi_0|)$, with i, j n.n. Using the computed ground states, Betts and Otimaa also obtain the correlation functions for i, j *not* n.n. Two spins correlate positively when on the same sublattice, but negatively when on different sublattices. For the antiferromagnetic XY model, the X and Y correlations are of course equal, but the Z correlation function is distinct. Calculation of both $(\phi_0|S_i^x S_j^x|\phi_0)$ and $(\phi_0|S_i^z S_j^z|\phi_0)$ show them to decrease surprisingly slowly with distance. These results are evident in Table 5.3 taken with suitable modification from Oitmaa *et al.*[40] Betts and Oitmaa concluded that the long-range order persists in 2D in the ground states; plotting $2/N \sum_{i,j}(-1)^{i-j}\langle S_i^x S_j^x \rangle$ versus $1/N$, they obtained the lines in Fig. 5.9 which extrapolate to LRO $= 0.116$ for the XY model and 0.059 for the Heisenberg antiferromagnet (compared to a possible maximum of 0.25). At finite T, however, where the XY model has a *bona-fide* phase transition at $T_c \approx 0.9$ J, the Heisenberg model "melts" at $T = 0^+$ and has no phase transition in 2D.

LSW theory on the 2D and 3D XY models is extremely simple, although highly inaccurate for $s = 1/2$ (see, however, the following section). Starting with the exact Hamiltonian

$$\mathcal{H} = \frac{1}{4}\sum_{i,j}(S_i^+ S_j^- + \text{H.c.}) \tag{5.154}$$

we use the linear approximation to the Holstein-Primakoff operators, and write the above as

$$\mathcal{H} = \frac{1}{4}s\sum_{i,j}(a_i^* a_j + \text{H.c.}) \tag{5.155}$$

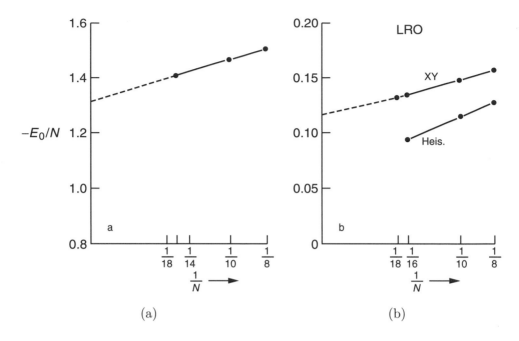

(a)   (b)

Fig. 5.9. (a) Ground state energy of the $s = \frac{1}{2}$ Heisenberg antiferromagnet on a square lattice. (b) Long-range order parameter for XY and Heisenberg antiferro-magnets, implying the existence of LRO in the ground state.[40]

which, after the phase change a_i, $a_i^* \to -a_i$, a_i^* on the B sublattice alone, yields

$$\mathscr{H} = -\frac{1}{4}s\sum_{i,j}(a_i^* a_j + \text{H.c.})\qquad(5.156)$$

and after transformation to running wave operators, (5.112, 5.113), yields

$$\mathscr{H} = -\frac{1}{4}s\sum_{\mathbf{k}}(\cos k_x a + \cos k_y a)\mathbf{n_k}\qquad(5.157)$$

in 2D, with an analogous expression in 3D.

In the ground state, a symmetry principle requires the eigenvalue of the operator S_{tot}^z to be zero,[43] thus for $s = \frac{1}{2}$ the magnon population is:

$$\mathbf{n_{op}} = \sum_{\mathbf{k}}\mathbf{n_k} = \frac{1}{2}N\qquad(5.158)$$

---

[43] D. Mattis, *Phys. Rev. Lett.* **42**, 1503 (1979).

in the ground state. Evidently, the ground state is achieved by populating the $\mathbf{k} = \mathbf{0}$ mode only, thus $n_0 = \frac{1}{2}N$ and $n_{\mathbf{k}} = 0$ for $\mathbf{k} \neq \mathbf{0}$. This yields for the $s = \frac{1}{2}$ correlation functions

$$4(\phi|S_i^x S_j^x|\phi) = (-1)^{i-j}(\phi|ca_i^* a_j + a_i^* a_i + a_i a_j + a_i^* a_j^*)|\phi)$$

$$= (-1)^{i-j}\frac{1}{N}\sum_{\mathbf{k}}(2\cos \mathbf{k} \cdot R_{ij})n_{\mathbf{k}} \tag{5.159}$$

in any eigenstate ϕ. In the ground state ϕ_0, this sum is N, so (5.159) is ± 1 depending on whether the two spins are on the same sublattice or not. This result is twice as large as the computed values in Table 5.3 and does not exhibit the rather mild, but quite definite fall-off with distance of the computed results. Similarly, the other $s = \frac{1}{2}$ correlation function is

$$4(\phi_0|S_i^z S_i^z|\phi_0) = 4(\phi_0|a_i^* a_j|\phi_0)(\phi_0|a_j^* a_i|\phi_0) = 1, \tag{5.160}$$

regardless of sublattice, having the correct sign but some 5 times too large! It seems that LSW is fortuitously good in the Heisenberg model, for the accuracy evaporates in the XY model, although it *does* explain LRO.

Turning to 3D, we compare LSW with other methods in Table 5.4, and in Table 5.5, present the correlation functions in the XY and Heisenberg antiferromagnets computed by the cell method.[44] The LSW correlation functions for the XY model are the same in 3D as in 2D and remain in quantitative disagreement with the tabulated values. A striking result of Table 5.4 is that the spin deviation is larger by a factor of 3 or 4 in the cell method than in the other calculations. Experimental tests to establish the correct value of this parameter in Nature have not yet been devised.

Table 5.4. Ground state energy parameter γ and spin deviation in 3D Heis. A.-F. spin one-half.[40]

Method	Reference	s.c.$(z = 6)$		b.c.c.$(z = 8)$	
		γ	$\langle \delta S_i^z \rangle$	γ	$\langle \delta S_i^z \rangle$
LSW	37	0.58	0.078	0.58	0.059
Perturbation theory	39	0.60	0.064	0.60	0.047
Variational	41	0.54	0.045	0.54	0.033
Variational	42	0.60	0.059	0.60	0.051
Cellular	44	0.4 ± 0.4	0.20 ± 0.02	0.4 ± 0.4	0.19 ± 0.02

[44] J. Oitmaa and D. Betts, *Phys. Lett.* **68A**, 450 (1978).

Table 5.5.   Ground state correlations for $s = \frac{1}{2}$ antiferromagnets in 3D, for a 16 spin cell in two distinct lattices.[44]

R_{ij}/a	XY		Heis.
	$4\langle S_i^z S_j^z \rangle$	$4\langle S_i^x S_j^x \rangle$	$4\langle S_i^z S_j^z \rangle$
Simple cubic (s.c.)			
1	0.1230	−0.5464	−0.4327
$2^{1/2}$	0.0297	0.5054	0.3141
$3^{1/2}$	0.0246	−0.4926	−0.3107
2	0.0347	0.5174	0.3333
Body-centered cubic (b.c.c.)			
$(3/4)^{1/2}$	0.0965	−0.5395	−0.4167
$1, 2^{1/2}, 3^{1/2}$	0.0326	0.5163	0.3333

5.9.   Ferrimagnetism

Ferrimagnets are generally insulators containing localized spins which are antiferromagnetically coupled. One example pointed by Néel, who coined the term *ferrimagnetism* (because of its existence in the *ferrites*, of which the lodestone is an example) occurs if the spins on the A and B sublattices of the previous section are of unequal magnitudes $s_A \neq s_B$. Other, vastly more complicated examples of ferrimagnetism exist in theory and in nature,[45] but their study is a complex and specialized field. In the present section, we just want to display the spin-wave Hamiltonian for the simple two-sublattice model of ferrimagnetism.

We also show how the same expression, which for unequal spins gives a magnon energy $\sim k^2$ at long wavelengths, will yield the antiferromagnon energy $\sim k$. The mathematics is based on work by Nakamura and Bloch[47] and consists of a straightforward expansion of the square roots in the Holstein-Primakoff representation. For large spins, agreement with the classical equations of motion gives some confidence in this procedure, for which there is no other formal mathematical justification.

Let $s_A \geq s_B$, let there be N of each, and put

$$s_A = (1 + \alpha)s, \quad s_B = (1 - \alpha)s. \tag{5.161}$$

---

[45] For a review and references to the literature, see Wolf's review.[46]
[46] W. P. Wolf, *Rep. Prog. Phys.* **24**, 212 (1961).
[47] T. Nakamura and M. Bloch, *Phys. Rev.* **132**, 2528 (1963).

Except for the unequal spins, the Hamiltonian is precisely that of (5.144) in the preceding section. We keep only the leading term in an expansion of the Hamiltonian in powers of s:

$$\mathcal{H} = \mathcal{H}_0 + \mathcal{H}_1 + \mathcal{H}_2 + O(s^{-2}).$$
(5.162)

It is,

$$\mathcal{H}_0 = \sum [\gamma_0(s_B a_{\mathbf{k}}^* a_{\mathbf{k}} + s_A b_{\mathbf{k}}^* b_{\mathbf{k}})$$

$$+ \sqrt{s_A s_B}\gamma_{\mathbf{k}}(a_{\mathbf{k}} b_{\mathbf{k}} + \text{H.c.})] - N\frac{z}{2}s_A s_B$$
(5.163)

where

$$a_{\mathbf{k}} = \frac{1}{\sqrt{N}} \sum_{j \leq A} a_j e^{-i\mathbf{k} \cdot \mathbf{R}_j} \quad \text{and} \quad b_{\mathbf{k}} = \frac{1}{\sqrt{N}} \sum_{i \leq B} b_i e^{+i\mathbf{k} \cdot \mathbf{R}_i}$$
(5.164)

and

$$\gamma_{\mathbf{k}} = \sum_{\delta} e^{i\mathbf{k} \cdot \delta}, \quad \gamma_0 = z.$$
(5.165)

Next, $\mathcal{H}_0$ is diagonalized by the Bogolubov transformation,

$$a_{\mathbf{k}} \to a_{\mathbf{k}} \cosh u_{\mathbf{k}} + b_{\mathbf{k}}^* \sinh u_{\mathbf{k}}$$

$$b_{\mathbf{k}} \to a_{\mathbf{k}}^* \sinh u_{\mathbf{k}} + b_{\mathbf{k}} \cosh u_{\mathbf{k}}$$
(5.166)

mixing operators on the A and B sites. The degeneracy of the magnon spectrum in antiferromagnetism is lifted for $\alpha \neq 0$, and the following *two* magnon branches are found:

$$\varepsilon_b(\mathbf{k}) = \gamma_0 s(f_{\mathbf{k}} - \alpha) \quad \text{and} \quad \varepsilon_b(\mathbf{k}) = \gamma_0 s(f_{\mathbf{k}} + \alpha)$$
(5.167)

where

$$f_{\mathbf{k}} = \sqrt{1 - (1 - \alpha^2)\left(\frac{\gamma_{\mathbf{k}}}{\gamma_0}\right)^2}$$
(5.168)

provided the transformation parameter, $u_{\mathbf{k}}$, chosen so as to eliminate $a_{\mathbf{k}} b_{\mathbf{k}} +$ H.c. from the Hamiltonian, takes the value

$$\tanh 2u_{\mathbf{k}} = -\frac{\gamma_{\mathbf{k}}}{\gamma_0}\sqrt{1 - \alpha^2}.$$
(5.169)

In the long wavelength approximation, the magnon energies are

$$\varepsilon_a(\mathbf{k}) \approx (\sqrt{\alpha^2 + (1 - \alpha^2)(ka)^2} - \alpha)\gamma_0 s \propto \frac{1 - \alpha^2}{2\alpha}(ka)^2$$

$$\varepsilon_b(\mathbf{k}) \approx [\sqrt{\alpha^2 + (1 - \alpha^2)(ka)^2} + \alpha]\gamma_0 s$$

(5.170)

from which it is easy to see the range (dependent on α), over which the lower branch is quadratic, as in ferromagnets, before becoming approximately linear, as in antiferromagnets. The ground-state energy, in this linear approximation, is

$$E_0 = -\frac{z}{2}(2N)s^2(1 - \alpha^2)\left(1 + \frac{\gamma}{zs}\right)$$

(5.171)

(it must be noted that there is a total of $2N$ spins in the present calculation) with

$$\gamma = zs\left(1 - \frac{1}{N}\sum f_{\mathbf{k}}\right).$$

(5.172)

This quantum-mechanical correction is smaller in ferrimagnets than in antiferromagnets, which is not too surprising in view of the resemblance with ferromagnets, for which $\gamma \equiv 0$.

5.10.   Some Rigorous Notions in Antiferro- and Ferri-Magnetism

The following concerns the symmetries in the ground state of antiferromagnets and ferrimagnets, to the extent that these can be modeled by spins on a bipartite lattice (a lattice that can be decomposed into A and B sublattices, such that spins in A interact only with those in B and vice versa.) These interactions are supposed to be of the Heisenberg type, e.g., $+J\mathbf{S}_{i(A)} \cdot \mathbf{S}_{j(B)}$. The lattice has a bipartite bond structure but it is otherwise arbitrary; it could even be random, quasicrystalline, or whatever.

What we wish to show is that if the numbers and magnitudes of the spins on B differs from those on A, there will be a net *ferrimagnetic* moment attributable to a net total spin angular momentum in the ground state. The magnitude of this net spin is $S_T = |N_A s_A - N_B s_B|$, in which s_A is the magnitude of spins on the A sublattice and s_B on the B sublattice (i.e., $S_{i(A)}^2 = s_A(s_A + 1)$) in units $\hbar = 1$. This includes antiferromagnets as a special case. In simple antiferromagnets, both $N_A = N_B$ and $s_A = s_B$, thus quite evidently $S_T = 0$. But one may use this theorem to generalize the very *notion* of antiferromagnetism to a larger class of magnetic systems, those for which both $N_A \neq N_B$ and $s_A \neq s_B$ but $N_A s_A = N_B s_B$ (e.g., twice as many spins one-half on one sublattice as spins 1 on the other.)

The proof[48] *is simple.* It involves rotating each spin on the B sublattice $180°$ about its z-axis, such that the x and y components of the spin vector acquire a $(-)$ sign. Then *each* individual bond is transformed as follows:

$$JS_i \cdot S_j \Rightarrow -\frac{J}{2}(S_i^+ S_j^- + S_j^+ S_i^-) + JS_i^z S_j^z . \tag{5.173}$$

The total Hamiltonian is just the sum over all such bonds. It commutes with total S_z so we can express the energies in subspaces identified by M, the eigenvalue of total S_z. When expressed in the Hilbert space of all the spin operators within each subspace, H is a matrix with *all negative* off-diagonal elements. The eigenstate of lowest energy in each subspace is, unambiguously, nodeless. We denote it $E_0(M)$. But while we have a unique (nondegenerate) ground state for each M, to what value of S_T does it actually belong? How can we identify it, aside from the knowledge that $|M| \leq S_T$?

To find this, we next construct a special, soluble, example that satisfies all the conditions as this ferrimagnet in question, but is a *mean-field-type* example. That is, *every* spin on the A sublattice has an equal interaction with *every* spin on the B sublattice. This mean-field Hamiltonian is exactly soluble. Its Hamiltonian,

$$\mathcal{H}_{M-F} = J \sum_{\text{all } i(A)} S_i \cdot \sum_{\text{all } j(B)} S_j$$

$$\Rightarrow \sum_i \sum_j \left(-\frac{1}{2}(S_i^+ S_j^- + S_j^+ S_i^-) + S_i^z S_j^z \right), \tag{5.174}$$

shares the property that each ground state energy $E_0(M)$ belong to a nodeless eigenvector.

Alternatively, we can write this Hamiltonian in its diagonal form,

$$\mathcal{H}_{M-F} = \frac{J}{2}(S_T^2 - S_A^2 - S_B^2)$$

$$= \frac{J}{2}(S_T(S_T + 1) - S_A(S_A + 1) - S_B(S_B + 1)) \tag{5.175}$$

in which S_A and S_B range from 0 (or 1/2) to $N_A s_A$ and $N_B s_B$, and

---

[48] See E. H. Lieb and D. C. Mattis, *J. Math. Phys.* **3**, 749 (1962).

$$|N_A s_A - N_B s_B| \le S_T \le |N_A s_A + N_B s_B|. \tag{5.176}$$

Quite clearly, the lowest energy is attained if S_A and S_B each have their *maximum* values (i.e., all the spins *within* each sublattice are parallel to each other) while S_T has its *minimum* value, which is $S_T = |N_A s_A - N_B s_B|$. The ground state thus exhibits a sort of ferromagnetism within each sublattice and antiferromagnetism between sublattices.

The ground state of $\mathscr{H}_{M-F}$ belongs to $S_T = |N_A s_A - N_B s_B|$ and it is nodeless; hence it is not orthogonal to the ground state of any other ferrimagnet on this bipartite lattice that shares the same value of the parameters N_A, N_B, s_A and s_B. What is more, in any subspace of given M, the ground state of *any* such lattice belongs to the same common value of S_T.

As we vary M we can take advantage of the inequality $S_T \ge |M|$. Thus, as more spinwaves are emitted and S_T is increased from its minimum value, one finds a monotonic increase in ground state energy reflected in the inequality:

$$E_0(S_T + 1) \ge E_0(S_T). \tag{5.177}$$

This is, of course, verifiable by inspection in the particular case of Eq. (5.175).

5.11.   Vortices

Just as dislocations in solids cannot be studied using the theory of small amplitude vibrations, neither can the topological disorder known as a vortex in the magnetic problem be described in terms of spinwaves or magnon excitations. Vortices have to date been studied only in the classical ($s \to \infty$) limit. They were introduced into magnetism by Kosterlitz and Thouless[49] in the context of the 2D classical XY model. They exemplify one type of topological singularity.

Following those authors, let us consider classical vectors with ferromagnetic nearest-neighbor interactions, on a square lattice as illustrated in Fig. 5.12. There we picture a single counter-clockwise vortex of charge +1. (A similar but clockwise vortex has charge −1.) It extends throughout the lattice and because spins are practically parallel in any neighborhood far removed from the core, a vortex cannot be "cured" by any finite number of small re-orientations of the spins, nor can it be caused by them. Once introduced, it is a stable state, although one of higher energy than the ground

---

[49] J. Kosterlitz and D. Thouless, *J. Phys.* **C6**, 1181 (1973).

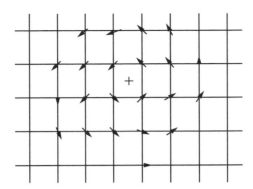

Fig. 5.10. A single vortex of "charge" +1. At large distances from the core, the spins are locally almost parallel but there is still circulation: a spin far to the left and one far to the right are antiparallel. The self-energy is infinite.

state. The vortex represents an extra degree of freedom of the spin system, independent of those associated with the magnons.

The excitation energy of a single vortex is easily estimated. In 2D one reasons as follows: the misalignment of 2 neighboring spins is a small quantity, say $\delta\theta$. Starting at some spin at a radial distance R from the core of the vortex and moving along the circumference in a ccw direction, we must wind through a total angle $\theta = (2\pi R/a)\delta\theta$ which is required to be an integer multiple of 2π, say $2\pi n$; for the case illustrated in Fig. 5.10, $n = +1$. Thus,

$$\delta\theta = na/R. \tag{5.178}$$

The energy of misalignment at each bond is of the order of $J(\delta\theta)^2$, with J the exchange parameter. Thus the total excitation energy or "self-energy" of the vortex is

$$\Delta E = 2\pi \int_a^L dR R a^{-2} J(na/R)^2$$

$$= 2\pi n^2 J \ln(L/a) \tag{5.179}$$

where we take the lattice to be circular of radius L and area πL^2, centered about the core. Because the ln is a slowly varying function one does not expect any dependence on sample geometry and the most convenient one is used.

Two vortices of opposite sign (say a pair $+n$, $-n$) at a distance $\mathbf{R}_{ij}$ apart have an energy which is independent of L. Their fields cancel at large

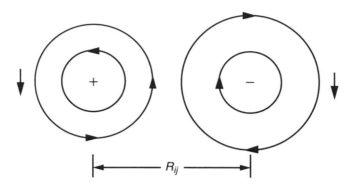

Fig. 5.11.   A vortex dipole: effects on spins far from the cores is much reduced: the energy is finite.

distances as shown in Fig. 5.11; and all that remains is the dipolar near-fields, which result in

$$\Delta E_{ij} = 2\pi n^2 J \ln\left(\frac{|\mathbf{R}_{ij}| + a}{a}\right). \tag{5.180}$$

Such vortex-antivortex (dipole) pairs can have a strong effect on the finite temperature thermodynamics.[49,50]

One might wonder whether the effect depends on dimensionality. In 1D, for a chain of N spins we have

$$N\delta\theta = n2\pi \tag{5.181}$$

and a self-energy

$$\Delta E = JN(2\pi n/N)^2 = J(2\pi n)^2/N \tag{5.182}$$

a negligible value. Moreover, the interaction energy of a vortex-antivortex pair separated by m sites is

$$\Delta E_{i,j} = |m|(2\pi n/N)^2 J. \tag{5.183}$$

What is interesting here is the similarity to the interaction energy of two sheets of charge in 1D, and of course, the energetics of the vortices is analogous to the flux calculus of classical electrostatics. Thus, in any number of dimensions, the study of magnetic vortices parallels that of charged fluids.

---

[50] J. Kosterlitz, *J. Phys.* **C7**, 1046 (1974).

5.12.   Solitons and Bloch Domain Walls: Introductory Material

It has been observed throughout this chapter that various problems in magnetism become simpler as s, the magnitude of the individual spin becomes larger and the quantum fluctuations become correspondingly smaller. A century of progress has recently culminated in the exact solution of many problems in classical nonlinear dynamics in one spatial dimension, including those of greatest relevance to this chapter. A common thread is the identification of at least two different types of propagating excitations: the ordinary wave (here, *magnon*) and the solitary wave, or *soliton*. The soliton is a particle-like manifestation that preserves its shape and identity after a long course of travel or period of time, and after collisions with stationary defects or other solitons. To quote its discoverer:[51]

> I was observing the motion of a boat which was rapidly drawn along a narrow channel by a pair of horses, when the boat suddenly stopped — not so the mass of water in the channel which it had put in motion — it accumulated round the prow of the vessel in a state of violent agitation, then suddenly leaving it behind, rolled forward with a great velocity, assuming the form of a large solitary elevation, a rounded, smooth and well-defined heap of water, which continued its course along the channel apparently without change of form or diminution of speed. I followed it on horseback, and overtook it still rolling on a rate of some eight or nine miles an hour, preserving its original figure some thirty feet long and a foot to a foot and a half in height. Its height gradually diminished and after a chase of one of two miles I lost it in the windings of the channel. Such, in the month of August 1834, was my first chance interview with that singular and beautiful phenomenon ...

To relate this remarkable phenomenon to magnetism, some concepts will have to be developed. Chief among these are Heisenberg's equations of motion for the spin dynamics, such as

$$\frac{\partial}{\partial t}\mathbf{S}_n = \frac{i}{\hbar}[\mathcal{H},\mathbf{S}_n]$$
(5.184)

---

[51] J. Scott Russell, *Proc. Roy. Soc. Edinburgh* (1844), p. 319.

with $\mathscr{H}$ the Hamiltonian of (5.1). Abbreviating $\partial/\partial t$ by $(\,\cdot\,)$, we find

$$\dot{\mathbf{S}}_n = -\frac{i}{\hbar}\sum_m J_{mn}[\mathbf{S}_m\cdot\mathbf{S}_n,\mathbf{S}_n] - \frac{i}{\hbar}g\mu_{\mathrm{B}}[\mathbf{H}\cdot\mathbf{S}_n,\mathbf{S}_n]. \tag{5.185}$$

The Zeeman contribution is easy to compute by the familiar commutation relations [Chapter 3, (3.5, 3.6)]. The bracket coefficient of J_{mn} has been previously calculated in [Chapter 3, (3.55)]. Combining the results, we find

$$\dot{\mathbf{S}}_n = \mathbf{S}_n \times \left(\sum_m J_{mn}\mathbf{S}_m + g\mu_{\mathrm{B}}\mathbf{H}\right). \tag{5.186}$$

Summing over all spins, we see that the cross terms cancel and the total spin merely precesses about the external field

$$\mathbf{S}_T = \mathbf{S}_T \times g\mu_{\mathrm{B}}\mathbf{H}. \tag{5.187}$$

As the quantum unit of action $\hbar$ does not appear explicitly in the equations of motion (5.186), it follows they are valid without modification in the classical limit $s \to \infty$. (In proceeding to this limit, it is necessary to scale J and μ_{B} as $1/s$ to keep the results finite, while setting $\hbar = 1$ for dimensional convenience.) We can now represent a spin by a classical unit vector: $\mathbf{S}_n = s\mathbf{u}_n$. Assuming a simple lattice [s.c., or sq., or n.n. linear chain, having primitive unit vectors $\boldsymbol{\delta}$ with $\delta = a$ the lattice parameter], the equations of motion take the form of difference equations

$$\dot{\mathbf{u}}(\mathbf{R}_n) = \mathbf{u}(\mathbf{R}_n) \times \left\{Js\sum_\delta [\mathbf{u}(\mathbf{R}_n + \boldsymbol{\delta}) - \mathbf{u}(\mathbf{R}_n)] + \mathbf{h}\right\} \tag{5.188}$$

with $\mathbf{h} = g\mu_{\mathrm{B}}\mathbf{H} = (0,0,h)$. For all long wavelength phenomena, whether linear or not, a lattice may be approximated by a continuum. In (5.188) we have subtracted terms such as $\mathbf{u}_n \times \mathbf{u}_n$ that vanish in the classical limit, to obtain a form in which the continuum limit is easily taken. One expands in a Taylor series

$$\mathbf{u}(\mathbf{R}_n + \boldsymbol{\delta}) = \mathbf{u}(\mathbf{R}_n) + \boldsymbol{\delta}\cdot\nabla\mathbf{u}(\mathbf{R}_n) + \frac{1}{2}(\boldsymbol{\delta}\cdot\nabla)^2\mathbf{u}(\mathbf{R}_n) + \cdots. \tag{5.189}$$

Terms in δ_α or in $\delta_\alpha\delta_\beta$ $(\alpha \neq \beta)$ vanish by symmetry, leaving, in leading order, only the following:

$$\dot{\mathbf{u}}(\mathbf{R}) = Jsa^2\mathbf{u}(\mathbf{R}) \times [\nabla^2\mathbf{u}(\mathbf{R})] + \mathbf{u}(\mathbf{R}) \times \mathbf{h}. \tag{5.190}$$

This is better written in terms of $u^\pm = u_x \pm i u_y$ and u_z:

$$\dot{u}^+ = iJsa^2(u_z\nabla^2 u^+ - u^+\nabla^2 u_z) - iu^+ h \qquad (5.191a)$$

with u^- just the complex conjugate of u^+, and:

$$\dot{u}_z = -i\frac{1}{2}Jsa^2(u^-\nabla^2 u^+ - u^+\nabla^2 u^-). \qquad (5.191b)$$

From our earlier studies we already know the form of a spinwave: it is a solution of these equations with constant amplitude and plane wave character. Thus, inserting $u^\pm = A\exp[\pm i(\mathbf{k}\cdot\mathbf{R} - \omega t)]$ in (5.191) yields

$$\omega = Jsa^2 u_z \mathbf{k}^2 + h \qquad (5.192a)$$

and

$$\dot{u}_z = 0, \quad \text{hence} \quad u_z = p_A, \quad \text{a constant.} \qquad (5.192b)$$

As $\hat{\mathbf{u}}$ is a unit vector, the amplitude A must be $\sqrt{(1 - p_A^2)}$. These results agree with (5.16) for $A \to 0$, $p_A \to 1$. At finite amplitude there is a "softening" of the frequency. But at finite amplitude, the total energy of the spinwave is extensive, i.e., it is infinite! For a finite amplitude spinwave is the superposition of $O(N)$ magnons. (Thus the momentum and angular momentum are also infinite.) This contrasts with the soliton, a localized excitation whose energy, momentum and angular momentum are all finite (1D) or proportional to the cross-sectional area of the wave front (in 2D or 3D). Cf. Fig. 5.12. To study it further, we shall need explicit expressions for the excitation energy, momentum and angular momentum. It will also be convenient to specialize to one dimension.

The Hamiltonian (5.1) can be written

$$\mathcal{H} = \frac{1}{2}J\sum_{n=1}^{Na}(\mathbf{S}_{n+1} - \mathbf{S}_n)^2 - h\sum_{n=1}^{N}(\mathbf{S}_n^z - s) + E_0$$

$$\to \frac{1}{2}Js^2a\int_0^{Na}dx\left[\frac{d\hat{\mathbf{u}}(x)}{dx}\right]^2$$

$$+ (hs/a)\int_0^{Na}dx[1 - u_z(x)] + E_0 \qquad (5.193)$$

proceeding to both classical and continuum limits. As unit vectors need only two parameters to specify their direction, we can introduce the two,

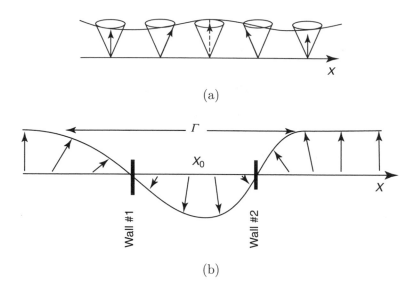

(a)

(b)

Fig. 5.12. (a) Spin wave (S_z = constant) extends to $x = \pm\infty$. (b) Soliton is a moving pulse centered about $x_0 = vt$. S_x and $S_y \rightarrow 0$ for $|x - x_0| > \Gamma$. The characteristic pulse width Γ and shape are obtained in the text. It actually consists of 2 domain walls, defined by the 180° change in direction of the magnetization near each.

canonically conjugate, dynamical variables p, q through the Villain representation [see Chapter 3, (3.87–3.89)] in the classical limit

$$u^+ = e^{iq}\sqrt{1 - u_z^2}, \quad u^- = (u^+)^+, \quad \text{and} \quad u_z = p. \tag{5.194}$$

Thus, the excitation Hamiltonian, in the natural units ($Js^2a = 1$) is

$$\mathbf{H} = \mathscr{H}_0 - E_0 = \frac{1}{2} \int dx \left[\frac{1}{1 - p^2}(dp/dx)^2 + (1 - p^2)(dq/dx)^2 \right]$$

$$+ h' \int dx(1 - p) \tag{5.195}$$

writing hs/a as h'. The equations of motion (5.191) now take on the aspect

$$-\dot{q}(x, t) = \frac{1}{1 - p^2}(d^2p/dx^2)$$

$$+ \frac{p}{(1 - p^2)^2}(dp/dx)^2 + p(dq/dx)^2 + h' \tag{5.196a}$$

and

$$\dot{p}(x,t) = (1 - p^2)(d^2q/dx^2) - 2p\frac{dp}{dx}\frac{dq}{dx}$$

$$= \frac{d}{dx}\left[(1 - p^2)\frac{dq}{dx}\right]. \tag{5.196b}$$

Along with the Hamiltonian, we identify two conserved quantities[52] as, the total momentum,

$$P = \frac{s}{a}\int_a^s dx(1 - p)\frac{dq}{dx} \tag{5.197}$$

and the z component of the angular momentum carried by the excitations,

$$\mathbb{M}_z = \frac{s}{a}\int_a^s dx(p - 1) \tag{5.198}$$

respectively. The latter is naturally negative, as the maximum spin of a ferromagnet occurs in the ground state.

5.13.   Solitary Wave Solution

Here we set out to construct the most general solutions permitted by the equations of motion, having a "permanent profile" character, $p(x - vt)$ and $q(x - vt)$. To qualify as *solitons* they must moreover be stable against small perturbations, against collisions with other solitons, etc., i.e., they must possess some quality of permanence.

With time dependence arising only through $x - vt$ and v an adjustable parameter, (5.196) simplifies

$$v\frac{dq}{dx} = \frac{1}{1 - p^2}\frac{d^2p}{dx^2} + \frac{p}{(1 - p^2)^2}(dp/dx)^2 + p(dq/dx)^2 + h' \tag{5.199a}$$

and

$$v(dp/dx) = \frac{d}{dx}\left[(p^2 - 1)\frac{dq}{dx}\right]. \tag{5.199b}$$

(b) is integrated at once, yielding

$$dq/dx = v\frac{p_0 - p}{1 - p^2} \tag{5.200}$$

---

[52] J. Tjon and J. Wright, *Phys. Rev.* **B15**, 3470 (1977).

in which p_0 is a constant of integration. Insertion into (a) now yields a nonlinear, second-order differential equation for p

$$v^2 \frac{p_0 - p}{1 - p^2} = \frac{1}{1 - p^2} \frac{d^2 p}{dx^2} + \frac{p}{(1 - p^2)^2} (dp/dx)^2$$

$$+ pv^2 \frac{(p_0 - p)^2}{(1 - p^2)^2} + h' . \tag{5.201}$$

This can nonetheless be solved, by the device of setting $(dp/dx)^2 \equiv F(p)$, so that $d^2 p/dx^2 = \frac{1}{2} dF/dp$. The above then turns into a linear, first-order differential equation in the new unknown, F. It is readily solved, yielding

$$(dp/dx)^2 = F(p) = 2h' p(p^2 - 1)$$

$$- v^2 (1 + p_0^2 - 2p_0 p) - p_1 (p^2 - 1) . \tag{5.202}$$

Here p_1 is the arbitrary constant of integration. It should be remarked that this equation is that of a classical particle of mass $\frac{1}{2}$ and energy E, in a potential well $V = E - F$. There exists the additional constraint, that $|p| \le 1$. Aside from this, the motion is confined to the positive region of F, which is a cubic polynomial with asymptotic behavior $2h' p^3$. The constants p_0, p_1 must thus be adjusted so that the solution is physically allowed. It is possible to integrate (5.202) and analyze the resulting elliptic functions. It is, however, much simpler to study F for the behavior between the turning points, from which one concludes that p_0, p_1 must be adjusted to allow one of three distinct patterns shown as (a), (b) and (c) in Fig. 5.13.

The first curve, (a), satisfies the conditions appropriate to a spinwave of amplitude p_A, previously analyzed. Only $p = p_A$ is allowed, hence (5.200) has the solution $q = k(x - vt) + q_0$.

The second curve describes the soliton superposed onto an otherwise perfect ferromagnetic background. The asymptotic value of p is $p_C = 1$, and the largest deviation from the asymptotic value is at p_A which can lie anywhere in the range $-1 \le p_A < +1$, by suitable adjustment of the parameters p_0, p_1.

The final curve (c) leads to a periodic repetition of the soliton pulse, i.e., to a sort of wavetrain. With p_B and p_C very close and straddling 1, each pulse in the infinite train resembles the solution (b), although the period is finite rather than infinite. With p_A and p_B very close, the solution reduces to the spinwave of (a). We thus see the spinwave and the soliton as two rather opposite limiting cases of the general behavior exemplified by (c).

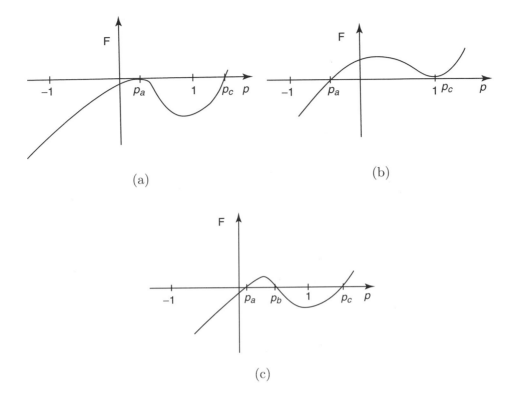

Fig. 5.13.   (a) Double root at p_a, single root outside allowed interval (spinwave). (b) Single root at $-1 < p_a < 1$, double root at $p = 1$ (soliton). (c) p oscillates between p_a and p_b; this describes a wavetrain, with characteristic of (a) if p_a and p_b are very close and of (b) if p_b and p_c are very close.

The soliton solution of this problem was first obtained by Nakamura and Sasada,[53] later by Lakshmanan et al.,[54] and had its stability tested by Tjon and Wright.[52] The construction of an F having the shape (b) in Fig. 5.13 requires the right-hand side of (5.202) to have a double root at $p = 1$. $F(1) = 0$ implies $p_0 = 1$; then $(dF/dp)_{p=1} = 0$ yields $p_1 = v^2 + 2h'$. Writing $p = 1 - 2\sin^2\frac{1}{2}\theta$ and defining a new variable $y = (1 - v^2/4h')^{-1/2}\sin\frac{1}{2}\theta$, and a new independent variable $t = x(h' - \frac{1}{2}v^2)^{1/2}$, we obtain

$$(dy/dt)^2 = y^2(1 - y^2) \tag{5.203}$$

[53] K. Nakamura and T. Sasada, *Phys. Lett.* **A48**, 321 (1974).
[54] M. Lakshmanan, T. Ruijgrok and C. Thompson, *Physica* **84A**, 577 (1976).

a well known ("canonical") equation, known to have the solution $y = \text{sech}\, t$. It is then a simple matter to substitute the original variables

$$p = 1 - 2\left(1 - \frac{1}{4}v^2/h'\right)\text{sech}^2\left(\frac{x - vt - x_0}{\Gamma}\right) \qquad (5.204)$$

where $\Gamma \equiv (h' - \frac{1}{2}v^2)^{-1/2}$, is the width of the pulse. This pulse is centered at x_0 (an arbitrary constant, chosen to satisfy initial conditions) at $t = 0$, travels to the left ($v < 0$) or right ($v > 0$) with the given speed in the range $0 \le |v| < 2(h')^{1/2}$. Integration of the equation for q proceeds by similar substitutions, to yield

$$q = \phi_0 + \frac{1}{2}v(x - vt - x_0)$$

$$+ \tan^{-1}\left[\left(\frac{4h'}{v^2} - 1\right)^{1/2}\tanh\left(\frac{x - vt - x_0}{\Gamma}\right)\right]. \qquad (5.205)$$

The distortions associated with the soliton are maximal at $v = 0$, yielding the narrowest possible pulse. Conversely, at $v^2 \to 4h'$, the pulse is infinitely broad and the local distortions infinitesimal.

Several quantities are of interest: the phase shift, the momentum, angular momentum, and energy. From (5.205) we find, for the total phase shift,

$$\Delta q = 2\tan^{-1}\left(\frac{4h'}{v^2} - 1\right)^{1/2} \qquad (5.206)$$

which varies from a maximum of π at $v = 0$ to zero as $v^2 \to 4h'$.

Using (5.195) we find the energy density to be

$$\mathbf{H}(x, t) = \frac{4}{\Gamma^2}\,\text{sech}^2\left(\frac{x - vt - x_0}{\Gamma}\right). \qquad (5.207)$$

The total energy is time independent, of course. Integrating the above, we find it to be

$$\Delta E = 8/\Gamma.$$

Restoring the original units

$$\Delta E = 4Js^2a/\Gamma. \qquad (5.208)$$

Similarly, the momentum is

$$P = \frac{4s}{a}\sin^{-1}\left(1 - \frac{1}{4}v^2/h'\right)^{1/2} \qquad (5.209)$$

and

$$\mathbb{M}_z = -4s/h\Gamma a \,. \tag{5.210}$$

Combining these gives a more perspicuous relation

$$\Delta E = \frac{16 J s^3}{|\mathbb{M}_z|} \sin^2(Pa/4s) \,. \tag{5.211}$$

For the spin one-half two-magnon bound state, $|\mathbb{M}_z| = 2$, $s = \frac{1}{2}$, this formula agrees *precisely* with the energy previously derived in (5.85) by quantum-mechanical analysis. The above also demonstrates the further lowering of the energy upon binding 3, 4, or more magnons into a bound state "soliton" of a given total momentum.

Tjon and Wright[52] studied the collision of two solitons numerically, some of their results being displayed in Fig. 5.14. But the crucial progress has occurred more recently. First Takhtajan[55] and then Fogedby[56] have come to the realization that the nonlinear equations of motion (5.190) can be replaced by a set of coupled linear eigenvalue equations. Difficult though the linear problem may be to bring to a closed form solution, there is the fact that the spectrum of eigenfunctions and eigenvalues is complete. This spectrum maps, in a one-to-one correspondence, onto the magnons, solitons and wavetrains we have encountered. It is therefore no accident that two solitons, having a collision described in detail in Fig. 5.14, survive this collision with their identities unimpaired. This is a necessary consequence of the large number of constraints — constants of the motion — that the equations of motion imply and that a solution must satisfy.

The interested reader will find the analysis elsewhere[52–55,56] for the mathematics are beyond the scope of the present volume. Recent applications of this theory include the XY model, modified somewhat to allow slight oscillations out of the plane[57] as required for the experimental applications; future applications promise to explain the outstanding mysteries about one-dimensional magnets: their response to neutrons, to time-dependent forces and fields, etc.

---

[55] L. A. Takhtajan, *Phys. Lett.* **64A**, 235 (1977).
[56] H. Fogedby, "Theoretical Aspects of Mainly Low-Dimensional Magnetic Systems," notes from the Inst. Laue-Langevin, Grenoble, France (March 1979); *J. Phys.* **A13**, 1467 (1980).
[57] H. Mikeska, *J. Phys.* **C11**, L29 (1978); **13**, 2913 (1980); K. Leung, D. Hone, D. Mills, P. Riseborough and S. Trullinger, *Phys. Rev.* **B21**, 4017 (1980); J. José and P. Sahni, *Phys. Rev. Lett.* **43**, 78 (1978), *erratum* **43**, 1843 (1978).

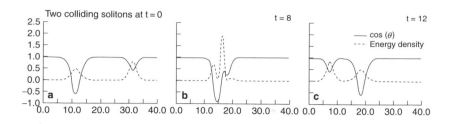

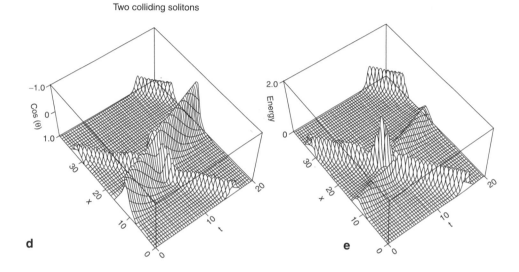

Fig. 5.14.   Two colliding solitons: one has velocity $v = \frac{1}{2}$ and amplitude $b = \sin\frac{1}{2}\theta = 0.8$, the other $v = -2$, $b = 0.3$.[51]
(a) $t = 0$ (before collision)
(b) $t = 8$ (impact)
(c) $t = 12$ (after collision)
(d) $p = \cos\theta$ versus x and t
(e) Energy density versus x and t

Although there is no reason to exclude solitons from consideration in 2D and 3D systems, where indeed they must exist, their energy which is proportional to the area of the wave front (N^{D-1}) will be too large for them to be included among those elementary excitations that are found spontaneously at low T, although bound complexes of them (such as the vortex–antivortex pair of a preceding section) may be.

In a later chapter on thermodynamics, we shall see how much simpler is the equilibrium statistical mechanics of the linear chain when compared with its dynamics, of which the reader has had a glimpse in these pages, or compared with statistical mechanics in 2D and 3D.

5.14.   More on Magnetic Domains

The soliton theory can be directly adapted to the study of the formation of magnetic domains, the shape of domain walls, their energy and their motion under external forces.

Domains exist thermodynamically to minimize electromagnetic energy, although the energetic cost associated with the creation, motion or elimination of *each wall* contributes to the total loss by hysteresis in each cycle.

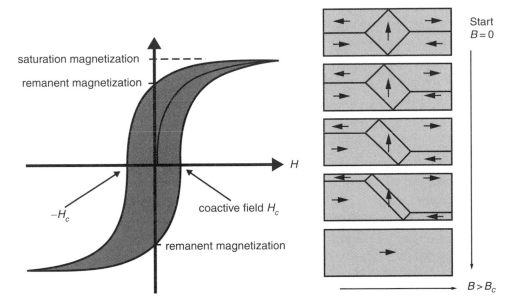

Fig. 5.15.   Hysteresis in the Magnetization of Oblong Ferromagnet.
Initially unmagnetized (note the assumed 90° domain structure at $B = 0$), an elongated sample acquire a magnetic moment as the domains deform and move in increasing field, as shown, until M_{sat} is achieved asymptotically at large, positive, H. Then, on the return path, upon reversing the external field, a finite positive ("remanent") magnetization persists at $H = 0$ and finally disappears at $-H_c$; then, as oppositely oriented domains become energetically favorable the magnetization decreases further to $-M_{\text{sat}}$. The shaded area in the hysteresis loop $\approx 4M_s H_c$ is a measure of the irreversible energy loss in each cycle. It is causally related to the irreversible motion of the domain walls.

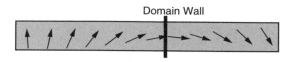

Domain Wall

Fig. 5.16.   Néel domain wall rotates the spins primarily perpendicular to the wall.

Assuming the process of magnetization to be performed as slowly as possible (adiabatically), we know from the previous section that the energy stored in the domain structure is approximately $4/\Gamma$ per domain wall, per unit area, with Γ no greater than the dimension of the sample, L. Does the lowering of magnetostatic energy (proportional to the volume of the sample) justify this cost? Without detailed calculation, it is easy to see that by shrinking the three dimensions at constant aspect ratios (shape), the *cost* in energy *per unit area* varies as $1/L$ while the magnetostatic energy *savings* are $\propto L$ *per unit area*. It follows there exists a *critical size* below which domain formation is energetically discouraged *ab initio*.

Figure 5.16 shows 90° domain walls, in which the magnetization is rotated by that angle through the wall. Alternatively, 180° walls are also stable. Traditionally these come in two types. In the first ("Bloch walls,") the magnetization rotates primarily in the plane of the wall. In the second ("Néel walls," illustrated above) the magnetization rotates perpendicular to the wall.

However, this entire field of *micromagnetics* is beyond our purview. The reader who wishes to examine it in detail can find a number of texts outlining simple and detailed theories for both isotropic and anisotropic materials.[58]

5.15.   Time-Dependent Phenomena

Interactions among spins are peculiar, due to the ability of the constituents to rotate. Let us spend a few pages on this aspect of the dynamics, some of which is counter-intuitive.

Suppose a spin $1/2$ is embedded in a solid. Its environment consists of normal modes of frequency ω that can, somehow, interact with it. The spin is in

---

[58] A good first choice: Stephen Blundell, *Magnetism in Condensed Matter*, Oxford Univ. Pess, Oxford, 2001, or L. D. Landau and E. M. Lifshitz, *Electrodynamics of Continuous Media*, Pergamon, Oxford, 1960, Sec. 39. A. Aharony's *Introduction to the Theory of Ferromagnetism*, Oxford Univ. Press, Oxford, 2000 (2nd Edition) will especially appeal to physicists interested in micromagnetics.

a magnetic field B and may be assumed to be in thermodynamic equilibrium at $T \approx 0$, its ground state energy $E_0 = Bg\mu_B$. Then, shortly after $t = 0$, suppose that the magnetic field is suddenly reversed from $(0, 0, B) \to (0, 0, -B)$. How does the spin regain its equilibrium now that it is suddenly in a high-energy configuration and how long does it take? The Hamiltonian at $t > 0$ is,

$$H = -Bg\mu_B \sigma_z + \lambda \cos \omega t \sigma_x \qquad (5.212)$$

and the state of the spin is initially $\begin{bmatrix} 0 \\ 1 \end{bmatrix}$, with energy $+Bg\mu_B$. This Hamiltonian is not conservative and we do not keep track of the energy once it is donated to the environment (to which it is connected by the coupling parameter λ).

This problem is exactly soluble by means of the time-dependent Schrödinger equation, $H\Psi = i\hbar \partial \Psi / \partial t$. *One finds this type of coupling does not allow real transitions.* The explicit solution that takes the initial condition into account is:

$$\begin{bmatrix} \phi_\uparrow \\ \phi_\downarrow \end{bmatrix} = e^{-\frac{i}{\hbar} \int_0^t dt' (-Bg\mu_B \sigma_z + g \cos \omega t' \sigma_x)} \begin{bmatrix} 0 \\ 1 \end{bmatrix}. \qquad (5.213)$$

Recalling the definitions of the Pauli matrices, one readily obtains:

$$|\phi_\uparrow|^2 = \frac{\lambda^2 \sin^2 \omega t}{\lambda^2 \sin^2 \omega t + (Bg\mu_B \omega t)^2}$$

$$\times \sin^2 \left(\frac{\sqrt{\lambda^2 \sin^2 \omega t + (Bg\mu_B \omega t)^2}}{\hbar \omega} \right). \qquad (5.214)$$

This formula is plotted in Fig. 5.17 at two values of the parameters.

Problem 5.3. Using the properties of the Pauli matrices, obtain $\phi_\uparrow$ and $\phi_\downarrow$, verify this last result and show that the spinor (5.213) remains normalized at all times $t > 0$.

Here we examined fully quantum spin $s = 1/2$ interacting with a classical oscillator. A generalization to a quantum oscillator is suggested in Problem 5.4 below. But does quantization speed the return to thermal equilibrium? Next we examine a spin of arbitrary magnitude s interacting with just such a quantized oscillator.

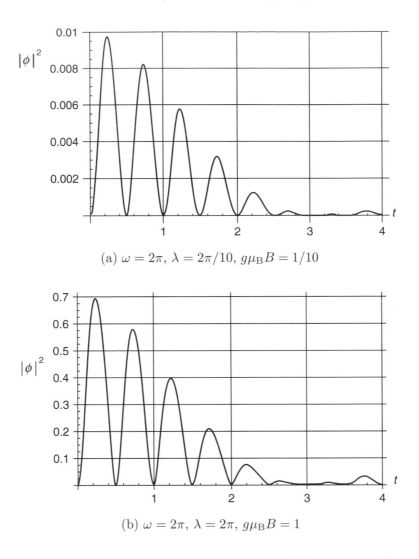

(a) $\omega = 2\pi$, $\lambda = 2\pi/10$, $g\mu_{\rm B}B = 1/10$

(b) $\omega = 2\pi$, $\lambda = 2\pi$, $g\mu_{\rm B}B = 1$

Fig. 5.17.   Probability of Return to Ground State after Time t.
In part (a), the probability of returning into the ground state never exceeds 1%, while in (b) it does reach a substantial value (70%) at first, but then oscillates and decays. Note the scaling similarity of the two curves.

The "toy" model to study next is that of a spin of magnitude s originally in its ground state at $S_z = s$ when the external magnetic field is suddenly reversed at $t = 0$. Approximating the off-diagonal matrix elements in a number-conserving approximation and retaining only expressions quadratic in *bosons* a and b, for $t > 0$ $\mathscr{H}$ takes the form,

$$\mathcal{H} = -Bg\mu_{\mathrm{B}}(s - b^+b) + \lambda(ab^+ + ba^+) + \hbar\omega_o(a^+a). \qquad (5.215)$$

The time-dependent $|\Psi(t)\rangle\rangle$ is subject to the initial condition: $|\Psi(0)\rangle\rangle = |2s\rangle \otimes |0\rangle$. (Here the state of the spin is designated as $|m\rangle$ while that of the oscillator is $|n\rangle$.)

Suppose we deal with spin $s = 2$. The Hilbert space consists of the following 5 states:

$$|\Phi_1\rangle\rangle = |4\rangle \otimes |0\rangle, \quad |\Phi_2\rangle\rangle = |3\rangle \otimes |1\rangle, \quad |\Phi_3\rangle\rangle = |2\rangle \otimes |2\rangle,$$

$$|\Phi_4\rangle\rangle = |1\rangle \otimes |3\rangle, \quad \text{and} \quad |\Phi_5\rangle\rangle = |0\rangle \otimes |4\rangle.$$

Instead of exponentiating the operators as in the preceding treatment, we adopt a different but equivalent approach to the time-dependent problem.

First, we construct the eigenvectors $|\xi_j\rangle\rangle$ using the basis set of 5 states; second, we expand the desired solution in the eigenstates: $|\Psi(t)\rangle\rangle = \sum_{j=1}^5 A_j|\xi_j\rangle\rangle e^{-iE_jt/\hbar}$, determining the coefficients A_j from the initial condition. Finally, we set $Bg\mu_{\mathrm{B}} - \hbar\omega_o = \Omega$ as a measure of how close one is to "resonance." The energy eigenvalue E is measured relative to the initial energy $(+Bg\mu_{\mathrm{B}}s)$. The 5×5 Hamiltonian matrix for $s = 2$ is,

$$\begin{pmatrix} -E & \sqrt{4} & 0 & 0 & 0 \\ \sqrt{4} & \Omega - E & \sqrt{2(4-1)} & 0 & 0 \\ 0 & \sqrt{2(4-1)} & 2\Omega - E & \sqrt{3(4-2)} & 0 \\ 0 & 0 & \sqrt{3(4-2)} & 3\Omega - E & \sqrt{4(4-3)} \\ 0 & 0 & 0 & \sqrt{4(4-3)} & 4\Omega - E \end{pmatrix}. \qquad (5.216)$$

The eigenvalue spectrum in that example is,

$$E = 2\Omega, 2\Omega - \sqrt{4 + \Omega^2}, -2(-\Omega + \sqrt{4 + \Omega^2}),$$

$$2(\Omega + \sqrt{4 + \Omega^2}), 2\Omega + \sqrt{4 + \Omega^2}.$$

If we set $\Omega = 0$ (go *on resonance*), the solution is relatively simple. After some algebra it yields the following plot for $P(t) =$ the probability of being the new ground state $|\Phi_5\rangle\rangle$.

Problem 5.4.   In the case $s = 1/2$, the Hamiltonian in (5.215) is 2×2. Find the corresponding $P(t)$ analytically and examine any significant consequences on the time-dependent behavior of being on- or off-resonance.

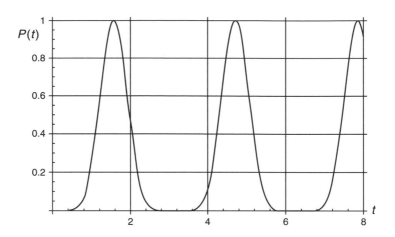

Fig. 5.18. Probability of Attaining the New Ground State $P(t)$.
For the $s = 2$ system on resonance, the new ground state is *approached* quickly and
periodically. On time-average, however, the probability of being in the ground state
is $P = 0.5$.

The fact is, the excess energy placed into the system by the reversal of
external field cannot be shed, either in the example illustrated in Fig. 5.18 (or
in the simpler example of Problem 5.4), regardless of resonance. The $t > 0$
Hamiltonian of Eq. (5.215) is *conservative* and the energy is shared among
very few degrees of freedom. Nor does it help to use a classical oscillator, as
we saw in the example of the preceding Fig. 5.17.

How does thermalization occur? The answer has to be, *into a reservoir
that contains a large number of normal modes, not just the one.* To simulate
"friction," the final states must have vastly higher entropy than the initial
one ...

Which presents us with a much more difficult model to dissect, as it
combines inherent nonlinearities (of the spin) with a large number of degrees
of freedom. It is the type of problem in which Green function techniques have
been most successful. Pending this, let us examine a generalization of (5.215),
using ordinary scattering theory.

We shall show that to leading order in the coupling λ, the decay of the
localized excited state connecting to N delocalized modes is no longer oscillatory, but strictly exponential. The effective lifetime is found to be inversely
proportional to the "density of states" of the delocalized modes resonating
at the excited state's energy. For spin $s = 1/2$ the multi-mode Hamiltonian
is simply,

$$\mathcal{H} = -Bg\mu_{\rm B}\left(\frac{1}{2} - b^+b\right) + \frac{\lambda}{\sqrt{N}}\sum_k(a_kb^+ + ba_k^+) + \sum_k\hbar\omega_k(a_k^+a_k). \quad (5.217)$$

For present purposes let us assume λ to be $\ll |Bg\mu_{\rm B}|$ so λ need be retained only to leading order. Then the normal modes' new eigenstates are,

$$|k)) = |k\rangle + \frac{\lambda/\sqrt{N}}{\hbar\omega_k - \Delta}b^+|0\rangle + O(\lambda^2) \qquad (5.218a)$$

where $|k\rangle$ is short-hand for $a_k^+|0\rangle$ and Δ for $|Bg\mu_{\rm B}|$. The decaying state at or near $\Delta/2$ is,

$$|b)) = \left(b^+|0\rangle + \sum_k\frac{\lambda/\sqrt{N}}{E_b - \hbar\omega_k}|k\rangle + O(\lambda^2)\right)e^{-iE_bt/\hbar}, \qquad (5.218b)$$

as a function of t. Its energy eigenvalue is,

$$E_b = \Delta + \frac{\lambda^2}{N}\sum_k\frac{1}{E_b - \hbar\omega_k}$$

$$= \Delta + \lambda^2(R(E_b) + i\pi\rho(\Delta)) = \Delta + i\pi\lambda^2\rho(\Delta). \qquad (5.218c)$$

Here we assumed the normal modes to be evenly distributed about Δ, causing R to vanish; this simplifies the result somewhat.[59] According to (5.218b), one then finds the decay of the probability of the initial spin being up to be $\propto \exp -t/\tau$, where

$$\tau = (2\pi\lambda^2\rho(\Delta))^{-1} \qquad (5.219)$$

in perfect accordance with Fermi's well-known "Golden Rule."

The knotty issue of spin reversals and decay (the tunneling can be surprisingly easy!) is the subject of a recent monograph by Chudnovsky and Tejada, to which we refer the reader.[60]

5.16.   Majumdar-Ghosh Model and "Quantum Frustration"

In an antiferromagnet, nearest-neighbor (n.-n.) spins have a tendency to be antiparallel while next-nearest-neighbor (n.-n.-n.) spins correlate naturally.

---

[59] Otherwise there has to be a slight shift $(O(\lambda^2))$ in the argument of ρ.
[60] E. M. Chudnovsky and J. Tejada, *Macroscopic Quantum Tunneling of the Magnetic Moment*, Cambridge Univ. Press, 1998.

Whenever this tendency is frustrated, e.g., by n.-n.-n. antiferromagnetic interactions,[61] there can appear a new quantum phase from which long-range
spin order simply disappears. It can be confusing to solve the Majumdar-
Ghosh model *ab initio* so we study it constructively for half-odd integer
spins, $s = 1/2, 3/2, \ldots$.

Write $\mathscr{H}$ as a sum over all inner products of clusters of three half-odd-
integer spins,

$$
\begin{aligned}
(S_1 &+ S_2 + S_3)^2 + (S_2 + S_3 + S_4)^2 + (S_3 + S_4 + S_5)^2 + \ldots \\
&= (2S_1 \cdot S_2 + 2S_1 \cdot S_3 + 2S_2 \cdot S_3) + 3s(s+1) \\
&\quad + (2S_2 \cdot S_3 + 2S_2 \cdot S_4 + 2S_3 \cdot S_4) + 3s(s+1) \\
&\quad + (2S_3 \cdot S_4 + 2S_3 \cdot S_5 + 2S_4 \cdot S_5) + 3s(s+1) \ldots .
\end{aligned}
\tag{5.220}
$$

From the right-hand side of this identity we infer an Heisenberg antiferromagnet in which the n.-n.-n. bonds are *half* as strong as the n.-n. bonds. By
inspection of the left-hand side, we find this system supports a novel phase,
obtained by pairing spins #2 and 3 in a singlet state, #4 and 5 similarly,
etc. For spins $s = 1/2$, the energy of this paired state is (by inspection!)[62]
$E_0 = N3/4$, each parenthesis having been reduced to a single spin of magnitude $1/2$. This value is the exact ground state energy, for it coincides with
a lower bound to the energy spectrum obtained by adding the minimum
energy of each parenthesis.

What is more, the ground state is degenerate as it supports two orthogonal pairings. (We could have paired spin #1 with 2, #3 with 4, #5 with
6, etc.; although in (5.220) there appears to be a trivial end correction, it
disappears if the chain on the left-hand side of (5.220) is cyclic and ends in
a term $(S_N + S_1 + S_2)^2$.) Given this ground-state degeneracy, a well-known
theorem[63] states that the excitation spectrum has a gap and that it is mas-
sive (i.e., the dispersion above the gap $\propto k^2$ rather than $\propto |k|$ as in ordinary
antiferromagnets.)

Now for spins $1/2$ let us consider a more general Hamiltonian,

$$
\mathscr{H} = \sum_n S_n \cdot S_{n+1} + \lambda \sum_n S_n \cdot S_{n+2}
\tag{5.221}
$$

[61] C. K. Majumdar and D. K. Ghosh, *J. Math. Phys.* **10**, 1388 and 1399 (1969).
[62] In a singlet state, the vector $S_2 + S_3$ simply vanishes.
[63] E. Lieb, T. Schultz and D. Mattis, *Ann. Phys. (N.Y.)* **16**, 407 (1962).

with variable n.-n.-n. coupling. At $\lambda = 0$ we know its ground state (and it is unique) and its magnon spectrum (fermionic and massless, with dispersion $|k|$ at long wavelengths), from the Bethe ansatz solution. At $\lambda = 0.5$ this reduces to (5.220). Therefore, at some critical value $0 \leq \lambda_c < 0.5$ there has to occur a sort of "quantum phase transition" from the massless phase with long-range order ("LRO") to the massive phase without LRO, as the result of the frustration parameter λ.

Problem 5.5. What is the ground state energy of (5.221) at $\lambda = 1/2$, for spins $s = 3/2$?

Let us now generalize (5.220), again just for N half-odd-integer spins,[64] by including a parameter g in a range to be determined.

$$\mathscr{H}(g) = \{\ldots (S_n + S_{n+1} + gS_{n+2})^2 + (gS_{n+1} + S_{n+2} + S_{n+3})^2$$
$$+ (S_{n+2} + S_{n+3} + gS_{n+4})^2 + (gS_{n+3} + S_{n+4} + S_{n+5})^2 + \ldots\}.$$
(5.222a)

Within this range of g the ground state energy is precisely $E_0(g) = Ng^2 3/4$ for $s = 1/2$. (Proof: this value corresponds to pairing n with $n+1$ into a singlet, and similarly for $n+2$ with $n+3$, $n+4$ with $n+5$, etc.; hence $E_0(g)$ is a variational *upper bound*. But if g is not *too* negative, $3g^2/4$ is also a *lower bound* on each parenthesis, hence $E_0(g)$ is a lower bound on the total energy. QED.) Let us re-arrange $\mathscr{H}(g)$ into the more perspicuous form:

$$\mathscr{H}(g) = 2 \sum_n (S_{2n} + S_{2n+1})^2$$

$$+ g \sum_n (S_{2n} + S_{2n+1}) \cdot (S_{2n-1} + S_{2n+2}) + Ng^2 s(s+1). \quad (5.222b)$$

The first sum favors singlet over triplet bonds $(2n, 2n+1)$, but does this hold for all g? Compare the ground state energy of the singlet state with that of a ferromagnetic state, say: all spins "up." Then $E_F = N(1 + g/2)^2$. This is lower than $E_0(g)$ if $(2+g)^2 < 3g^2$. Thus, the disordered singlet state is stable against LRO and ferromagnetism for g within a range $-\sqrt{3}+1 < g < \sqrt{3}+1$, with the symmetry point $g = 1$ in the middle of the range.

---

[64] Assuming cyclic boundary conditions and N divisible by 6.

5.17.   Integer Spins

We were limited to half-odd integer spins in the preceding discussions because the lower bound of the expression $(S_1 + S_2 + S_3)^2$ is zero for any three identical integer spins (and one lacks the advantage of a lower bound that actually coincides with an upper bound, allowing to determine the ground state energy). Thus, we saw that integer spins behave differently. Suppose we had a lattice of spins $s = 1$ that interacted with their neighbors antiferromagnetically? Haldane[65] first conjectured that, *even without frustration*, the ground state of any *one-dimensional* antiferromagnet of integer spins is degenerate and has no LRO. More importantly, the excitation spectrum is massive with an energy gap now known as the "Haldane gap." In these details and more, the integer-spin linear chain antiferromagnet differs profoundly from its half-odd-integer counterpart.

Although originally predicated on the continuum field-theoretic "nonlinear sigma model," Haldane's arguments have by now been confirmed, but only in 1D. Profound differences between one-dimensional magnetic materials composed of half-integer and those of integer spins have been documented.[66] The most startling contrast is between spins $s = 1/2$ and $s = 1$. One example: the two ends of an antiferromagnetic chain consisting of magnetic atoms of spin 1 are predicted — and actually observed by magnetic resonance techniques — to behave like Kramers doublets (i.e., artificial spins $1/2$).[66] By way of contrast, the ends of an antiferromagnet of spins $1/2$ are nonmagnetic. Another difference: the excitation spectrum of an isotropic antiferromagnet of half-integer spins is gapless, while that of spins 1 has a reasonably large "Haldane gap."

Because in the $\lim s \to \infty$, the quasiclassical expansion of Sec. 5.8 does not discriminate between integer or half-integer spins, such distinctions must be related to quantum fluctuations. Thus they must vanish as $1/s$ at large s. Also, there is no hint of corresponding anomalies in two- and three-dimensions, where spin wave theory appears reliable. That is what

---

[65] F. D. M. Haldane, *Phys. Lett.* **A93**, 454 (1983) and *Phys. Rev. Lett.* **50**, 1153 (1983). The relation between sigma models and spin chains is investigated further in D. Controzzi and E. Hawkins, *Int. J. Mod. Phys.* **B9**, 4449 (2005).

[66] For example, the simple proof of a gapless excitation spectrum in the one-dimensional Heisenberg model (see previous footnote 63) that can be derived for *any* half-integer spins *cannot* be extended to integer spins. This is not mere technical glitch! See I. Affleck and E. H. Lieb, *Lett. Math. Phys.* **12**, 57 (1986). End states were measured by M. Kenzelmann *et al.*, *Phys. Rev. Lett.* **90**, 087202-1 (2003).

motivates our present study of the *one-dimensional* Heisenberg antiferromagnet for spins $s = 1$. The first task is to determine an acceptable ground state wave function. Consider then,

$$\mathscr{H} = \sum_{n=1}^{N} S_n \cdot S_{n+1}$$

$$= \sum_{n=1}^{N} \left(\frac{1}{2}(S_n^+ S_{n+1}^- + \text{H.c.}) + S_n^z S_{n+1}^z \right) \qquad (5.223)$$

where each $S_n^2 = s(s+1) = 2$.

First, the Néel state ($\uparrow\downarrow\uparrow\downarrow \ldots$) has ground state energy $E_{\text{Néel}} = -N$. If we include second-order perturbation corrections totaling $\approx N/3$, this phase has a total energy $\approx -(4/3)N$ *and LRO*. But numerical investigations show that the Néel state is not even close to the ground state and the true ground state energy is close to $-3/2N$. Nor is spinwave theory a satisfactory way to deal with the fluctuations in this model, insomuch as the nonlinearities are large and the linearization produces any preconceived result — without justification! Beyond spin waves there is Bethe's ansatz, which has been the subject of several published but ultimately invalid attempts to adapt Bethe's ansatz to spins $s \geq 1$.[67]

This conundrum suggests we examine the *bonds* rather than the *spins*. Schwinger's coupled boson representation introduced in Chapter 3 handily serves the purpose. Recall,

$$S^+ = a^+ b, \quad S^- = b^+ a,$$

$$S^z = \frac{1}{2}(a^+ a - b^+ b), \quad s = \frac{1}{2}(a^+ a + b^+ b) = 1.$$

In the language of the boson operators, each bond

$$\frac{1}{2}(S_n^+ S_{n+1}^- + \text{H.c.}) + S_n^z S_{n+1}^z$$

$$= \frac{1}{2}(a_n^+ b_n b_{n+1}^+ a_{n+1} + \text{H.c.})$$

$$+ \frac{1}{4}(a_n^+ a_n - b_n^+ b_n)(a_{n+1}^+ a_{n+1} - b_{n+1}^+ b_{n+1}) \qquad (5.224)$$

---

[67] The ansatz fails because the Baxter-Yang criterion (simply put, multiple scatterings of three particles must not cancel one another) is not satisfied. Thus, the two-body scattering theory cannot be exended to three or more magnons — and we need N of them! See the discussion in Schlottmann[36].

is resolutely *quartic* in the fields. The fields are, moreover, severely constrained by the above condition $s_n = 1$ for all n. We can actually use this constraint to our advantage — to simplify the form of the bond operators — by adding in $(s_n s_{n+1} - 1) \equiv 0$ at each bond. After some trivial manipulations, the $\mathscr{H}$ of (5.223) takes a new form,

$$\mathscr{H} = \frac{-1}{2} \sum_n (\Omega_n^+ \Omega_n - 2), \quad \text{where} \quad \Omega_n = (b_n a_{n+1} - a_n b_{n+1}). \quad (5.225)$$

This Ω_n is a bond operator on $(n, n+1)$. It is also a *"half-singlet operator,"* in the sense that $\Omega_n^+ |0\rangle = (a_{n+1}^+ b_n^+ - b_{n+1}^+ a_n^+)|0\rangle$ is a *singlet state* of angular momentum but *contains only half of each spin.*[68] With only Ω and its Hermitean conjugate present, the above is undoubtedly the simplest possible form in which to cast $\mathscr{H}$.

Both $\Omega_n^+ |0\rangle$ and $(\Omega_n^+)^2 |0\rangle$ are singlet states annihilated by the total spin raising operator of the 2 spins: $S_{n,n+1}^+ = a_n^+ b_n + a_{n+1}^+ b_{n+1}$, as well as by the corresponding lowering operator $S_{n,n+1}^- = b_n^+ a_n + b_{n+1}^+ a_{n+1}$ and $S_{n,n+1}^z = \frac{1}{2}(a_n^+ a_n - b_n^+ b_n + a_{n+1}^+ a_{n+1} - b_{n+1}^+ b_{n+1})$.

Now the singlet ground state and ground state energy (-2) of *any one* bond in this Hamiltonian could be found without recourse to such fancy operators. Just complete the square and express the energy in terms of total spin: $S_n \cdot S_{n+1} = \frac{1}{2}((S_n + S_{n+1})^2 - 2 \times 2) = 1$ (for quintet states of total spin 2), -1 (for the triplet state, total spin 1) and finally, -2 (total spin zero). We examined the projection operators more systematically in Sec. 3.15. This result does give us a rigorous lower bound on the total energy eigenvalue of $\mathscr{H}$, $E_0 > -2N$ (i.e., $\approx 2/3N$ lower than the Néel state or about $1/2N$ lower than the best numerical results), where $N = $ total number of bonds for $N+1$ spins.

Let us verify whether the new notation also yields the above results. Of course the half-singlet state is somewhat deficient, having an insufficient number of factors (each spin requires 2 a's, 2 b's or one a and one b per site to be a legitimate $s = 1$, for a total 4 *per* bond). So consider instead the *true* singlet state $|\phi\rangle_{n,n+1} = \frac{1}{\sqrt{C}}(\Omega_n^+)^2 |0\rangle$ of two neighboring spins, where C is an appropriate normalization constant. If it is to be an eigenstate of the "one-bond Schrödinger equation," it must satisfy:

---

[68] This shows why use of the product wavefunction (5.227) is not practicable in 2D, for the bonds could not be treated isotropically (4 half-bond operators are 2 too many). But it suggests that a 2D sq antiferromagnet made of spins $s = 2$ should be attempted as should a *sc* antiferromagnet made of spins $s = 3$.

$$-\frac{1}{2}\{(a_{n+1}^+b_n^+ - b_{n+1}^+a_n^+)(b_n a_{n+1} - a_n b_{n+1}) - 2\}(a_{n+1}^+b_n^+ - b_{n+1}^+a_n^+)^2|0\rangle$$

$$= E(a_{n+1}^+b_n^+ - b_{n+1}^+a_n^+)^2|0\rangle. \tag{5.226}$$

Problem 5.6. Show that $[\Omega, \Omega^{+2}]|0\rangle = 6\Omega^+|0\rangle$ with Ω defined in (5.225).

Using a property of commutators: $AB = [A, B] + BA$, with $A = \Omega$ and $B = \Omega^{+2}$ the above simplifies to, $-\frac{1}{2}\{\Omega^+([\Omega, \Omega^{+2}] + \Omega^{+2}\Omega) - 2\Omega^{+2}|0\rangle\} = E\Omega^{+2}|0\rangle$. Note that $\Omega|0\rangle = 0$, so with the solution of Problem 5.6, all surviving terms are $\propto \Omega^{+2}|0\rangle$. Equating coefficients yields an eigenvalue $E = -(6 - 2)/2 = -2$, as expected.

The singlet solution comes easily. We try as a variational ground state the product state of half-singlets with extra linear operators at either end to satisfy the "$s = 1$ rule."

$$|\Phi_0\rangle = (a_1^+ \text{ or } b_1^+) \cdot \Omega_1^+ \Omega_2^+ \cdots \Omega_N^+ \cdot (a_{N+1}^+ \text{ or } b_{N+1}^+)|0\rangle. \tag{5.227}$$

To within "surface energy" terms of $O(1)$, the variational energy is N times the energy of a typical bond (somewhere in the middle). That is,

$$E_0(\text{var.}) = -\frac{1}{2}N\frac{(\Phi_0|\Omega_n^+\Omega_n - 2)|\Phi_0)}{(\Phi_0|\Phi_0)}. \tag{5.228}$$

We can calculate this similarly to the 1 bond problem solved above, but the results turn out no better than the Néel state. So let us generalize the ansatz. If it is clear that the $(n, n+1)$ bond is a singlet, then the neighboring $(n + 1, n + 2)$ and $(n - 1, n)$ bonds can interfere with it and break it up into triplet components, while preserving $S_{\text{total}} = 0$. (These interferences are reciprocated by the effects of the $(n, n + 1)$ bond onto its neighbors.) So let us try for a more general — indeed the best possible — product state approximation,

$$|\Phi(\lambda)\rangle = (a_1^+ \text{ or } b_1^+)\Gamma_1^+(\lambda)\Gamma_2^+(\lambda)\ldots\Gamma_N^+(\lambda)(a_{N+1}^+ \text{ or } b_{N+1}^+)|0\rangle \tag{5.229}$$

with which to evaluate the variational ground state energy:

$$E_0(\lambda) = -\frac{1}{2}N\frac{(\Phi_0(\lambda)|(\Omega_n^+\Omega_n - 2)|\Phi_0(\lambda))}{(\Phi_0(\lambda)|\Phi_0(\lambda))}, \tag{5.230}$$

where $\Gamma_n^+ = a_{n+1}^+b_n^+ - b_{n+1}^+a_n^+ + \lambda(a_{n+1}^+a_n^+ - b_{n+1}^+b_n^+)$.

The first two terms comprise the half-singlet Ω^+ while the next two terms comprise a half-triplet bond state, with λ a parameter with which to adjust the optimal mix. However this state lacks correlations among neighboring half-triplets to make an overall singlet, therefore even the optimized one cannot be an eigenstate of $\mathscr{H}$, strictly speaking.

This optimal $E_0(\lambda)$ wrt λ yields an acceptable result $\approx -1.4N$ at the lowest computational cost. Because nearest neighbor correlations are automatically built into this wavefunction it takes advantage of the exponentially decaying correlations presumed to exist in any massive gap phase. Specific results are given elsewhere.[69]

5.18.   The AKLT Spin One Chain

A slight modification of Heisenberg's nearest-neighbor antiferromagnet proposed by Affleck, Kennedy, Lieb and Tasaki (AKLT)[70] turns it into an exactly solvable model. These authors used a coupled fermion representation useful only for spins $s = 1$. Here we rewrite their model using Schwinger's coupled bosons as in the preceding section, allowing the theory to be extended for any integer spin. The idea is to express $\mathscr{H}$ using $Q^{(2)}$ projection operators of Chapter 3. By construction, the model exhibits the following property: (1) any eigenstate of $\mathscr{H}$ devoid of bonds with spin angular momentum 2 has energy zero and (2) except for end effects, the ground state of the infinite chain is unique, and is separated from the excitation spectrum by an energy gap (Haldane gap).

Recall $Q^{(2)} = (2 + 3S_1 \cdot S_2 + (S_1 \cdot S_2)^2)/6 = 1$ when acting on the 5 states of the pair $(1, 2)$ of total spin 2, and is zero otherwise. It is therefore a non-negative operator, from which we can deduce that the ground state energy of the following $\mathscr{H}$ is ≥ 0,

$$\mathscr{H}_{\text{AKLT}} = \sum_n Q^{(2)}_{n,n+1} . \tag{5.231}$$

Now, using $S_1 \cdot S_2 = \frac{-1}{2}(\Omega_1^+ \Omega_1 - 2)$ we obtain:

---

[69] D. P. Arovas and A. Auerbach, *Phys. Rev.* **B38**, 316 (1988).

[70] I. Affleck, T. Kennedy, E. H. Lieb and H. Tasaki, *Phys. Rev. Lett.* **59**, 799 (1987) and *Commun. Math. Phys.* **115**, 477 (1988). Numerical results are in L. Venuti *et al.*, *Int. J. Mod. Phys.* **B16**, 1363 (2002), *inter alia*.

$$\mathscr{H}_{\text{AKLT}} = N - \frac{5}{12} \sum_n \left(\Omega_n^+ \Omega_n - \frac{1}{10} (\Omega_n^+ \Omega_n)^2 \right), \tag{5.232}$$

a slight but meaningful modification of the Heisenberg antiferromagnet.

The state (5.227) easily satisfies the AKLT condition $Q_{n,n+1}^{(2)} |\Phi_0) = 0$ at each bond and it is unique (all half-singlets) hence its eigenvalue for $\mathscr{H}_{\text{AKLT}}$ is 0 and it is the ground state. (The generalized states $\Phi(\lambda)$ in (5.229) might not be useful here, as they are not explicitly rotationally invariant.)

For the excitation spectrum we might construct a "spin wave" in the usual way,

$$|k) = \sum_n e^{ikn} |n) \equiv \sum_n e^{ikn} \Omega_1^+ \Omega_2^+ \dots \Omega_{n-1}^+ \kappa_n^+ \Omega_{n+1}^+ \dots |0) \tag{5.233}$$

in which each term $|n)$ is orthogonal to the ground state. Obviously it "costs" energy to replace an Ω by a κ but the gap could disappear *if* the overlap matrix $(n+1|H_{\text{AKLT}}|n)$ were sufficiently large.

For further conjectures concerning the existence of an energy gap and extensions of the present calculations to similar cases in 2 and 3 dimensions (e.g., to spins 3/2 on a lattice with coördination number 3, etc.), the story, albeit similar, becomes less simple. The interested reader will find it helpful to examine the surprisingly readable original papers of AKLT.[70]

5.19. Defective Antiferromagnets

It is difficult to maintain perfect chemical structure (stoichiometry) in high temperature superconductors based on planar CuO_2. In principle, the native material is antiferromagnetic. But, typically, some magnetic cells are replaced by nonmagnetic ones. The Néel temperature is a rapidly decreasing function of the impurity concentration and of electron doping and may vanish at a few percent doping concentration. This is surprising, for if one or more spins were missing (or replaced by a nonmagnetic entity) in a *ferromagnet* there would hardly be any effect on the magnetic or thermodynamic behavior of the remaining spins.

However, recent studies have shown that the absence of one or more spins from one and the same sublattice in an Heisenberg quantum *antiferromagnet* of spin S entails many unexpected consequences: the remaining spins create, collectively, an effective magnetic moment centered at the site of each defect but having great reach. Experimentally, local magnetic moments can be measured *via* shifts in the nuclear magnetic resonance frequency (NMR) due to the polarization (or lack thereof) of the closest s-state electrons or spins, an effect known as the *Knight shift*. Thus there is basis for comparison of theory and experiment.

Supposing the total magnetic susceptibility of the antiferromagnet χ (an extensive quantity) originally has the value χ_0. In the presence of a single defect, this changes to χ_1. It has been shown[71] that the *difference* is independent of the size of the system (intensive). For the 2D Heisenberg model the correction takes the form of an expansion,

$$\chi_{\text{imp}} = \chi_1 - \chi_0 = \frac{S^2}{3T} + \frac{1}{3\pi\rho_s} \log \frac{T_0}{T} + \cdots \tag{5.234}$$

In these units a single noninteracting quantum spin satisfies Curie's Law with $\chi = S(S+1)/T$. Thus, the defect acts as a *classical spin* oriented isotropically on a unit sphere, with logarithmic corrections. The quantity ρ, is the "spin stiffness' coefficient and T_0 is an appropriate constant.

The most recent work of Anfuso and Eggert[71] is based on their quantum Monte Carlo simulations on a lattice of 100×100 sites. They studied the effects of defects on the neighborhood and on the entire system, as well as on their interactions. They examined the consequences of having a number of impurities, on the staggered susceptibility as well as on the total susceptibility. If spins are missing from distinct sublattices the collective moments tend to cancel and the effects decay quickly at large distances from the defects. Interestingly, if the number of defects on one sublattice exceeds the number on the other sublattice by an amount ΔN_I ($\Delta N_I \ll N$), the Curie term, unlike the logarithmic contributions, increases proportional to $(\Delta N_I)^2$. Consequently, the defects' spins appear to be additive in their contribution to Curie Law paramagnetism ("superparamagnetism") and the logarithmic corrections incoherent. For more details one should consult a rapidly evolving literature.[71]

---

[71] N. Nagosa, Y. Hatsugai and M. Imada, *J. Phys. Soc. Jpn.* **58**, 978 (1989); S. Sachdev, C. Buragohain and M. Voijta, *Science* **286**, 2479 (1999); K. H. Höglund and A. W. Sandvik, *Phys. Rev.* **B70**, 24406 (2004); O. P. Sushkov, *Phys. Rev.* **B68**, 094426 (2003); S. Sachdev, *Phys. Rev.* **B68**, 064419 (2003); F. Anfuso and S. Eggert, cond-mat/0511001 v1, 31 Oct. 2005.

Chapter 6

Magnetism in Metals

In which we discuss miscellaneous topics relevant to the theory of magnetism in metals

Not all the electrons in a metal participate in the electrical conduction, nor in other physical, chemical, or magnetic processes; some are core electrons, belonging to filled shells tightly bound to the ion and unaware of the metallic environment. Electrons in unfilled shells have a range of behavior intermediate between such tightly bound localized electrons and that of quasifree particles experiencing the smooth periodic atomic potential and participating fully in the electrical conductivity. The geometry plays a big role, either because it affects the band structure or because it affects the electron-electron interactions. Body-centered cubic Iron (Fe) has a curie temperature 0 (1000 K); face-centered Iron is *anti*ferromagnetic, $T_N \approx 70$ K.

It is easiest to discuss the theory of magnetism in such metals where the electrons can be clearly divided into distinct sets of tightly bound and quasifree particles. Hopefully, the results still have qualitative merit when the carriers have properties intermediate between these two extremes.

The simplest model of magnetism in metals is the following: electrons in well-localized magnetic d or f shells interact with one another via a Heisenberg nearest neighbor exchange mechanism, whilst an entirely distinct set of (quasifree) electrons in Bloch states accounts for the metallic properties without partaking of the magnetic ones. *Unfortunately, this model is purely fictional*; x-ray data, optical experiments, measurements of specific heat all indicate that the d electrons in the iron transition series metals are largely conduction electrons, in bands an electron volt wide. Whereas in the rare earths, on the contrary, the magnetic f shells are *so* well localized (~ 0.3 Å) that the overlap between them (at a distance of approximately 3 Å) must be negligibly small, and there is therefore *no* Heisenberg nearest-neighbor exchange, to a good approximation. The observed magnetism in such a case must involve the *s*-band conduction electrons, which alone are capable of sustaining correlations over several interatomic distances: "indirect exchange."

For these and other reasons,[1] it is not possible to ignore the electrons' itineracy in any sensible theory of magnetism in metals.[2] In the present chapter we start with an elementary review of the band theory in the one-electron approximation with emphasis on *tight-binding*, the simplest approximation of any value in the investigation of magnetic properties. Proceeding from there we shall see what gives rise to strong magnetic properties such as ferromagnetic or antiferromagnetic (AF) behavior. We shall even show that the Heisenberg Hamiltonian for insulators can be derived on the basis of a band picture entirely analogous to the band theory of metals. We shall also derive the theory of magnons in metals for the various models considered.

The case of a half-filled band is easiest. To the intratomic mechanisms (intra-atomic Coulomb repulsion, exchange corrections thereto, etc.) we can add the inter-atomic "transfer" matrix elements responsible for the motion of the electrons through the solid. One or the other of these mechanisms can be mathematically turned off, to recover the well-known atomic or free electron limits.

On the other hand, the case of partially-filled (low density) magnetic bands is far more intriguing. The electrons repel, and are not constrained by density to occupy the same atomic sites. Turning off the transfer matrix elements would result in an inhomogeneous localized array. We shall treat this case by a multiple scattering formalism that is exact in the dilute limit, and yields interesting criteria for the occurrence of magnetic moments in metals. We study it for clues to the eventual ferromagnetism or AF of the electron fluid.

This chapter affords a preliminary treatment of the magnetic impurity problem: how does an impurity atom acquire a magnetic moment, and how do these moments interact with one another without additional mathematics? In a later chapter we expand on this topic.

---

[1] See the indictment of the Heisenberg and Heitler-London theories in J. C. Slater, *Rev. Mod. Phys.* **25**, 199 (1953).

[2] For a comprehensive treatment see C. Herring, "Exchange Interactions among Intinerant Electrons," in *Magnetism*, G. Rado and H. Suhl, eds., Academic Press, New York, 1966; T. Moriya, *J. Magn. Magn. Mat.* **14**, 1 (1979).

6.1.   Bloch and Wannier States

Consider the one-electron Hamiltonian in a lattice where the atoms are located at fixed, regular points $a(n_i, m_i, l_i) \equiv \mathbf{R}_i$:

$$\mathcal{H} = \frac{\mathbf{p}^2}{2m} + \sum_i V(\mathbf{r} - \mathbf{R}_i) \tag{6.1}$$

in which $V(\mathbf{r} - \mathbf{R}_i)$ is the averaged potential due to the nucleus and all other electrons except the one under consideration. In the simple cubic structure, lattice spacing a, the translation $\mathbf{r} \to \mathbf{r} + a(n_1, n_2, n_3)$ leaves $\mathcal{H}$ invariant, and so the translation operator can be used to provide the eigenfunctions of $\mathcal{H}$ with an important quantum number, the crystal momentum $\mathbf{k}$. In other lattices, the translations $\mathbf{R}_\alpha$ take a different form, but a crystal momentum can always be defined and, together with the band index t, it provides the quantum numbers for the *Bloch functions*,

$$\psi_{t,\mathbf{k}}(\mathbf{r}) = e^{i\mathbf{k}\cdot\mathbf{r}} u_{t,\mathbf{k}}(\mathbf{r}) \tag{6.2}$$

which are the eigenfunctions of $\mathcal{H}$. The function $u_{t,\mathbf{k}}(\mathbf{r})$ has periodicity of the lattice and satisfies the eigenvalue equation,

$$e^{-i\mathbf{k}\cdot\mathbf{r}} \mathcal{H} e^{+i\mathbf{k}\cdot\mathbf{r}} u_{t,\mathbf{k}}(\mathbf{r}) = \left[\frac{(\mathbf{p} + \hbar\mathbf{k})^2}{2m} + \sum_i V(\mathbf{r} - \mathbf{R}_i) \right] u_{t,\mathbf{k}}(\mathbf{r})$$

$$= E_t(\mathbf{k}) u_{t,\mathbf{k}}(\mathbf{r}) \tag{6.3}$$

subject to the boundary condition $u_{t,\mathbf{k}}(\mathbf{r} + \mathbf{R}_\alpha) = u_{t,\mathbf{k}}(\mathbf{r})$, with $\mathbf{R}_\alpha \equiv a$ translation vector of the lattice (see below), denoted $\boldsymbol{\delta}$ sometimes in other contexts.

The meaning of the band index t is best understood in connection with the Fourier transform of the Bloch functions, viz., the *Wannier functions*

$$\psi_{t,i}(\mathbf{r}) = \frac{1}{\sqrt{N}} \sum_\mathbf{k} e^{-i\mathbf{k}\cdot\mathbf{R}_i} \psi_{t,\mathbf{k}}(\mathbf{r})$$

$$= \psi_t(\mathbf{r} - \mathbf{R}_i) \tag{6.4}$$

which like the Bloch functions, form a complete, orthonormal set of functions in the Hilbert space of the Hamiltonian, (6.1). The sum over $\mathbf{k}$ is restricted to the *first Brillouin zone*, i.e., to N values of $\mathbf{k}$ in the region

$$|\mathbf{k}\cdot\mathbf{R}_\alpha| < \pi$$

where $\mathbf{R}_\alpha$ = any one of the smallest translation vectors of the lattice (primitive translation vectors). In the limit of infinite interatomic separation, the Wannier functions reduce to ordinary atomic orbitals. In that limit, i identifies the atom, and t the set of atomic quantum numbers (principal, orbital, azimuthal, spin; the use of a single index is for typographical simplicity). When atoms are brought close together, the atomic levels identified by t broaden into a band, unless, as in the f shell of the rare earths and the $1s$ helium core common to all metallic atoms, the electrons are still so tightly bound to the nucleus at the observed interatomic separation that the very concept of one-electron bands remains inapplicable. But this is not the case of the $3d$ states, and we note that the first metal to have a filled $3d$ band (Cu), and the elements immediately following it in the periodic table (Zn, Ga, etc.) are nonmagnetic; whereas the iron series just preceding these, noted for the unfilled d shell in the atom and d band in the metal, form materials with varied and interesting magnetic properties. One may rightly suspect the d-band electrons of being particularly important in the study of magnetism, and the unfilled d band of containing "magnetically active" electrons.

Before making these notions more precise it is necessary to review some of the properties shared by *all* the electrons, including nonmagnetic ones. Some of these can be studied in the "plane wave approximation," in which we set $V(\mathbf{r} - \mathbf{R}_i) = 0$ and $u_{t,\mathbf{k}} = 1$. But we do not wish to sacrifice the band structure, the qualitative features of which are retained in the tight-binding approximation which we study next.

6.2.   Tight-Binding

The basic premise of this theory is that it is easier to estimate matrix elements involving Wannier functions (because of their supposed localization about specified atoms) than to solve the differential equations for the Bloch functions. We illustrate this, using Hamiltonian $\mathscr{H}$ defined in (6.1), and form the Wannier matrix elements within the nth band,

$$H(\mathbf{R}_{ij})_{n,n} \equiv \int \psi_{n,i}^*(\mathbf{r})\mathscr{H}\,\psi_{n,j}(\mathbf{r})\,d^3r \qquad (6.5)$$

so that Schrödinger's equation reduces to the determinantal eigenvalue problem,

$$\det\|H(\mathbf{R}_{ij})_{n,n} - E\delta_{ij}\| = 0\,. \qquad (6.6)$$

Evidently it does not matter in which representation we solve Schrödinger's equation and the eigenvalues E will coincide exactly with the Bloch energies $E_n(\mathbf{k})$. Moreover, the eigenfunctions necessarily turn out to be precisely the proper linear combination,

$$\psi_{n,\mathbf{k}}(\mathbf{r}) = \frac{1}{\sqrt{N}} \sum_{i=1}^{N} e^{i\mathbf{k}\cdot\mathbf{R}_i} \psi_{n,i}(\mathbf{r}) \tag{6.7}$$

which make up the Bloch functions, (6.4).

If one uses *approximate* Wannier functions instead, the *interband* $(n \neq m)$ matrix elements $H(\mathbf{R}_{ij})_{n,m}$ need not vanish. It is common practice to use atomic orbitals instead of Wannier orbitals as a first approximation, and therefore this method is often known as the LCAO method, acronym of *linear combination of atomic orbitals*. The mixing of orbitals of different angular momenta to form the bands in the solid expresses the well-known fact that angular momentum is "quenched" (not conserved) due to conflict with the translational and point-group symmetries.

It is standard practice to limit the matrix elements $H(\mathbf{R}_{ij})_{n,m}$ to nearest-neighboring (in extreme cases, perhaps as far as second- and third-nearest-neighboring) atomic distances $\mathbf{R}_{ij}$. It is not sensible to consider more distant interactions, for if they become important the tight-binding procedure itself becomes unwieldy, and other methods such as the quasifree electron approximation are then simpler and more appropriate.

Consider the band structures derived in the following simple examples. The determinantal equation yields the energy at all points in $\mathbf{k}$-space, with only the constant parameters (overlap integrals) required to be numerically calculated. And if we do not know the atomic orbitals, nor trust them in the particular crystal under consideration, these constants may be taken as adjustable parameters to be fitted either by experiment or by comparison with a few calculated points given by more accurate band structure calculations.

As our first example, *assume a simple cubic structure* and consider:

s bands: By symmetry, the matrix elements to the six nearest neighbors are all equal, so that only two parameters enter the problem:

$$A \equiv H(\mathbf{0}) = \int \psi_i^*(\mathbf{r}) \mathcal{H} \psi_i(\mathbf{r}) \, d^3r$$

and

$$-B \equiv H(0,0,a) = \cdots = H(a,0,0)$$

$$= \int \psi_i^* (|\mathbf{r} + (0,0,a)|) \mathscr{H} \psi_i(\mathbf{r}) \, d^3r \,.$$

In terms of these, the energy eigenvalues are

$$E(\mathbf{k}) = A - 2B(\cos k_x a + \cos k_y a + \cos k_z a) \,. \tag{6.8}$$

Problem 6.1.   (a) Assuming nearest-neighbor overlap, prove that in the body-centered cubic structure the s bands have the form

$$E(\mathbf{k}) = A - B \cos k_x a \cos k_y a \cos k_z a$$

and that in the face-centered cubic structure the appropriate formula is

$$E(\mathbf{k}) = A - B(\cos k_x a \cos k_y a + \cdots + \cos k_y a \cos k_z a)$$

(b) Derive the s-band structure in the hexagonal close-packed lattice.

---

p *bands*: In cubic structures, with nearest-neighbor interactions only, the threefold degeneracy of the atomic orbitals is not lifted. We write the orbitals as $\psi_n(\mathbf{r})$, $n = 1$, 2 or 3:

$$\psi_1(\mathbf{r}) = x\phi(r)\,, \quad \psi_2(\mathbf{r}) = y\phi(r) \quad \text{and} \quad \psi_3(\mathbf{r}) = z\phi(r)$$

using $l = 1$ functions from Table 3.1. A first parameter,

$$A \equiv H(0)_{n,n} = \int x\phi^*(\mathbf{r})\mathscr{H} x\phi(\mathbf{r}) \, d^3r$$

is the same for all three bands. A second parameter,

$$-B = \int x\phi^*(|\mathbf{r} + (0,0,a)|) \mathscr{H} x\phi(\mathbf{r}) \, d^3r$$

$$= \int y\phi^*(|\mathbf{r} + (0,0,a)|) \mathscr{H} y\phi(\mathbf{r}) \, d^3r$$

and finally a third one

$$-C = \int (z + a)\phi^*(|\mathbf{r} + (0,0,a)|) \mathscr{H} z\phi(\mathbf{r}) \, d^3r$$

are required. All other integrals may be obtained from the above, except those for $m \neq n$. We set these $= 0$ by symmetry. When the Hamiltonian eigenvalue equation (6.6) is finally solved, we find three degenerate bands at $l = 1$:

$$E_1(\mathbf{k}) = A - 2B(\cos k_x a + \cos k_y a) - 2C(\cos k_z a) \tag{6.9a}$$

$$E_2(\mathbf{k}) = A - 2B(\cos k_y a + \cos k_z a) - 2C(\cos k_x a) \quad \text{and} \tag{6.9b}$$

$$E_3(\mathbf{k}) = A - 2B(\cos k_z a + \cos k_x a) - 2C(\cos k_y a). \tag{6.9c}$$

The three p bands are related to one another by those rotations in $\mathbf{k}$-space, which permute the Cartesian components $k_{x,y,z}$. In each band, the symmetry of contours of constant energy is lower than cubic and even at small $\mathbf{k}$, these contours are not spherical but are ellipsoids of revolution with principal non-equivalent axes along the $k_{x,y,z}$ directions.

Had we chosen the axes of quantization of the three p functions along other than a crystal axis, the interband matrix elements $H(0,0,a)_{n,m}$ for $n \neq m$ would not have vanished so conveniently. The eigenvalue equation, (6.6), which held for the exact Wannier functions, would have to be replaced in the general case by the r-dimensional equation

$$\det\|H(\mathbf{k})_{n,m} - E\delta_{n,m}\| = 0, \tag{6.10}$$

where $r = $ number of interacting bands; $n, m = 1, 2, \ldots, r$. Here we have taken advantage of translational invariance to form the Fourier transforms, and to define

$$(n/m) \equiv H(\mathbf{k})_{n,m}$$

$$= \frac{1}{N} \sum_{i,j} e^{i\mathbf{k} \cdot \mathbf{R}_{ij}} \int \psi_{n,i}^* \mathcal{H} \psi_{m,j} \, d^3 r. \tag{6.11}$$

Table 6.1 reproduces results by Slater and Koster, who have calculated these matrix elements out to third nearest neighbor in the simple cubic structure. This is sufficient also for obtaining nearest-neighbor interactions in face- and body-centered cubic structures (and with some manipulations, next nearest neighbors also, but we shall not be interested in these). Note that $H(\mathbf{k})_{n,m}$ is abbreviated (n/m) in Table 6.1 for typographical simplicity. The matrix elements $H(\mathbf{k})_{n,m}$ are given, abbreviated as (n/m), among s-like functions (s); p-like functions $(x, y, \text{ and } z)$; and d-like functions $(xy, xz, yz, 3z^2 - r^2, \text{ and } x^2 - y^2)$.

It is sensible to consider these matrix elements as empirical constants; they can be estimated by performing the appropriate two-center integrals.

Table 6.1. Matrix elements of (6.10) and (6.11) in tight-binding (LCAO) approximation.[3]

(s/s)	$H(000)_{s,s} + 2H(100)_{s,s}(X + Y + Z)$
	$\quad + 4H(110)_{s,s}(XY + XZ + YZ) + 8H(111)_{s,s}XYZ$
(s/x)	$2iH(100)_{s,x}\tilde{X} + 4iH(110)_{s,x}\tilde{X}(Y + Z) + 8iH(111)_{s,x}\tilde{X}YZ$
(s/xy)	$-4H(110)_{s,xy}\tilde{X}\tilde{Y} - 8H(111)_{s,xy}\tilde{X}\tilde{Y}Z$
$(s/x^2 - y^2)$	$\sqrt{3}H(001)_{s,3z^2-r^2}(X - Y) + 2\sqrt{3}H(110)_{s,3z^2-r^2}(Y - X)Z$
$(s/3z^2 - r^2)$	$H(001)_{s,3z^2-r^2}(2Z - X - Y)$
	$\quad - 2H(110)_{s,3z^2-r^2}(-2XY + XZ + YZ)$
(x/x)	$H(000)_{x,x} + 2H(100)_{x,x}X + 2H(100)_{y,y}(Y + Z)$
	$\quad + 4H(110)_{x,x}X(Y + Z) + 4H(011)_{x,x}YZ$
	$\quad + 8H(111)_{x,x}XYZ$
(x/y)	$-4H(110)_{x,y}\tilde{X}\tilde{Y} - 8H(111)_{x,y}\tilde{X}\tilde{Y}Z$
(x/xy)	$2iH(010)_{x,xy}\tilde{Y} + 4iH(110)_{x,xy}X\tilde{Y} + 4iH(011)_{x,xy}\tilde{Y}Z$
	$\quad + 8iH(111)_{x,xy}\tilde{Y}XZ$
(x/yz)	$-8iH(111)_{x,yz}\tilde{X}\tilde{Y}\tilde{Z}$
$(x/x^2 - y^2)$	$\sqrt{3}iH(001)_{z,3z^2-r^2}\tilde{X} + 2\sqrt{3}iH(011)_{z,3z^2-r^2}(\tilde{X}Y + \tilde{X}Z)$
	$\quad + 2iH(011)_{z,x^2-y^2}\tilde{X}(Y - Z) + 8iH(111)_{x,x^2-y^2}\tilde{X}YZ$
$(x/3z^2 - r^2)$	$-iH(001)_{z,3z^2-r^2}\tilde{X} - 2iH(011)_{z,3z^2-r^2}\tilde{X}(Y + Z)$
	$\quad + 2\sqrt{3}iH(011)_{z,x^2-y^2}\tilde{X}(Y - Z)$
	$\quad - (8/\sqrt{3})H(111)_{x,x^2-y^2}\tilde{X}YZ$
$(z/3z^2 - r^2)$	$2iH(001)_{z,3z^2-r^2}\tilde{Z} + 4iH(011)_{z,3z^2-r^2}\tilde{Z}(X + Y)$
	$\quad + (16/\sqrt{3})iH(111)_{x,x^2-y^2}XY\tilde{Z}$
(xy/xy)	$H(000)_{xy,xy} + 2H(100)_{xy,xy}(X + Y) + 2H(001)_{xy,xy}Z$
	$\quad + 4H(110)_{xy,xy}XY + 4H(011)_{xy,xy}(X + Y)Z$
	$\quad + 8H(111)_{xy,xy}XYZ$
(xy/xz)	$-4H(011)_{xy,xz}\tilde{Y}\tilde{Z} - 8H(111)_{xy,xz}X\tilde{Y}\tilde{Z}$
$(xy/x^2 - y^2)$	Zero

---

[3] J. Slater and G. Koster, *Phys. Rev.* **94**, 1498 (1954).

Table 6.1.   (*Continued*)

$(xy/3z^2 - r^2)$	$-4H(110)_{xy,3z^2-r^2}\tilde{X}\tilde{Y} - 8H(111)_{xy,3z^2-r^2}\tilde{X}\tilde{Y}Z$
$(xz/x^2 - y^2)$	$2\sqrt{3}H(110)_{xy,3z^2-r^2}\tilde{X}\tilde{Z} + 4\sqrt{3}H(111)_{xy,3z^2-r^2}\tilde{X}Y\tilde{Z}$
$(xz/3z^2 - r^2)$	$2H(110)_{xy,3z^2-r^2}\tilde{X}\tilde{Z} + 4H(111)_{xy,3z^2-r^2}\tilde{X}\tilde{Z}Y$

$(x^2 - y^2/x^2 - y^2) \quad H(000)^{\mathrm{a}} + \dfrac{3}{2}H(001)^{\mathrm{a}}(X+Y)$

$$+ 2H(001)^{\mathrm{b}}\left(\frac{1}{4}X + \frac{1}{4}Y + Z\right) + 3H(110)^{\mathrm{a}}(X+Y)Z$$

$$+ 4H(110)^{\mathrm{b}}\left(XY + \frac{1}{4}XZ + \frac{1}{4}YZ\right) + 8H(111)^{\mathrm{a}}XYZ$$

$(3z^2 - r^2/3z^2 - r^2) \quad H(000)^{\mathrm{a}} + 2H(001)^{\mathrm{a}}\left(\frac{1}{4}X + \frac{1}{4}Y + Z\right)$

$$+ \frac{3}{2}H(001)^{\mathrm{b}}(X+Y)$$

$$+ 4H(110)^{\mathrm{a}}\left(XY + \frac{1}{4}XZ + \frac{1}{4}YZ\right)$$

$$+ 3H(110)^{\mathrm{b}}(XZ + YZ) + 8H(111)^{\mathrm{a}}XYZ$$

$(x^2 - y^2/3z^2 - r^2) \quad \dfrac{1}{2}\sqrt{3}H(001)^{\mathrm{a}}(-X+Y) - \dfrac{1}{2}\sqrt{3}H(001)^{\mathrm{b}}(-X+Y)$

$$+ \sqrt{3}H(110)^{\mathrm{a}}(X-Y)Z - \sqrt{3}H(110)^{\mathrm{b}}(X-Y)Z$$

---

[a] $H(LMN)_{3z^2-r^2,3z^2-r^2}$.

[b] $H(LMN)_{x^2-y^2,x^2-y^2}$.

Note: Assuming nearest-neighbor interactions only, in the simple cubic structure retain only terms in (100), (010), and (001). For face-centered cubic retain only (110), (011), and (101). For bodycentered cubic retain only (111).

$$Key: \quad X = \cos k_x a \qquad \tilde{X} = \sin k_x a$$
$$Y = \cos k_y a \qquad \tilde{Y} = \sin k_y a$$
$$Z = \cos k_z a \qquad \tilde{Z} = \sin k_z a$$

The band parameter constants are the integrals

$$H(LMN)_{m,n} = \int \psi_m^*[\mathbf{r}_m + a(L,M,N)]\mathscr{H}\psi_n(\mathbf{r})\, d^3r$$

Let us consider five d bands in the simple cubic structure, ignoring s and p bands as well as non-nearest-neighbor interactions. Looking up Table 6.1, we extract the following special case of the eigenvalue equation, (6.10):

$$
\det
\begin{array}{c}
\\ (xy) \\ (xz) \\ (yz) \\ \\ x^2-y^2 \\ 3z^2-r^2
\end{array}
\left\|
\begin{array}{ccc:cc}
(xy) & (xz) & (yz) & x^2-y^2 & 3z^2-r^2 \\[4pt]
F_1(k)-E & 0 & 0 & 0 & 0 \\
0 & F_2(k)-E & 0 & 0 & 0 \\
0 & 0 & F_3(k)-E & 0 & 0 \\ \hdashline
0 & 0 & 0 & F_4(k)-E & V(k) \\
0 & 0 & 0 & V(k) & F_5(k)-E
\end{array}
\right\| = 0
$$

$$(6.12)$$

The definitions of the F''s and $V(k)$, and some features of the solutions, are examined in Problem 6.2.

Problem 6.2.   (a) In the example of (6.12) in the text, find $F(k)$ and $V(k)$ by referring to Table 6.1. Show that the solutions of the eigenvalue equation describe three degenerate d bands with ellipsoidal contours of constant energy much like the p bands of (6.9) and two nondegenerate s-like bands. Obtain the contours of energy of the latter near $k = 0$. Plot the energy as a function of k along the three principal directions, (100), (110), and (111).

(b) Calculate the first-order effect of an infinitesimal next-nearest-neighbor interaction on the band structures calculated in part (a).

Making some assumptions about the relative and absolute magnitudes of the band-structure parameters, Slater and Koster calculated the histogram *density of states* curve (number of eigenvalues per unit energy) which includes all five d bands in the body-centered cubic structure, such as in Fe. This is reproduced in Fig. 6.1. The lower peak belongs to the *bonding orbitals* in the chemical terminology, and, according to Slater and Koster, explains the anomalously great binding energies of some metals in the first half of the iron transition series.

What the histogram could not show are the so-called *Van Hove singularities*. Whenever an $E(\mathbf{k})$ curve has a minimum, maximum, or simply a saddle point, its contribution to the density of states becomes excessive. [For example, a totally constant $E(\mathbf{k}) = E_0$ contributes a delta-function singularity to the density of states function.] Some years after the Slater-Koster work,

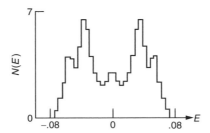

Fig. 6.1. Original tight-binding density-of-states histogram for d bands in bcc structure. $N(E)$ is plotted E (Rydbergs), and is equal to the number of eigenvalues $E_n(k)$ within ± 1.025 Ry of the energy E, normalized such that the total area under the curve equals 5.[3]

Wohlfarth and Conwell published the curve reproduced in Fig. 6.2, giving the density of states

$$N(E) \propto \sum_{n,\mathbf{k}} \delta[E - E_n(\mathbf{k})] \tag{6.13}$$

as the output of a computer calculation in which the Van Hove singularities were scrupulously preserved. The peaky nature of this new curve is evident, with the principal maxima occurring whenever an energy band touches the Brillouin zone. Note also the revision upwards of the estimated effective width of the bands over the earlier work, by a factor of approximately 3.

As we shall see subsequently, it is of crucial importance in the theory of magnetism in metals[4] whether the density of states is high or low, particularly in the vicinity of the Fermi level μ.

6.3. Weak Magnetic Properties

All metals share some weak magnetic characteristics. The study of these has enjoyed great vogue, because of the very detailed information on the band structure and Fermi surface parameters which it divulges. As some examples of important dynamic properties, we list the Hall effect, magneto-

---

[4] The band structure of some magnetic metals such as nickel, is now known fairly well. See H. Ehrenreich, H. Phillipp and D. Olechna, *Phys. Rev.* **131**, 2469 (1963); J. C. Phillips, *Phys. Rev.* **133**, A1020 (1964); L. F. Mattheiss, *Phys. Rev.* **134**, A970 (1964); C. Wang and J. Callaway, *Phys. Rev.* **B15**, 298 (1977); D. R. Penn, *Phys. Rev. Lett.* **42**, 921 (1979); L. Kleinman *et al.*, *Phys. Rev.* **B22**, 1105 (1980). The most up-to-date book at the present time is by R. M. Martin, *Electronic Structure*, Cambridge Univ. Press, 2004.

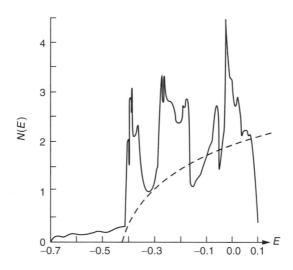

Fig. 6.2. More accurate density-of-states curve $N(E)$ versus E (Rydbergs) for bcc iron. Dashed line is average free-electron approximation $N(E) \propto \sqrt{E}$; and note how at several points (for example, $E = 0$ or -0.4) the computed density of states can exceed the average curve by a large factor.[5]

resistance, and cyclotron resonance. Some important static properties of the electrons include Pauli spin paramagnetism, Landau diamagnetism, and the De Haas–Van Alphen effect. All these are well known and abundantly discussed in standard texts on solid-state or metal physics.[6] Therefore, here we shall be content with a qualitative discussion of the physical basis of the static phenomena listed, without emphasizing the mathematics which can become rather complicated. The principal purpose is to display the important role of the density of states function $N(E)$ in the magnetic properties of electrons, and to introduce the concepts of Fermi energy and Fermi distribution.

In the ground state of a normal metal ($T = 0$ K) all the one-electron states of energy less than μ are occupied, all those above are empty. μ is the chemical potential or Fermi energy. Moreover, in the absence of magnetic fields or spin-orbit coupling to lift the Kramers' degeneracy, every state of given $(n, \mathbf{k})$ below the Fermi level is *doubly* occupied, by an electron with spin up and one with spin down.

[5] E. Wohlfarth and J. Cornwell, *Phys. Rev. Lett.* **7**, 342 (1961).

[6] Less known, but also interesting, is the subject of high fields in metals, treated by Fawcett in E. Fawcett, *Adv. Phys.* (*Philos. Mag. Suppl.*) **13**, 139 (1964).

At finite temperature, states within $\pm kT$ of the Fermi level are partly occupied, as may be seen from the Fermi distribution function

$$f(E_\mathbf{k}) = \frac{1}{e^{(E_\mathbf{k}-\mu)/kT} + 1} \tag{6.14}$$

which gives the thermal-average probability that the state of energy $E_\mathbf{k}$ is occupied (absorbing band and spin indices into $\mathbf{k}$). In a weak magnetic field, the spins of the electrons within the $\pm kT$ neighborhood of the Fermi energy will be free to orient themselves parallel to the field; and according to the laws of Langevin and Curie, each will contribute a magnetization proportional to the applied field, to Curie's constant C, and to the inverse temperature, viz.,

$$\delta\mathcal{M} \sim H\frac{C}{T}.$$

The number of participating electrons is $\sim 2kTN(\mu)$, and therefore the total paramagnetic spin susceptibility is

$$\mathcal{M}/H \equiv \chi_\mathrm{p} = 2CkN(\mu). \tag{6.15}$$

This agrees with more rigorous derivations of Pauli's spin paramagnetism of free electrons and is correct to $O[(kT/\mu)^2]$ at finite temperatures. ($kT/\mu \ll 10^{-2}$ at room temperature for most nonmagnetic metals.) Note the dependence on $N(\mu)$, the density of states *at* the Fermi energy. This susceptibility is smaller by a factor $2kTN(\mu)/\mathcal{N}$ than that of $\mathcal{N}$ free spins.

The De Haas–Van Alphen effect is not so easy to explain nor to understand; nevertheless it also reflects the dependence of the thermodynamic properties of the metal on the density of states. In this case, the *density of states is affected* by a magnetic field, and therefore the thermodynamic functions will depend on the field. For illustrative purposes, consider the wavefunctions in the free-electron approximation,

$$\psi_\mathbf{k} = e^{i\mathbf{k}\cdot\mathbf{r}}, \quad \text{where } \mathcal{H} = \frac{\mathbf{p}^2}{2m^*} = -\frac{(\hbar\nabla)^2}{2m^*}. \tag{6.16}$$

The effective mass m^* may differ from the free-electron mass $m_0 = 9.1 \times 10^{-28}$ gram by one or more orders of magnitude (greater or smaller). The effective mass approximation used here may be quite successful for describing s bands, but it does not lead to a realistic density of states for the d bands, as seen in Fig. 6.2; so the following derivation is merely illustrative.

In a weak electromagnetic field, described by the vector potential $\mathbf{A}(\mathbf{r}, t)$, the electron momenta $\mathbf{p}$ become $\mathbf{p} - e\mathbf{A}/c$, and the time-dependent Schrödinger equation becomes

$$\frac{[\mathbf{p} - (e/c)\mathbf{A}]^2}{2m^*}\psi(\mathbf{r}, t) = \hbar i \frac{\partial}{\partial t}\psi(\mathbf{r}, t). \tag{6.17}$$

For a static magnetic field, $\mathbf{A}(\mathbf{r}) = (0, Hx, 0)$ does not depend on the time and satisfies the two equations

$$\nabla \times \mathbf{A} = (0, 0, H) \quad \text{and} \quad -\frac{1}{c}\frac{\partial \mathbf{A}}{\partial t} = \mathscr{E}(\mathbf{r}, t) = 0. \tag{6.18}$$

Therefore, writing $\psi(\mathbf{r}, t) = \exp[i(k_y y + k_z z)]\phi(x)\exp(-iEt/\hbar)$, we find that $\phi(x)$ obeys a harmonic-oscillator equation, so that the total energy is given by

$$E = \frac{\hbar^2 k_z^2}{2m^*} + \left(n + \frac{1}{2}\right)\hbar\omega_c \quad n = 0, 1, 2, \ldots \tag{6.19}$$

with the "cyclotron frequency" ω_c defined by

$$\omega_c = \frac{eH}{m^*c}. \tag{6.20}$$

Problem 6.3.   Derive (6.19, 6.20) of the text, by solving Schrödinger's equation in the manner described.

The expression for the energy may be interpreted as the result of quantization on the classical circular motion of a charge in a magnetic field.

The density of states is obtained by differentiating the function which gives the total number of states lying below energy E. Thus, with a factor 2 for the two values of spin

$$N(E) \propto 2\frac{d}{dE}\sum_{m=1}^{M(E)}\int_0^{\sqrt{E-(m+1/2)\hbar\omega_c}} \hbar\, dk_z (2m^*)^{-1/2}$$

$$\propto \sum_{m=1}^{M(E)}\frac{1}{\sqrt{E - (m + 1/2)\hbar\omega_c}} \tag{6.21}$$

where $M(E) = $ largest positive integer for which the radicand is positive. A plot of this function is given in Fig. 6.3a; it is similar to the free-electron

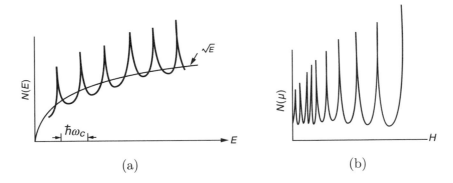

(a)   (b)

Fig. 6.3. (a) Density-of-states $N(E)$ versus E in constant magnetic field. Averaging the singularities would *lower* curve *below* its zero-field value $\sqrt{E}$. Therefore the energy of a constant number of electrons in weak magnetic fields is *higher* than in zero field, and the Landau motional susceptibility is diamagnetic. (b) Density of states at the Fermi energy, $N(\mu)$, as function of strong applied magnetic field. The oscillatory behavior results in De Haas–Van Alphen effect. Note that μ is itself a function of H, determined by the requirement $\mathscr{N} = \int^{\mu} dE N(E) = \text{constant}$.

function $N(E) \sim E^{1/2}$ except for narrow, integrable, square-root singularities at half-odd-integer multiples of the cyclotron energy $\hbar\omega_c$.

The most interesting behavior occurs in the neighborhood of the Fermi energy μ as the magnitude of H is increased. Whenever $(\mu - \frac{1}{2}\hbar\omega_c)$ becomes an integer multiple of $\hbar\omega_c$, the above sums acquire a new integer $M(\mu)$, and the density of states $N(\mu)$ itself acquires a square-root singularity. This is shown in Fig. 6.3b. Clearly, this must lead to fluctuating, quasiperiodic behavior of all the thermodynamic properties of the metal: the specific heat, magnetic susceptibility, electrical resistance, etc., must all be oscillatory functions of the maximum integer, $M(\mu) \propto 1/H$.

The weak-field limit occurs either when $\hbar\omega_c \ll kT$ or when scattering results in a mean free path smaller than the radius of the cyclotron orbit. In either case it is permissible to expand the free energy and other thermodynamic quantities in powers of H.

A useful method due to Peierls[7] is based on the Poisson summation formula for evaluating (6.21):

$$N(E) \propto \sum_{p=-\infty}^{+\infty} (-1)^p \int_0^{E/\hbar\omega_c} \frac{e^{2\pi i p x}}{\sqrt{E - x\hbar\omega_c}}\, dx$$

---

[7] R. E. Peierls, *Quantum Theory of Solids*, Oxford Univ. Press, Oxford, 1955, p. 148.

and some partial integrations to evaluate the leading terms in the free energy.[8] As an example, the internal energy to leading order, as calculated using this formula is:

$$E_{\text{tot}}(H) = \int dE N(E) E f(E) \cong E_{\text{tot}}(0) + \frac{1}{2}\chi_{\text{d}}H^2 \qquad (6.22)$$

with χ_{d} a positive quantity. Because the increase in energy results in a force tending to repel the material from an applied field, this is a diamagnetic susceptibility.

Finally, the total susceptibility, combining the Pauli spin paramagnetism with the Landau orbital diamagnetism can be shown to have the value

$$\chi = 2CkN(\mu)\left[1 - \frac{1}{3}\left(\frac{m_0}{m^*}\right)^2\right] \qquad (6.23)$$

which explains the weak net paramagnetism of most metals (where $m^* \sim m_0$) and, on the other hand, the strong diamagnetism of bismuth, in which $m^* = O(0.01m_0)$, $(m_0/m^*)^2 = O(10^4)$.

In common with other magnetic phenomena studied in this book, the Landau diamagnetism is a purely quantum mechanical effect, which disappears in the correspondence limit by virtue of the oft-invoked Bohr-Van Leeuwen theorem. Also the diamagnetic increase in energy, (6.22), is extensive, i.e., every unit volume of the material contributes equally to the diamagnetism current density. In crossed electric and magnetic fields the conductivity is a tensor, the components of which yield magneto-resistance and the Hall effect.[9] In 2D the latter becomes the "quantum" Hall effect.[10]

Note: It is possible to view the nonvanishing diamagnetism as a direct consequence of the uncertainty principle; for if the electrons have perfectly sharp momenta $\mathbf{p}$, the vector potential $\mathbf{A} = (0, Hx, 0)$ cannot be simultaneously specified nor removed by a gauge transformation and vice versa. Therefore for small H, the energy is raised above the ground state value it had in the absence of the field. This point of view has been implemented by Van Vleck[11] in a review of the weak magnetic properties of metals and of the effects of exchange and correlations, through an expansion of the free

---

[8] M. Glasser, *J. Math. Phys.* **5**, 1150 (1964); *Erratum: J. Math. Phys.* **7**, 1340 (1966); A. W. Saenz and R. O'Rourke, *Rev. Mod. Phys.* **27**, 381 (1955).
[9] A. Smith, J. Janak and R. Adler, *Electronic Conduction in Solids*, McGraw-Hill, New York, 1967.
[10] *The Quantum Hall Effect*, 2nd ed., R. E. Prange and S. M. Girvin, eds., Springer-Verlag, Berlin, 1990; *Quantum Hall Effect*, M. Stone, ed., World Scientific, 1992.
[11] J. H. Van Vleck, *Nuovo Cimento* **6** (Ser. X, Suppl. 3) 857 (1957); D. Kojima and A. Isihara, *Phys. Rev.* **B20**, 489 (1979).

energy in powers of $\hbar$. Many-body effects have been thoroughly explored by Isihara and Kojima.[11]

An inaccurate but frequently heard explanation of the quantum diamagnetism is that it is caused by the inability of surface currents to cancel volume currents, due to quantum mechanical effects. But this is only a half-truth; for it obscures the physically significant fact that χ_d depends on the bulk properties, and is independent of the surface geometry, boundary conditions, scattering, etc.

6.4. Exchange in Solids: Construction of a Model Hamiltonian

In solids as in atoms, the really strong magnetic phenomena are *electrostatic* in origin, the powerful Coulomb forces being "triggered" by the spins of the electrons under the regulation of the Pauli principle. This is well demonstrated in second quantization, introduced in Chapter 4. Specific applications to insulators and metals will be the subject of the remainder of the chapter. In second quantization, it is possible for a unique Hamiltonian to be applicable to all the various sorts of solids, with only numerical parameters and the occupation of the various bands remaining to be specified. Thus there is no need to deal differently with insulators or metals at the present stage.

First, let us reformulate the theory of noninteracting electrons, starting with the operator $c_{j,n,m}$ which destroys an electron at jth Wannier site, in the nth band, with spin index $m(=\uparrow$ or $\downarrow)$. The operator that creates an electron in precisely the same state is the Hermitean conjugate operator $c_{j,n,m}^{\dagger}$. The band Hamiltonian of (6.1) can be written in terms of these operators as

$$\mathscr{H}_0 = \sum_{i,i,n,m} H(\mathbf{R}_{ij})_{n,n} c_{i,n,m}^{\dagger} c_{j,n,m} \, . \tag{6.24}$$

This represents quite graphically the "hopping" or transfer of an electron from site j to site i, with the matrix element previously calculated in (6.5), and displays the conservation laws obeyed by true Wannier functions in the present (one-electron) approximation. These are the conservation of the spin index, and of the band index (which could be considered as an "isotopic" spin), ensuring that the one-electron bands are well defined.

The Fermion operators above obey the usual *anti*commutation relations

$$c_r c_s + c_s c_r \equiv \{c_r, c_s\} = 0 \qquad \{c_r^{\dagger}, c_s^{\dagger}\} = 0, \quad \text{therefore}$$
$$(c_r)^2 = (c_r^{\dagger})^2 = 0 \quad \text{and} \quad \{c_r, c_s^{\dagger}\} = \delta_{r,s} \tag{6.25}$$

and the occupation number operator $n_r = c_r^\dagger c_r$ has eigenvalues 0, 1 only; here, r or s stand for any *set* of quantum numbers, including the electron's spin, m.

The band Hamiltonian is diagonal in the Bloch representation. We show this by means of a canonical transformation, which in turn is equivalent to choosing the plane wave combination of Wannier operators, as follows:

$$c_{\mathbf{k},n,m} = \frac{1}{\sqrt{N}} \sum_{i=1}^{N} e^{-i\mathbf{k}\cdot\mathbf{R}_i} c_{i,n,m} \quad \text{and}$$

$$c_{\mathbf{k},n,m}^\dagger = \frac{1}{\sqrt{N}} \sum_{i=1}^{N} e^{+i\mathbf{k}\cdot\mathbf{R}_i} c_{i,n,m}^\dagger . \tag{6.26}$$

The reader can easily verify that the c_k's and $c_k^\dagger$'s also satisfy anticommutation relations, (6.25). The inverse linear combinations are simply

$$c_{i,n,m} = \frac{1}{\sqrt{N}} \sum_{\substack{\mathbf{k}\,\text{in}\\ \text{first BZ}}} e^{i\mathbf{k}\cdot\mathbf{R}_i} c_{\mathbf{k},n,m} \quad \text{and}$$

$$c_{i,n,m}^\dagger = \frac{1}{\sqrt{N}} \sum_{\substack{\mathbf{k}\,\text{in}\\ \text{first BZ}}} e^{-i\mathbf{k}\cdot\mathbf{R}_i} c_{\mathbf{k},n,m}^\dagger . \tag{6.26a}$$

Therefore let us substitute these expressions into $\mathcal{H}_0$, and obtain

$$\begin{aligned}
\mathcal{H}_0 &= \frac{1}{N} \sum_i e^{i(\mathbf{k}-\mathbf{k}')\cdot\mathbf{R}_j} \sum_i H(\mathbf{R}_{ij})_{n,n} e^{i\mathbf{k}\cdot\mathbf{R}_{ij}} \sum_{n,m} c_{\mathbf{k}',n,m}^\dagger c_{\mathbf{k},n,m} \\
&= \sum_{\mathbf{k},n,m} E_n(\mathbf{k}) c_{\mathbf{k},n,m}^\dagger c_{\mathbf{k},n,m} \\
&= \sum_{\mathbf{k},n,m} E_n(\mathbf{k}) n_{\mathbf{k},n,m}
\end{aligned} \tag{6.27}$$

using the definition of the energy of a Bloch electron $E_n(\mathbf{k})$ given previously.

Because $\mathcal{H}_0$ is diagonal in the Bloch operator representation (the eigenvalue of the number operator $n_{\mathbf{k},n,m}$ is $n_{\mathbf{k},n,m} = 0$ or 1), as eigenstate of this Hamiltonian can be specified merely by stating which $\mathbf{k}$, n, m are occupied, and which are not. For example, the no-particle *vacuum* state $|0)$ is annihilated by *every* $c_\mathbf{k}$: $c_\mathbf{k}|0) = 0$ therefore $n_\mathbf{k}|0) = 0$; and $\mathcal{H}_0$ must also have zero eigenvalue in this state. A more important eigenfunction is the *Fermi sea*, defined to be the state of lowest energy among all the eigenstates containing precisely $\mathcal{N}$ electrons. In terms of the Fermi energy μ (below which there are precisely $\mathcal{N}$ one-electron states $\mathbf{k}, n, m$) the Fermi sea can be written as

$$|F) \equiv \prod c_{\mathbf{k},n,m}^{\dagger}|0) \tag{6.28}$$

where the product extends over all $\mathbf{k}$, n, m for which $E_n(\mathbf{k}) < \mu$.

The eigenvalue of $\mathscr{H}_0$ in this state will be "unperturbed" ground state energy W_0,

$$W_0 = \sum E_n(\mathbf{k}) = \int_{-\infty}^{\mu} dE N(E) E \quad \text{where } \mathcal{N} \equiv \int_{-\infty}^{\mu} dE N(E) \tag{6.29}$$

and where the sum over $\mathbf{k}$, n, m again extends only over the states in the Fermi sea. It is the principal object of the theory of magnetism in metals to explain precisely how the electronic interactions modify the Fermi sea, and perturb the ground state energy.

One possible result of the interactions, and of thermal excitations as well, is to create any number of *elementary excitations*. These are constructed by removing a single electron from the Fermi sea and placing it above in one of the unoccupied states. For example, letting b stand for a set of labels $\mathbf{k}$, n, m *within* $|F)$, and a for a set of such labels *outside* $|F)$, the eigenfunction and eigenvalue of a single elementary excitation are

$$\psi_{ab}^{+} = c_a^{\dagger} c_b |F) \quad \text{and} \quad W_{ab} = W_0 + E_a - E_b. \tag{6.30}$$

As an alternative to the above description, we may conceive the elementary excitation of (6.30) as the creation of two *quasiparticles*: both a quasielectron of energy $E_a - \mu$, and a quasihole of energy $\mu - E_b$ are added to the ground state. The energy of each quasiparticle, and of the elementary excitation as well, must be positive (by definition of the *ground* state).

The elementary excitations occupy a continuum (in the limit $L \to \infty$, naturally) of energy levels even when restricted to a specific total momentum. In the free electron approximation $E(\mathbf{k}) = \mathbf{k}^2$ in some appropriate units, with the Fermi level at $\mu = k_F^2$, the Fermi sea is represented in $\mathbf{k}$ space as a sphere of radius k_F (Fig. 6.4), with total momentum, total current, total spin all zero. The elementary excitations are $c_{\mathbf{k}+\mathbf{q}}^{\dagger} c_{\mathbf{k}} |F)$, with $k < k_F$ and $|\mathbf{k}+\mathbf{q}| > k_F$, omitting spin indices. Even if $\mathbf{q}$ is fixed, there is a continuum of elementary excitations corresponding to the possible angles between $\mathbf{k}$ and $\mathbf{q}$, and the energy of these is bounded by two parabolas and the horizontal axis, as shown in Fig. 6.5. Brillouin zone and magnetic field effects on the spectrum of elementary excitations are discussed in Problem 6.4.

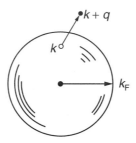

Fig. 6.4. Fermi sphere of radius k_F with an elementary excitation indicated: electron taken to $(\mathbf{k} + \mathbf{q})$, leaving hole at $\mathbf{k}$.

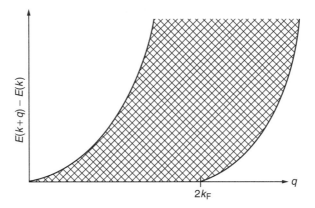

Fig. 6.5. Continuum of elementary excitations in nonmagnetic Fermi sea of *non-interacting* electrons.

Problem 6.4. (a) Discuss the double spectrum of elementary excitations + spin flip,

$$\psi_{\mathbf{k},\mathbf{q}}^{+} = c_{\mathbf{k}+\mathbf{q}\uparrow}^{\dagger} c_{\mathbf{k}\downarrow} |F_H) \quad \text{and} \quad c_{\mathbf{k}+\mathbf{q}\downarrow}^{\dagger} c_{\mathbf{k}\uparrow} |F_H)$$

for free electrons whose spins, only, are interacting with a magnetic field (e.g., an exchange field); that is, for which $E_m(\mathbf{k}) = \mathbf{k}^2 + m\mu_B H$, where $m = \pm 1$, and $|F_H)$ is the Fermi sea appropriate to this situation. Plot the continua in the manner of Fig. 6.5, and notice that since $|F_H)$ is the ground state, all excitation energies are required to be positive. Consider H both small, *and* large.

(b) Neglecting spin (as in the text above), show the effects of the Brillouin zone by plotting qualitatively the elementary excitation spectrum of a half-filled s band in the simple cubic structure. Pay special attention to the effects of *umklapp* ($\mathbf{k}$ is necessarily in first B.z. but $\mathbf{k} + \mathbf{q}$ is not), and to the maximum energy cutoff in the spectrum.

Besides all one-body potentials, the band-theoretic Hamiltonian $\mathscr{H}_0$ can include the averaged effects of two-body forces. Let us see why this is the case; the most general two-body matrix element is, in the Wannier representation,

$$\mathscr{H}' = \sum V(i,j,n',n;i',j',t',t)c^{\dagger}_{i't'm'}c^{\dagger}_{in'm}c_{jnm}c_{j'tm'} \qquad (6.31)$$

where

$$V(i,j,n',n;i',j',t',t)$$
$$= \frac{1}{2}\int d^3r \int d^3r' \psi^*_{t',i'}(\mathbf{r}')\psi_{t,j'}(\mathbf{r}')\frac{e^2}{|\mathbf{r}-\mathbf{r}'|}\psi^*_{n',i}(\mathbf{r})\psi_{n,j}(\mathbf{r}). \qquad (6.32)$$

These matrix elements will connect states that differ only by a change in the quantum numbers of two electrons, from jnm and $j'tm'$ to $in'm$ and $i't'm'$, and so long as we use orthogonal Wannier functions and two-body potentials (such as the physically important Coulomb repulsion), *there are no further matrix elements*, and *the total Hamiltonian consists of $\mathscr{H}_{\text{tot}} = \mathscr{H}_0 + \mathscr{H}'$.* As stated above, some of the terms in $\mathscr{H}'$ can be incorporated in $\mathscr{H}_0$; consider as one possible example the terms with $i' = j'$ and $t' = t$,

$$\left[\sum V(i,j,n',n;i',j',t,t)\mathsf{n}_{i'tm'}\right]c^{\dagger}_{in'm}c_{jnm}. \qquad (6.33)$$

Although the factor which multiplies $c^{\dagger}_{in'm}c_{jnm}$ is an operator, its average value in the Fermi sea serves as a useful estimate. Thus, we should incorporate into $\mathscr{H}_0$ the terms,

$$\delta\mathscr{H}_0 = \sum \delta H(\mathbf{R}_{ij})_{n,n'}c^{\dagger}_{in'm}c_{jnm}$$

with

$$\delta H(\mathbf{R}_{ij})_{n,n'} \equiv \sum_{i'tm'} V\cdot(F|\mathsf{n}_{i'tm'}|F) \qquad (6.34)$$

and subtract them from $\mathscr{H}'$. This is the sort of procedure which was already anticipated in Eqs. (6.1) and ff., when it was stated that $V(\mathbf{r} - \mathbf{R}_i)$ is the *averaged* potential due to the nucleus, and all other electrons except the one under consideration. Thus the transfer of all the averaged effects of $\mathscr{H}'$ into

$\mathscr{H}_0$ has the result that what remains of the former has vanishing expectation value in the Fermi sea

$$(F|\mathscr{H}' - (F|\mathscr{H}'|F)F) \equiv 0\,.$$

This procedure "renormalizes" the band structure $E_n(\mathbf{k})$ in a self-consistent way. Note that the self-consistent one-electron band picture must be altered somewhat if we consider a different state, such as the ferromagnetic state instead of $|F)$; but the resultant changes should be small if charge neutrality, the prime consideration in the energy balance, is maintained. We may therefore imagine $\mathscr{H}_0$ to have constant parameters, to include *a priori* all the important electron-electron interactions (on the average), as well as all the interactions of the electrons with nuclei and their kinetic energy. From $\mathscr{H}'$ we shall extract for present consideration only a significant subset of two-body terms. Using as a guide, that we should consider only those interactions that matter most when the atoms are very far apart (the intra-atomic terms) or when they are very close (the ubiquitous Coulomb interaction), we can extract from $\mathscr{H}_{\text{tot}} = \mathscr{H}_0 + \mathscr{H}'$ the features of greatest physical significance and provide a starting point for later more ambitious investigations.

The first of the interactions to be retained is the direct two-particle *Coulomb repulsion* obtained from $\mathscr{H}'$ by setting $in' = jn$ and $i't' = j't$. After a change of dummy indices, it is

$$\mathscr{H}_{\text{c}} = \frac{1}{2} \sum_{\substack{i,j,n,t,\\ m,m'}} V(\mathbf{R}_{ij})_{n,t} \mathfrak{n}_{i,n,m}[\mathfrak{n}_{j,t,m'} - (F|\mathfrak{n}_{i't'm'}|F)] \qquad (6.35)$$

where the simplified notation is used,

$$V(\mathbf{R}_{ij}) \equiv V(i,i,n,n;j,j,t,t)$$

$$= \int d^3r \int d^3r' \psi_{t,j}^*(\mathbf{r}')\psi_{t,j}(\mathbf{r}')\frac{e^2}{|\mathbf{r}-\mathbf{r}'|}\psi_{n,i}^*(\mathbf{r})\psi_{n,t}(\mathbf{r}) \qquad (6.36)$$

At large distances the leading term in a multipole expansion of this integral is

$$V(\mathbf{R}_{ij})_{n,t} \sim \frac{e^2}{R_{ij}+a} \qquad (6.37)$$

modified to be accurate even as close as nearest-neighbor separation. The terms subtracted in (6.35) ensure that we have electrical neutrality,

$$(F|\mathscr{H}_{\text{c}}'|F) = 0$$

as required.

The next important class of terms are the *exchange interactions*, which are the subset of terms in (6.31) for which $i = j'$, $n' = t$ and $i' = j$, $t' = n$. Because $c_{in'm}^{\dagger}$ and c_{jnm} *anticommute*, the interchange of these two operators to properly associate the pairs in the manner indicated, results in a minus sign. Except for the operator formalism, this then is the ordinary exchange interaction discussed repeatedly in this volume. It acts as a correction to the Coulomb repulsion of *two electrons in a relative triplet state*, and lowers the energy of this state as compared to the singlet configuration. By grouping pairs of Fermions into spin operators, we can include in our Hamiltonian the natural and obvious generalization to the exchange operator $\frac{1}{2}(1 + 4\mathbf{S}_i\cdot\mathbf{S}_j)$, including the variable occupation-number feature of the second quantization. Associating the indices as prescribed, one extracts from (6.31) the exchange Hamiltonian,

$$\mathscr{H}_{\text{ex}} = - \sum_{\substack{n,n', \\ i,j \\ (n,i)\neq(n',j)}} J(\mathbf{R}_{ij})_{n,n'} \left\{ \mathbf{S}_{n,j}\cdot\mathbf{S}_{n',i} + \frac{1}{4}\mathfrak{n}_{n,j}[\mathfrak{n}_{n'i} - (F|\mathfrak{n}_{n',i}|F)] \right\}$$

$$(6.38)$$

where

$$\mathfrak{n}_{n,j} \equiv \mathfrak{n}_{n,i,j,\uparrow} + \mathfrak{n}_{n,j,\downarrow} \tag{6.39}$$

and the spin operators,

$$S_p^z = \frac{1}{2}(\mathfrak{n}_{p\uparrow} - \mathfrak{n}_{p\downarrow}) = \frac{1}{2}(c_{p\uparrow}^{\dagger}c_{p\uparrow} - c_{p\downarrow}^{\dagger}c_{p\downarrow})$$

$$S_p^+ = S_p^x + iS_p^y = c_{p\uparrow}^{\dagger}c_{p\downarrow} \quad \text{and} \quad S_p^- = S_p^x - iS_p^y = c_{p\downarrow}^{\dagger}c_{p\uparrow} \tag{6.40}$$

are recognized as one of the Fermion representations of the spin one-half operators, with $\hbar = 1$, and $p \equiv (n, j)$.

The exchange constant, in the simplified notation, is a *positive-definite* integral that favors parallel spins:

$$J(\mathbf{R}_{ij})_{n,n'} = 2V(i, j, n', n; j, i, n, n')$$

$$= \int d^3r \int d^3r' \psi_{n,j}^*(\mathbf{r}')\psi_{n',i}(\mathbf{r}')\frac{e^2}{|\mathbf{r} - \mathbf{r}'|}\psi_{n',i}^*(\mathbf{r})\psi_{n,j}(\mathbf{r}). \tag{6.41}$$

The largest such exchange integral is the familiar *intra*-atomic Hund's rule integral to which we assign a special symbol,

$$J_{n,n'}^{\text{Hu}} \equiv J(\mathbf{0})_{n,n'}$$

$$= \int d^3r \int d^3r' \psi_{n,i}^*(\mathbf{r}')\psi_{n',i}(\mathbf{r}') \frac{e^2}{|\mathbf{r} - \mathbf{r}'|} \psi_{n',i}^*(\mathbf{r})\psi_{n,i}(\mathbf{r}) \qquad (6.42)$$

and in most cases it will not be necessary to consider any exchange contribution other than $J(0)$. The reason for this is that $J(\mathbf{R}_{ij})$ decreases *much*
faster than the Coulomb integral $V(\mathbf{R}_{ij})$, so that unless (6.42) vanishes by
symmetry or for any other reason, even the nearest-neighbor exchange will
be small in comparison, and the non-nearest-neighbor contributions completely negligible. (An exception: for only one band, two electrons can be in
a triplet state only if they are on different Wannier sites; and the leading
exchange integral is (6.41) at nearest-neighbor distances, with $n = n'$.)

The designation of the exchange forces has a certain degree of arbitrari-
ness, so it is useful to conclude this section with a review of the choice which
was exercised. One traditional approach which was implicitly discarded, was
the one-band, free electron approximation, with Bloch electrons interacting
solely via the long-range Coulomb repulsion, (6.35). While this idea might
have appeared plausible at the time Bloch[12] first proposed it, it had to
be discarded in the face of strong theoretical and experimental evidence.
"... those causes of the magnetical motions ... we relinquish to the moths
and the worms" (see [Ref. 1.5]). Obviously, experiment cannot disprove a
correct theory; the fact that metals with simple nondegenerate bands are
never found to possess strong magnetic properties should merely have been
a spur to the experimentalists to look harder, had it not been for wiser
counsel, notably Wigner's calculation[13] showing the correlations among the
charged particles keep electrons both of parallel spin *and* of antiparallel spin
apart (see also Chapter 1).

We discussed the counter-example in dilute "jellium" in Chapter 4. That
example aside, *the key to magnetic behavior at normal density must therefore
be the existence of degenerate bands* permitting a large number of electrons
to be on the same or neighboring atoms. In such cases they cannot escape
the Coulomb repulsion even by elaborate correlations. The fact is that the
same forces which are present in the individual atoms to produce Hund's rule
"atomic magnetism" cannot suddenly be nullified in the solid state.[14] Compelling as these arguments may be, they are not rigorous. A mathematical

---

[12] F. Bloch, *Zeit. Phys.* **57**, 545 (1929).

[13] E. P. Wigner, *Trans. Faraday Soc.* **205**, 678 (1938).

[14] The credit for this idea belongs to Slater, in J. C. Slater, *Phys. Rev.* **49**, 537, 931 (1936);
ibid. **52**, 198 (1937).

proof is available only in one dimension, where we have shown[15] (see also Chapter 4) that the ground state of $\mathscr{H}_0$ + any interaction Hamiltonian $\mathscr{H}_c$, is *always* a nonmagnetic state, regardless of the number of electrons or the nature of the potential energy. Our proof breaks down, of course, in the face of degeneracy of three-dimensional atoms, as symbolized by the exchange Hamiltonian of (6.38). This adds plausibility to the choice of a single effective Hamiltonian for use in all solids which includes (6.38), viz.,

$$\mathscr{H}_{\text{eff}} = \mathscr{H}_0 + \mathscr{H}_c + \mathscr{H}_{\text{ex}} \tag{6.43}$$

defined in (6.27, 6.35, 6.38). *The retention of that part of $\mathscr{H}'$ that is explicitly spin-dependent, namely $\mathscr{H}_{\text{ex}}$, enhances the possibility of magnetic behavior.*

Unfortunately, the eigenfunctions and eigenvalues of $\mathscr{H}_{\text{eff}}$ cannot be found exactly because the three constituent terms do not commute with one another and therefore cannot be simultaneously diagonalized. $\mathscr{H}_{\text{eff}}$ itself is only part of the total Hamiltonian,

$$\mathscr{H}_{\text{tot}} = \mathscr{H}_0 + \mathscr{H}' \tag{6.44}$$

defined in (6.27, 6.31), which contains all manners of complicated interactions, some of which have spin-dependent consequences and some not. Although we cannot assess their importance in any specific metal, we discard them because they are not required in the high-density limit for which $\mathscr{H}_0$ suffices, nor in the low-density limit where $\mathscr{H}_c + \mathscr{H}_{\text{ex}}$ will do. Thus, (6.43), containing these essential terms and only these terms, will be our starting point.

6.5. Perturbation-Theoretic Derivation of Heisenberg Hamiltonian

To illustrate the uses of $\mathscr{H}_{\text{eff}}$, we shall obtain the Heisenberg Hamiltonian valid in an insulating magnetic material.[16] This derivation can only be suggestive of what is required in a general theory of magnetism in insulators, a complicated and rather specialized subject when investigated in appropriate detail. According to the theory, the larger contributions will mostly be the antiferromagnetic ones, in accordance with experiment; although there exist exceptions to this prevalent antiferromagnetism in insulators:

---

[15] E. Lieb and D. Mattis, *Phys. Rev.* **125**, 164 (1962).
[16] The many-body nature of the "insulating state" is investigated in W. Kohn, *Phys. Rev.* **133**, A171 (1964).

... a curopium oxide of the formula EuO becomes truly *ferromagnetic* at 77 K with a saturation moment of close to 7 Bohr magnetons. This is thus the first rare earth oxide to be found to become ferromagnetic, and with the exception of CrO_2 the only oxide to our knowledge that has true ferromagnetic coupling.[17]

In the magnetic insulator, $\mathcal{H}_c + \mathcal{H}_{ex}$ is the *big* part of $\mathcal{H}_{eff}$, and the "hopping" part, $\mathcal{H}_0$ is the perturbation. However, $\mathcal{H}_{ex}$ may be further decomposed; the Hund's rule part is retained in lowest order, whereas the nearest-neighbor exchange is treated by first-order perturbation theory. The starting Hamiltonian is, therefore,

$$\mathcal{H} = \mathcal{H}_0 + \mathcal{H}_{ex}^{Hu} \tag{6.45}$$

describing the internal dynamics of N *noninteracting* Wannier "atoms," each endowed with a net spin $\mathbf{S}_i$ of magnitude $s_i = \frac{1}{2}$ times the number of electrons is unfilled shell. (It is assumed that a crystal field quenches the angular momentum, or else the appropriate quantities to use here are $\mathbf{J}_i$ and j_i.) The ground-state eigenfunctions of $\mathcal{H}$ are highly degenerate, in fact they number N_0,

$$N_0 = (2s + 1)^N$$

and it is the perturbation $\mathcal{H}_0$ which will lift this degeneracy. The perturbation $\mathcal{H}_0$ can be seen from (6.24) to transfer an electron from one site to a neighboring site, this virtual transition occurring in second-order perturbation theory with matrix element $H(\mathbf{R}_{ij})_{n,n}$ and an increase of energy $U > 0$ in the intermediate state due to the creation of two ionized sites: a positively ionized ion at $\mathbf{R}_i$ and a negatively ionized one at $\mathbf{R}_j$. Let one of the neutral ground states be $|\alpha\rangle$ and one of the excited ionized states be $|\beta\rangle$; then $U \equiv (\beta|\mathcal{H}|\beta) - (\alpha|\mathcal{H}|\alpha)$ and the second-order perturbation theoretic change in the energy of $|\alpha\rangle$ equals

$$\delta E_\alpha = -\sum_{\beta \neq \alpha} \frac{|(\beta|\mathcal{H}_0|\alpha)|^2}{U} = \frac{(\alpha|\mathcal{H}_0|\alpha)^2 - (\alpha|\mathcal{H}_0^2|\alpha)}{U} \tag{6.46}$$

with the second line obtained by closure $[\sum_\beta |\beta)(\beta| \equiv \mathbf{1}]$. $\mathcal{H}_0$ has vanishing ground state expectation value in the insulator (as opposed to a metal)

---

[17] B. T. Matthias, R. Bozorth and J. H. Van Vleck, *Phys. Rev. Lett.* **7**, 160 (1961).

and therefore $(\alpha|\mathcal{H}_0|\alpha)$ vanishes. In writing out the nonvanishing terms, one collects such terms as

$$-H^2(\mathbf{R}_{ij})_{n,n}c_{in\uparrow}^\dagger c_{in\downarrow}c_{jn\downarrow}^\dagger c_{jn\uparrow} = -H^2(\mathbf{R}_{ij})_{n,n}S_{i,n}^+S_{j,n}^- \qquad (6.47)$$

and others corresponding to $S_i^z S_j^z$, as well as terms which do not depend on the relative orientation of the two spins. The latter have the same magnitude for all states $|\alpha)$ and contribute a constant shift in their energy δE_α. The degeneracy of the ground state and relative ordering of the energy levels is given entirely by the subset of spin-dependent terms we lump together into an effective Hamiltonian

$$\mathcal{H}_{\mathrm{Heis}} = \sum_i \sum_{j\neq i}\left[\frac{2}{U}H^2(\mathbf{R}_{ij}) - J(\mathbf{R}_{ij})\right]\mathbf{S}_i\cdot\mathbf{S}_j$$

$$= -\sum_{j\neq i}J_{\mathrm{eff}}\mathbf{S}_i\cdot\mathbf{S}_j . \qquad (6.48)$$

It is important to note that the vectors $\mathbf{S}_i$ are the *total atomic* spin operators and *not* individual electron components such as $\mathbf{S}_{i,n}$, which cannot be specified in any of the N_0 ground states. Therefore the quantities $H^2(\mathbf{R}_{ij})$ and $J(\mathbf{R}_{ij})$ are appropriate averages of the corresponding quantities in the various bands. Note that the non-Hund's rule nearest-neighbor exchange $J(\mathbf{R}_{ij})$ has been reintroduced at this point by means of first-order perturbation theory. The Hund's rule parameter $J(\mathbf{0})$ is absent from (6.48). Its role is just to create the atomic spins $\mathbf{S}_i$. The total Heisenberg Hamiltonian in nonconducting media must be considered as the sum of two effects: the first, an antiferromagnetic interaction due to the virtual "hopping" of an electron from $\mathbf{R}_i$ to $\mathbf{R}_j$ (and back), which is a "one-body" or "kinetic" exchange mechanism. It is antiferromagnetic because the hopping is enhanced when the spins are antiparallel, the exclusion principle prohibiting certain hops when the spins are parallel. The second, ferromagnetic, contribution arises from the exchange of two electrons not on the same atom; its matrix elements among the degenerate ground states may be described by first-order perturbation theory; it is *always* ferromagnetic.

The same arguments hold for the interactions between two distinct partly occupied shells on the *same* atom, e.g., the valence and magnetic shells on a transition or rare earth atom or ion. The net coupling, J_{eff} given by the square bracket in (6.48), gives the tendency of the two shells to have their spins parallel or antiparallel, with important consequences in the Kondo effect studied later.

In real materials (e.g., magnetite) the hopping occurs via an intermediary nonmagnetic ion, such as O^{-2}. This is called *superexchange*, a mechanism first proposed by Kramers[18] 70 years ago and previously analyzed here in Sec. 4.13. For more details, the reader will wish to consult H. Zeiger and G. W. Pratt's comprehensive monograph on magnetic interactions, appropriately named *Magnetic Interactions in Solids*, Clarendon Press, Oxford, 1973.

6.6. Heisenberg Hamiltonian in Metals

The "indirect exchange theory" of magnetism in metals is another example where an effective interatomic Heisenberg Hamiltonian can be derived by second-order perturbation theory. The relative importance of the various energies is here completely reversed over the previous section, for in a metal $\mathscr{H}_0$ is the principal Hamiltonian, and $\mathscr{H}_c$ and $\mathscr{H}_{ex}$ the perturbations.

The theory developed below was first invented in connection with nuclear magnetic resonance by Ruderman and Kittel[19] and independently, by Bloembergen and Rowland[20]; these authors studied the effective long-ranged interaction between nuclear spins due to the hyperfine coupling with the common sea of conduction electrons. The extension of their analysis to the s–d or s–f interaction[21] permitted the explanation of some significant experiments by Zimmerman on the long-ranged interaction between Mn atoms[22] dissolved in Cu. Briefly, the Mn impurity atom retains part of its Hund's rule magnetization in the solute state and by J_{eff}, (6.48), polarizes the spins of the conduction electrons in its neighborhood. The conduction electrons, constrained by the Pauli exclusion principle, respond with a characteristic wavelength

$$\lambda_F = \pi/k_F \qquad (6.49)$$

and the resultant spin polarization is not well localized in the vicinity of the impurity but is oscillatory and long-ranged. A second manganese atom at an

[18] H. A. Kramers, *Physica* **1**, 182 (1934).

[19] M. A. Ruderman and C. Kittel, *Phys. Rev.* **96**, 99 (1954).

[20] N. Bloembergen and T. J. Rowland, *Phys. Rev.* **97**, 1679 (1955).

[21] K. Yosida, *Phys. Rev.* **106**, 893 (1957).

[22] See review of theory and experiments in footnotes 23–25, and discussions on the "spin glass" elsewhere in this volume.

[23] W. Marshall, T. Cranshaw, C. Johnson and M. Ridout, *Rev. Mod. Phys.* **36**, 399 (1964).

[24] F. Smith, *Phys. Rev. Lett.* **36**, 1221 (1976).

[25] G. S. Rushbrooke, *J. Math. Phys.* **5**, 1106 (1964).

arbitrary distance from the first, suffers a ferromagnetic or an antiferromagnetic interaction with it — depending upon whether it is in the trough or on the crest of the polarization wave. The strength of the interaction gradually decreases with distance, in a manner we shall calculate.

Assume a pair of solute magnetic atoms at $\mathbf{R}_1$ and $\mathbf{R}_2$, in an otherwise ideal nonmagnetic metal characterized by an s-band Hamiltonian $\mathscr{H}_0$. Internal Hund's rule coupling maintains the magnitudes of the solutes' spins fixed at s_1 and s_2, respectively, but the *relative orientation* of the two spins will be governed by the interaction derived below. The exchange coupling of the localized electrons with the conduction electrons is the perturbation,

$$\mathscr{H}_{\mathrm{ex}} = -J_{\mathrm{eff}}[\mathbf{S}_1 \cdot \mathbf{s}_c(\mathbf{R}_1) + \mathbf{S}_2 \cdot \mathbf{s}_c(\mathbf{R}_2)] \tag{6.50}$$

with the conduction-band spin operators $\mathbf{s}_c(\mathbf{R}_i)$ given by (6.40). The substitution of Bloch operators for the Wannier operators, given in (6.26a), results in the following:

$$\mathbf{s}_c(\mathbf{R}_i)^z = \frac{1}{2N} \sum_{\mathbf{k},\mathbf{q}} e^{-i\mathbf{q}\cdot\mathbf{R}_i} (c^{\dagger}_{\mathbf{k}+\mathbf{q}\uparrow} c_{\mathbf{k}\uparrow} - c^{\dagger}_{\mathbf{k}+\mathbf{q}\downarrow} c_{\mathbf{k}\downarrow})$$

$$\mathbf{s}_c(\mathbf{R}_i)^+ = \frac{1}{N} \sum_{\mathbf{k},\mathbf{q}} e^{-i\mathbf{q}\cdot\mathbf{R}_i} (c^{\dagger}_{\mathbf{k}+\mathbf{q}\uparrow} c_{\mathbf{k}\downarrow}) \quad \text{and} \tag{6.51}$$

$$\mathbf{s}_c(\mathbf{R}_i)^- = \frac{1}{N} \sum_{\mathbf{k},\mathbf{q}} e^{-i\mathbf{q}\cdot\mathbf{R}_i} (c^{\dagger}_{\mathbf{k}+\mathbf{q}\downarrow} c_{\mathbf{k}\uparrow}).$$

With this definition in mind, let us calculate the eigenvalues and eigenfunctions of

$$\mathscr{H} = \mathscr{H}_0 + \mathscr{H}_{\mathrm{ex}} = \sum_{\substack{m=\uparrow,\downarrow \\ \mathbf{k}}} E(\mathbf{k}) \mathfrak{n}_{\mathbf{k},m}$$

$$- J_{\mathrm{eff}} \sum_{i=1}^{2} \left\{ s_c(\mathbf{R}_i)^z S_i^z + \frac{1}{2}[s_c(\mathbf{R}_i)^+ S_i^- + \text{H.c.}] \right\} \tag{6.52}$$

by ordinary perturbation theory. In particular, we wish to see how the perturbation lifts the degeneracy of the $r \equiv (2s_1 + 1) \times (2s_2 + 1)$ states of orientations of the two solute spins. The conduction electrons are assumed to be initially in the ground state, a singlet (except for the resultant polarization effects).

The first-order correction to the energy,

$$\delta E^{(1)} = (t; F|\mathscr{H}_{\text{ex}}|F; t) = 0 \tag{6.53}$$

is seen to vanish. We use the notation $|F; t)$ to indicate the product state of Fermi sea with the two solute spins, with the index t spanning the range $t = 1, \ldots, r$ of degenerate orientations of the two spins.

The conduction-band spin operators $c^\dagger c$ create elementary excitations of energy $E(\mathbf{k} + \mathbf{q}) - E(\mathbf{k})$; their matrix elements are unity if $k < k_{\text{F}}$ and $|\mathbf{k} + \mathbf{q}| > k_{\text{F}}$, and zero otherwise. Therefore, by second-order perturbation theory

$$\delta E_t^{(2)} = -\frac{J_{\text{eff}}^2}{2N^2} \sum_{\substack{t', k < k_{\text{F}} \\ |\mathbf{k}+\mathbf{q}| > k_{\text{F}}}} \frac{(t|e^{i\mathbf{q}\cdot\mathbf{R}_1}\mathbf{S}_1 + e^{i\mathbf{q}\cdot\mathbf{R}_2}\mathbf{S}_2|t')\cdot(t'|e^{-i\mathbf{q}\cdot\mathbf{R}_1}\mathbf{S}_1 + e^{-i\mathbf{q}\cdot\mathbf{R}_2}\mathbf{S}_2|t)}{E(\mathbf{k} + \mathbf{q}) - E(\mathbf{k})}$$

$$= -\frac{J_{\text{eff}}^2}{2N^2} \sum_{\substack{k < k_{\text{F}} \\ |\mathbf{k}+\mathbf{q}| > k_{\text{F}}}} \frac{(t|s_1(s_1 + 1) + s_2(s_2 + 1) + 2\mathbf{S}_1\cdot\mathbf{S}_2 \cos \mathbf{q}\cdot\mathbf{R}_{12}|t)}{E(\mathbf{k} + \mathbf{q}) - E(\mathbf{k})}. \tag{6.54}$$

The second line is the result of using closure on the intermediate states $|t')$. It is convenient to separate this formula into two parts: a self-energy

$$\delta E^{(2)} = -K \sum_i s_i(s_i + 1) \tag{6.55}$$

with

$$K = \frac{J_{\text{eff}}^2}{2N^2} \sum_{\substack{k < k_{\text{F}} \\ |\mathbf{k}+\mathbf{q}| < k_{\text{F}}}} \frac{1}{E(\mathbf{k} + \mathbf{q}) - E(\mathbf{k})} \tag{6.55a}$$

and an interaction energy, which is the eigenvalue of the effective Hamiltonian

$$\mathscr{H}_{\text{IE}} = -\sum_{(i,j)} J(\mathbf{R}_{ij})_{\text{IE}} \mathbf{S}_i\cdot\mathbf{S}_j \tag{6.56}$$

where the indirect exchange coupling constant is

$$J(\mathbf{R}_{ij})_{\text{IE}} = \left(\frac{J_{\text{eff}}}{N}\right)^2 \sum_{\substack{k < k_{\text{F}} \\ |\mathbf{k}+\mathbf{q}| > k_{\text{F}}}} \frac{\cos \mathbf{q}\cdot\mathbf{R}_{ij}}{E(\mathbf{k} + \mathbf{q}) - E(\mathbf{k})}. \tag{6.56a}$$

If instead of only two impurities there are N_I, the sum in (6.55) runs over all N_I spins and the interaction, (6.56), over all $\frac{1}{2}N_I(N_I - 1)$ distinct pairs.

For an estimate of the indirect exchange coupling, one uses the effective mass approximation, $E(\mathbf{k}) = \hbar^2 \mathbf{k}^2/2m^*$ and $\mu = \hbar^2 k_F^2/2m^*$, and introduces an imaginary part $\hbar i/\tau$ to the denominator to account for a finite electronic mean free path.

$$
\begin{aligned}
J(\mathbf{R}_{ij})_{\mathrm{IE}} &= J_{\mathrm{eff}}^2 \left(\frac{a_0}{2\pi}\right)^6 \int_{[E(k)<\mu]} d^3k \int d^3k' \frac{\cos(\mathbf{k} - \mathbf{k}')\cdot\mathbf{R}_{ij}}{(\hbar^2/2m^*)(\mathbf{k}'^2 - \mathbf{k}^2 + i2m^*/\hbar\tau)} \\
&= J_{\mathrm{eff}}^2 \left(\frac{a_0}{2\pi}\right)^6 \frac{m^*}{2\hbar^2}\left(\frac{4\pi}{R_{ij}}\right)^2 \int_0^{k_F} dk\,k \int_{-\infty}^{\infty} dk'\,k' \frac{\sin kR_{ij} \sin k'R_{ij}}{k'^2 - k^2 + i2m^*/\hbar\tau} \\
&= \frac{-J_{\mathrm{eff}}^2}{\mu} \frac{(k_F a_0/2)^6}{\pi^3}\left[\frac{\sin 2k_F R_{ij} - 2k_F R_{ij} \cos 2k_F R_{ij}}{(2k_F R_{ij})^4}\right] e^{-R_{ij}/\lambda}
\end{aligned}
$$

$$(6.56b)$$

where λ = mean free path = $\hbar k_F \tau/m^*$ and a_0^3 = volume of unit cell. This formula (without the mean free path factor) was first published by Ruderman and Kittel[19] and is referred to as the Ruderman-Kittel interaction or "RKKY."[26] The mean free path factor originally introduced in the first edition of the present book has since been measured.[27]

6.7.   Ordered Magnetic Metals: Deriving the Ground State

Some of the most interesting applications of the indirect exchange theory are to metals containing elements in the gadolinium rare-earth series (*lanthanides*). Generally the *f*-shell radii of the rare-earth atoms are so small that even nearest neighboring atoms do not have significant direct overlap and the interaction is assumed to be principally the Ruderman-Kittel indirect exchange mechanism derived in the preceding section (with crystal field anisotropy and hybridization i.e., band-mixing, the principal correction).[28] Ordered alloys containing transition elements are also described by this theory if the magnetic atoms are sufficiently far apart for the *d*-shell overlaps to be unimportant.

---

[26] For Ruderman-Kittel-Kasuya-Yosida, the last two also contributors to the theory.
[27] A. Heeger, A. Klein and P. Tu, *Phys. Rev. Lett.* **17**, 803 (1966).
[28] R. J. Elliott, *Phys. Rev.* **124**, 346 (1961); H. Miwa and K. Yosida, *Prog. Theor. Phys.* (*Kyoto*) **26**, 693 (1961); T. A. Kaplan, *Phys. Rev.* **124**, 329 (1961).

Therefore we are led to consider the Hamiltonian

$$\mathscr{H}_{\mathrm{IE}} = -\sum J_{ij}\mathbf{S}_i\cdot\mathbf{S}_j \quad [J_{ij} \equiv J(\mathbf{R}_{ij})_{\mathrm{IE}}] \tag{6.57}$$

and to calculate its eigenstates and eigenvalues in cases where the spins $\mathbf{S}_i$ occupy points on a regular lattice. For convenience, we shall assume it to be one of the Bravais lattices, and particularly one of the three principal cubic lattices. There is no exact method known to obtain the ground state, but the following procedure avoids unnecessary complications, and appears reliable. First, we construct the product wavefunction

$$\psi = \prod \phi_i \tag{6.58}$$

in which the ϕ_i are, as yet, unspecified but normalized states of the spins $\mathbf{S}_i$. The variational energy in this configuration is

$$E = (\Psi|\mathscr{H}_{\mathrm{IE}}|\Psi)$$

$$= -\sum J_{ij}(\phi_i|\mathbf{S}_i|\phi_i)\cdot(\phi_j|\mathbf{S}_j|\phi_j) \tag{6.59}$$

where

$$(\phi_i|\mathbf{S}_i|\phi_i)^2 \le \mathbf{s}_i^2 . \tag{6.60}$$

In the remainder, assume all N_{I} magnetic atoms to belong to the same species, so that $s_i = s$. The *method of Luttinger and Tisza*, similar to the "spherical model" discussed later, can then be used to find the lowest energy attainable with a trial function of the type in (6.58). This method works only on a Bravais lattice, i.e. where there is just one spin per cell. It consists mainly of relaxing the inequality above, requiring only

$$\sum_i (\phi_i|\mathbf{S}_i|\phi_i)^2 \le N_{\mathrm{I}}s^2 \tag{6.61}$$

which is a weaker constraint. *The lowest energy (6.59) subject to the weaker constraint (6.61) must be lower than the lowest energy (6.59) subject to the more rigorous constraint (6.60).* E is calculated by Fourier transforms, letting $\mathbf{S}_{\mathbf{k}}$ be defined by

$$\mathbf{S}_{\mathbf{k}} \equiv \frac{1}{N_{\mathrm{I}}}\sum_i e^{-i\mathbf{k}\cdot\mathbf{R}_j}(\phi_i|\mathbf{S}_i|\phi_i) \quad \text{and}$$

$$\tag{6.62}$$

$$(\phi_i|\mathbf{S}_i|\phi_i) = \sum_{\mathbf{k}} e^{i\mathbf{k}\cdot\mathbf{R}_i}\mathbf{S}_{\mathbf{k}} ,$$

substituting it into (6.59) and obtaining,

$$E = -N_I \sum_{\mathbf{k}} J(\mathbf{k})|\mathbf{S_k}|^2 \tag{6.63}$$

where

$$J(\mathbf{k}) = \frac{1}{2N_I} \sum_{i,j} J_{ij} e^{i\mathbf{k}\cdot\mathbf{R}_{ij}} . \tag{6.64}$$

The weak inequality (6.61), by substitution of the Fourier transform, is fully equivalent to,

$$\sum_{\mathbf{k}} |\mathbf{S_k}|^2 \leq s^2 \tag{6.65}$$

and if we define $\mathbf{q}_0$ to be the wave vector for which $J(\mathbf{q}_0)$ attains its largest value (or one of the wave vectors which have this property if there are more than one) then let $\mathbf{S_{q_0}} = \mathbf{s}$ and all other $\mathbf{S_k} = 0$, it is obvious that

$$E(\mathbf{q}_0) = -N_I J(\mathbf{q}_0) s^2 \tag{6.66}$$

is the lowest energy, subject to the weak constraint. Now choose the wave-functions ϕ_i such that

$$(\phi_i|\mathbf{S}_i|\phi_i) = s(\cos \mathbf{q}_0\cdot\mathbf{R}_i, \sin \mathbf{q}_0\cdot\mathbf{R}_i, 0) \tag{6.67}$$

which is called a *spiral* configuration of pitch $\mathbf{q}_0$. Inserting this variational *ansatz* into (6.59) leads to precisely the energy $E(\mathbf{q}_0)$ calculated above; and being an upper bound as well as a lower bound to the ground-state energy (in the Hartree product wavefunction approximation), $E(\mathbf{q}_0)$ must itself be the Hartree ground-state energy.

When $\mathbf{q}_0 = 0$, all spins are parallel and the ground state is ferromagnetic. When $\mathbf{q}_0$ is a wavevector on one of the points of symmetry of the Brillouin zone boundary, for example, $(\pi/a)(1\pm,\pm1,\pm1)$ in the simple cubic structure, the ground state is an antiferromagnetic configuration of some sort (the Néel state in the example given). If $\mathbf{q}_0$ is none of these special wave vectors, the ground state is a "spiral spin configuration," as these screw structures are usually identified. The various possible configurations can be found from group theory or by matrix methods, even in non-Bravais lattices, particularly in insulators when $\mathbf{J}_{ij}$ is limited to the Heisenberg nearest-neighbor forces.[29,30] Note that in present derivation the range of the interaction does not matter, therefore the proof holds equally well for the short-ranged Heisenberg interaction in insulators. In the case of the long-ranged Ruderman-Kittel interaction, numerical calculation is required to obtain $J(\mathbf{k})$, and to determine $\mathbf{q}_0$.

The numerical calculation of $J(\mathbf{0})$ is particularly interesting, because according to the theory developed in a later chapter, *it is proportional to the paramagnetic Curie temperature* θ. A negative θ necessarily precludes ferromagnetism; a positive θ is a likely indication of ferromagnetism, but it is still possible for a spiral configuration belonging to some $\mathbf{q}_0 \neq 0$ to be the stable ground state at low temperatures. Therefore it is also necessary to study the spin-wave spectrum, or magnon energy

$$\hbar\omega(\mathbf{k}) = 2s[J(\mathbf{0}) - J(\mathbf{k})] \tag{6.68}$$

to determine that all $\hbar\omega(\mathbf{k})$ are positive, if the ferromagnetic state is to be stable. This is a necessary but *not sufficient* condition for ferromagnetism.[31] The minimizing wavevector $\mathbf{q}_0$ can also be found at the minimum of the function $\hbar\omega(\mathbf{q}_0)$.

As an example, we calculate $\hbar\omega(\mathbf{k})$ for the Ruderman-Kittel interaction, in the continuum limit $k_F a \to 0$, where the lattice sums can be replaced by integrals. Let us definite ($x_i \equiv 2k_F R$, a = separation of n.n. *magnetic* atoms):

$$\varepsilon(\mathbf{k}) \equiv \sum a^3 \frac{\sin x_i - x_i \cos x_i}{5x_i R_i^3} e^{-R_i/\lambda}(1 - \cos \mathbf{k}\cdot\mathbf{R}_i) \propto \hbar\omega(\mathbf{k}) \tag{6.69}$$

which is (6.68), the magnon energy, with unimportant constant factors eliminated, for tabular convenience. Also,

$$\theta \equiv \sum_{R_i \neq 0} a^3 \frac{\sin x_i - x_i \cos x_i}{5x_i R_i^3} e^{-R_i/\lambda} \tag{6.70}$$

[29] D. Lyons and T. Kaplan, *Phys. Rev.* **120**, 1580 (1960); D. Lyons, T. Kaplan, K. Dwight and N. Menyuk, *Phys. Rev.* **126**, 540 (1962).

[30] E. F. Bertaut, in *Magnetism*, Vol. 3, G. Rado and H. Suhl, eds., Academic Press, New York, 1963, Chap. 4.

[31] Cf. Mattis[32], in which it is shown that in some exceptional instances the ground state may be nonferromagnetic, due to quantum fluctuations *even though* the ferromagnetic state is stable against the emission of (any finite number of) spin waves. This analysis has been extended, see Bader and Schilling[33], and ongoing work. As for the practical task of fitting the formulas to experiment, we may cite Reitz and Stearns[34] for the Heusler alloys.

[32] D. Mattis, *Phys. Rev.* **130**, 76 (1963).

[33] H. Bader and R. Schilling, *Phys. Rev.* **B19**, 3556 (1979); **20**, 1977 (1979).

[34] J. Reitz and M. B. Stearns, *J. Appl. Phys.* **50** (3), 2066 (1979).

is the paramagnetic Curie temperature with similar constant factors removed. The limiting values are elementary integrals

$$\lim_{k_F \to 0} \theta = \frac{8\pi}{10} \left(1 + \frac{1}{32 k_F \lambda} - \frac{1}{16\pi k_F \lambda} \tan^{-1} \frac{1}{2k_F \lambda} \right) \tag{6.71}$$

and

$$\lim_{k_F \to 0} \varepsilon(\mathbf{k}) = \frac{4\pi}{10} \left\{ 1 - \frac{(2k_F \lambda)^2 + 1 - (k\lambda)^2}{(4k\lambda)(2k_F \lambda)} \ln \frac{(2k_F \lambda + k\lambda)^2 + 1}{(2k_F \lambda - k\lambda)^2 + 1} \right.$$
$$\left. + \frac{1}{2k_F \lambda} \left[\tan^{-1} \frac{4k_F \lambda}{(2k_F \lambda)^2 - 1} - \tan^{-1} \frac{4k_F \lambda}{(2k_F \lambda)^2 - (k\lambda)^2 - 1} \right] \right\}. \tag{6.72}$$

These are the magnetic parameters for a "continuum" or "jellium" lattice or for a low density electron gas. Note that they are both positive, indicating ferromagnetism, agreeing with the prediction of the "double-exchange" mechanism.

Closely associated with k_F is the dimensionless parameter $n_{c/a} \equiv$ number of conduction electrons per magnetic atom. (If there is one conduction electron per unit cell, but a magnetic atom is present only in every other cell, $n_{c/a} = 2$, etc. The "jellium" limit is equivalent to $n_{c/a} \to 0$.) As $n_{c/a}$ is raised, the paramagnetic Curie temperature goes through zero at approximately $n_{c/a} = 1/4$ for the three principal cubic lattices. There is then an antiferromagnetic region as $n_{c/a}$ is increased further, until it exceeds the value $3/2$ for the bcc and fcc lattices, and $5/2$ for the sc lattice. At this point the paramagnetic Curie temperature becomes positive once more until $n_{c/a}$ is approximately doubled, whereupon a new antiferromagnetic region is encountered, etc.

A plot of θ as a function of k_F and $n_{c/a}$ which displays the features just discussed, is given in Fig. 6.6 for the simple cubic lattice. In Fig. 6.6 we also highlight the spin-wave stable regions where $\hbar\omega(\mathbf{k}) > 0$; these are seen to be somewhat smaller than the regions of positive paramagnetic Curie temperature. To scale the results so that they correspond to the Ruderman-Kittel interaction of (6.56) with constant J, it is necessary to multiply the plotted and tabulated values by appropriate functions of k_F, a, and $a_0 = $ nonmagnetic metal lattice parameter. In the calculations, the value $a = 1$ is taken for convenience. Therefore, comparing (6.56b, 6.68, 6.69), we obtain the physical parameters $\tilde{\theta}$ and $\hbar\omega(\mathbf{k})$

$$\frac{\tilde{\theta}}{\theta} = \frac{\hbar\omega(\mathbf{k})}{\varepsilon(\mathbf{k})} \propto J^2 (k_F a_0)(k_F a)^3 \left(\frac{a_0}{a} \right)^6 \tag{6.73}$$

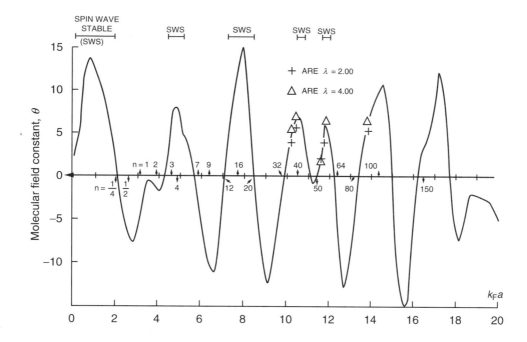

Fig. 6.6. Θ versus $k_F a$ in the sc lattice (bcc, fcc are similar). Values of electron density $n = (8\pi/3) \times (k_F a/2\pi)^3 =$ number of conduction electrons *per* magnetic atom, are indicated by arrows. Main curve is for $\lambda = 3$ (some points at $\lambda = 2$ or 4 are shown to indicate the insensitivity of the results). Spinwave-stable regions indicated are seen to correlate well although not *exactly*, with $\Theta > 0$.

in terms of θ and $\varepsilon(\mathbf{k})$, the computed functions.

In Fig. 6.7 we reproduce experimental results of Methfessel and collaborators,[35] on the paramagnetic Curie temperature of certain ordered Eu-Gd-Se alloys, in which the electronic concentration could be varied from insulator to metallic, corresponding to the range $0 < k_F a < 2$ (the upper value is an estimate). These results are in qualitative agreement with the applicable portions of the theoretical curves.

A plot of $J(\mathbf{k})$ in the simple cubic lattice is given in Fig. 6.8. One observes the ferromagnetic state characterized by $\mathbf{q}_0 = \mathbf{0}$ being succeeded by antiferromagnetic configurations as k_F is increased from frame (a) to (b). The resulting configuration consists of alternating planes of parallel spins, the alternation being in the (100) direction. As one proceeds to frames (d), (e), and (f), he sees the alternation going into the (110) direction, and finally the (111) direction, the last being the Néel state.

---

[35] S. Methfessel, private communication.

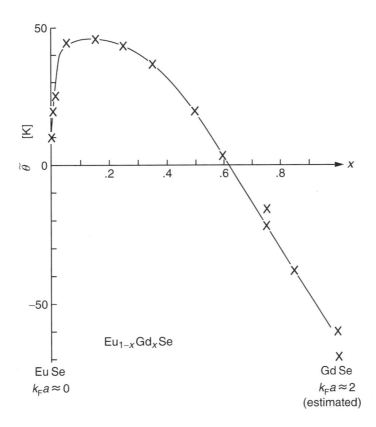

Fig. 6.7. Experimental values of $\tilde{\Theta}$ (paramagnetic Curie temperature) versus composition (roughly, k_F^3) at constant a, obtained by Methfessel *et al.*[35] in a series of ordered rare-earth alloys. An order-of-magnitude estimate is that $k_F a$ varies in the range 0–2 over the range of compositions, and these results are in qualitative agreement with Ruderman-Kittel theory (cf. Fig. 6.6 over same range of k_F).

Note that if the Ruderman-Kittel interaction is valid at small k_F, then the statistical mechanics will be very well described by the molecular field theory which we develop in a later chapter. For then, the interaction is long-ranged and practically nodeless (J_{ij} is ferromagnetic out to distances $\sim 1/k_F$ and is very small beyond) and the criteria of the molecular field theory are met at all but the lowest temperatures, where spin-wave theory is applicable.

Because the indirect exchange theory is used in the description of the magnetic properties of the rare-earth metals and alloys,[36] it is interesting to note that in many cases the angular momentum of the f shell in these ions is not quenched, and the total $\mathbf{J}_i$ angular momentum must be specified and

---

[36] T. A. Kaplan and D. H. Lyons, *Phys. Rev.* **129**, 2072 (1963).

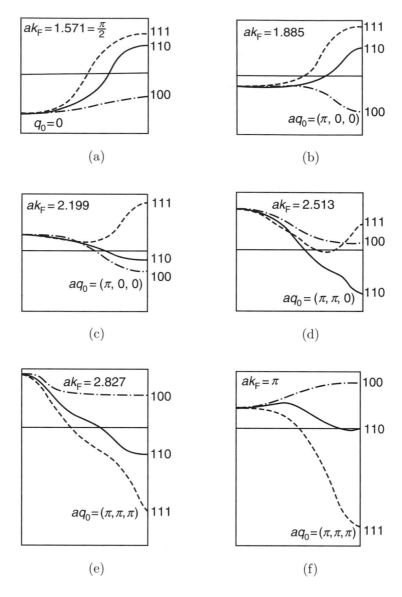

Fig. 6.8. Plot of $-J(\mathbf{k})$. In all these examples, $\mathbf{q}_0$ lies at a BZ surface point such that the spiral is commensurate with the lattice. If and when $\mathbf{q}_0$ lies at some arbitrary point *inside* the BZ, the magnetic order is incommensurate with the lattice for the Ruderman-Kittel interaction versus $\mathbf{k}$ in the three principal directions, sc lattice. In the range $0 < k_F a \leq \pi/2$ the ground state is ferromagnetic. However, as k_F is increased in frames (b) to (f) various antiferromagnetic states become stable, indicated by nonzero values of spiral pitch parameter $\mathbf{q}_0$. ($J(\mathbf{k})$ is defined in (6.56, 6.64).)

Table 6.2. g-factor and angular momenta of rare earths.

Number of electrons in f shell	Symbol	s	l	j	$g-1$
0	La	0	0	0	$\cdots$
1	Ce	$\frac{1}{2}$	3	$\frac{5}{2}$	$-\frac{1}{7}$
2	Pr	1	5	4	$-\frac{1}{5}$
3	Nd	$\frac{3}{2}$	6	$\frac{9}{2}$	$-\frac{3}{11}$
4	Pm	2	6	4	$-\frac{2}{5}$
5	Sm	$\frac{5}{2}$	5	$\frac{5}{2}$	$-\frac{5}{7}$
6	Eu	3	3	0	$\cdots$
7	Gd	$\frac{7}{2}$	0	$\frac{7}{2}$	$+1$
8	Tb	3	3	6	$+\frac{1}{2}$
9	Dy	$\frac{5}{2}$	5	$\frac{15}{2}$	$+\frac{1}{3}$
10	Ho	2	6	8	$+\frac{1}{4}$
11	Er	$\frac{3}{2}$	6	$\frac{15}{2}$	$+\frac{1}{5}$
12	Tm	1	5	6	$+\frac{1}{6}$
13	Yb	$\frac{1}{2}$	3	$\frac{7}{2}$	$+\frac{1}{7}$
14	Lu	0	0	0	$\cdots$

Note: From the following formula: $g - 1 = [j(j+1) + s(s+1) - l(l+1)]/2j(j+1)$.

not just the total spin $\mathbf{S}_i$. That is, the magnetic degrees of freedom of each rare earth are described by $2j_i + 1$ eigenfunctions, and not by $2s_i + 1$. But this is easily taken into account by using the definition of the Landé g factor. In the subspace of the $2j_i + 1$ eigenfunctions, the following equality defines the Landé factor g_i,

$$\mathbf{M}_i = \mathbf{J}_i + \mathbf{S}_i = g_i\mathbf{J}_i \tag{6.74}$$

where $\mathbf{M}_i$ is the magnetic moment operator of the ion. Subtracting $\mathbf{J}_i$ from both sides of the equation, we obtain the prescription useful in the present case:

$$\text{replace } \mathbf{S}_i \text{ by } (g_i - 1)\mathbf{J}_i \tag{6.75}$$

in $\mathscr{H}_{\text{IE}}$ for all rare earths *except* when $j = 0$, as in the case of Eu in some states. Because $(g_i - 1)$ can be positive or negative, a sort of "charge" is introduced into the indirect exchange theory: ions with opposite signs of $(g_i - 1)$ will interact antiferromagnetically for ferromagnetic J_{ij}, and vice versa, so that this gives to mixed rare-earth alloys yet another degree of freedom.

The indirect exchange Hamiltonian is thus,

$$\mathscr{H}_{\text{IE}} = -\sum J_{ij}(g_i - 1)(g_j - 1)\mathbf{J}_i\cdot\mathbf{J}_j \tag{6.76}$$

except when $\mathbf{J}_i = 0$, when $\mathbf{S}_i$ is used. Table 6.2 lists the rare earths and their effective "spin charge" $g_i - 1$.

In transition metal ions the angular momentum is quenched because of strong crystal field effects on the relatively extensive d orbitals. This is reflected in experimentally measured g factors close to 2, the theoretical spin-only value. In those cases, the correct low-lying states are designated by m_s, and the correct vector operator is just $\mathbf{S}_i$.

6.8.   Kondo Effect

In the preceding, a large number of spins arranged in a regular way in a metallic matrix were found to correlate their orientations via the Ruderman-Kittel interaction mediated by the nonmagnetic conduction electrons. The symmetry of the ground state depended on geometric factors such as the ratio of distance between magnetic atoms to the deBroglie wavelength of an electron at the Fermi surface.

In this section, we give the reader a brief introduction to the case of random, dilute, magnetic alloys, in which the concentration of spins may be as few as parts per million (ppm). In the extremely dilute limit where the self-energy of each spin is the dominant factor, one finds the Kondo effect[37]: an anomaly in the conduction electrons in the vicinity of the impurity spin. In the moderately dilute case, when interactions among neighboring spins become more significant than the self-energy of each, the spin-glass phase — a condensed phase in the presence of a high degree of disorder — will yield the presumed ground state. In either case, the internal degrees of freedom peculiar to spins are the cause of highly unusual phenomena, which theorists, assisted by a variety of experimental facts, have been seeking to understand in ever finer detail.

One goes back to the 1930's for the first inklings of these phenomena. Meissner and Voigt[38] first noted an anomaly in the resistivity of nominally

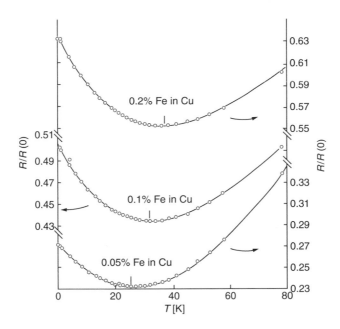

Fig. 6.9.   Resistance as a function of temperature for dilute magnetic alloys; note various scales.[39]

[37] J. Kondo, *Prog. Theor. Phys.* (*Kyoto*) **32**, 37 (1964).

[38] W. Meissner and G. Voigt, *Ann. Physik* **7**, 761, 892 (1930).

[39] J. P. Franck, F. D. Manchester and D. L. Martin, *Proc. Roy. Soc.* (*London*) **A263**, 494 (1961).

pure gold samples, in reality containing trace impurities of iron from the manufacturing process. These showed an increasing resistance as the temperature was lowered below some 10 K. Such behavior is in diametric opposition to that of ordinary metals and alloys, which have their minimal resistance at $T = 0$ when all thermal vibrations have ceased. A recent confirmation of the Meissner-Voigt discovery is shown in Fig. 6.9. The effects are highlighted in Fig. 6.10, from which the resistivity contributed by lattice vibrations has been subtracted out. A few ppm iron in Cu or Au are seen to be capable of substantially affecting the resistivity and, as it turns out, a variety of other thermodynamic properties as well. Between the low-temperature saturation of the resistivity (at 1 K in the example of Fig. 6.10) and the high temperatures at which the magnetic contributions become negligible, there exists a wide range over which the resistivity follows an approximately *logarithmic* behavior. This is seen in Fig. 6.11 for several samples, which satisfy a universal law, once the temperature is expressed in terms of suitable units. The scaling temperature (the unit) is constant for a given alloy but varies from

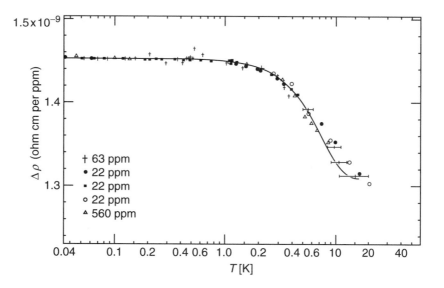

Fig. 6.10.   Low temperature resistivity of CuFe, *per* iron atom; with phonon contribution subtracted out.[40] Note logarithmic scale.

---

[40] M. D. Daybell and W. Steyert, *Phys. Rev. Lett.* **18**, 398 (1967).

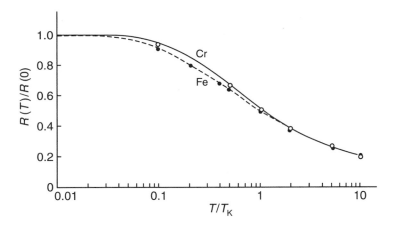

Fig. 6.11.   The normalized resistivity $R(T)/R(0)$ for a variety of samples, plotted as function of $\log(T/T_{\rm K})$. The results suggest a universal curve.[41]

one to the other. It is most appropriately denoted the "Kondo temperature" $T_{\rm K}$, given as:

$$T_{\rm K} = (D/k_{\rm B})\exp(-D/2|J|) \qquad (6.77)$$

with D = bandwidth, J the exchange parameter ($J > 0$ for ferromagnetic s–d coupling, < 0 for AF coupling), $k_{\rm B}$ = Boltzmann's constant. Kondo first gave an expression for the resistivity[37]:

$$R = R_0(T) + c(J^2/D)s(s+1)A\left[1 - \frac{2J}{D}\ln(D/k_{\rm B}T)\right] \qquad (6.78)$$

in which R_0 is the nonmagnetic contribution, A is a lumped constant, s = spin of the impurity, and the other parameters have already been given; the Kondo temperature $T_{\rm K}$ signals the onset of peculiar behavior in (6.78) at $T \leq T_{\rm K}$. In fact, (6.78), derived by perturbation theory, ceases to be valid at low temperature. Since the enunciation of the logarithmic law, it has been observed in a number of laboratories and Fig. 6.11 is typical Kondo behavior.

It is also generally accepted that this behavior is the consequence of a large "soft" cloud of conduction electrons' spin polarization resonating about

---

[41] A. J. Heeger, "Localized Moments and Nonmoments in Metals: the Kondo Effect," in *Solid State Physics*, Vol. 23, F. Seitz, D. Turnbull and H. Ehrenreich, eds., Academic Press, New York, 1969, p. 284 and Fig. 28 on p. 380.

the impurity's local moment, so that the role of temperature is to break up this cloud and reduce the strong scattering that such a resonance produces. It should then be possible to predict at which concentration of magnetic impurities the overlap of neighboring clouds becomes more significant than the energy binding each to its local moment. At such a value of the concentration, the *spin glass* phase must become important; at yet higher concentrations, the *ordered magnetic alloys* studied in previous sections become relevant.

The one-spin interaction Hamiltonian appropriate to the very dilute Kondo limit (ppm) is precisely half of what we considered in (6.50, 6.52), viz.

$$\mathscr{H}_{\text{ex}} = -J\mathbf{S}\cdot\mathbf{s}_{\text{c}}(0), \quad \text{with} \quad J = J_{\text{eff}} \quad \text{of (6.48)}. \tag{6.79}$$

The isolated spin is taken to be located at the origin, without loss of generality.

Now, however weak the interaction parameter J may be compared with the other parameters: Fermi energy ε_{F}, electron bandwidth D, etc., it cannot be considered a small perturbation because of a singularity at $J = 0$. Specifically, one can show[42]:

In any large region surrounding the magnetic defect, the ground state spin (impurity + conduction band) has the value $S_{\text{tot}} = s - \frac{1}{2}$ for antiferromagnetic interactions and $S_{\text{tot}} = s$ or $s + \frac{1}{2}$ for ferromagnetic interactions.

The proof of this follows the arguments in Chapter 4, by comparing the ground state of the present system with that of a reference system for which the ground state quantum numbers are known. It is a tricky proof, however, as the energy levels of the conduction band form a continuum and they are so dense that their fluctuations can overwhelm the impurity. Cragg and Lloyd[42] have investigated this problem by numerical means and use of the renormalization group, finding for $J > 0$ (ferromagnetic coupling) a dependence of the ground state symmetry on the number of iterative steps in their procedure. For $J < 0$ (AF coupling), however, the predicted $S_{\text{tot}} = s - \frac{1}{2}$ is obtained without ambiguity. In any event, the significant feature is the change in symmetry as the sign of J is varied; this indicates that perturbation theory in powers of J has a vanishing radius of convergence, as the point $J = 0$ is singular.

---

[42] D. Cragg and P. Lloyd, *J. Phys.* **C12**, L215 (1979); D. Mattis, *Phys. Rev. Lett.* **19**, 1478 (1967).

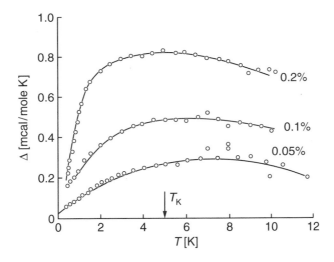

Fig. 6.12.   The excess specific heat for the CuFe alloys, corresponding to a total entropy of precisely $k_B \ln 2$ per spin.[39]

It is known from thermodynamics that the entropy $\mathscr{S}(T)$ relative to the value at high temperature $\mathscr{S}_\infty$ is given by

$$\mathscr{S}(T) = \mathscr{S}_\infty - \int_T^\infty dT' c(T')/T'$$

with $\mathscr{S}_\infty = k_B \ln 2$ for spin one-half. In Fig. 6.12 the specific heat curves of dilute copper-iron alloys are shown; the area under the suitable integral yields $k_B \ln 2$, implying $s = 1/2$ on the iron and confirming that the ground state entropy is zero. Such experimental evidence of a singlet ground state is paralleled by the variational estimates concerning the ground state of the system for antiferromagnetic coupling[43] and by the above-mentioned theorem[42] applied to $s = 1/2$. A variety of approximate methods[44] have all shown the expected decrease in total moment as the temperature is lowered ($J < 0$) or the increase ($J > 0$).

In (6.55) we have already obtained the self-energy of a magnetic ion in a metal, seen to be finite and symmetric in J. It is thus the higher-order terms that must lead to the expected anomaly, third- or higher-order in the

[43] K. Yosida, *Phys. Rev.* **147**, 223 (1966).
[44] See the various chapters of Rado and Suhl series.[45]
[45] K. Yosida and A. Yoshimori, in *Magnetism*, Vol. 5, G. Rado and H. Suhl, eds., Academic Press, New York, 1973, p. 253.

energy or second- or higher-order in the Born expansion of the scattering amplitude. To pinpoint the origin of the difficulties, consider what occurs when one attempts to eliminate the perturbing Hamiltonian $\mathscr{H}_{ex}$ by a unitary transformation, even to lowest order in the "small parameter" J. First,

$$\mathscr{H} = \mathscr{H}_0 + \mathscr{H}_{ex}$$

$$\mathscr{H}_0 = \sum_{km} \varepsilon_k c_{km}^\dagger c_{km}$$

$$\mathscr{H}_{ex} = (-J/2N)\mathbf{S}\cdot \sum_{\substack{kk'\\mm'}} c_{km}^\dagger \sigma_{mm'} c_{k'm'}$$

$$(6.80)$$

in which $\boldsymbol{\sigma} = (\sigma_x, \sigma_y, \sigma_z)$ is a vector with 2×2 matrix components, the Pauli spin matrices inserted into (6.79); cf. (3.79). Following a common procedure,[46] pick an operator Ω such that

$$[\Omega, \mathscr{H}_0] = \mathscr{H}_{ex}.$$

$$(6.81)$$

It will be $O(J)$. A canonical transformation $\mathscr{H} \to e^{-\Omega}\mathscr{H} e^{\Omega}$ produces

$$\left.\begin{array}{l} \mathscr{H}_0 \to \mathscr{H}_0 - [\Omega, \mathscr{H}_0] + \dfrac{1}{2!}[\Omega, [\Omega, \mathscr{H}_0]] + \cdots \\[2mm] \mathscr{H}_{ex} \to \mathscr{H}_{ex} - [\Omega, \mathscr{H}_{ex}] + \cdots \end{array}\right\}$$

$$(6.82)$$

and

$$\mathscr{H} \to \mathscr{H}_0 - \frac{1}{2}[\Omega, \mathscr{H}_{ex}] + O(J^3).$$

$$(6.83)$$

The operator linear in J has been transformed away (except for the small part "on the energy shell" which leads to scattering and thus cannot be physically removed). The Ω that achieves this and solves (6.81) is,

$$\Omega \equiv JS\cdot\boldsymbol{\lambda}_c, \quad \boldsymbol{\lambda}_c = (1/2N) \sum_{\substack{\varepsilon \neq \varepsilon'\\mm'}} \frac{c_{km}^\dagger \sigma_{mm'} c_{k'm'}}{\varepsilon_k - \varepsilon_{k'}}.$$

$$(6.84)$$

It is well defined except on the energy shell $(\varepsilon = \varepsilon')$. Writing $\mathbf{S}\cdot\mathbf{s}_c(0)$ as $S_\alpha s_c^\alpha$ and $\mathbf{S}\cdot\boldsymbol{\lambda}_c$ as $S_\beta \lambda_c^\beta$ using a summation convention on repeated indices, we find, to new leading order in J, a Hamiltonian

$$\mathscr{H} = \mathscr{H}_0 - \frac{1}{2}J^2\{S_\alpha S_\beta[\lambda_c^\alpha, s_c^\beta] + [S_\alpha, S_\beta]s_c^\beta \lambda_c^\alpha\},$$

$$(6.85)$$

---

[46] A simplified version of time-ordered perturbation theory.

in which we have used the fact that $\mathbf{S}$ commutes with all components $\mathbf{s}_c$ of and $\boldsymbol{\lambda}_c$. For classical spins, the components of $\mathbf{S}$ commute with each other and the second commutator vanishes. This commutator is also absent in ordinary potential scattering. The first commutator is representative of conventional scattering. $\boldsymbol{\lambda}_c$ and $\mathbf{s}_c$ are both quadratic forms in fermion operators; their commutator is, again, quadratic. The contribution from this term to the scattering amplitude (which, in first order, was J) is $O(J^2/D)$. It is negligible for small J.

For quantum mechanical spins, the second commutator does *not* vanish. In terms of components, the spin commutation relations of Chapter 3 are

$$[S_\alpha, S_\beta] = i\varepsilon^{\alpha\beta\gamma} S_\gamma \tag{6.86}$$

(with $\hbar = 1$, $\varepsilon^{\alpha\beta\gamma} = \pm 1$ for even/odd permutations of x, y, z and 0 otherwise). The operator $s_c^\alpha \lambda_c^\beta$ is *quartic* in the fermion operators. Among the contributions to electron scattering on the energy shell, we find such new terms as

$$\frac{J^2}{2N} \sum \frac{1 - 2f_{k'}}{\varepsilon_k - \varepsilon_{k'}} \tag{6.87}$$

in which $f_{k'} = \langle c_{k'}^\dagger c_{k'} \rangle$ is the Fermi function. The principal part of (6.87) diverges logarithmically as $\varepsilon_k \to \varepsilon_F$ and $T \to 0$; the Fermi function discontinuity at ε_F is sharper at low temperature.

Indeed, the calculation by Kondo showed that the scattering amplitude of an electron $(\mathbf{k})$ including constructive interference between first and second orders became

$$\sim J \left(1 + \frac{J}{D} \ln \frac{|\varepsilon_k - \varepsilon_F|}{D} \right) \tag{6.88}$$

at $T = 0$. The scattering cross section should then vary as

$$\sim J^2 \left(1 + \frac{2J}{D} \ln \frac{|\varepsilon_k - \varepsilon_F|}{D} \right) \tag{6.89}$$

to the stated order. A typical energy is $\varepsilon_k = \varepsilon_F + kT$; then, the resistivity is:

$$R = R_0 + R_1 \left[1 - \frac{2J}{D} \ln(D/kT) \right] \tag{6.90}$$

in which R_0 contains the temperature-independent contributions, R_1 is a lumped constant. The result is the Kondo formula quoted earlier, (6.78).

The magnitudes of T_K estimated from resistivity measurements range from less than 1 K (MgMn, CuMn, CdMn, etc.) to greater than 1000 K

(AuTi, CuNi)[47] while typical conduction electron bandwidths D range from 10^3 to $10^4 \times k_B$. From this we may assume that even at the largest T_K's the Kondo problem remains within the realm of "weak coupling," however peculiar the ground state might appear. We return to this topic in a later chapter.

6.9.   Spin Glasses

As the polarization cloud surrounding a magnetic ion effectively screens it from further interactions with other magnetic ions, there must be a phase transition to an interactive phase when the interaction energy of two neighboring spins becomes more favorable than the Kondo binding energy of each. For spins ferromagnetically coupled to the conduction sea, this question does not pose itself; the Ruderman-Kittel interaction will connect them at all concentrations c. We can estimate c for the AF coupled ions by recalling that the interaction energy is $\sim (J^2/D)(a/R_{ij})^3 \cos(2k_F R_{ij})$. With $c \sim (a/R_{ij})^3$, this yields

$$c_{\text{crit}} = A(D/J)^2 \exp(-D/2|J|) \qquad (6.91)$$

where A is a constant. For J/D much less than 0.1, the critical concentration drops exponentially to zero. Much of the current research in this topic is centered on the distinction, and competition, between Kondo and spin glass phases. It is clear that the spin glass (SG) phase is the more common.

In the SG phase, the spins interacting *via* the Ruderman-Kittel interactions, have either ferromagnetic or AF couplings depending on the phase of $\cos(2k_F R_{ij})$, an uncertain quantity at large distances R_{ij}. Later, we shall see that the principal requirements for the existence of a SG phase concern the AF bonds: the concentration of AF bonds, disposed at random, must exceed a critical value, the magnitude of which depends on the topology of the lattice and the effective range of the interaction. We shall return to these considerations in due course, but first, point out the results of numerical experiments by Binder and Schröder on the statistical properties of some half a million bonds J_{ij}, as summarized in Fig. 6.13. Note the near symmetry in the distribution of positive and negative bonds.[48]

---

[47] M. Daybell, "Thermal and Transport Properties," in *Solid State Physics.*[41]
[48] K. Binder and K. Schröder, *Phys. Rev.* **B14**, 2142 (1976).

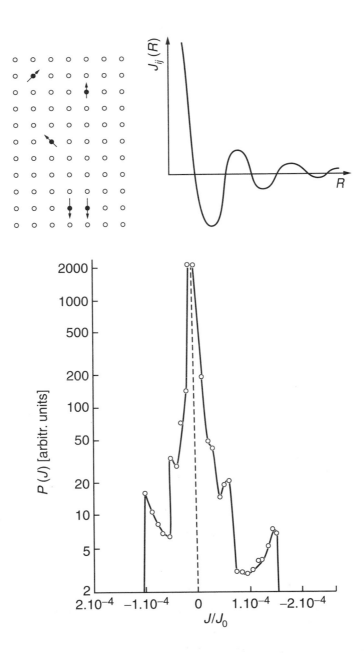

Fig. 6.13. Interactions leading to Spin Glass phase in dilute magnetic alloy, as computed in Binder and Schröder[48]. Top figures represent RKKY interactions among spins randomly distributed in dilute alloy. Lower figure shows computed distribution of J_{ji}.

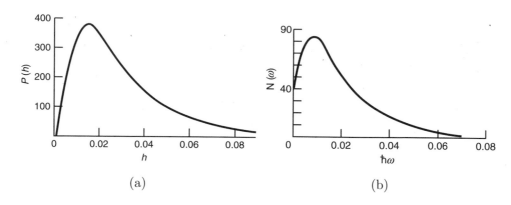

Fig. 6.14. Properties of computer simulation of RKKY spin-glass. (a) Distribution of internal fields and (b) density of elementary excitations. Curves shown are smoothed approximations to the histograms in Walker and Walstedt[49].

Given a particular set of bonds, the spin orientations in the ground state have to minimize the energy. While presenting the appearance of total disorder, the spin-glass ground state is, in fact, highly correlated. It is parametrized by the internal fields (also called "molecular fields") felt by each spin. The distribution $P(h)$ of molecular fields h characterizes the ground state uniquely. On the other hand, the ground state itself is not unique, for there may be many different configurations yielding the same energy; all, however, with presumably the same distribution function $P(h)$.

A computer simulation of the ground state and elementary excitation spectrum of the Ruderman-Kittel SG has been worked out by Walker and Walstedt,[49] with the S_i taken to be classical vectors. The spectrum of elementary excitations is found by quantizing each spin along the direction in which the classical spin points, then diagonalizing leading terms in the Holstein-Primakoff LSW approximation, the quadratic Hamiltonian discussed in Chapter 5. The results of their computations, confirmed by semiempirical theoretical considerations, lead to the curves of Fig. 6.14 obtained for 324 spins on a fcc lattice, $c = 0.003$. It should be noted that the elementary excitations (magnons) are extended states for the entire low-frequency spectrum, although the states in the high-frequency "tail" are localized at

---

[49] L. R. Walker and R. E. Walstedt, *Phys. Rev. Lett.* **38**, 514 (1977); *Phys. Rev.* **B22**, 3816 (1980).

the various strongly interacting closest neighbor pairs. We return to this model in a later discussion of the finite temperature properties of the SG phase, but now fix our attention on other popular models.

The simplest "separable," model first introduced by the present author,[50] takes the signs of the bonds to be separately random

$$J_{ij} = \varepsilon_i \varepsilon_j F(\mathbf{R}_{ij}) \tag{6.92}$$

with the $\varepsilon_i = \pm 1$ random variables, distributed independently according to a specified probability distribution. The $F(\mathbf{R}_{ij})$ can be chosen to mimic the Ruderman-Kittel long-range interaction, but two simpler choices can be made: $F(\mathbf{R}_{ij}) \neq 0$ for $\mathbf{R}_{ij}$ a nearest-neighbor vector only, or $F(\mathbf{R}_{ij}) =$ constant, independent of distance. The former leads straightaway to the ordered n.n. models which have been extensively studied, and are found in the various chapters of the present book. The long-range version[51] is soluble in closed form by mean-field theory. The trick leading to the solution of the separable models is the transformation from spins $\mathbf{S}_i$ to new spin variables σ_i, as follows: $\mathbf{S}_i \equiv \varepsilon_i \sigma_i \ (= \pm \sigma_i)$. Only for a quantum Heisenberg model is this transformation *in*admissible, for it is forbidden to invert all *three* components of a spin vector [recall the example following (3.6) of Chapter 3]. For classical spins, or quantum XY or Ising models, the transformation brings the Hamiltonian into a simpler form

$$\mathscr{H}_{\text{sep}} = -\sum_{i,j} J_{ij} \mathbf{S}_i \cdot \mathbf{S}_j$$

$$= -\sum_{i,j} F(\mathbf{R}_{ij}) \sigma_i \cdot \sigma_j \, . \tag{6.93}$$

The latter is the type of non-random Hamiltonian we previously considered in Chapter 5. Now, obtaining correlated $\sigma_i \cdot \sigma_j$ in the ground state (e.g., parallel spins in the case $F > 0$) does not mean the physical spins $\mathbf{S}_i$ have an *obvious* correlation, for the quantity $\mathbf{S}_i \cdot \mathbf{S}_j = \varepsilon_i \varepsilon_j \sigma_i \cdot \sigma_j$ will have the sign of $\varepsilon_i \varepsilon_j$ and the *appearance* of a random variable. Nevertheless, it is obvious that the **S**'s are as strongly correlated as the σ's, *for a given set* of ε. *But these correlations are random and, in principle, un-knowable*. The long-range

[50] D. Mattis, *Phys. Lett.* **56A**, 421 (1976); err. **60A**, 492 (1977); D. Sherrington, *Phys. Rev. Lett.* **41**, 1321 (1978); W. Y. Ching and D. L. Huber, *Phys. Rev.* **B20**, 4721 (1979). We return to this model in Sec. 8.14 of Chap. 8.
[51] J. M. Luttinger, *Phys. Rev. Lett.* **37**, 778 (1976).

version of this model is remarkably simple; we illustrate for the ground state, replacing F by its average

$$\mathscr{H}_{\text{sep L.R.}} = -\frac{\langle F \rangle}{N} \left(\sum_i \sigma_i \right)^2. \tag{6.94}$$

For $\langle F \rangle < 0$, we have $\sum \sigma_i = 0$, a condition satisfied by almost any random configuration of the σ's. The ground state is then totally disordered, just as for N noninteracting spins, regardless of the magnitude of F! This implies also the lack of an order-disorder phase transition at finite T, as we shall determine later. In the absence of correlations, the SG phase does not exist, so these spins are either perfectly free or are in their Kondo ground states. At the opposite extreme is $\langle F \rangle > 0$, for which the ground state σ_i's are *all* parallel.

The physical problem studied numerically by Walker and Walstedt does not have all bonds positive or negative, but rather a mixture of both. There have been several attempts to introduce models that combine this element together with some simplification, sufficient to render the analysis tractable. Notably, Edwards and Anderson[52] and Sherrington and Kirkpatrick[53] have introduced models with such features, but these have also, to a great extent, resisted analysis. Although such studies have attracted much attention lately, the nature of the ground state is still not clear, not to mention the question of the existence of a thermodynamic phase transition at finite temperature! Recent studies indicate[54] that $T_c = 0$ in less than 3 dimensions (if the bonds are random with zero mean, and restricted to nearest-neighbors). Nevertheless, the study of ground state and low-lying states in such models is of the greatest interest.

Take, for example, a model in which Ising spins $\mathbf{S}_i = \pm 1$ are regularly arrayed at the vectices of a square lattice with nearest-neighbor bonds $J_{ij} = \pm 1$. Let p be the probability of an AF bond, in the range $0 \leq p \leq 1/2$ ($p > 1/2$ can be obtained by symmetry). The ground state energy, E_0, is easily found numerically: for p in the range $0 < p \leq p_c$ ($p_c \sim 1/4$) the energy increases monotonically from $-2N$ to approximately $-2^{1/2}N$. From p_c on,

[52] S. F. Edwards and P. W. Anderson, *J. Phys.* **F5**, 965 (1975) (nearest-neighbor model).

[53] D. Sherrington and S. Kirkpatrick, *Phys. Rev. Lett.* **35**, 1792 (1975); *Phys. Rev.* **B17**, 4384 (1978) (long-range interaction model).

[54] P. Reed, *J. Phys.* **C12**, L799 (1979) and references therein; L. Morgenstern and K. Binder, *Phys. Rev. Lett.* **43**, 1615 (1979) (for 2D). See also K. Binder and W. Kob, *Glassy Materials and Disordered Solids*, World Scientific, 2005, pp. 221ff.

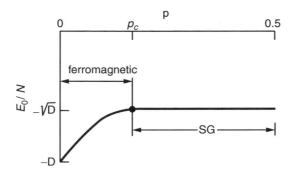

Fig. 6.15.   Ground state energy per spin in n.n. model, bonds ± 1, with $p =$ probability of AF bond. Energy is independent of p in the SG phase, and takes the values $-D^{1/2}$ for $D = 1, 2, 3$ (l. c., sq. and sc); p_c varies according to D.

E_0 is independent of p and maintains the value $-2^{1/2}N$; this is suggestive of a distinct phase. And indeed, in 3D, E_0 starts at $-3N$ at $p = 0$ and rises to approximately $-3^{1/2}N$ at p_c, staying constant thereafter. In 1D, we of course have the trivial result: $-1N$ and $-1^{1/2}N$, with $p_c = 0$. It is not known whether in 4D the numbers are 4 and $4^{1/2}$, respectively, but this is certainly suggestive; the typical random-walk like behavior being shown in Fig. 6.15. The ground state properties have been studied, in a variety of models, by a variety of techniques too numerous to list.[55]

A somewhat more tractable model SG is based on the "spherical model"; each spin interacts with every other spin, the $N(N-1)/2$ bonds $J_{ij} = \pm 1/N^{1/2}$ being randomly selected. If the spins are treated in the spherical model, i.e., $\sum_i S_i^2 = N$ is required (and there is no constraint on the individual S_i^2) the ground state and thermodynamic properties can be calculated[59] from the eigenvalue spectrum of the J_{ij} matrix. Fortunately, this spectrum is known exactly.[59,60] For example, the ground state energy is precisely $E_0 = -N$.

We return to these topics in a later chapter on statistical mechanics.

---

[55] Ground state properties, particularly the problems of "frustration" were pioneered in Toulouse[56]. The quantum features are treated in Marland and Betts[57] and also in Sherrington[50]; the S-K model[53] in Palmer and Pond[58].

[56] G. Toulouse, *Commun. Phys.* **2**, 115 (1977); S. Kirkpatrick, *Phys. Rev.* **B16**, 4630 (1977); E. Fradkin *et al.*, *Phys. Rev.* **B18**, 4789 (1978).

[57] L. Marland and D. Betts, *Phys. Rev. Lett.* **43**, 1618 (1979).

[58] R. Palmer and C. Pond, *J. Phys.* **F9**, 1451 (1979).

[59] J. M. Kosterlitz, D. J. Thouless and R. C. Jones, *Phys. Rev. Lett.* **36**, 1217 (1976).

[60] M. L. Mehta, *Random Matrices*, Academic Press, New York, 1967; D. C. Mattis and R. Raghavan, *Phys. Lett.* **75A**, 313 (1980).

6.10. Magnetism without Localized Spins: Preliminaries

The strongly magnetic properties of the iron transition series metals and of their alloys must be explained by band theory. This would be required for the following two reasons alone: the number of Bohr magnetons per atom is generally far from being an integer, and both calculated and observed bandwidths are of the order of electron volts. Neither of these facts could be accorded with a scheme based on localized spins, and together with such additional evidence as the abnormally high specific heat (which can only be explained by a continuum of states for the magnetic carriers) they point instead to the need for associating the uncompensated spins with Bloch electrons rather than with their Wannier counterparts in the first approximation.

The band theory of magnetism is currently a very active field of research, albeit far less developed than the relatively straightforward Heisenberg theory for insulators; but the physical mechanisms are established and some of the consequences therefrom approximately understood. Let us start by visualizing, inaccurately perhaps, localized spins in the indirect exchange theory being gradually modified so as to allow some overlap and the formation "magnetic" electron bands. The energy gap against the excitation of electrons in the magnetic states disappears and a finite density of states appears at the Fermi surface. This now allows charge fluctuations, hence the conduction of electricity by magnetic electrons; it permits them to contribute to the electronic specific heat, microwave absorption, etc. On the other hand, there is no particular reason why the inherently *magnetic* properties — such as the magnon spectrum — should be affected to leading order. As with indirect exchange, there is no universally valid reason that the long-range order in the ground state be ferromagnetic, and spiral or other antiferromagnetic configurations, such as are actually observed in Mn and in Cr, are not difficult to reconcile with this approach.

One salient difference with the indirect-exchange theory, is that nonmagnetic *s*-like bands can be ignored in the first approximation. Doubtless some indirect exchange still goes on, but the dominant, primary interactions are *internal* to the magnetic bands themselves. An interesting point of view has it that the *d* electrons (e.g., in Fe) effectively split into two sub-bands; one of which is narrow, so that the electrons it contains are well localized like the rare-earth *f* states, and the other is more like an ordinary conduction band. It is quite likely, that even were such a model describable by the indirect exchange Hamiltonian of (6.52), the perturbation-theoretic

Table 6.3.   Scale of energies in descending order.

~ 10 eV	(a) Atomic Coulomb integrals U
$\wr$	(b) Hund's rule exchange energy, J
$\wr$	(c) Energy of electronic excitations violating Hund's rule
$\downarrow$	(d) Electronic band widths, W (also denoted D)
~ 1 eV	(e) $\mathcal{N} \div$ (density of states at Fermi Surface)
0.1–1.0 eV	(f) Crystal field splittings
10^{-2}–10^{-1} eV	(g) Spin-orbit coupling
	kT_C or kT_N
10^{-4} eV	(h) Magnetic spin-spin coupling
	Interaction of a spin with external field 10 kG
10^{-6}–10^{-5} eV	(i) Hyperfine electron-nuclear coupling.

solution on which the Ruderman-Kittel interaction is based, would not be valid. The most spectacular feature of the band theory is that with neglect of the nonmagnetic electrons the remaining magnetic ones do not number any particular rational multiple of the number of atoms. Therefore, the occupation number of the magnetic bands is a vital, and often an adjustable, parameter in the theory.

For definiteness in this study, we shall wish to demonstrate the existence of a threshold magnitude for the interactions, that *below* a certain magnitude of the Coulomb repulsion and Hund's rule coupling parameters, *there is no metallic magnetism*. To establish the possible orders of magnitude, we list in Table 6.3 the relative strengths of various plausible physical forces.

In the creation of magnetic moments in the metal, the competition is between the "kinetic energy" terms (d) and (e), and the potential energy terms (a)–(c), representing forces internal to each incomplete shell in each atom of the solid. For isolated atoms the kinetic terms are zero and the magnetization of the incomplete shell proceeds according to Hund's rules, detailed and proved in an earlier chapter. But what happens when kinetic energy is introduced?

And once the criteria for the existence of magnetic moments are satisfied, what is their alignment? We have already observed in connection with the indirect-exchange theory, that the geometric factor $k_F a$ plays an important role, i.e., the filling of the band. For an almost full or almost empty band, $k_F \to 0$ and long wavelengths predominate — i.e., ferromagnetism. For nearly half-filled bands, some species of AF are possible. The use of a variable axis of quantization for local spin polarization as an adjunct in the study of electron interactions in magnetic metals has become increasingly popular in the recent literature.[61]

Much of the commonplace understanding about itinerant ferromagnets comes from the example of nickel, a ferromagnetic metal having 0.6 Bohr magnetons *per* atom, with 0.6 holes per atom in the *d* band. Alloying with copper, having a similar band structure but a higher Fermi level, "plugged up" the holes and quenched the ferromagnetism in direct proportion to the added number of electrons. Now, the important discovery of metallic ferromagnetic alloys made out of ordinarily nonmagnetic constituents[62] — e.g., $ZrZn_2$ — has made the need for a reliable theory of itinerant electron ferromagnetism more urgent. Neutron diffraction has shown that some of the magnetic moment in such an alloy resides between the atoms and not on them, i.e., is identified with the bonding, mobile electrons.[63]

There is one common but mystifying finding that comes out of practically all band structure — and local density approximation — (*LDA*) calculations. The majority-spin band is invariably *narrower* than the partly filled minority-spin band. The simplest explanation is the following: in building up the solid one inserts the $N_\uparrow$ electrons first, therefore they experience an ionic potential that is more attractive (in the amount of ≈ 1 charge *per* atom) than what the succeeding $N_\downarrow$ will experience. This potential draws the Wannier orbitals of the $\uparrow$ electrons closer to the individual nuclei than the Wannier orbitals of the $\downarrow$ particles. It follows that the majority band structure is narrower because the overlap of a function centered about $\mathbf{R}$, with sites at $\mathbf{R} + \boldsymbol{\delta}$ and $\mathbf{R} + \boldsymbol{\delta} + \boldsymbol{\delta}'$ (its n.-n. and n.-n.-n. sites) *shrinks accordingly.*[64]

In a magnetic solid, the lower energy band will be completely filled first. Then, its width or narrowness becomes, in fact, irrelevant, insofar as electrical transport and charge fluctuations are concerned. These are the exclusive domain of the minority band, unless the majority band is only partially filled.

One complicating factor: LRO is not always ferromagnetic, as the axis of polarization could be defined differently at each atomic site (cf. the spiral

[61] S. H. Liu, *Phys. Rev.* **B17**, 3629 (1978); V. Korenman, J. Murray and R. Prange, *Phys. Rev.* **B16**, 4032, 4048, 4058 (1977); J. Hubbard, *Phys. Rev.* **B19**, 2626; **20**, 4584 (1979).
[62] B. Matthias and R. Bozorth, *Phys. Rev.* **109**, 604 (1958).
[63] S. Pickart, H. Alperin, G. Shirane and R. Nathans, *Phys. Rev. Lett.* **12**, 444 (1964).
[64] We could also invoke an analogy to semiconductors — to the extent that higher mass implies narrower bands. It is well known that the mass of holes in the valence bands of Ge, Si, GaAs, etc., is typically a factor $O(5-10)$ times greater than the mass of electrons at the bottom of the (higher energy) conduction band.

antiferromagnets). So first we just look at the creation of local moments in Fe and similar metals, to see what differentiates them from ordinary paramagnetic electronic fluids such as in Cu or Al.

The following "toy model" serves only to elucidate the tendency to *filling a majority band*, not the narrowing effect discussed above. We consider $N_{elec} = 5$ and 6 on $N = 4$ sites, at the corners of a square. Let us assume that — either because of on-site Coulomb repulsions, Hund's rule mechanisms or the n.-n. exchange mechanism favored by Hirsch[65] — there exists on overall preference for parallel spins, as expressed by the mean-field parameter J in the following Hamiltonian:

$$\mathcal{H} = -t \left(\sum_{j=1,\sigma}^{3} (c_{j,\sigma}^{\dagger} c_{j+1,\sigma} + \text{H.c.}) + c_{4,\sigma}^{\dagger} c_{1,\sigma} + \text{H.c.} \right) - \frac{J}{2} S_{tot}^2 . \quad (6.95)$$

Plane wave states have wave-vectors $k = 0$, $\pm\pi/2$ and π, the corresponding "Bloch energies" being $\varepsilon(k) = -2t \cos k = -2t$, ± 0 and $+2t$.

$N_{elec} = 5$, $s = 3/2$ (maximum spin for 5 electrons on 4 sites). Now if 4 electrons have spin ↑ (the maximum), the total ground state energy for the 5 electrons is $E_0 = -2t - (15/4)J$ for a rotationally invariant state of total spin 3/2. Excited states belonging to the same total spin are $E_1 = E_2 = -(15/4)J$ and $E_3 = +2t - (15/4)J$, all associated with the motion of the minority spin electron.

Next, how does that compare with $s = 1/2$?

$N_{elec} = 5$, $s = 1/2$ (minimum spin). Three electrons of spin ↑, two of spin ↓. *This is as close to nonmagnetic as one gets with an odd number of electrons.* The ground state $E_0 = -4t - (3/4)J$ lies below $s = 3/2$ only if $J < (2/3)t$. Here, the bandwidths for the electrons of spin ↑ and ↓ are identical, each $4t$. (The lowest excited states' energies include $E_1 = \cdots = E_4 = -2t - (3/4)J$ etc., the result of motion of either majority of minority particles.)

$N_{elec} = 6$, $s = 1$ (maximum spin). Four electrons of spin ↑ are immobilized, 2 "minority" spins ↓ can move. The lowest energy is $E_0 = -2t - 2J$, the highest $+2t - 2J$.

$N_{elec} = 6$, $s = 0$ (minimum spin). $E_0 = -4t$. This lies lower than the preceding only if $J < t$.

---

[65] J. Hirsch, *J. Appl. Phys.* **67**, 4549 (1990).

Recapitulating: For 5 electrons on 4 sites, the ground state has maximum spin 3/2 if $J > (2/3)t$. For 6 electrons, the ground state has maximum spin 1 if $J > t$. Moreover, in that case both conditions are met, such that in *both* instances the majority spins ↑ are immobilized while the minority spins ↓ have a full bandwidth available for their motional degrees of freedom. So if we were to make a phase diagram, $0 < J/t < 2/3$ describes one phase, $2/3 < J/t < 1$ a second phase and $J/t > 1$ a third phase. This is taken up in more detail in the Problem.

Problem 6.5.   Consider a three-dimensional $2 \times 2 \times 2$ cube with atomic sites at each corner connected to 3 n.-n. sites by a matrix element $-t$. The exchange term will be again be taken as $-JS_{\text{tot}}^2$, i.e., $-Js(s+1)$, where $s =$ total spin number. For 10 electrons on 8 sites, s ranges from 0 to 3. Identify stable phases as a function of J/t by the value of s in the ground state. The possibilities are: $s = 3$ (maximum magnetization, (a clear winner if J is sufficiently large), partial ferromagnetism $s = 2$ and 1, and the nonmagnetic singlet phase at $s = 0$).

Identify and analyze the excited states also, according to whether they correspond to charge fluctuations (S_z is invariant) or spinwave excitations (S_z changes by ± 1) from the ground state.

6.11. Degenerate Bands and Intra-Atomic Exchange Forces

To assess the role of intra-atomic exchange, we here turn to the special case of two degenerate d-like bands, a and b, with various Coulomb and exchange integrals. We then assess the situation in the iron series ferromagnetic metals, to see which are the physical parameters to which theory must address itself. The calculations are an extension of the theory we first broached in Sec. 4.15.

The interaction of a singlet pair within each band is $U_{aa} = U_{bb}$ and the t matrix that ensues is approximated by the constant

$$\tilde{t}_{aa}(\rho) = \frac{U_{aa}}{1 + U_{aa}(1 - \rho^{1/3})/W} \,. \tag{6.96}$$

For a singlet pair of which one member occupies one band and one the other, there is a similar expression with U_{ab} replacing U_{aa}. It is a simple mathematical identity for Coulomb integrals given in (6.32) to prove $U_{aa} > U_{ab} > 0$. (Intuitively, electrons in two distinct orbitals overlap less, hence have a smaller Coulomb repulsion, than two electron in identical orbitals.) Moreover, for triplet pairs (e.g., electrons of parallel spins) in two different bands, this Coulomb integral is further reduced by an exchange correction, (6.41), and becomes $U_{ab} - J_{ab} = U'_{ab}$, which we also write as $U_{ab}(1 - j)$. The parameter j is thus the fractional inter-band exchange parameter, $0 < j < 1$. The ground state singlet energy thus includes 3 distinct interactions:

$$E_0 = N\rho^{5/3}\left\{\frac{18}{5}W + 2\rho^{1/3}[\tilde{t}_{aa}(\rho) + \tilde{t}_{ab}(\rho) + \tilde{t}'_{ab}(\rho)]\right\}. \tag{6.97}$$

It is to be compared to the ferromagnetic state, with half the particles in band a and half in band b but all with spin "up," which has energy

$$E_f = N\rho^{5/3}\left[2^{2/3}\frac{18}{5}W + 4\rho^{1/3}\tilde{t}'_{ab}(2\rho)\right], \tag{6.98}$$

and with the ferromagnetic state in which all the particles, of spin up, are also in a single band, say a (this represents spin *and* orbital magnetism, and

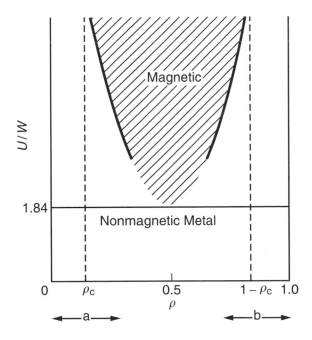

Fig. 6.16. Phase diagram of 1-band Hubbard model at $T = 0$, based on low-density theory. Region (a) is low density for electrons and (b) for holes. The high density results near 0.5 are unreliable, due to limitations of the theory. Potentially magnetic region is shaded, and $\rho_c = 0.136$ in 3D.

should occur when the perturbations of the solid are too weak to quench the orbital moments of the individual atoms, as in f shells of the rare earths). This state has energy

$$E_{f,0} = N\rho^{5/3} \left(\frac{9}{5} 4^{5/3} W \right). \qquad (6.99)$$

The case of maximal interband exchange $j = 1$ is special; we compare the three energies and conclude that (6.99), representing spin + orbital magnetism, can never lie lowest. The phase diagram showing the region where the spin-only magnetic moments form is remarkably similar to Fig. 6.16, with only the numerical value of parameters $\rho_c = 0.04$ and $U_{min}/W = 1$ being different. Doubling the number of bands allows magnetic moment formation at much lower densities and interaction parameter U than previously.

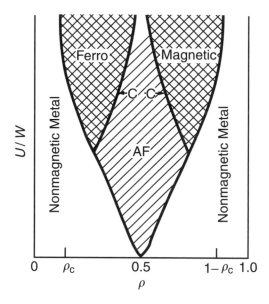

Fig. 6.17. Same as preceding, with magnetic ordering taken into account. The half-filled band is most easily susceptible to AF ordering (Examples: NiO, and also 1D Hubbard model[66]) but at strong enough coupling in 2D or 3D (not 1D) gives way to the ferromagnetic phase (cross-hatching). Comparison of the energies in AF and ferromagnetic phases produces curves C: see Eq. (4.114).

---

[66] E. Lieb and F. Wu, *Phys. Rev. Lett.* **20**, 1445 (1968).

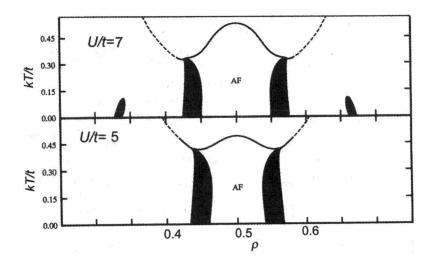

Fig. 6.18. Phase diagram at finite T in 3D Hubbard model, near half-filling, for two values of U in weak coupling. Shaded areas represent Berker's τ-phase. [Adapted from a figure kindly provided by Dr. Nihat Berker, private communication, cond-mat/0503226 v2 15 Mar. 2005.]

Effects of temperature on the phase diagram of the one-band Hubbard model have been examined by N. Berker, in several publications based on quantum Monte Carlo numerical calculations. Figure 6.18 is representative of the results. At various values of U (note that $U/t = 6$ is small, equivalent to $U/W = 0.5$ on the sc lattice) the AF phase that is stable at half-occupancy gives way to a disordered phase at higher temperatures. But most remarkably, new phases, shaded here and called the τ-phases, appear in the diagram. There is no other theoretical justification for, or description of, the τ-phases. These are possibly artifacts of the method. However, the very *existence* of high-temperature superconductivity should make one reluctant to dismiss these so-far unidentified states.

The physically plausible cases of partial Hund's rule exchange $j = J_{ab}/U_{ab} < 1$ may be of interest. We compared (6.97–6.99) by numerical calculation, assuming $U_{ab} = \frac{1}{2}U_{aa} \equiv U$, and found: *spin + orbital magnetism never occurs* for $U/W < 20$ nor for densities far from 1/2, so this case may be eliminated from practical considerations even though it is good to know it exists (in principle) in the atomic limit $W \to 0$. Spin magnetism has itself a restricted range of stability, which depends strongly on j. For $j = 0.5$ or greater, the situation is qualitatively similar to $j = 1$, whereas for smaller values of j (0.2 or 0.1) the regions of stability shrink rapidly and become nonexistent at $j = 0$. Thus, regardless of the strength of the interaction parameter U, the *existence of a magnetic moment* ultimately *depends on the Hund's rule exchange* parameter, as a stabilizing factor.

The spatial ordering of the moments, once they are created, is a delicate competition between several mechanisms already considered in this chapter, principally, the indirect exchange and Nagaoka mechanisms. But before proceeding, it is prudent to consider to what extent electrons in real materials satisfy the various simplifying assumptions we have made, for example, the effective-mass approximation $E(k) = \hbar^2 k^2/2m$ with $m = m^*$ the band structure effective mass before interactions and $= \tilde{m}$ the total mass after the interactions (which are necessarily strong and thus nontrivial) have been incorporated.

The hypothesis of two kinds of d electrons, has been given added credence lately by Stearns. Her "95% local" and "5%" itinerant model counters the pure itinerant or pure Heisenberg models, against which substantial evidence has been accumulating. In her view,[67] most (95%) of the d electrons lie in the relatively flat portions of the band structure, where the high density of states (or small W, large U/W) promotes magnetic moment formation, with the small residual fraction occupying states well described in the effective mass approximation, with effective masses $\sim m_{\text{el}}$. To see this, let us examine

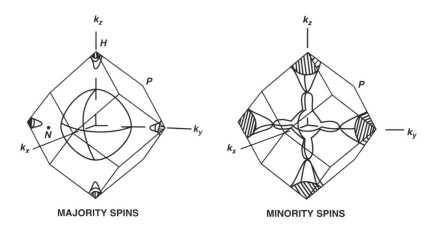

MAJORITY SPINS   **MINORITY SPINS**

Fig. 6.19. Fermi surface of iron, as given in Gold *et al.*[68], here adapted from Stearns[67].

[67] M. B. Stearns, *Phys. Today* (April 1978), pp. 34–39.
[68] A. V. Gold, L. Hodges, P. Panousis and R. Stone, *Int. J. Magn.* **2**, 357 (1971).

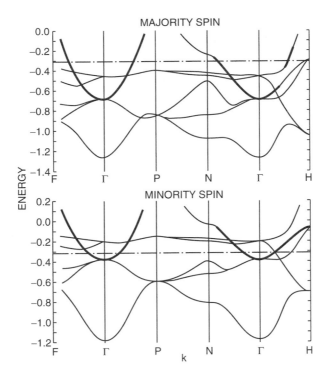

Fig. 6.20.   Band structure of iron, for majority and minority electrons as calculated in Duff and Das[69] and reproduced in Stearns[67].

the Hartree-Fock band structure of iron, as shown in Figs. 6.19 and 6.20 reproduced from her article, "Why is iron magnetic?"[67] There are shown the flat parts of the d bands and the itinerant parts, drawn in heavy lines. The latter have a curvature corresponding to approximately the electron free mass m_{el}, Fermi wave vectors and occupation numbers $n\pm$ that have been given as[70,71]

	$ak_{F+}/2\pi$	$n+$	$ak_{F-}/2\pi$	$n-$	
Fe	0.50	4.4	0.19	2.2	
Ni	0.65	4.6	0.53	4.0	(6.100)

in agreement with tunneling experiments. Various other experiments indicate that the s electrons are only weakly spin polarized, and that the nodes in the

[69] K. Duff and T. Das, *Phys. Rev.* **B3**, 192, 2294 (1971).

[70] M. B. Stearns, *J. Magnetism Magn. Mater.* **5**, 167 (1977).

[71] J. Kuebler, *Theory of Itinerant Electron Magnetism*, Clarendon Press, Oxford, 2000, p. 193.

Ruderman-Kittel oscillation make this weak indirect exchange mechanism favor antiferromagnetism. It follows that the ferromagnetic alignment arises from the d electrons themselves; but the above wave vectors are all indicative of AF (see Fig. 6.8(f), or Fig. 6.6 for the expected ground state symmetry at the above values of k_F), except the minority one for Fe, k_{F-}. There is thus a mystery to be resolved. Could it be that nearest-neighbor ferromagnetic exchange is sufficiently strong that Heisenberg-like ferromagnetism coexists with the itinerant electron minority? The answer: nearest-neighbor exchange is too weak by one order of magnitude,[72] so this will not help.

A plausible explanation lies in a combination of physical effects. First, the density of itinerant d electrons is small, and the interactions are large — placing the parameters in the ferromagnetic region of the phase diagram, Fig. 6.17. Second, the indirect exchange mechanism is no longer given by the Ruderman-Kittel formula of (6.56) with R_{ij} an interatomic distance, for the local moments are not point-like but rather spread out, occupying a large fraction of the volume in each unit cell. The paramagnetic Curie temperature measuring the strength of the indirect exchange and the spin wave spectrum are then better given by the jellium model, (6.71, 6.72), in which both are positive, guaranteeing ferromagnetism. The only remaining question concerns the applicability of what is basically second-order perturbation theory to the interactions between itinerant and localized d-electrons, of equal or greater strength than the relevant energy denominators. The only practical procedure is to calculate the Hartree-Fock energy first, as deduced from the band structure, Fig. 6.20, and add thereto the interaction energies as computed from the $\tilde{t}$'s, as in the 2-band case analyzed above. This total energy is then to be compared to the energy of varying states of polarization, including the antiferromagnetic ordered magnetic phases and the nonmagnetic state. The qualitative arguments given here indicate that a proper calculation will favor the ferromagnetic state of iron. The magnon spectrum, such as we calculate next, should confirm this.

6.12.   Magnons in Metals

Once Coulomb and exchange forces have created a magnetic solid, they profoundly affect the elementary excitations. To the quasi-particles and collective modes which are common to all metals, one must add magnons for the

---

[72] R. Stuart and W. Marshall, *Phys. Rev.* **120**, 353 (1960).

magnetic metals. These are the quantized spinwaves which are the property of an ordered magnetic medium. This section is devoted to their properties, especially in ferromagnetic metals.

With the exception of the strong-coupled one-band model, our knowledge of the itinerant ground state is imperfect even for ferromagnets. We cannot simply proceed by computing an excited state, subtracting the ground state energy therefrom, as even small errors of $O(N)$ will obscure effects $O(1)$. A more direct method is required, that of equation of motion. In this procedure, the scattering of a certain type of excitation is treated exactly but other terms are neglected ("random phase approximation"). The neglected terms vanish at low density, and the procedure gives satisfactory answers at long wavelengths, therefore one can accept the results with the same confidence (or skepticism!) as for magnons in 3D Heisenberg antiferromagnets — as a semiquantitative, systematic approximation. A similar procedure brings the "plasmons," quantized charge fluctuations, into play. They are ignored in the present book but are discussed in any textbook on metal physics.

For simplicity we treat the case of a single band in some detail, including the cases — so far ignored — of partial magnetization. We then indicate the extensions of the theory to multi-band cases and to questions of stability of the assumed ferromagnetic ground state, such as previously examined in connection with the indirect exchange mechanism.

So, let U and ρ be in the range shown in Fig. 6.17 for which the ground state is ferromagnetic. In the Heisenberg model, translational invariance was sufficient to construct the one-magnon states uniquely. Now, it is no longer enough. The degrees of freedom in a metal are sufficiently numerous, that if we only specify that the spin angular momentum must decrease by one unit and the total momentum wave vector must be $\mathbf{q}$, there will be a very large number of excitations to satisfy this requirement. We therefore add to these requirements that of a very long or infinite lifetime, and find then that the magnon is a bound state, a linear combination of all the states in which a hole is created in the majority-spin band (taken to be spin "down" henceforth) while a particle, with extra momentum $\hbar\mathbf{q}$, is added to the minority spin ("up") sub-band.

The quasi-particle energies are,

$$E_m(\mathbf{k}) = E(\mathbf{k}) + \frac{1}{N}\sum_{\mathbf{k}'}\tilde{t}(\mathbf{k}\mathbf{k}')f_{-m}(\mathbf{k}')$$

$$\equiv E(\mathbf{k}) + \Delta_m(\mathbf{k}) \tag{6.101}$$

where the Fermi function $f_m(\mathbf{k})$ is 1 if $E_m(\mathbf{k}) \leq E_m(\mathbf{k}_{Fm})$ and zero otherwise, as discussed previously. Thus the elementary unit of magnetic excitation must be $a^+_{\mathbf{k}+\mathbf{q}\uparrow} a_{\mathbf{k}\downarrow} |F)$, where $|F)$ is the ground state, and $a_{\mathbf{k}m}$ destroys a particle (or creates a hole) in spin sub-band m, and $a^+_{\mathbf{k}m}$ creates one. The one-magnon state must be $|q) \equiv \Omega_{\mathbf{q}} |F)$, where $\Omega_{\mathbf{q}}$ is the operator

$$\Omega_{\mathbf{q}} = N^{-1/2} \sum_{\mathbf{k}} F_{\mathbf{k}} a^+_{\mathbf{k}+\mathbf{q}\uparrow} a_{\mathbf{k}\downarrow} . \qquad (6.102)$$

The energy of each unit must be,

$$E_{\mathbf{kq}} = \frac{\hbar^2}{2m^*} (2\mathbf{k}\cdot\mathbf{q} + \mathbf{q}^2) + \Delta_\uparrow(\mathbf{k} + \mathbf{q}) - \Delta_\downarrow(\mathbf{k}) \qquad (6.103)$$

in the effective mass approximation. Each scatters into the others with a strength proportional to $\tilde{t}(\mathbf{kk}')$ and to the occupation-number factor. We determine this from the equations of motion; that is, if $\Omega_{\mathbf{q}}$ is a raising operator for $\mathscr{H}$ by an energy $\hbar\omega_{\mathbf{q}}$, then

$$[\mathscr{H}, \Omega_{\mathbf{q}}] = \hbar\omega_{\mathbf{q}}\Omega_{\mathbf{q}} . \qquad (6.104)$$

The kinetic energy contribution to this commutator yields

$$[\mathscr{H}_0, \Omega_{\mathbf{q}}] = N^{-1/2} \sum_{\mathbf{k}} F_{\mathbf{k}}(E_{\mathbf{kq}}) a^+_{\mathbf{k}+\mathbf{q}\uparrow} a_{\mathbf{k}\downarrow} \qquad (6.105)$$

whereas the scattering part yields

$$[\mathscr{H}', \Omega_{\mathbf{q}}] = -N^{-3/2} \sum_{\mathbf{k},\mathbf{k}'} F_{\mathbf{k}'}\tilde{t}(\mathbf{kk}')[f_\downarrow(\mathbf{k})' - f_\uparrow(\mathbf{k} + \mathbf{q})] a^+_{\mathbf{k}+\mathbf{q}\uparrow} a_{\mathbf{k}\downarrow} \qquad (6.106)$$

omitting terms such as $a^+_{\mathbf{k}'+\mathbf{q}'m} a_{\mathbf{k}'m} a^+_{\mathbf{k}+\mathbf{q}-\mathbf{q}'\uparrow} a_{\mathbf{k}\uparrow}$ which are smaller by some power of the density ρ and are neglected in the "random-phase approximation." We have identified $a^+_{\mathbf{k}m} a_{\mathbf{k}m} = n_{\mathbf{k}m}$ with its average, $f_m(\mathbf{k})$, the Fermi function.

Equating terms on both sides of the equation of motion (6.104) we obtain the equations of the amplitudes $F_{\mathbf{k}}$

$$F_{\mathbf{k}}(E_{\mathbf{kq}} - \hbar\omega_{\mathbf{q}}) = N^{-1} \sum_{\mathbf{k}'} \tilde{t}(\mathbf{kk}')[f_\downarrow(\mathbf{k}') - f_\uparrow(\mathbf{k}' + \mathbf{q})] F_{\mathbf{k}'} . \qquad (6.107)$$

This equation is exactly soluble at $\mathbf{q} = 0$, by the choice $F_{\mathbf{k}} = $ const., and yields $\omega_0 = 0$. This is an exact result, reflecting the rotational invariance of the magnetic ground state.

For $\mathbf{q} \neq \mathbf{0}$, (6.107) is a transcendental equation that must generally be solved graphically or numerically. However, in the special case of $\tilde{t}(\mathbf{k}\mathbf{k}') = \tilde{t} = $ const., there is a very simple solution $F_{\mathbf{k}} = A(E_{\mathbf{kq}} - \hbar\omega_{\mathbf{q}})^{-1}$ which leads to the secular equation for the eigenvalue

$$1 = N^{-1} \sum_{\mathbf{k}} \frac{\tilde{t}}{E_{\mathbf{kq}} - \hbar\omega_{\mathbf{q}}} . \qquad (6.108)$$

This yields an approximately parabolic magnon $\hbar\omega_{\mathbf{q}} = D\mathbf{q}^2$ for $\mathbf{q} < \mathbf{q}_{max}$; the maximum q is not at the edge of the BZ but occurs when the denominator of (6.108) becomes complex. This signifies the onset of the scattering regime, in which the bound state does not exist; an attempt to excite a magnon at $\mathbf{q} > \mathbf{q}_{max}$ will result only in broad absorption. Rather than detail this calculation, we examine the case of *two* degenerate bands, which we do calculate in detail. This case has an interesting feature: the elementary units in (6.102) can be $a^+_{\mathbf{k}+\mathbf{q},a,\uparrow} a_{\mathbf{k},a,\downarrow} \pm a^+_{\mathbf{k}+\mathbf{q},b,\uparrow} a_{\mathbf{k},b,\downarrow}$, where a, b are the two bands in question. The $(+)$ combination yields $\mathbf{S}^+_{tot}$ at $\mathbf{q} = \mathbf{0}$, and thus commutes with the Hamiltonian. It is identified as the zero energy mode which expresses the rotational invariance of our approximation of the magnetic ground state; the $(+)$ mode at $\mathbf{q} \neq \mathbf{0}$ is identical to that of a single band worked out above. These are denoted the "acoustic magnons" by analogy with lattice vibrations, where the low-lying spectrum are the acoustic modes. Correspondingly there exist the "optical magnons," the $(-)$ branch, which do not exist for a single band ferromagnet. The optical spectrum start at $2(\Delta_\uparrow - \Delta_\downarrow)$ at $\mathbf{q} = \mathbf{0}$ in the two-band case, and increases slightly at higher wave vectors. The equations (with $2m^* = 1$, $\hbar = 1$, and $\mathbf{k}_F = 1$) are given below, and the calculated results plotted in Fig. 6.21. [The acoustic branch is identical to the solution of the one-band case, (6.108).] We replace the parameter $\Delta_\uparrow - \Delta_\downarrow$ by the single parameter Δ, and separately consider the cases when $\Delta < 1$ (weak-coupling and partial magnetization), $= 1$ (intermediate), and > 1 (strong-coupling and saturation magnetization).

When $\Delta \geqslant 1$, $n_{\mathbf{k}\uparrow} = 0$ and we find the following transcendental equation:

$$\pm \frac{2}{3\Delta} = \frac{1}{2q} L(Q) \qquad (6.109)$$

where

$$Q = \frac{2q}{\Delta + q^2 - \hbar\omega_q} \qquad (6.110a)$$

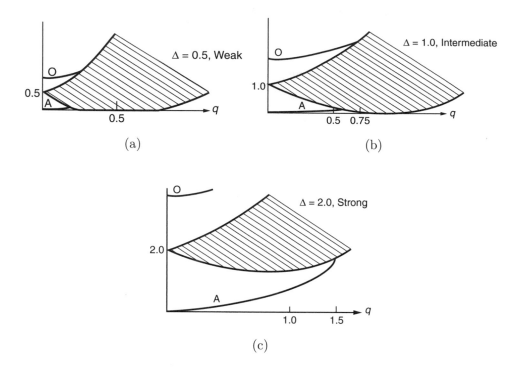

Fig. 6.21. Acoustic (A) and optical (O) magnons in the band theory, for various values of Δ, the "Stoner gap" parameter. Continuum indicated (*shading*) is for elementary excitations with spin flip; continuum for elementary excitations without spin flip remains the same as shown in Fig. 6.5.

and

$$L(Q) = \frac{1}{Q^2}\left[\frac{1}{2}(Q^2 - 1)\ln\left|\frac{Q+1}{Q-1}\right| + Q\right]. \tag{6.110b}$$

For $\Delta \leq 1$, $n_{\mathbf{k}\uparrow} \neq 0$, and we obtain a slightly more complicated equation:

$$\pm\frac{2}{3}\left[\frac{1 - (1-\Delta)^{3/2}}{\Delta}\right] = \frac{1}{2q}[L(Q) - (1-\Delta)L(Q')] \tag{6.111a}$$

where

$$Q' = \frac{2q(1-\Delta)^{1/2}}{\Delta - q^2 - \hbar\omega_q} \tag{6.111b}$$

and (6.110) defines the other parameter and the function $L(Q)$.

The optical magnon mode $(-)$ starts at 2Δ for $\mathbf{q} = 0$ and increases some-what before merging with the continuum. The acoustic $(+)$ branch starts at $\hbar\omega_0 = 0$ for $\mathbf{q} = 0$ and increases approximately $\sim Dq^2 + O(q^4)$, with the parabolic approximation $\hbar\omega_q \sim Dq^2$ improving in relative accuracy as Δ is increased. Expansion of the equations leads to a formula for D

$$D = \frac{1 + (1-\Delta)^{3/2} - 4/5[1 - (1-\Delta)^{5/2}]/\Delta}{1 - (1-\Delta)^{3/2}} \quad \text{for } \Delta \leq 1 \qquad (6.112a)$$

and

$$D = 1 - \frac{4}{5\Delta} \quad \text{for } \Delta \geqslant 1. \qquad (6.112b)$$

The acoustic branch enters the continuum and "dies" at $q_{max} = 0.75\Delta$ for $\Delta \geq 1$.

Antiferromagnetism occurs when spin-down electrons start to fill the Bril-louin zone, as may be seen in the following demonstration for strong cou-pling. Assume *that $\Delta \geq 1$*, and *every state in the spin-down zone is filled, while every state in the spin-up zone is empty.* The eigenvalue equation can be expanded in power of $E(\mathbf{k} + \mathbf{q}) - E(\mathbf{k}) \equiv w(\mathbf{k}, \mathbf{q})$, a procedure cer-tainly valid for small $\mathbf{q}$. We make use of the assumption $E(-\mathbf{k}) = E(\mathbf{k})$ to prove,

$$\sum_{\mathbf{k}} w^{2p+1} \equiv 0 \quad p = 0, 1, 2, \dots, \quad \text{all } \mathbf{q}, \qquad (6.113)$$

which together with the identity $\sum w^{2p} \geq 0$ readily establishes the desired result

$$\hbar\omega_0 = 0 \geq \hbar\omega_{\mathbf{q}} \quad \text{all } \mathbf{q} \neq 0. \qquad (6.114)$$

Not only does a collective magnon mode exist at every $\mathbf{q}$ in the Brillouin zone, but the $\mathbf{q} = 0$ mode is a *maximum*, and the ferromagnetic state must be unstable against the emission of any number and any type of magnons. The new ground state must then be an antiferromagnetic, or a spiral spin configuration of the type previously discussed.

The antiferromagnetic behavior sets in even before the spin-down Bril-louin zone is completely filled by electrons, in the neighborhood of a half-filled zone, although the precise point at which it occurs must be calculated numerically. It is akin to the antiferromagnetism of insulators. Whereas the exclusion principle has the effect in the ferromagnetic state of preventing the kinetic motion of electrons from atom to atom, such motion is not prohibited in the antiferromagnetic configurations. The difference with insulators, is that the "hopping" (band, or kinetic, energy) represents real transitions in the metallic state, but only virtual transitions in the insulator.

Connection between the band theory and the indirect exchange theory is very natural and easy to establish. For when one band is supposed narrow, the Hund's rule splitting of Kramers' degeneracy usually will send the entire spin-up band above the Fermi level, whereas the entire spin-down band will remain below this energy. This explains why narrow bands most likely lead to integral numbers of Bohr magnetons, just as in a localized electron theory. If this narrow band interacts with a broader band, the broader band will be only slightly poloarized. Whenever perturbation theory is applicable, *something like* the Ruderman-Kittel formula must result therefrom. If perturbation theory is not applicable, then the formulas derived in this section and the previous one provide a fair initial approximation to a strong-coupling theory, and should be used instead. Instead of two degenerated bands, one should consider the two dissimilar bands — one very narrow and the other delocalized, and the resultant magnon spectrum. The magnons are constituted out of the pair excitations,

$$a^+_{\mathbf{k}+\mathbf{q},r',\uparrow} a_{\mathbf{k},r,\downarrow} \qquad\qquad (6.115)$$

with r, r' labeling the respective bands; there are thus 4 such operators for each $\mathbf{k}$, $\mathbf{q}$; but the linear combinations are no longer the obvious $\pm$ ones, due to lack of degeneracy. Nevertheless, the coefficients can be readily obtained and the eigenvalue problem solved if the relevant scattering amplitudes $\tilde{t}_{r,r}$, are constants, independent of $\mathbf{k}$, $\mathbf{k}'$. Such a calculation, with an accurate band structure such as in Fig. 6.20, is required for a semi-quantitative comparison of theory with such experiments as have been performed on the iron series metals.

It should be mentioned that the collective charge oscillations "plasmons" also exist in magnetic metals. For Ω_q in Eq. (6.104) one chooses the spin independent operator, $\sum_{k,\sigma} g(k,q) a^+_{\mathbf{k}+\mathbf{q},\sigma} a_{\mathbf{k},\sigma}$, and the rest of the calculation is pretty much similar to what we have done to obtain the magnon spectra. The magnons lie *above* the continuum of elementary excitations.

6.13.   Marginal Magnetism of Impurities

A magnetic impurity atom does not necessarily retain its spin or any finite fraction thereof when it is immersed in a nonmagnetic metal, although
the tendency to do so is greatest when the magnetic orbitals are isolated
from the host metal by virtue of small size (the case for f-shell magnetism) or disparity in energies (lack of resonance with the conduction
band). Consequently, there have been studies to determine the properties
of marginally magnetic solute atoms, to establish the key physical parameters involved in the preservation of a moment. This work, initiated by
Friedel, Anderson, Wolff and taken up by many others[73] has been evolving
over many decades. The question to be answered concerns the magnitude
of the overlap between the magnetic shell and the nonmagnetic conduction band, which so reduces the "effective" interaction $t(U, \rho)$ that the spin
disappears.

   This question is somewhat confused by the Kondo effect, for if a moment
were to appear, it would so couple to the conduction band electrons that the
resultant spin might remain zero. So it is not clear under what circumstances
one expects a discontinuity in the ground state parameters with a change in
the interaction parameters, and under what circumstances one does not. In
this section, we shall indicate some of the methods for calculating the criteria
for the appearance of a localized moment. The Kondo effect, principally
interesting because of the finite T effects, will be treated in the appropriate
chapter.

   The method we present below is illustrative of the simplifications that
can be made in the study of an impurity in a 3D solid. First, it is only
required to study s waves, for the higher angular momenta have little if any
overlap with a point impurity at the origin. Second, the conduction band
can be treated in the leading approximation, in which the correct density of
states is preserved near the band edges. This is achieved by assuming that
the matrix element connecting one quasi-spherical shell about the origin
to the next distant shell is a constant, $-W/4$. For example, if the central
atom is a regular host atom, the Hamiltonian matrix for the s states is
merely

---

[73] The topic is reviewed from many points of view in the compendium *Magnetism*[45]. An
earlier, still valuable review was given by Heeger in *Solid State Physics*[41].

$$\mathscr{H}_0 = \begin{vmatrix} 0 & -\dfrac{1}{4}W & 0 & \cdots & \\ -\dfrac{1}{4}W & 0 & -\dfrac{1}{4}W & \cdots & 0 \\ 0 & -\dfrac{1}{4}W & 0 & -\dfrac{1}{4}W & \cdots \\ \vdots & 0 & \vdots & \vdots & \end{vmatrix} \tag{6.116}$$

taking the center-of-gravity of the band to be at zero. It is, in fact, possible to prove that $\mathscr{H}_0$ can always be put into tri-diagonal form such as shown above for *arbitrary* band structure,[74] but the matrix element connecting the central $n = 0$ site to the first $n = 1$ shell is not necessarily equal to that which connects the $n = 1$ to the $n = 2$, etc. However, for *any* band structure in 3D, the matrix elements become $-W/4$ asymptotically. It is a great simplification to consider them constant down to $n = 0$, so let us examine this case in detail. By inspection, the eigenvectors of (6.116) have components $u_n = (2/N)^{1/2} \sin k(n+1)$, i.e.,

$$\mathbf{u} = \left(\frac{2}{N}\right)^{1/2} (\sin k, \sin 2k, \ldots, \sin Nk) \tag{6.117}$$

with $n = 0$ labeling the origin, and the sample assumed to be spherical in shape, of radius Na. Thus, the k's are determined by the vanishing of $\sin(N+1)k$, and are

$$k = p\pi/(N+1), \quad p = 1, 2, \ldots \tag{6.118}$$

and the energies are,

$$E_k = -\frac{1}{2}W \cos k \tag{6.119}$$

with the bandwidth W, as desired. The local density of states, as measured

---

[74] R. Haydock, V. Heine and M. Kelly, *J. Phys.* **C5**, 2845 (1972).

at the origin, is

$$N(E) = \frac{2}{N} \sum_k \sin^2 k \, \delta \left(E + \frac{1}{2} W \cos k \right) \qquad (6.120a)$$

$$= (2/\pi) \int_0^\pi dk \, \sin^2 k \, \delta \left(E + \frac{1}{2} W \cos k \right) \qquad (6.120b)$$

$$= (8/\pi W^2) \left(\frac{1}{4} W^2 - E^2 \right)^{1/2}. \qquad (6.120c)$$

It is semicircular in shape, and has the proper square root dependence on the energy near the band edges, characteristic of all energy bands in 3D; the approximation of constant matrix element sacrifices only the van Hove singularities. The units of the density of states are $1/W$.

Problem 6.6. Derive the density of states (6.120c) from (6.120a), filling in the indicated steps and integration.

It requires only a slight modification to incorporate the changes due to an impurity at the origin. The $n = 0$ to 1 matrix element can be written $-\lambda W/4$, with λ generally $\ll 1$ for a compact magnetic shell; this feature (small λ) is the essence of Anderson's model magnetic impurity,[75] whereas the Wolff model[76,77] takes $\lambda = 1$; both also include a two-body potential at the $n = 0$ site. In the Hartree-Fock approximation, we can even take this into account by an effective potential $\varepsilon_m \equiv V + U \langle n_{0,-m} \rangle$ acting only on the central site, where $V =$ one body potential (localizing the orbital with respect to the c.o.g. of the conduction band), U the parameter characterizing the two-body forces, and $\langle n_{0,m} \rangle$, with $m = \uparrow$ or $\downarrow$ (or $\pm 1/2$), the occupation of the impurity orbital, averaged in the ground state. This is to be determined self-consistently as we shall see. The Hamiltonian now has the matrix representation

---

[75] P. W. Anderson, *Phys. Rev.* **124**, 41 (1961).
[76] P. A. Wolff, *Phys. Rev.* **124**, 1030 (1961).
[77] J. Friedel, *Canad. J. Phys.* **34**, 1190 (1956).

$$\mathcal{H} = \begin{vmatrix} \varepsilon & -\lambda\frac{1}{4}W & \cdots\cdots \\ -\lambda\frac{1}{4}W & 0 & -\frac{1}{4}W & 0 \\ 0 & -\frac{1}{4}W & 0 & \cdots \\ & \vdots & \vdots & \end{vmatrix}. \tag{6.121}$$

If, as a further idealization we assume a half-filled band, the energy ε at the impurity site has the effect of lifting an existing symmetry between holes and electrons and "charging up" the impurity orbital relative to the host atoms. A compensating charge must thus exist in the conduction band of the metal, in the vicinity of the impurity. Similarly, if the impurity acquires a net spin magnetic moment by the mechanism we shall shortly investigate, then a compensating spin polarization of the conduction electrons in the neighborhood of the impurity should cancel it, resulting in a net singlet state. The calculation of these compensations requires many-body techniques, whereas (6.121) does not.

The eigenstates of (6.121) are specified by a phase shift for $n \geq 1$. That leaves the amplitude at $n = 0$ undetermined. Assuming there is no bound state, the $n \geq 2$ amplitudes are found, using the band energies $-\frac{1}{2}W \cos k$ as before, together with

$$u_n = (2/N)^{1/2} \sin(kn + \theta_k), \quad n \geq 1 \tag{6.122}$$

and $u_0 = (2/N)^{1/2} A_k$, with A_k and θ_k to be determined by the equations at $n = 0$ and 1. At $N + 1$, the requirement $\sin(kN + \theta_k) = 0$ yields the slightly shifted k's

$$k = \frac{p\pi - \theta_k}{N} = \frac{p\pi - (\theta_k - k)}{N + 1} = k_0 - \frac{\delta k}{N + 1} \tag{6.123}$$

referring to the evenly spaced k's previously found in (6.118) as k_0, and defining the conventional *phase shift* $\delta_k = \theta_k - k$.

The eigenvalue equation at $n = 1$ yields

$$-\lambda\frac{1}{4}W A_k - \frac{1}{4}W \sin(2k + \theta_k) = -\frac{1}{2}W \cos k \sin(k + \theta_k)$$

which is solved by inspection by setting

$$-\frac{1}{4}\lambda W A_k = -\frac{1}{4}W \sin\theta_k. \tag{6.124}$$

The equation at $n = 0$ is

$$\varepsilon A_k - \lambda \frac{1}{4} W \sin(k + \theta_k) = -\frac{1}{2} W \cos k A_k .\qquad(6.125)$$

These two equations in the two unknowns A and θ are to be solved, with the resulting θ then used in (6.123) to yield the new k's.

It is more convenient to use $e^{ik} \equiv z$ than k as the independent variable and $e^{i\theta}$ than θ as the dependent variable, for then the above simply become binomials in the exponentiated functions. Thus, complex variables impose themselves in this problem.

$$e^{2i\theta(z)} = \frac{4\varepsilon z + W(z^2 + 1) - \lambda^2 W}{4\varepsilon z + W(z^2 + 1) - \lambda^2 W z^2}\qquad(6.126)$$

or

$$\theta_k = \tan^{-1}\left[\frac{\lambda^2 W \sin k}{4\varepsilon + W(2 - \lambda^2)\cos k}\right] \quad \text{and}$$

$$A_k = \frac{\lambda W \sin k}{[(4\varepsilon + W(2 - \lambda^2)\cos k)^2 + (\lambda^2 W \sin k)^2]^{1/2}} .\qquad(6.127)$$

There can be *no* bound state as long as $[4\varepsilon + W(2 - \lambda^2)\cos k]$ does not vanish, i.e., as long as the inequality

$$2|\varepsilon| + \frac{1}{2} W \lambda^2 < W\qquad(6.128)$$

is satisfied.

The quantity of greatest interest is the occupation of each spin component at the impurity, given by

$$\langle n_{0m} \rangle = \frac{2}{N} \sum_{k < k_F} (A_k)^2$$

$$= (2/\pi) \int_0^{k_F} dk \frac{(\lambda W \sin k)^2}{(4\varepsilon_m + W(2 - \lambda^2)\cos k)^2 + (\lambda^2 W \sin k)^2}$$

$$= (2\lambda^2 \pi)^{-1} \int_{-k_F}^{+k_F} dk(1 - \cos 2\theta_k)\qquad(6.129)$$

in which we use the fact that θ is odd in k, and $\varepsilon_m \equiv V + U \langle n_{0,-m} \rangle$. For the special case of a 1/2-filled band this integral is exactly calculated by complex variables. With $\supset$ a contour of half the unit circle, we have

$$\langle n_{0m} \rangle = (2\lambda^2 \pi i)^{-1} \int \frac{dz}{z}[1 - e^{2i\theta(z)}] .\qquad(6.130)$$

As there are no poles within the unit circle with the exception of $z = 0$, this contour can be deformed into the line integral up the imaginary axis $\uparrow$ with the origin excluded, and evaluated in terms of logarithms. One finds

$$\langle n_{0m} \rangle = \frac{1}{2} - \frac{1}{2\pi i(1 - \lambda^2)} \int_0^1 \frac{dy}{y} \left[\frac{1 + y^2}{y^2 - c_m iy - b} - \text{c.c.} \right] \qquad (6.131)$$

where

$$c_m = \frac{4}{W(1 - \lambda^2)}(V + U\langle n_{0,-m} \rangle), \qquad b = 1/(1 - \lambda^2). \qquad (6.132)$$

The Wolff model[76] corresponds to $\lambda = 1$, Anderson's model[75] to the limit $\lambda \to 0$. Let us discuss Wolff's model first. The physics is particularly simple. One atom differs from the others only in the energy of its atomic orbital (V) relative to the others, and in the magnitude of the intra-atomic Coulomb potential U (assumed zero on the rest of the lattice). The bond connecting it to its neighbors is unexceptional, hence $\lambda = 1$. In the absence of the potentials, the occupation of the "impurity" is $\langle n_{0m} \rangle = 1/2$ for both values of m. So now let us define the "electron occupation number defect" δn_m:

$$\delta n_m \equiv \frac{1}{2} - \langle n_{0m} \rangle, \qquad (6.133)$$

put $\lambda^2 = 1$ in (6.131) and obtain a somewhat simpler relation

$$\delta n_m = +\frac{1}{2\pi i} \int_0^1 \frac{dy}{y} \left[\frac{1 + y^2}{1 + \frac{4}{W}(V + \frac{1}{2}U - U\delta n_{-m})iy} - \text{c.c.} \right]. \qquad (6.134)$$

Performing the integrals, one obtains two equations in the unknowns $\delta n_\uparrow$ and $\delta n_\downarrow$

$$\delta n_m = \frac{1}{\pi}[a_m^{-1} + (1 - a_m^{-2}) \tan^{-1} a_m] \qquad (6.135)$$

with

$$a_m = \frac{4}{W} \left(V + \frac{1}{2}U \right) - \frac{4U}{W}\delta n_{-m}$$

$$= \frac{4U}{W}(\mu - \delta n_{-m}), \qquad \mu = \left(V + \frac{1}{2}U \right) \Big/ U \qquad (6.136)$$

defining a new parameter μ (not to be confused with the chemical potential!).

Although derived under the assumption that no bound state exists, these formulas are now serendipitously valid over the entire range of parameters (by analytical continuation). Thus the appearance or disappearance of bound states turns out not to be a significant consideration in the appearance or disappearance of a local moment! As for the magnetic moment: nonmagnetic solutions have $\delta n_\uparrow = \delta n_\downarrow$, magnetic solutions have $m = \delta n_\uparrow - \delta n_\uparrow \neq 0$. At $\mu = 0$ symmetry dictates $\delta n_\uparrow = -\delta n_\downarrow = \pm m/2$ for the magnetic solution; insert this into (6.135) to obtain

$$(\mu = 0) \quad x = \frac{2}{\pi} \left(\frac{2U}{W} \right) [x^{-1} + (1 - x^{-2}) \tan^{-1} x] \tag{6.137}$$

writing $x = 2Um/W$. This equation always has the trivial solution $x = 0$, but in addition it may have a nontrivial solution if the slope of the right-hand side exceeds 1 at the origin. This yields, at $\mu = 0$,

$$\frac{4}{3\pi} \cdot \frac{4U}{W} > 1 \quad \text{or} \quad 4U/W > 3\pi/4 = 2.3562 \tag{6.138}$$

as the necessary condition for a local moment to form. At $\mu \neq 0$ one requires a larger ratio $4U/W$ for the moment to persist. Expanding (6.135) to third order in δn_m yields a parabolic relation valid for small μ

$$(|\mu| \ll 1) \quad \frac{4U}{W} > \frac{3\pi}{4} \left[1 + \frac{6}{5} \left(\frac{3\pi}{8} \mu \right)^2 + \cdots \right]. \tag{6.139}$$

Numerically, one may solve for the simultaneous solutions of (6.135) or one may converge to them by *relaxation*: insert an initial value for $\delta n_\downarrow$, calculate $\delta n_\uparrow$ by (6.135), reinsert the result to obtain a new $\delta n_\downarrow$, etc. This procedure is found to always converge to the magnetic solution whenever it exists. When there is no magnetic solution, it quickly converges to the nonmagnetic solution $\delta n_\uparrow = \delta n_\downarrow$. This type of behavior characterizes the magnetic solution as a "stable fixed point" of the coupled nonlinear equations, and the trivial solution as an "unstable fixed point." An independent confirmation can be found in a calculation of the energy associated with each solution. Where a magnetic solution exists, the energy is a minimum at the proper value of m (and $-m$), and a local maximum at $m = 0$.

The results of numerical calculation of (6.135) can then be put in graphical form, as shown in Fig. 6.22. In the figure we also indicate the regions where bound states form: above the band for $\mu > 0$ and below it for $\mu < 0$. Such bound states give the magnetic impurity the aspect of an isolated one-electron atom. But there is ample parameter space for a magnetic solution without any bound state detaching itself from the continuum, and this is of course the new feature.

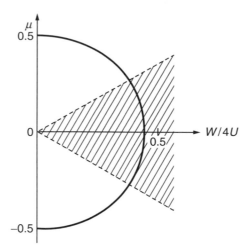

Fig. 6.22.   Parameter space in the Wolff model. Magnetic solutions occupy the approximately semicircular region. Shading indicates regions free of one-body bound states. Above this, *one* level will detach itself and lie above $\frac{1}{2}W$, the maximum permissible energy in the continuum; below this, it will lie below $-\frac{1}{2}W$, the lowest energy of the continuum.

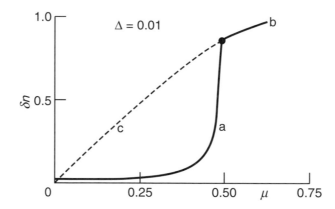

Fig. 6.23.   Electron deficit (or charge accumulation) $\delta n = \delta n_\uparrow + \delta n_\downarrow$ on the impurity's orbitals as a function of the next effective potential μ defined in the text. The magnetic solution (*curve a*) is stable until approximately $\mu = \frac{1}{2}$; at larger μ the nonmagnetic solutions yield curve b. The nonmagnetic solution at smaller potentials, curve c, is unstable. The magnetic solution shows a great stability against charge accumulation compared to the (unstable) nonmagnetic solution. This rather typical curve was calculated for $\Delta = 0.01$. For $\mu \to -\mu$, $\delta n \to -\delta n$.

Anderson's model is characterized by two parameters, one of which is the same μ and the other is denoted Δ

$$\Delta \equiv \frac{1}{4} W \lambda^2 / U . \tag{6.140}$$

We solve it first, then provide explanations. In the limit $\lambda^2 \to 0$, the amplitudes (6.127) are seen to yield a Lorentzian probability centered at the impurity level ε; only that part lying below the Fermi level will contribute to the occupation of the impurity orbital. The result:

$$\delta n_m = \frac{1}{\pi} \tan^{-1} \left(\frac{\mu - \delta n_{-m}}{\Delta} \right) . \tag{6.141}$$

We calculate some of the physical properties from the numerical solution of these coupled equations, much as we did for (6.135). Some of the quantities of interest are: the net electron occupation-number defect $\delta n = \delta n_\uparrow + \delta n_\downarrow$ in the magnetic and nonmagnetic states, shown in Fig. 6.23; the magnetization as a function of μ for several values of Δ shown in Fig. 6.24; and noting that m is a maximum at $\mu = 0$, the dependence of m on Δ at $\mu = 0$ in Fig. 6.25. The symmetry of the solutions with respect to μ is maintained throughout, of course, as it was in the Wolff model previously. $\mu = 0$ is called "symmetric Anderson model."

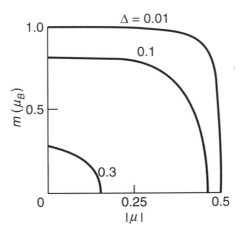

Fig. 6.24. Magnetization (in Bohr magnetons) as function of effective potential parameter μ defined in the text. For $\mu \gtrsim \frac{1}{2}$ there is no magnetic solution, at smaller values the curves $m(\mu)$ depend on the kinetic energy parameter Δ, the general tendency being shown by the three curves for increasing Δ.

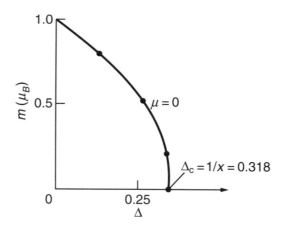

Fig. 6.25. The curve $m(\Delta)$ at the most favorable potential, $\mu = 0$, i.e., $V + \frac{1}{2}v = 0$, $v > 0$, the so-called "symmetric Anderson model."

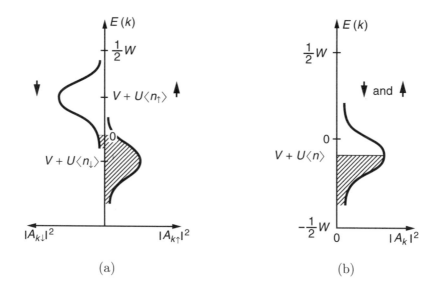

Fig. 6.26. (Amplitude)^2 of waves at impurity in cases (a) moment exists and (b) no magnetic moment for the same Anderson model impurity. The approximately Lorentzian shaped curves apply for the case $\lambda \ll 1$. Magnetic case (a) is degenerate: a solution with $\uparrow$ and $\downarrow$ interchanged is equally valid. *Note:* $V < 0$, $V + U > 0$ are assumed.

The electron probability distribution at the impurity site, the Lorentzian distribution characterizing Anderson's model, is shown in Fig. 6.26. The asymmetric magnetic configuration is degenerate: one can interchange spins up and down to obtain a distinct, yet degenerate state; a magnetic field applied to the impurity site only, would lift this degeneracy by an amount linear in the applied field. Thus, we know that a magnetic moment is associated with the state. This behavior mimics, on the scale of a single atom, what we observed for a whole band in a ferromagnet.

The original physical basis for Anderson's model was the following: a magnetic d-orbital (shell index $n = 0$ in our nomenclature) is weakly connected ($\frac{1}{4}\lambda W$) to an s-orbital (shell index $n = 1$) which then connects to the rest of the lattice in the usual way. The 2-body forces and the "weak link" characterize the magnetic shell. After the exact Hartree-Fock solution given here, we seek some insight into 2-body correlations in the following section.

6.14.   Correlations and Equivalence to s–d Model

By solving the Hartree-Fock equations, we determined the range of parameters over which an impurity atom, imbedded in a nonmagnetic metal, acquires a localized moment. Now we consider the correlations required to re-establish the rotational invariance of the model and to understand the concomitant polarization of the conduction electrons near the localized moment. For the symmetry-breaking Hartree-Fock solutions, which offer us a choice of two ground states (Fig. 6.26a with majority spins "up" or "down"), are obviously just the first step toward the correct singlet ground state.

De Gennes[78] and Schrieffer and Wolff[79] have established the correspondence between the magnetic solution of Anderson's model and the so-called s–d exchange model, have shown that the effective exchange parameter is AF in sign and determined its magnitude from the microscopic parameters such as λ, W, U, etc. Their principal motivation: the s–d model being explicitly rotationally invariant, no further demonstrations are required. Moreover, the Kondo effect and other thermodynamic and dynamic phenomena are fairly

---

[78] P. G. de Gennes, *J. Phys.* **23**, 510 (1962).
[79] J. R. Schrieffer and P. A. Wolff, *Phys. Rev.* **194**, 491 (1966).

well understood for the s–d model but have not yet been thoroughly studied for the models of the preceding section for which one must go to much higher orders in the perturbation theory. In fact, a direct calculation of the Kondo effect starting with Anderson's model requires an accuracy better than $O(\lambda^6)$ and is reasonably complicated.[80]

For maximum simplicity, we treat only the symmetrical model $\mu = 0$, $(V + \frac{1}{2}U) = 0$, but the reader can proceed to the more general case with no essential modifications. We assume U is large so that $m \sim 1$, and follow the Schrieffer-Wolff type analysis. We then show how to extract the effective exchange parameter, not just in the Anderson model, but for arbitrary λ including the Wolff limit $\lambda = 1$.

To start, suppose there is precisely one electron of spin "up" on the impurity site (the zeroth shell, denoted $r = 0$) with the rest of the electrons in the normal Fermi sea on the shells $r = 1, 2, \ldots$. Then the matrix element for the scattering of an electron $k \downarrow$, of spin "down" from near the Fermi level, into the localized $r = 0$ shell, is $\lambda W (2/N)^{1/2}$, at the cost of an additional excitation energy $\frac{1}{2}U$. In second order, this process restores the electron to the conduction band at k'. Thus, a scattering from $k \downarrow$ to $k' \downarrow$ has occurred, with amplitude $-(\lambda W)^2 (2/N)/(U/2)$. Conduction electrons of spin "up" do not benefit from this scattering mechanism, which is forbidden to them by the exclusion principle. Thus, we have found a spin-dependent scattering mechanism that can be written as the sum of a potential and a spin term

$$V_{\text{eff}} (2/N) \sum_{kk'} (c_{k\uparrow}^{\dagger} c_{k'\uparrow} + c_{k\downarrow}^{\dagger} c_{k'\downarrow}) - J_z S_z \frac{1}{2} (2/N) \sum_{kk'} (c_{k\uparrow}^{\dagger} c_{k'\uparrow} - c_{k\downarrow}^{\dagger} c_{k'\downarrow}) \quad (6.142)$$

with $2/N$ instead of the usual $1/N$ to take care of the different amplitudes in our shell formalism, and the k, k' restricted to the neighborhood of the Fermi $k_F = \frac{1}{2}\pi$. S_z is a counter: $+1/2$ for an electron with spin "up" at $r = 0$, and $-1/2$ for spin "down." For no $k \uparrow$ scattering to occur, we require

$$V_{\text{eff}} - \frac{1}{4} J_z = 0$$

[80] D. Hamann, *Phys. Rev.* **B2**, 1373 (1970); P. B. Wiegmann, *Phys. Lett.* **80A**, 163 (1980); H. Fukuyama and A. Sakurai, *Progr. Theor. Phys.* **62**, 595 (1979); H. Krishna-Murthy and J. Wilkins; K. G. Wilson, *Phys. Rev.* **B21**, 1003–1083 (1980).

and for the $k\downarrow$ scattering to yield the correct result, we need

$$V_{\text{eff}} + \frac{1}{4}J_z = -(\lambda W)^2 \Big/ \frac{1}{2}U \,.$$

This leads to the following value for the exchange constant and effective potential

$$J_z = -4\lambda^2 W^2/U = -16W\Delta \quad \text{and} \quad V_{\text{eff}} = -4W\Delta \,. \tag{6.143}$$

A similar reasoning yields the spin-flip scattering terms. In the intermediate state, 2 electrons of opposite spin occupy the $r = 0$ shell. If instead of the spin "down" electron returning to the Fermi sea, the spin "up" electron did so, we would have a scattering from $k\downarrow$ to $k'\uparrow$. This is described by

$$-J_\perp S^- c^\dagger_{k'\uparrow} c_{k\downarrow}(2/N)$$

with S^- indicating that the counter has gone from $+1/2$ to $-1/2$, and the negative sign from the Pauli principle (the ordering of two electrons has been interchanged). The magnitude of $J_\perp$ is again given by second-order perturbation theory: $-16W\Delta$, and is identical to J_z. The inverse process provides us with the Hermitean conjugate of the above term and completes our effective s–d exchange Hamiltonian

$$\mathcal{H} = \text{K.E.} + J\mathbf{S}\cdot\mathbf{s_c} - \frac{1}{4}J\mathbf{n}\,, \quad \text{with} \quad J > 0\,, \tag{6.144}$$

where $\mathbf{s_c}$ and $\mathbf{n}$ are the local spin-density and particle-density operators, restricted to the neighborhood of k_F (more or less). A solution of this Hamiltonian is attempted in Chapter 9.

For a more quantitative study at finite λ, we can start by the Hartree-Fock solution of the preceding section and then add in higher-order terms if desired. The proper of an AF exchange parameter is that the conduction sea close to the local moment is polarized antiparallel to it. Then the Friedel (i.e., Ruderman-Kittel) oscillations periodically reverse the polarization as one proceeds further. We shall verify these features explicitly.

At the rth shell, $r = 1, 2, \ldots$ the magnetization $m(r)$ is

$$m(r) = \frac{2}{N} \sum_{k<k_F} [\sin^2(rk + \theta_{k\uparrow}) - \sin^2(rk + \theta_{k\downarrow})]$$

$$= \frac{1}{N} \text{Re}\left\{ \sum e^{2irk}(e^{2i\theta_{k\downarrow}} - e^{2i\theta_{k\uparrow}}) \right\}. \tag{6.145}$$

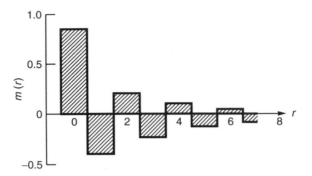

Fig. 6.27. Schematic Hartree-Fock result: magnetization $m(r)$ of the rth shell relative to magnetic shell at $r = 0$, according to Eq. (6.146).

With $z = \exp(ik)$ and $\theta(z)$ given in (6.126) we follow the procedure of (6.130) et seq. to evaluate the above as the following integral:

$$m(r) = \frac{\lambda^2(-1)^r}{2\pi i(1 - \lambda^2)} \int_0^1 \frac{dy}{y} y^{2r}(1 + y^2)$$

$$\times \{[(y^2 - c_\downarrow iy - b)^{-1} - (y^2 - c_\uparrow iy - b)^{-1}] - \text{c.c.}\} \qquad (6.146)$$

with c_m, b defined in (6.131). The main difference with the $r = 0$ calculation is the newly appeared factor λ^2 and the $(-y^2)^r$. The results, plotted in Fig. 6.27, show that the first shell is polarized antiparallel to the $r = 0$ impurity shell, therefore imply the effective AF coupling. The change of sign at each successive shell reflects the half-filled band, $k_F = \pi/2$. To obtain J_{eff} by this procedure, evaluate the spin oscillations that would be caused by an equivalent s–d interaction [restricted to $J_z S^z s_c^z(0)$ for fair comparison with the Hartree-Fock]. The latter is also a point-scatterer which is positive for spins up and negative for spins down, and can be analyzed by the same set of equations above. Thus, we re-evaluate $m(r)$ which is once more given in (6.146), with however a different set of parameters: $\lambda = 1$ and $c_\uparrow = 0$, $c_\downarrow = \frac{1}{2}J_z$, corresponding to the appropriate exchange and potential scattering. Denote the result, $m_{sd}(r)$. The best fit to the equation

$$m_{sd}(r - 1) = m(r), \quad \text{for } r > 1, \qquad (6.147)$$

will yield the value of J_z (*alias* J_{eff}) in terms of the more "microscopic" parameters V, U, λ, W.

In summary we establish the equivalence of physically distinct situations that can be mapped onto each other. The subsequent calculation of the Kondo "state," (including the matter of a resonance pinned at the Fermi level), the anomalous scattering, the formation of a singlet bound state, etc., are all concerns that are not amenable to the Hartree-Fock theory. They are to be taken up later.

Secondary issues have also been addressed: how far does the polarization cloud extend, how do various impurities interact magnetically (as in RKKY) if their total spin is zero, what happens at finite T, etc.? These can be studied by looking not just at the ground state but also at the low-lying excitations of the spin-conduction electron complex. A first approximation consists of carving out a state for an extra electron just above the Fermi surface, and one for holes just below it, in the general form:

$$A_\sigma^\dagger(R) = \frac{1}{\sqrt{N}} \sum_{k>k_F} \phi_k e^{ik\cdot R} c_{k,\sigma}^\dagger \quad \text{and}$$

$$B_\sigma^\dagger(R) = \frac{1}{\sqrt{N}} \sum_{k<k_F} \chi_{-k} e^{ik\cdot R} c_{-k,-\sigma} \tag{6.148}$$

The impurity atom sports a one-body zero-range potential $V(n_\uparrow + n_\downarrow)$ and a two-body repulsive potential located at the same site, $Un_\uparrow n_\downarrow$, to which the A and B states are subject. The result is a 16×16 Hamiltonian matrix to be diagonalized, with the functions ϕ and χ to be optimized variationally. Amazingly, this latter is easily accomplished and enforces a functional form:

$$\phi, \chi = \frac{C_\pm}{|\varepsilon_k - \varepsilon_F| + \Delta_\pm}, \quad \text{with } C_\pm \text{ the respective norms.} \tag{6.149}$$

The 16-dimensional matrix is easily reduced into singlets, doublets and a triplet. Of course the singlet is *always* the ground state. At a temperature T such that the doublets may be excited, one finds the triplet lying very close to the low-lying doublets (regardless of model parameters such as filling factor, etc.). The small excitation energies mean that magnetic excitations can range over very large distances (this is determined by Δ in (6.142)), and are preferred when two or more impurities interact. For example, the following energies E and wave-function radii ξ are found at $\varepsilon_F = 0.7$, $V = 1$ and $U = 1.2$ (in units of bandwidth) for spins zero, $1/2$ and 1[81]:

$$E_0 = -0.848\ldots \text{(a.u.)}, E_{1/2} = -0.828\ldots, E_1 = -0.828\ldots$$

$$\text{while} \quad \xi_0 = O(1), \xi_{1/2} > 100 \text{ and } \xi_1 > 10^3.$$

[81] D. C. Mattis, "Nonmagnetic Phase of a Transition Series Alloy," *Mod. Phys. Lett.* **B12**, 655 (1998).

6.15.   Periodic Anderson Model

In its original formulation the Anderson model[82] was just another micro-scopic version of the s–d model: a localized f-state of fixed energy ε_f inter-acting with the conduction d-band of dispersion $\varepsilon(k)$. The extended states in the latter were supposed to have little or no Coulomb or Hubbard inter-actions, whereas the localized state had a Hubbard-like interaction Hamil-tonian $Un_{f\uparrow}n_{f\downarrow}$. The band hybridization term $V_{k,f}$, the energy ε_f relative to the Fermi energy in the dispersive band, ε_F, and the bandwidth W were the parameters in the model, which turned out to be neither simpler nor more complicated than the Wolff model or the s–d version, all purportedly describing the same magnetic impurity.

But interesting properties of metals and alloys of the rare-earth and ura-nium series have transformed the *periodic* version of this Hamiltonian into a fairly realistic and appropriate subject of study.[83] An otherwise mysterious volume collapse at a certain $(\alpha \rightarrow \gamma)$ phase transition in Cerium compounds could only be explained as the massive transfer of electrons from one band to the other.

In the "periodic Anderson model" both bands are allowed a measure of dispersion (even though f-shells barely overlap) but the Hubbard-like interactions are limited to the f's. Thus,

$$\mathscr{H} = \sum_{k,\sigma} (\varepsilon_k^d n_{k,d,\sigma} + \varepsilon_k^f n_{k,f,\sigma} + V_k(d_{k,\sigma}^\dagger c_{k,\sigma} + \text{H.c.}))$$

$$+ U \sum_i n_{k,f,\uparrow} n_{k,f,\downarrow} \qquad\qquad (6.150)$$

where typically the f-band lies well below the d-band, although the U in-teractions may force particles out of the one into the other, virtually or permanently. Note that *momentum conservation* is a simplifying feature of the extended model. These and other properties of the model have been widely discussed in the contemporary literature.[84]

Recently, however, Gulácsi and Vollhardt[85] have made a major step for-ward, as they have been able to construct exact ground-states of this model

[82] P. W. Anderson, *Phys. Rev.* **164**, 352 (1967).
[83] C. M. Varma and Y. Yafet, *Phys. Rev.* **B13**, 2950 (1976); O. Gunnarson and K. Schönhammer, *Phys. Rev.* **B31**, 4815 (1985).
[84] P. A. Lee *et al.*, *Comments Condens. Matter Phys.* **12**, 99 (1986).
[85] Z. Gulácsi and D. Vollhardt, "Exact Ground States of the Periodic Anderson Model," arXiv:cond-mat/0504174v1 (7 April 2005), in press.

in 3D for filling factor $\geq 3/4$, finding for the most part *nonferromagnetic,
non-Fermi liquid* type ground states. Different circumstances obtain at small
filling factors — indeed, they find exact *ferromagnetic* ground states at filling
factor $\leq 1/4$.

At the time of writing, it is too early to tell whether their construction
can be parlayed into a better understanding of strongly interacting metal
physics in general, or whether it shall stand as an isolated achievement.

Chapter 7

Statistical Thermodynamics

This chapter is intended as a self-contained introduction to the statistical thermodynamics of magnetic systems. To the extent possible, all concepts and quantities are derived and calculated from first principles. Within the context of simple spin systems various basic quantities emerge: the free energy F, its derivatives the internal energy U and the entropy $\mathscr{S}$, and higher derivatives such as the specific heat c and magnetic susceptibility χ. As usual, the first hurdle in generalizing the study of physical systems to finite temperature is the very definition of temperature itself. It emerges in the following section, devoted to the simplest possible model of a thermodynamic system.

7.1. Spins in a Magnetic Field

This study begins with the magnetic analogue of an ideal gas, a single spin interacting only with a specified external magnetic field. If it were experimental, the experiment would be repeated a number of times and the results subjected to statistical analysis. It is advantageous for theory to emulate this procedure; therefore we consider not one, but an assembly of N similar spins, each statistically and dynamically independent of the others, and calculate the properties *on the average*, in the limit $N \to \infty$. This limit is usually denoted the *thermodynamic limit*.

Let the energy of each spin pointing parallel to the field be $-h$ and of each spin antiparallel to the field be $+h$, where this Zeeman splitting parameter is

$$h = \frac{1}{2}g\mu_b B \,. \tag{7.1}$$

[The Landé factor $g = 2$ for spin and 1, for orbital angular momentum. Bohr magneton $\mu_b = e\hbar/2mc = 0.9274 \times 10^{-20}$ erg/Gauss. Thus in a magnetic field 1 Tesla $= 10,000$ Gauss, the energy splitting $2h$ is equivalent to 0.0001 eV, or (dividing by Boltzmann's constant k_B) to a temperature $T \approx 1$ K.]

If n spins are parallel and $N - n$ antiparallel to the field, the average energy u of each spin is related to the total energy E as follows:

$$u = \frac{E}{N} = \frac{1}{N}(N - 2n)h = (1 - 2p)h, \quad \text{with } p \equiv n/N. \tag{7.2}$$

To substantiate the notion of average we need to specify the counting procedure. But, at first, let us calculate the *most probable* values of p and u. This is not only the most convenient procedure, it also serves to illustrate the methodology of statistical thermodynamics.

The probability of attaining the configuration of (7.2) is $Q(n)$,

$$Q(n) = P(E) \times \frac{N!}{n!(N-n)!} \times 2^{-N}. \tag{7.3}$$

It is a *relative* probability — un-normalized as yet. $P(E)$ is the yet-to-be-determined thermal (Maxwell–Boltzmann) factor. The binomial coefficient counts the number of distinct arrangements of N spins that can lead to precisely the energy E of (7.2). The last factor relates to the normalization: 2^N is the total number of distinct configurations, or partitions, of N binary variables. In this manner, we have postulated that $Q(n)$ is the product of two distinct probabilities: $P(E)$, the thermal *a priori* probability of the system having energy E and the remaining factors being the purely statistical probability of achieving the value of n required to yield E according to (7.2).

The factorials can be estimated with the use of Stirling's expansion:

$$\ln N! \sim N \ln N - N + \frac{1}{2}\ln(2\pi N) + \cdots . \tag{7.4}$$

Not only is this formula asymptotically exact at large N, but the error is negligible even for small N (0.3% for $N = 5$). The statistical factor thus depends exponentially on N in leading approximation. Unless there occurs unexpected cancellation the logarithm of Q must therefore also scale with N, the size of the system.

Quantities proportional to N are denoted *extensive* to distinguish them from the *intensive* variables (temperature, pressure, field-strength, etc.) that are independent of N. Thus, $\ln Q$ will be extensive, the sum of two terms: the first, related to the energy and the second to statistics.

We calculate $\ln Q$:

$$\ln Q = \ln P(E) - N[\ln 2 + p \ln p + (1-p) \ln(1-p)] - \frac{1}{2} \ln(2\pi) \qquad (7.5)$$

setting $p = n/N$. The last term is smaller by a factor N than the rest, and is discarded. The remainder of the statistical contribution is extensive. This suggests the four reasonable criteria which $P(E)$ must satisfy for $\ln Q$ to be useful in the determination of thermodynamic properties:

(1) $\ln P(E)$ must be extensive, otherwise it would not be commensurable with the statistical term in the thermodynamic limit $(N \to \infty)$. [Hence, $\ln Q$ will be extensive.]
(2) $\ln P(E)$ must be dimensionless, else a change of units would modify it, while leaving the statistical term invariant. [Hence, $\ln Q$, too, will be dimensionless.]
(3) P, hence $\ln P$, must be maximal in the ground state, i.e., must favor low energies.

   Requirement (1) is met by setting $\ln P = -\beta E$, where the (intensive) coefficient β must have units $(\text{energy})^{-1}$ to satisfy criterion (2). The negative sign is chosen so that the lowest energies have higher probability than the highest ones.

   The result is a rather familiar expression:

$$P(E) = \exp(-\beta E). \qquad (7.6)$$

[Other logical possibilities (such as $\ln P(E) = -\beta E^3/N^2$, with β now having dimensions $(\text{energy})^{-3}$) will not be considered because, ultimately, they fail to agree with experiment.] Finally:

(4) To attain the most probable p and u, we must maximize Q.

   The coefficient β is identified with temperature as follows:

$$\beta \equiv 1/k_B T \qquad (7.7)$$

where Boltzmann's constant $k_B = 1.3806 \times 10^{-16}$ erg/K for T in degrees Kelvin. (Of course, the choice of k_B is what determines the scale of T. Thus $1/\beta$ is the temperature in units $k_B = 1$, the scale most favored by theorists. Henceforth, we'll omit the subscript on k, to conform with usual notation.)

   Thermometry depends on being able to measure the temperature of a probe, assumed in thermal equilibrium with the system under investigation. Rudimentary though it be, the discussion above permits a proof that two or more systems in thermal equilibrium share the same temperature — that is, that a measurement of T on the one establishes T for all. The proof for spin systems is suggested in Problem 7.1.

Problem 7.1.   Separate the N spins into sets of $n_1, n_2, \ldots$ with $\sum n_i = N$. Assign temperatures $T_1, T_2, \ldots$ to these sets and a fixed total energy E. Show that Q is maximal for $T_1 = T_2 = \cdots = T$, i.e., show that at fixed total energy a constant temperature is most probable.

By analogy with $P(E)$ the logarithm of Q is an extensive, dimensionless quantity,

$$2^N Q = \exp(-\beta F) = \exp(-\beta N f) \tag{7.8}$$

which serves to define F, the free energy, and f, the free energy per spin. We use this definition, together with (7.6), to eliminate P and Q from (7.5), obtaining:

$$f = (1 - 2p)h + kT[p \ln p + (1 - p) \ln(1 - p)] \tag{7.9}$$

to $O(1/N)$. It remains to determine the correct value of p. As the fourth criterion on Eq. (7.5) requires f to be a minimum, $df/dp = 0$. Differentiation of (7.9) yields

$$p = \frac{1}{1 + \exp(-2\beta h)} \; . \tag{7.10}$$

With this thermal equilibrium value of p we can calculate the various thermodynamic properties, such as the *internal energy* u per spin defined in (7.2):

$$u = -h \tanh(\beta h), \tag{7.11}$$

and the free energy per spin f in (7.9):

$$f = -kT \ln[2 \cosh(\beta h)]. \tag{7.12}$$

Problem 7.2.   Derive (7.12) by inserting (7.10) into (7.9).

By the above arguments, one is led to the *most probable* properties of a thermodynamic ensemble of spins, although it was the *average* properties we originally set out to determine. We shall next establish that these two quantities coincide in the thermodynamic limit.

Let us denote thermal averaged quantities by $\langle\,\rangle$. For example, the average n is $\langle n \rangle$. The variance $\langle (n - \langle n \rangle)^2 \rangle$, written $\langle \Delta n^2 \rangle$ for typographical simplicity, is evaluated using $Q(n)$ as a weighing factor:

$$\langle \Delta n^2 \rangle = \sum_n (n - \langle n \rangle)^2 Q(n) \Big/ \sum_n Q(n) \tag{7.13}$$

as are *all* averaged quantities. To evaluate such averages, it is convenient to expand $\ln Q$ about its optimum value at pN, with p fixed by (7.10):

$$Q(n) = Q(pN)\exp\left[-\frac{1}{2}A_2(n - pN)^2 + \cdots\right] \tag{7.14}$$

where $(\cdots)$ indicates neglect of higher-order terms starting with $A_3(n - pN)^3/3!$ (they vanish in the thermodynamic limit, as the reader may wish to verify). Thus, $Q(n)$ is a sharply peaked Gaussian, centered on the optimum value pN. We use Stirlings's approximation to obtain A_2:

$$A_2 = |- d^2 \ln Q/dn^2|_{pN}$$

$$= \frac{d^2}{dn^2}[(N - 2n)h + n\ln n + (N - n)\ln(N - n)]_{pN}$$

$$= [Np(1 - p)]^{-1}. \tag{7.15}$$

Because $Q(n)$ in (7.14) is symmetric about pN, $\langle n \rangle = pN$ by inspection. To calculate (7.13) it is easiest to approximate the sums over the slowly varying summands by integrations.

$$\langle \Delta n^2 \rangle = \lim_{N\to\infty}\left[\frac{\int_{-Np}^{N(1-p)} dx\, x^2 \exp(-\frac{1}{2}A_2 x^2)}{\int_{-Np}^{N(1-p)} dx\, \exp(-\frac{1}{2}A_2 x^2)}\right]$$

$$= \frac{1}{A_2} = Np(1 - p). \tag{7.16}$$

The rms fluctuation $|\Delta n|$ is $N^{1/2}p^{1/2}(1 - p)^{1/2}$. Thus, the fluctuations per site $|\Delta n|/N = p^{1/2}(1 - p)^{1/2}N^{-1/2}$ vanish in the thermodynamic limit.

The above results imply that the intensive observable properties are virtually free of fluctuations — that for all practical purposes, the most probable, the optimum, and the average values all coincide. These claims are explored further in Problem 7.3, and are the basis for an axiomatic statistical mechanics developed in the next section.

Problem 7.3. Prove that the fluctuations in u about its thermodynamic equilibrium (7.11) are only $O(N^{-1/2})$ and, furthermore, that the fluctuations in f about (7.12) are smaller yet, $O(N^{-1})$.

7.2. The Partition Function

The results of the preceding section can be obtained more elegantly through the postulational statistical mechanics of Gibbs and Einstein. We start by separating the noninteracting particles into a number of sets, each with energy $e_i n_i$.

The free energy is the average of terms such as,

$$f_i = e_i p_i + kT p_i \ln p_i + \lambda p_i \tag{7.17}$$

with λ a Lagrange multiplier, chosen so as to maintain

$$\sum_i p_i = 1 \,. \tag{7.18}$$

Minimizing $F = \sum_i f_i$ with respect to the individual p_i:

$$p_i = \left[\exp\left(-\frac{\lambda + kT}{kT} \right) \right] \exp(-e_i/kT) \,. \tag{7.19}$$

The quantity in $[\,]$ is usually denoted $1/Z$, where Z (for *Zustandsumme*) is the *partition function*. From (7.18),

$$Z = \sum_i \exp(-\beta e_i) = \mathrm{Tr}\{\exp(-\beta \mathscr{H})\} \,. \tag{7.20}$$

Z, replacing the Lagrange multiplier, is thus fixed. The Hamiltonian $\mathscr{H}$ is hereby also introduced, assuming — as is usually the case — that the energies e_i can be written as the eigenvalues of a physically appropriate Hamiltonian $\mathscr{H}$. Tr, the *trace*, which is the sum over the diagonal values of an operator, is invariant and can be evaluated in any convenient representation including the one in which $\mathscr{H}$ is diagonal with eigenvalues e_i.

In the event the eigenvalues of $\mathscr{H}$ are not known, it is convenient to define the *density matrix* ρ:

$$\rho = Z^{-1} e^{-\beta \mathscr{H}} \,. \tag{7.21}$$

In this form, ρ satisfies the equation

$$-\partial\rho/\partial\beta = (\mathscr{H} - \langle\mathscr{H}\rangle_{\mathrm{TA}})\rho \tag{7.22}$$

and is normalized

$$\mathrm{Tr}\{\rho\} = 1\,.$$

Here we have introduced the notation for thermal average $\langle\,\rangle_{\mathrm{TA}}$. ρ can be used to compute the thermal average of any observable or operator, as follows:

$$\langle G\rangle_{\mathrm{TA}} = \mathrm{Tr}\{G\rho\} = \mathrm{Tr}\{\rho G\} \tag{7.23}$$

so that, in (7.22), $\langle\mathscr{H}\rangle_{\mathrm{TA}} = \mathrm{Tr}\{\mathscr{H}\rho\}$.

Intuitively, Z has a physical significance akin to Q of the preceding section. Indeed,

$$Z = e^{-\beta F} \equiv \mathrm{Tr}\{e^{-\beta H}\}\,, \tag{7.24}$$

which serves to define the free energy F and the partition function Z in terms of a given Hamiltonian $\mathscr{H}$ and temperature β^{-1}.

Define the internal energy $U \equiv \langle\mathscr{H}\rangle_{\mathrm{TA}}$. Comparison with (7.23) and (7.24) establishes that

$$U = \frac{\partial(\beta F)}{\partial\beta}\,, \tag{7.25}$$

a familiar thermodynamic identity. Suppose next, that in an applied field B, the Hamiltonian is given by $\mathscr{H} = \mathscr{H}_0 - BM$, where M is the *magnetization operator*. We want to know $\mathscr{M} = \langle M\rangle$, the thermal average magnetization. Again (7.23) and (7.24) yield the answer:

$$\mathscr{M} = -\frac{\partial F}{\partial B}\,, \tag{7.26}$$

another well-known thermodynamic identity.

Expanding (7.25) we find

$$U = F + \beta\frac{\partial F}{\partial\beta} = F - T\frac{\partial F}{\partial T}$$

and use this to define yet another derivative of F:

$$\mathscr{S} = \frac{\partial F}{\partial T} = (U - F)/T\,. \tag{7.27}$$

$\mathscr{S}$ is defined as the *entropy*, a measure of the thermal disorder. This aspect is explored further in Problem 7.4.

Problem 7.4. Using the definition of ρ in (7.21) show that: $\mathscr{S} = -k\,\mathrm{Tr}\{\rho\ln\rho\}$.

Problem 7.4 is related to the third law of thermodynamics: unless the ground state is macroscopically degenerate, the sum rule (7.20) guarantees that $\mathscr{S}/N$, as calculated in Problem 7.4, vanishes at $T = 0$.

For many magnetic systems, the quantities U, $\mathscr{M}$ and $\mathscr{S}$ exhaust the thermodynamic functions related to first derivatives of F. Turning to second derivatives, the *heat capacity* C is

$$C \equiv \frac{\partial U}{\partial T} = T\frac{\partial \mathscr{S}}{\partial T} = -T\frac{\partial^2 F}{\partial T^2}, \tag{7.28}$$

making use of (7.27). The heat capacity is alternatively expressible as a fluctuation in the total energy:

$$C = \frac{\partial U}{\partial T} = -k\beta^2\frac{\partial U}{\partial \beta} = -k\beta^2\frac{\partial}{\partial \beta}\left\{\frac{\mathrm{Tr}[\mathscr{H}\exp(-\beta\mathscr{H})]}{\mathrm{Tr}[\exp(-\beta\mathscr{H})]}\right\}$$

$$= k\beta^2[\langle\mathscr{H}^2\rangle - \langle\mathscr{H}\rangle^2] = k\beta^2\langle(\mathscr{H} - U)^2\rangle, \tag{7.29}$$

proving, incidentally, that C is non-negative.

In systems without spontaneous magnetization, when $B \to 0$, $\mathscr{M}$ will generally be proportional to the applied field B. The constant of proportionality is the *susceptibility*, χ, for which we offer a more general definition — valid for *all* systems:

$$\chi \equiv \frac{\partial \mathscr{M}(B, T, \ldots)}{\partial B}. \tag{7.30}$$

Again, because it is a second derivative of F, there exists an alternative formulation as a fluctuation — this time, of the total magnetization:

$$\chi = \frac{\partial}{\partial B}\left\{\frac{\mathrm{Tr}[M\exp(-\beta\mathscr{H})]}{\mathrm{Tr}[\exp(-\beta\mathscr{H})]}\right\} = \beta\langle(M - \mathscr{M})^2\rangle. \tag{7.31}$$

The quantities: $F, U, \mathscr{S}, \mathscr{M}, C, \ldots$, are all extensive. It is often convenient to divide them by N so they no longer depend on N in the thermodynamic limit. Thus:

$f = F/N$ is the free energy per spin

$u = U/N$ is the internal energy per spin

$s = \mathscr{S}/N$ is the entropy per spin,

σ or $m = \mathscr{M}/N$ is the magnetization per spin,

$C = \mathscr{C}/N$ is the *specific heat,* and

$x = \chi/N$ is the susceptibility per spin

(this last is used interchangeably with χ when there is no confusion possible).

A Simple Magnetic System. We now work out the properties of a spin in an external, fixed, field — the problem posed in the preceding section — using the new formalism.

The Hamiltonian of a spin one-half in an external field is a Pauli matrix, here given in a diagonal representation:

$$\mathbf{H} = - \begin{bmatrix} \dfrac{1}{2} & 0 \\ 0 & -\dfrac{1}{2} \end{bmatrix} g\mu_b B = \begin{bmatrix} -h & 0 \\ 0 & +h \end{bmatrix}. \tag{7.32}$$

The two eigenstates are

$$\text{``up''} : \begin{bmatrix} 1 \\ 0 \end{bmatrix} \quad \text{and} \quad \text{``down''} : \begin{bmatrix} 0 \\ 1 \end{bmatrix}. \tag{7.33}$$

To evaluate ρ, (7.21), we need to exponentiate the Pauli matrix. The simple identity is

$$\exp(K\boldsymbol{\sigma}_z) = \mathbf{1} \cos K + \boldsymbol{\sigma}_z \sinh K \tag{7.34}$$

with $K \equiv \beta h$. The matrices $\mathbf{1}$ and $\boldsymbol{\sigma}_z$ are

$$\mathbf{1} = \begin{bmatrix} 1 & 0 \\ 0 & 1 \end{bmatrix} \quad \text{and} \quad \boldsymbol{\sigma}_z = \begin{bmatrix} 1 & 0 \\ 0 & -1 \end{bmatrix}. \tag{7.35}$$

The properties of the Pauli matrices were previously derived in Chap. 3, where they can be reviewed. They are "unimodular." The identity (7.34) may be proved by summing all terms in the Taylor's series expansion, or else by verifying that both sides of the equation yield the same result on the complete set of states (7.33). Performing the trace over a single spin yields an ordinary function of K:

$$Z = \cosh K \, \text{Tr}\{\mathbf{1}\} + \sinh K \, \text{Tr}\{\boldsymbol{\sigma}_z\} = 2 \cosh K . \tag{7.36}$$

For N spins, the Hamiltonian is the sum of the individual Hamiltonians; and the collective trace is, as in multidimensional integrations, the product over individual traces. Thus, for N spins,

$$Z = (2 \cosh K)^N . \qquad (7.37)$$

The free energy and its derivative are given by

$$F = -kTN \ln(2 \cosh K) \quad \text{and} \qquad (7.38)$$

$$U = \frac{\partial(\beta F)}{\partial \beta} = -Nh \tanh \beta h$$

$$= -N\left(\frac{1}{2}g\mu_b B\right) \tanh\left(\frac{1}{2}\beta g\mu_b B\right) \qquad (7.39)$$

in exact agreement with the results of Sec. 7.1. Differentiation of F with respect to the external field B yields the magnetization:

$$\mathcal{M} = N\left(\frac{1}{2}g\mu_b\right) \tanh\left(\frac{1}{2}\beta g\mu_b B\right) \qquad (7.40)$$

and the susceptibility

$$\chi = \frac{\partial \mathcal{M}}{\partial B} = N\beta \left(\frac{1}{2}g\mu_b\right)^2 \operatorname{sech}^2\left(\frac{1}{2}\beta g\mu_b B\right) . \qquad (7.41)$$

In the limit of zero field, the zero-field susceptibility χ_0 yields Curie's law:

$$\chi_0 = \frac{\mathbb{C}_{1/2}}{T} \qquad (7.42)$$

in which $\mathbb{C}_{1/2}$, Curie's constant, is expressible in terms of known quantities:

$$\mathbb{C}_{1/2} = N\left(\frac{1}{2}g\mu_b\right)^2 \Big/ k . \qquad (7.43)$$

It is amusing that the order of the limits $T \to 0$ and $B \to 0$ may *not* be interchanged, for while χ_0 diverges at $T = 0$, the finite-field susceptibility χ (7.41) vanishes at $T = 0$ for any finite B, however small. We therefore note that $B = T = 0$ is a critical point in the phase space of noninteracting spins. For interacting systems we find nontrivial critical points elsewhere.

Problem 7.5.   Find $\mathcal{M}$ and $\mathbb{C}$ for the following cases:

(a) spins 1 (eigenvalues $e_i = hS_i$, with $S_i = 1, 0, -1$)

(b) classical dipoles (eigenvalues $e(\theta) = h\cos\theta$, with $-\pi \leqslant \theta \leqslant +\pi$)

The original calculation in case (b) dates back to 1905.[1]

7.3.   The Concept of the Molecular Field

Pierre Weiss invented the concept of a molecular field — or *mean* field — in 1907, by stretching an analogy with van der Waals' theory of nonideal gases. In his own words:

> In 1885, in his famous mémoire *On the Magnetic Properties of Bodies at Various Temperatures*, Pierre Curie gave the first experimental study of the magnetization of a ferromagnet, iron, as a function of field and temperature. He concluded from the curves he obtained, that 'by analogy with the hypotheses about fluids, the rapid increase of magnetization occurs when the magnetic intensity of the particles is sufficiently strong to permit them to interact'. But he also cautions against attaching too much importance to this similarity. There seems to be no reason to doubt the truth of Pierre Curie's idea, nor that many aspects of paramagnetism are to ferromagnetism what perfect gases are to dense fluids. The theory of the molecular field is an outgrowth of this idea ... modeled on van der Waals concept of internal pressure, yet different ... .[2]

The main proposition was that the interactions — of known or unknown origins — all added to provide a single molecular field B_m, such that the *total* force on each spin was the sum of the molecular field B_m plus any external, fixed, field B applied to the sample. The equation for determining B_m is the constitutive equation of *mean field theory* (abbreviated: MFT).

Weiss proceeded by means of a single assumption, without microscopic justification. His simple guess was that B_m would be proportional to the magnetization. For spins 1/2, (7.40) is applicable, and thus Weiss' assumption $B_m \propto \mathscr{M}$ was tantamount to:

$$B_m = B_0 \tanh[\beta b(B + B_m)] \qquad (7.44)$$

---

[1] M.P. Langevin, *J. Physique* **4**, 678 (1905).
[2] P. Weiss, *J. Physique* **6**, 661 (1907); *Proc. of 6th Solvay Congress*, 1930, Gauthier-Villars, Paris, 1932, pp. 281ff.

where B_0 is the constant of proportionality, and b is short-hand notation for $g\mu_b/2$, the ubiquitous coupling constant. Once B_m is determined by this equation, many of the results derived in the preceding sections can be taken over, the only change being the replacement of B in the various formulas by $B + B_m$. Experiment showed that the constant of proportionality B_0 and the resultant molecular field B_m could exceed the strongest laboratory fields (some 10^5 Gauss) by as much as two or three orders of magnitude!

In order to obtain the results of Weiss' assumption (7.44) it is useful to neglect the external field B altogether, and to solve first for all the thermo-dynamic properties in zero applied field. Later, we can re-introduce B into our considerations, for example, in the calculation of χ_0, the susceptibility calculated in the limit $B \to 0$.

In zero external field, (7.44) becomes

$$B_m = B_0 \tanh(\beta b B_m)\,. \tag{7.45}$$

This always possesses a trivial solution $B_m = 0$. For sufficiently large β it also admits the nontrivial solutions $\pm B_m(T)$ shown in Fig. 7.1. With increasing temperature, $B_m(T)$ decreases, and ultimately vanishes when the slope of $B_0 \tanh(\beta b B_m)$ drops below 1. This occurs at temperature

$$T_c = B_0 b/k = B_0(g\mu_b/2k)\,, \tag{7.46}$$

which we can identify as the Curie temperature.

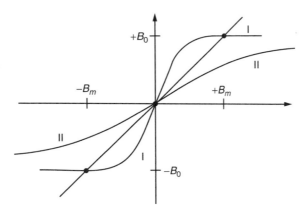

Fig. 7.1.   Graphical solution of (7.45): lhs and rhs of (7.45) vs. internal field para-meter B_m. 45° line is lhs, curve I is rhs at $T < T_c$ displaying the nontrivial solutions at $B_m(T)$, and the solution at $B_m = 0$. Curve II is the rhs at $T > T_c$, demonstrating only the trivial solution at $B_m = 0$.

How to choose between the trivial and nontrivial solutions for $T < T_c$? We can apply the criterion of maximum Z, i.e., lowest free energy F. This is one of the general principles of statistical mechanics, always invoked in case of doubt. While the trivial solution straightaway leads to

$$Z_{\text{triv.}} = 2^N \qquad (7.47)$$

for N spins, the nontrivial one yields

$$Z_{\text{nontriv.}} = (2 \cosh \beta b B_m)^N , \qquad (7.48)$$

according to (7.37). As $\cosh x > 1$ for any $x \neq 0$, the nontrivial solutions are clearly preferable over the entire temperature range below T_c.

In applying (7.39) for the internal energy to the present analysis we must introduce a factor $1/2$ to eliminate double-counting the interactions. The justification, from microscopic considerations, will come later. With this single modification, (7.39) becomes

$$U = -N(g\mu_b/2) B_m \tanh(\beta g \mu_b B_m / 2) . \qquad (7.49)$$

We eliminate the tanh factor by (7.45), express B_0 in terms of T_c by (7.46) to obtain a surprisingly simple quadratic:

$$U = -\frac{1}{2} N k T_c (B_m / B_0)^2 = -\frac{1}{2} N k T_c \sigma^2 . \qquad (7.50)$$

The physically significant quantities are: N the size of the system, T_c the experimentally determined Curie temperature, and $\sigma(T) \equiv B_m / B_0$, a function computed below.

Two other functions are of thermodynamic interest. The first is the specific heat

$$c = \frac{1}{N} \frac{\partial U}{\partial T} = -\frac{1}{2} k T_c \frac{\partial}{\partial T} \sigma^2 \qquad (7.51)$$

and the (spontaneous) magnetization per spin, $m(T) = \mathcal{M}/N$, proportional to B_m by hypothesis, hence satisfying

$$m(T) = m(0)\sigma(T) \quad \text{with} \qquad (7.52)$$

$$m(0) = \frac{1}{2} g\mu_b \qquad (7.53)$$

according to (7.40).

Rewriting (7.45) in terms of $\sigma = B_m(T)/B_0$, one obtains

$$\sigma = \tanh \frac{\sigma T_c}{T} , \tag{7.54}$$

a transcendental equation for σ. While such equations are solvable by numerical iteration, or graphically (7.54) can also be solved by specifying a value of σ in the range $-1 \leqslant \sigma \leqslant +1$ and calculating the resulting T/T_c. Specifically,

$$\frac{T}{T_c} = \frac{2\sigma}{\ln(\frac{1+\sigma}{1-\sigma})} = \frac{2\sigma}{2(\sigma + \frac{\sigma^3}{3} + \frac{\sigma^5}{5} + \cdots)} . \tag{7.55}$$

The right-hand side is invariant under the change $\sigma \to -\sigma$, yielding both nontrivial roots. The result is plotted in Fig. 7.2 where it is compared with the simple approximation $\sigma \sim 3^{1/2}(1 - T/T_c)^{1/2}$ valid near T_c, obtained from the two leading terms in Taylor's series expansion of $\ln(\ )$.

The result of the calculation for specific heat (7.51) is shown in Fig. 7.3, together with analogous curves for spins 1 and for classical dipoles, as previously introduced in Problem 7.5. Comparison with an experimental result on the particular substance $HoRh_4B_4$ is shown in Fig. 7.10 below. The conditions for a material to satisfy the premises of mean-field theory will be examined in Sec. 7.7.

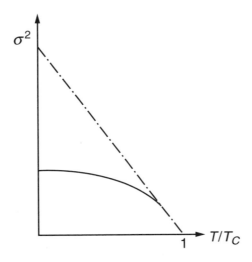

Fig. 7.2. (—): solution of (7.55) for σ^2 vs T/T_c. (- · - · - ·): linear approximation $\sigma^2 = 3(1 - T/T_c)$, valid near T_c. σ is the order parameter.

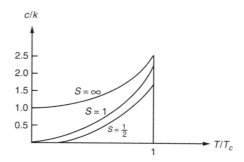

Fig. 7.3. Specific heat in molecular-field theory, for $s = 1/2$, 1 and ∞ (classical limit).

7.4. Discontinuity in Specific Heat

A striking result of mean-field theory is the discontinuity in specific heat at T_{c}. The free energy is continuous, as it is generally required to be, and all the first derivatives vary linearly near T_{c} and are continuous.

It is only the second derivatives — specific heat and, as we shall see, zero-field magnetic susceptibility — that are discontinuous. For this reason, the thermodynamic change of phase at T_{c} is denoted a *second-order phase transition*. It is also an *order–disorder* phase transition, from an ordered (magnetic) phase below T_{c} to a disordered phase above T_{c}. We may consider σ to be the *order parameter* as it vanishes at, and above T_{c}.

The magnitude of the jump discontinuity can be estimated from Fig. 7.3, or calculated by expanding the previous formulas for T near T_{c}, where $\sigma \sim 0$. It is convenient to define a dimensionless temperature t, measured from T_{c}:

$$t \equiv (T_{\mathrm{c}} - T)/T_{\mathrm{c}} \,. \tag{7.56}$$

Expansion of (7.54 or 7.55) to leading order yields:

$$\sigma = (3t)^{1/2} \,, \quad \text{thus} \quad U = -\frac{1}{2} N k T_{\mathrm{c}}(3t) \quad \text{and} \quad c = 3k/2 \,, \tag{7.57}$$

at small $t \geqslant 0$. At negative t, $\sigma = U = c = 0$. Thus, the jump discontinuity is $\Delta c = 3k/2$, independent of the model parameters T_{c} and B_0, but always in zero *external* field $B = 0$.

Although the jump discontinuity is generic to mean-field theories, the magnitude of the jump depends on such parameters as the magnitude of the spins (it varies from $\Delta c = 3k/2$ for spins one-half to $5k/2$ for spins $s \gg 1$) and the amount of correlations retained in the theory. Indeed, when *all* correlations are correctly taken into account in the more modern, microscopic theories which we examine later, the specific heat is seen to *diverge* near T_c as $c \sim |t|^{-\alpha}$, and it is with the calculation of such "critical exponents" as α that much of the modern research into statistical physics is concerned. Before examining these interesting new developments let us take Weiss' theory to its logical conclusions here.

We now investigate the form molecular-field theory must take when $s > 1/2$, and the manner in which a classical limit is attained as $s \gg 1/2$. For a spin of given s, the eigenvalue S_z along the internal-field direction may take on any of the values: $s, s-1, \ldots, -s+1, -s$. With $b = 1/2g\mu_b$ as before, the partition function of a single spin in a field B is exactly summable series:

$$Z = \exp(\beta 2bsB) + \exp[\beta 2b(s-1)B] + \cdots + \exp(-\beta 2bsB)$$
$$= \frac{\sinh[\beta bB(2s+1)]}{\sinh(\beta bB)} . \tag{7.58}$$

The magnetization (7.26) per spin is therefore given by

$$m(T) = \frac{\partial}{\partial B} \left\{ kT \ln \frac{\sinh[\beta bB(2s+1)]}{\sinh(\beta bB)} \right\} . \tag{7.59}$$

The molecular-field hypothesis replaces B in the above by $B + B_m$, with a molecular field $B_m = B_0 m(T)/m(0)$, and $m(0) = 2sb$. With $B = 0$, these substitutions result in the equation

$$B_m = B_0 \mathbb{B}_s(\beta 2sb B_m) \tag{7.60}$$

where $\mathbb{B}_s(y)$, following the differentiation indicated in (7.59), is the well-known *Brillouin function*.[3] The multiplicative constant is adjusted such that $\mathbb{B}_s(\infty) = 1$; thus the function in (7.60) is

$$\mathbb{B}_s(y) = \frac{1}{2s} \left[(2s+1) \coth \frac{2sy+y}{2s} - \coth \frac{y}{2s} \right] . \tag{7.61}$$

The leading terms in expansion in a small argument y are

$$\mathbb{B}_s(y) = \frac{s+1}{3s} y - \frac{s+1}{3s} \frac{2s^2 + 2s + 1}{30s^2} y^3 + O(y^5) . \tag{7.62}$$

In the classical limit, $s \to \infty$, the Brillouin function reduces to the so-called Langevin function.[1] This last was the function used by *Weiss* in his original analysis,[2] which predated the discovery of quantized spin by a decade. (We

---

[3] L. Brillouin, *J. Phys. Radium* **8**, 74 (1927).

have encountered the Langevin function in the guise of $\mathscr{M}$ (Problem 7.5b). The reader should check out this limit.) Taking $\mathbb{B}_s$ to the opposite limit, $s = 1/2$, yields the results of Sec. 7.3.

Near T_c the internal field vanishes, hence it is sufficient to equate terms linear in B_m in (7.60) in order to determine T_c. It is found to be

$$kT_c = \frac{2}{3}bB_0(s+1).$$ (7.63)

The expansion to $O(B_m^3)$ yields the behavior of the order parameter $\sigma = B_m/B_0$ (alternatively, $T = m(T)/m(0)$) near T_c. In terms of t (defined in Eq. (7.56) and small near T_c),

$$\sigma^2(T) = t\left(\frac{10}{3}\right)\frac{(s+1)^2}{s^2 + (s+1)^2} - O(t^3).$$ (7.64)

For $T > T_c$, t is negative and the nontrivial roots $\sigma(T)$ are imaginary, hence inadmissible. The trivial solution, $\sigma \equiv 0$, is thus mandated above T_c.

In the expression for the internal energy $u(T)$ one can eliminate the Brillouin function and obtain a quadratic form in $\sigma(T)$, just as was done previously for $s = 1/2$. The resulting formula for arbitrary s is

$$u = -(bsB_0)\sigma^2(T) = -(bsB_0)t\left(\frac{10}{3}\right)\frac{(s+1)^2}{s^2 + (s+1)^2} + \cdots.$$ (7.65)

Just below T_c, σ^2 is given by (7.64) and just above it, by $\sigma \equiv 0$. Thus, there is a jump in specific heat at T_c, of magnitude:

$$\Delta c = 5k\frac{s(s+1)}{s^2 + (s+1)^2}.$$ (7.66)

This discontinuity, which occurs only in zero external field, is independent of the parameters T_c and B_0 (but not of s). Δc varies from a minimum $3k/2$ for $s = 1/2$ to a maximum $5k/2$ in the classical limit $s \to \infty$. These findings are nicely universal. But there is a "glitch" in the classical limit: the molecular field theory predicts a specific heat that vanishes exponentially, as we approach $T = 0$. As the spin is increased, the region of exponential fall-off becomes less significant, until — in the limit $s = \infty$ — the specific heat remains finite at $T = 0$, as explored further in Problem 7.6.

Problem 7.6.   From the large y expansion of (7.61) show that in the molecular-field theory $c(T)$ approaches zero as $A\exp(-T_0/T)$ at low temperature. Express A and T_0 in terms of s and T_c. Obtain the anomalous behavior of these parameters in the classical limit (Fig. 7.3), and discuss whether the point $1/s = T = 0$ is in fact a pseudo-critical point of the same species as $B = T = 0$ for free spins.

The failure of $c(T)$ to vanish at absolute zero is tantamount to a failure of the third law of thermodynamics. This may be verified by integration of (7.28) near $T = 0$, to obtain the entropy $\mathscr{S}(T)$ in terms of the given heat capacity. It diverges, a symptom of the difficulties in classical statistical mechanics that could only be cured by the advent of quantum theory.

7.5. Magnetic Susceptibility and Spontaneous Magnetization

We now generalize mean-field theory to finite applied fields. Figure 7.4 shows the constitutive equation (7.44) plotted as function of the external field B, from which we note the salient features:

(a) At fixed $T < T_c$, the curve $\sigma(B/B_0)$ is multiple-valued. The regions where σ decreases with increasing B are unphysical: Equation (7.31) established $\chi \geqslant 0$. If the applied field is increased beyond a certain value, the magnetization must jump to the upper branch. This must occur for $|B| \leqslant |B_1|$ (B_1: external field value at which $\partial\sigma/\partial B = \infty$). Thus, even the primitive mean-field theory predicts *hysteresis* — an irreversible behavior as the external field is cycled. Unfortunately, however, we cannot use this theory to study physical hysteresis, which is related to the formation and evolution of domains (regions of opposing magnetization

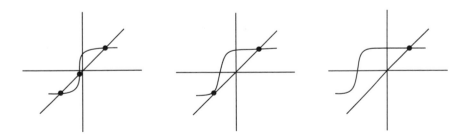

Fig. 7.4. Graphical solution of (7.44) in nonzero external fields. *Left curve*: weak field, the 3 solutions in Fig. 7.1 are slightly displaced. *Middle curve*: two roots merge at $B = B_1(T)$. *Right curve*: at fields B_1 only one root remains. See Fig. 7.6 for $B_1(T)$.

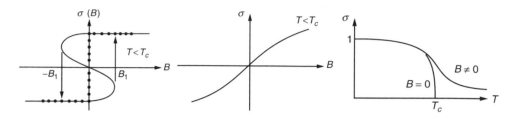

Fig. 7.5.   Magnetization m or order parameter σ as a function of B and T. Thermodynamic equilibrium curve is indicated by $(\cdots)$, arrows show possible extent of hysteresis when B changes with time along one branch or the other.

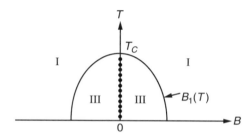

Fig. 7.6.   Order parameter σ is triple-valued in Regions III, single-valued in Region I although thermodynamic *equilibrium* value is unique everywhere *except* on vertical axis $B = 0$, $T < T_c$. Thus, 2 roots in Region III are always unstable or metastable.

which are created to minimize magnetostatic energy, totally ignored in the present treatment, but briefly discussed earlier in this book).

In thermodynamic equilibrium, the function $\sigma(B/B_0)$ has its discontinuity strictly at $B = 0$ and is reversible, as illustrated in Fig. 7.5.

(b) At fixed $B \neq 0$, $\sigma(T)$ is a smooth function, T_c being noted simply as a point of inflexion. Thermodynamic functions of magnetic systems are generally discontinuous only in zero external field. The neighborhood of $B = 0$, $T = T_c$ is denoted the critical region.

Many of these features are common to all known theories of ferromagnetism. The phase diagram we extract from mean-field theory is typical of all models that have phase transitions (not all do, as we shall see later) and is examined in Fig. 7.6. The line $\pm B_1$ indicates the extreme limit of hysteresis, although it has no significance in equilibrium thermodynamics. Figure 7.7 illustrates the zero-field susceptibility, $\chi_0 = \partial \mathcal{M}/\partial B|_{B=0}$, as a function of

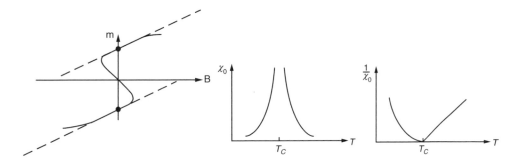

Fig. 7.7. Magnetization in a ferromagnet as a function of external field. *Left graph* shows that tangents are continuous at $B = 0$ even though m is discontinuous. The slope $\partial m/\partial B = \chi$ is the susceptibility. Plotted in *middle graph* is χ at $B = 0$, denoted χ_0, as a function of T. It is finite except at the critical temperature. For greater clarity, it is conventional to plot $1/\chi_0$ as in the *right figure*.

T. It diverges at T_c. More precisely, mean-field theory predicts $\chi_0 \sim 1/|t|$, as we shall see shortly. It may be interesting to note here, that other models of ferromagnetism also predict a singular χ_0 at T_c. In general, one writes $\chi_0 \sim |t|^{-\gamma}$, where the critical index $\gamma \geqslant 1$. The value $\gamma = 1$ is attained only for very long-ranged forces, and it will appear that, indeed, mean-field theory is valid near T_c only for very long-ranged interactions.

Let us proceed with the calculations. For $s = 1/2$ at temperatures $T > T_c$ we use (7.44), expanding to leading order in the presumably small quantities B and B_m:

$$B_m = B_0 \left(\frac{1}{2} \beta g \mu_b \right) (B + B_m) + O(B^3). \qquad (7.67)$$

With (7.46) for T_c, this yields

$$B_m = -B/t \quad \text{and} \quad B + B_m = B(1 - 1/t). \qquad (7.68)$$

The latter expression makes it easy to perform the derivative:

$$\chi_0 = \left| N \left(\frac{1}{2} g \mu_b \right) \frac{\partial}{\partial B} \tanh \left[\frac{1}{2} \beta g \mu_b (B + B_m) \right] \right|_{B=0}$$

$$= \frac{\mathbb{C}_{1/2}}{T - T_c} \quad (T > T_c), \qquad (7.69)$$

with Curie's constant $\mathbb{C}_{1/2}$ previously introduced in (7.43).

For $s > 1/2$ we start with

$$B_m = B_0 \mathbb{B}_s[\beta 2sb(B + B_m)], \tag{7.70}$$

and expand to first-order in the arguments, using (7.62). Again, this yields *precisely* (7.69), the s-dependence being entirely implicit in the definition of $T_c = 2bB_0(s+1)/3k$, (7.63). Now, with $\mathcal{M} = N2bsB_m/B_0$ we find

$$\chi_0 = \frac{\mathbb{C}_s}{T - T_c} \quad (T > T_c), \tag{7.71}$$

with

$$\mathbb{C}_s = Nb^2 \left[\frac{4}{3}s(s+1)\right] \Big/ k = \mathbb{C}_{1/2}\frac{4}{3}s(s+1),$$

the generalized Curie constant.

Below T_c one is interested not only in $\mathcal{M}_0(T)$, the spontaneous magnetization, but also in $\partial\mathcal{M}/\partial B|_0 \equiv \chi_0(T)$, its slope at zero field. The behavior of χ_0 is shown in Fig. 7.7 and derived in Problem 7.7.

Problem 7.7.   Extend χ_0 for $s = 1/2$ to the low-temperature region $T < T_c$ using the expansion $B_m = B_m^0 + \chi_0 B$ in (7.44), with B_m^0 = zero-field solution of (7.45). Show how to obtain:

$$\chi_0 \propto \frac{(1 - \sigma^2)}{T - T_c(1 - \sigma^2)} = \begin{cases} A(T_c - T)^{-1}, & T \lesssim T_c \\ \dfrac{D}{T}\exp(-T_0/T), & T \to 0 \end{cases}$$

and determine the constants A, D, T_0 in terms of T_c and $\mathbb{C}_{1/2}$. Next, obtain the analogous results for $s > 1/2$.

We briefly mentioned the unstable or metastable magnetization curves. Although they are typically neglected in texts on magnetism, the unstable solutions in the mean-field theory *are* examined in a companion book[4] on statistical mechanics. We shall not repeat the discussion here, except to mention that the sub-optimal solutions are the generalizations to interacting spins of "negative temperatures," a notion usually reserved for free spins. Negative temperature, in the context of quasi-free spins, is a useful tool in the study of magnetic resonance (ESR, NMR, EPR, ...), but little

---

[4] D.C. Mattis, *Statistical Mechanics Made Simple*, World Scientific, 2003, Chap. 2.

explored in strongly interacting spin systems. (Note that the idea is simplicity itself: at the moment an applied field B is suddenly reversed, spins
that previously were in thermal equilibrium now find themselves in highly
excited state(s). In establishing a new equilibrium they must interact with
their environment — and it is this interaction that reveals the nature of the
magnetic environment.)

7.6. Adiabatic Demagnetization

This is a second "exotic" topic that concerns quasi-free spins (frequently
embedded in a solid matrix, i.e. in a *magnetic salt*), and their use in low-T
thermometry. The entropy of N such spins one-half, in the presence of an
external magnetic field H is $\mathscr{S} = S/Nk = -\frac{1}{Nk}\frac{\partial F}{\partial T} = \log(2\cosh B/T) -$
$(B/T)\tanh(B/T)$, where $B = H\mu_B gs/k_B$ is the interaction energy with
the field $\div k$. Starting the system at some T in zero external field, point
a, in Fig. 7.8, we increase the field B isothermally to point b in diagram,
thence decrease it *adiabatically* to the point c.[5] Once thermal equilibrium is
recovered at c, the process can be repeated $(c \to d \to f)$, the temperature
halved once again, etc.

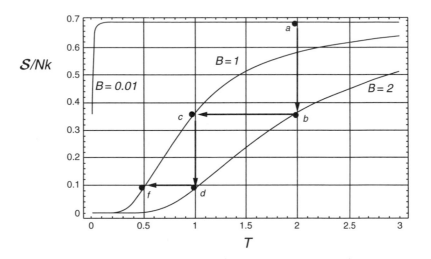

Fig. 7.8. Illustrating adiabatic demagnetization for spins $1/2$.

---

[5] For this illustration we have chosen, somewhat arbitrarily, $s = 1/2$ and $B_b = 2B_c$, which
leads to $T_b = 2T_c$, $T_d = 2T_f$, etc.

This process does not work well if the spins interact and are able thereby to lower their entropy by internal means. The external field then affects the total entropy to a much lesser extent, and prevents the dramatic lowering of T from occurring. Thus, once a temperature is achieved at which the residual spin-spin interactions are comparable to kT, this process stops.

7.7.   Antiferromagnetism

The discovery of antiferromagnetic materials tested the very foundations of molecular-field theory, for it introduced a geometrical, internal structure and a breaking of the homogeneous, isotropic symmetries of the Weiss field. Following the original explanation in terms two interpenetrating lattices put forth by Néel,[6] Bitter,[7] Van Vleck[8] and others, the concepts have been extended to include spiral-, canted-, triangular, and yet-more-complex-spin arrangements revealed by neutron scattering experiments. It is interesting to note how the structures were deduced prior to the developments of experimental methods which permit the direct observation of the spin correlations functions. We quote Van Vleck

> There is ... one class of materials known as 'antiferromagnetics' in which it is quite clear that the suppression of paramagnetism is to be identified with exchange coupling. These substances ... have a susceptibility which passes through a maximum as the temperature is raised ... explained theoretically in the following way. Suppose we have a crystal whose constituent atoms can be resolved into two sublattices A and B such that the nearest neighbors of the atoms of A are atoms of B, and vice versa. The simple-cubic and body-centered cubic lattices are both of this type ... With a negative exchange integral [connecting nearest-neighbors] the exchange energy of two atoms is a minimum if their spins are antiparallel. Hence the configuration of deepest energy for the crystal as a whole is that in which the spins of sublattice A all point northward, and those of B all southward or vice versa ... Consequently a "staggered" non-vanishing molecular field can exist.[9]

[6] L. Néel, *Ann. Physique* **17**, 64 (1932).
[7] F. Bitter, *Phys. Rev.* **54**, 79 (1937).
[8] J.H. Van Vleck, *J. Chem. Phys.* **9**, 85 (1941).
[9] J.H. Van Vleck, *Rev. Mod. Phys.* **17**, 27 (1945), pp. 45ff.

Let us pursue this idea to its logical development. For maximum simplicity, we specialize to spins $s = 1/2$, spins of type A interacting only with B's, and vice versa.

The linear constitutive equations for two sublattices of spins one-half are written,

$$\sigma_A = \tanh\left[\frac{T_N}{T}(H - \sigma_B)\right]$$

$$\sigma_B = \tanh\left[\frac{T_N}{T}(H - \sigma_A)\right]$$

(7.72)

in an obvious generalization of the ferromagnetic equation (7.44). The applied magnetic field (in appropriate units) is denoted H, to avoid confusion with the label on one of the sublattices. σ_A and σ_B, the fraction of maximum magnetization on each sublattice, are the *order parameters*. In the absence of an external field, the order parameters on each lattice are equal in magnitude and (7.72) has the obvious solution $\sigma_A = -\sigma_B = \sigma$. Setting this into (7.72) with $H = 0$ reduces it to (7.54), an equation previously solved: see (7.55) and Fig. 7.2 for the details.

Yet not all properties of the antiferromagnet can be related to the ferromagnet. The total magnetization, $N(\sigma_A + \sigma_B)/2$, now vanishes, so we define a staggered order parameter σ, $N(\sigma_A - \sigma_B)/2 = N\sigma$, vanishing only at or above the critical temperature. Appropriately enough, the last is denoted the *Néel temperature* T_N.

The parallel susceptibility is of great interest. It is the response to an external field along the directions of spontaneous sublattice magnetization. In a sufficiently weak external field, this quantity — denoted by $\chi_\|$ — is computed setting $\sigma_A = \sigma + H\chi_{\|A}$ and $\sigma_B = -\sigma + H\chi_{\|B}$, assuming σ is unaffected to $O(H^2)$. Subtracting the two equations (7.72), we readily verify $\chi_{\|A} = \chi_{\|B} \equiv \chi_\|$. Adding them yields

$$2H\chi_\| = \frac{\sigma + \frac{T_N}{T}H(1 - \chi_\|)}{1 + \frac{\sigma T_N}{T}H(1 - \chi_\|)} - \frac{\sigma - \frac{T_N}{T}H(1 - \chi_\|)}{1 - \frac{\sigma T_N}{T}H(1 - \chi_\|)}$$

$$= 2\frac{T_N}{T}H(1 - \chi_\|)(1 - \sigma^2) + O(H^3).$$

(7.73)

We divide by H, then take the limit $H = 0$ to obtain the zero-field susceptibility

$$\chi_{0\|} = \frac{\mathbb{C}(1 - \sigma^2)}{T + T_{\mathrm{N}}(1 - \sigma^2)} \tag{7.74}$$

after reëxpressing it in appropriate units, to agree with (7.41)–(7.43). σ is the molecular-field order parameter previously calculated in (7.55) and Fig. 7.2.

The parallel, zero-field susceptibility given above vanishes at absolute zero, increases to a maximum $\mathbb{C}/2T_{\mathrm{N}}$ at Néel's temperature, then decreases at high temperature, following a modified Curie law. These properties were originally documented by Van Vleck.

> ... At the absolute zero, the inhibiting effect of the powerful internal on any change in alignment due to a [parallel] weak external field is complete ...

In the same paper, he goes on to explain that an external field *perpendicular* to the existing sublattice magnetizations would have a more important effect:

> ... applied perpendicular to the alternating inner fields, the external field can still give rise to an outstanding moment, by twisting slightly the orientations of the elementary magnets.[9]

The response to such a perpendicular field is denoted $\chi_{\perp}$ the perpendicular susceptibility. We can calculate this quantity easily, by noting that the perpendicular magnetization is just $\mathcal{M} \propto \sigma H_{\perp}(H_{\perp}^2 + \sigma^2)^{-1/2}$, hence

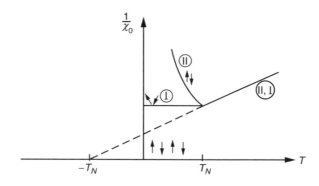

Fig. 7.9. Inverse of the zero-field longitudinal ($\|$) susceptibility and of the zero-field transverse susceptibility ($\perp$) in an antiferromagnet subject to Néel–Weiss molecular field theory. Both are equal and linear above T_{N}. Below T_{N} the transverse susceptibility is constant while the longitudinal vanishes (hence $1/\chi_{0\|}$ diverges at $T = 0$). (- - -) extrapolates high-temperature line to $-T_{\mathrm{N}}$.

$\chi_{0\parallel} = $ const. for $T \leqslant T_{\mathrm{N}}$. *Above* T_{N} the sublattice magnetizations vanish, hence χ must be isotropic and $\chi_{0\perp} = \chi_{0\parallel}$. By continuity we obtain $\chi_{0\perp} = \mathbb{C}/2T_{\mathrm{N}}$ for $T \leqslant T_{\mathrm{N}}$, as shown in Fig. 7.9. We shall see later that the separable spin-glass model has, coincidentally, exactly the same phase diagram!

7.8. Short-Ranged versus Long-Ranged Interactions

The microscopic forces in which the molecular field originates are quantum-mechanical in nature, coming in various guises depending on the specific mechanisms of electron transport in the solid. We have examined the several possibilities earlier in this book. Generally, interactions are quite short-ranged, so that taking nearest- and next-nearest neighbor forces into account suffices. But there are two instances of long-ranged interactions, the first being the ubiquitous dipole-dipole electromagnetic force — weak, long-ranged, and of course, anisotropic. Its principal effect is in ferromagnetism,[10] causing the shape of the sample to contribute to the overall energy (demagnetizing factor) and encouraging the creation of magnetic domains. It has no effect in causing a phase transition (the magnitude of the Curie temperaturature associated with purely electromagnetic forces would not exceed $O(1\ \mathrm{K})$), and has no bearing on the other magnetic structures: antiferromagnetic, spiral, spin-glass, etc. with which one is often concerned. The literature on this topic is comparatively sparse.[11] We shall not consider the magnetostatic forces at present, but rather, a second long-ranged mechanism, the Ruderman–Kittel–Kasuya–Yosida (RKKY) mechanism, also called the indirect-exchange mechanism.[12] It is operative only in metals and is generally oscillatory — spins at different distances R_{ij} may be connected by either ferromagnetic or antiferromagnetic bonds, depending on the magnitude of R_{ij}. At small concentrations of electrons, this interaction becomes essentially ferromagnetic and long-ranged.[13]

---

[10] Domains are thoroughly described in a number of texts: R.M. Bozorth, *Ferromagnetism*, Van Nostrand, Princeton, 1951; A.H. Morrish, *The Physical Principles of Magnetism*, Wiley, New York, 1964; S. Chikazumi, *Physics of Magnetism*, Wiley, New York, 1964; A.H. Eschenfelder, *Magnetic Bubble Technology*, 2nd ed., Springer Ser. Solid State Sci., Vol. 14, Springer, Berlin, Heidelberg, 1981.

[11] M.H. Cohen and F. Keffer, *Phys. Rev.* **99**, 1128 and 1135 (1955); R.B. Griffiths, *Phys. Rev.* **176**, 655 (1968); M.E. Fisher and D. Ruelle, *J. Math. Phys.* **7**, 260 (1966).

[12] M.A. Ruderman and C. Kittel, *Phys. Rev.* **96**, 99 (1954). For details see [Ref. 2.1, Chaps. 2 and 6].

[13] D.C. Mattis and W. Donath, unpublished IBM report (ca. 1961). For summary see [Ref. 2.1, p. 235 and Fig. 6.6].

It is important to know the range and nature of the magnetic interactions because *all* microscopic theories predict quite different thermodynamic properties for short-range interactions as distinguished from long-ranged ones. This is the topic we investigate in the present section. A typical Hamiltonian includes the interactions of all pairs of spins assumed to be at point R_i of a regular lattice. Let us assume there are N such points, and that the spins $S_i = \pm 1$ are Ising $s = 1/2$ spins. The interaction energy is thus

$$E = -\frac{1}{2}\sum_i \sum_j J_{ij} S_i S_j \,. \tag{7.75}$$

Translational invariance in an homogeneous medium requires that J_{ij} depend only on the distance, $J_{ij} = J(R_{ij})$. We distinguish the range of interactions as follows:

(a) Short-ranged and $\sum_j |J(R_{ij})| < \infty$

(b) Long-ranged: $\sum_j |J(R_{ij})| = \infty$.

If the bonds are ferromagnetic ($J \geqslant 0$) then there is no qualitative, substantive difference between the general short-range problem (a) and a simpler model, in which only nearest-neighbor bonds are retained. If some bonds are antiferromagnetic the geometry of the lattice plays an important role. Depending on the details of the interaction and the geometry of the lattice, it may be impossible to find a ground state configuration in which all bonds J are optimized, in which case we speak of frustration. For example, the properties of a ferromagnetic *triangular* lattice are more or less independent of the range of the forces, as long as it is finite. Whereas, for antiferromagnetic coupling, it is already impossible to satisfy all the bonds just for nearest-neighbor interactions. This phenomenon, first noted by Wannier,[14] is treated in the following chapter, and reappears in connection with spin glasses.[15]

Exercise: Find the ground state of 3 spins subject to (7.75) in the two cases: (i) all $J_{ij} > 0$ and (ii) all $J_{ij} < 0$. Note especially, the ground state degeneracy (the number of distinguishable configurations leading to the same energy) in the latter case, if $J_{12} = J_{13} = J_{23} = J < 0$. Contrast the two models: Ising and Heisenberg, and their correlations.

---

[14] G.H. Wannier, *Phys. Rev.* **79**, 357 (1950).
[15] G. Toulouse, *Commun. Phys.* **2**, 115 (1977); E. Fradkin, B. Huberman and S. Shenker, *Phys. Rev.* **B18**, 4789 (1978).

As a prototype of the long-ranged forces consider a *constant* interaction $J_{ij} = J_0/N$, each spin interacting with all the others with equal strength, the fact $1/N$ being included so that the total energy E remains extensive. Here,

$$E = -(J_0/2N)\left(\sum_i S_i\right)^2 + J_0 N/2. \tag{7.76}$$

The energy is a function of $\bar{S} = N^{-1}\sum_i S_i$ alone. We recall from Sec. 7.1: let n spins be "up" and $1 - n$ "down", such that $\bar{S} = 2p - 1$, with $p = n/N$. Thus, the energy

$$E = -NJ_0\bar{S}^2/2 + J_0 N/2 = -NJ_0(2p - 1)^2/2 + J_0 N/2 \tag{7.77}$$

can be written as a function of p, as can the spontaneous magnetization:

$$\mathcal{M} = N(2p - 1) \tag{7.78}$$

and the entropy (Problem 7.4):

$$\mathcal{S} = -kN[p\ln p + (1 - p)\ln(1 - p)]$$
$$= -kN\left[\frac{1}{2}(1 + \bar{S})\ln\frac{1}{2}(1 + \bar{S}) + \frac{1}{2}(1 - \bar{S})\ln\frac{1}{2}(1 - \bar{S})\right]. \tag{7.79}$$

Constructing $F(\bar{S}) = E - T\mathcal{S}$ and minimizing with respect to $\bar{S}$ (or, what is equivalent, p) we obtain a familiar result:

$$\bar{S} = \tanh\beta J_0\bar{S} \tag{7.80}$$

which, for positive J_0, is easily identifiable with results of the molecular-field theory (Sec. 7.3). We can define an internal field variable h_i, conjugate to the spin S_i:

$$h_i = -\frac{\partial E}{\partial S_i} = J_0\left(\bar{S} - \frac{S_i}{N}\right). \tag{7.81}$$

In the thermodynamic limit, all $h_i = J_0\bar{S}$. Thus, after we identify J_0/b in Sec. 7.3 as B_0, kT_c with $(J_0/b)b$, and $\bar{S}$ with σ, we recover *all* the results originally postulated by Weiss! The validity of this theory thus depends only on the range of the forces. Fortunately, there does exist an example of a metallic ferromagnet in which the range (proportional to π/k_F, the deBroglie wave-length of an electron at the Fermi surface[13]) may have become sufficiently long to satisfy the criteria of MFT. It is $HoRh_4B_4$, a model mean-field

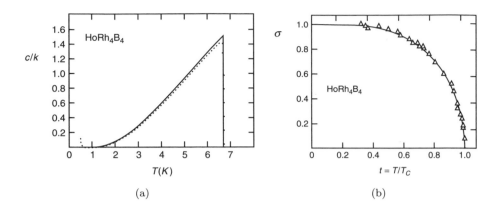

(a)   (b)

Fig. 7.10. An experimentally realized mean-field ferromagnetic substance showing excellent agreement between experiment and theory. (a) Temperature dependence of the specific heat [theory is (—), experimental data are the (· · · ·)].
(b) Magnetization or order parameter vs. T/T_c. Taken from Ott *et al.*,[16] where
additional graphs are given for magnetostriction and electrical resistance, all in
excellent accord with MFT predictions.

ferromagnet according to its discoverers.[16] The temperature-dependence of
its spontaneous magnetization and specific heat track the theoretical curves
with high precision, as shown in Fig. 7.10.

If all ferromagnets behaved as $HoRh_4B_4$, the theory of ferromagnetism
could be rather briefly disposed of. In fact, this *mean-field behavior is quite
exceptional*, and we shall determine that the properties near T_c are, and
should be, quite different for the more common case of short-ranged forces;
these depend on dimensionality, lattice geometry, etc.

So far, we have considered $J_0 > 0$, i.e., all ferromagnetic bonds. What
if J_0 were negative, would the case of all antiferromagnetic bonds lead to
Néel's theory of antiferromagnetism? The resultant equation,

$$\bar{S} = -\tanh(\beta|J_0|\bar{S}) \tag{7.82}$$

has the unique solution $\bar{S} = 0$ at all temperatures. Looking back at (7.76)
we see this is an example of *extreme frustration* — only half the bonds
have a negative energy, the remainder have positive energy. The energy is a
constant, independent of T, and one obtains $\mathscr{S}/N = k\ln 2$, the same entropy
as for free spins at all T! So, the answer is *no* — frustration is not the same
as antiferromagnetism.

---

[16] H.R. Ott *et al.*, *Phys. Rev.* **B25**, 477 (1982).

Although the long-ranged antiferromagnetic interaction does not lead to any significant cooperative behavior, it does modify the magnetic susceptibility. Extending (7.82) to include an applied external field B we find,

$$\bar{S} = \tanh \beta(B - |J_0|\bar{S}) \tag{7.83}$$

which now has a nontrivial solution $\bar{S}(B,T)$. Defining the Néel temperature $T_N = |J_0|/k$, we obtain for the zero-field susceptibility an expression:

$$\chi_0 = \frac{\mathbb{C}_{1/2}}{T + T_N} \tag{7.84}$$

which has the expected form (Fig. 7.9) above T_N. Unlike the results of Néel's theory, however, χ_0 here is continuous through T_N to absolute zero. Indeed, the solution of (7.83) is free of any singularity at T_N (or, indeed, at any other positive temperature!). We conclude that long-ranged $J_{ij} < 0$ result only in a disordered phase at all T, and not in any antiferromagnetic ordering.

Additional insight into this apparent asymmetry between positive and negative bonds is afforded by studying the eigenvalue spectrum of $J(R_{ij})$, which is given by its Fourier transform (we may think of J_{ij} as a cyclic matrix):

$$W_k \equiv \sum_n J(R_n) \exp(ik \cdot R_n) \,. \tag{7.85}$$

The amplitudes $X_k = X^*_{-k}$ of the normal modes $\exp(ik \cdot R)$ are

$$X_n = N^{-1/2} \sum_n S_n \exp(ik \cdot R_n) \,, \tag{7.86}$$

or, conversely, the original spins are themselves a sum over normal modes,

$$S_n = N^{-1/2} \sum_n X_k \exp(-ik \cdot R_n) \tag{7.87}$$

the sum being over the N points k of the first Brillouin zone. The total energy (7.75) is now brought into diagonal form,

$$E = -\frac{1}{2} \sum_k W_k |X_k|^2 \,. \tag{7.88}$$

We are still far from being able to evaluate the partition function because the N constraints $S_n = \pm 1$ (i.e., $S_n^2 = 1$) are not satisfied for arbitrary values of the normal mode amplitudes X_k. Nevertheless, it is easy to see that for long-ranged (non-integrable) ferromagnetic interactions W_k is singular at $k = 0$, thus the optimum energy comes from having X_0 as large as possible.

Conversely, for long-ranged antiferromagnetic interactions, W_k is *negative* at $k = 0$, thus $k = 0$ is singularly *disfavored* in comparison with all the other normal modes. In such a case, we expect maximum disorder and maximum entropy, precisely as we have found in the example (7.82).

Turning to short-ranged forces, we find the spectrum to be non-singular. It is no longer sufficient to consider one normal mode, or one order parameter. Take the example of simple-cubic lattice with nearest-neighbor bonds J_0:

$$W_k = -2J_0(\cos k_x a + \cos k_y a + \cos k_z a). \tag{7.89}$$

Even if $J_0 > 0$ and $k = 0$ yields the optimum W_k, there will be a continuum of nearby eigenvalues that contribute at finite temperature — the low-lying spinwaves. A single-parameter mean-field theory cannot then be expected to be accurate. If $J_0 < 0$, the optimum W_k is at $k = (\pi, \pi, \pi)/a$. The corresponding spin configuration extracted from (7.87) is:

$$S_n = (-1)^n \tag{7.90}$$

where $(-1)^n = +1$ on even-numbered sites and -1 on odd-numbered sites. This is precisely Néel's state, but the mean-field thermodynamics of the preceding section are not applicable, for the reason already stated: the existence of a continuum of eigenvalues in the neighborhood of the optimum value.

If we wish Néel's state to be favored, and mean-field theory to be applicable, we must contrive that (7.90) be uniquely, indeed singularly, favored. One such arrangement is:

$$J(R_n) = |J_0|(-1)^n/N \tag{7.91}$$

in which all bonds connecting even-numbered sites to one another are ferromagnetic, all bonds connecting odd-numbered sites to one another are also ferromagnetic, but all bonds connecting even- to odd-numbered sites are antiferromagnetic. It is intuitive that when the interactions have the structure of the desired ground state, the resulting order parameter is strongly favored. However, materials in which the interactions are long-ranged and oscillatory, in precisely this manner, are perhaps even rarer than those in which the bonds are ferromagnetic and long-ranged. To explain antiferromagnetism it is necessary to invoke short-range interactions.

For the most part, the remainder of this book is concerned with short-ranged interactions — usually, just nearest-neighbor interactions among localized entities that come in multiple varieties and internal symmetries.

As a result we have to learn about Ising models, Heisenberg models, Gaussian, spherical, Potts, and clock models, etc. Some that belong to the same symmetry class can be lumped together, others cannot.

7.9. Fermions, Bosons, and All That

Before turning to microscopic theories of cooperative phenomena, it is necessary to study the statistical mechanics of idealized particles — noninteracting fermions, bosons, as well as of spins (which are neither fermions nor bosons), and of classical systems as well. We illustrate with a couple of examples that could prove useful later on.

Fermions

No more than 1 fermion may be in any one-particle quantum state. If a given state has energy e_k, this may also be written $n_k e_k$, with $n_k = 0$ if the state is unoccupied and $n_k = 1$ when it is occupied. The partition function Z is therefore of the form

$$Z = Z_0 \, \text{tr}_{(k)} \, \exp(-\beta e_k n_k) = Z_0[1 + \exp(-\beta e_k)] \tag{7.92}$$

with Z_0 the partition function for all states other than k, and $\text{tr}_{(k)}$ refers to the sum over the two possibilities: $n_k = 0$ and 1. One may include all other one-particle states by induction,

$$Z = \prod_k [1 + \exp(-\beta e_k)], \tag{7.93}$$

where the product is over *all* values of k. This formula is not applicable if the particles interact, as we have not included the interaction energies.

The thermal averaged occupation of the kth state is

$$\langle n_k \rangle = \frac{0 + 1 \exp(-\beta e_k)}{1 + \exp(-\beta e_k)} = [1 + \exp(\beta e_k)]^{-1} \tag{7.94}$$

the familiar *Fermi function*.

Bosons

We need to understand Bose–Einstein (boson) statistics for spin waves. As many bosons as desired may occupy any given one-particle state. Thus, if this state has energy e_k when occupied by a single particle, it will have energy $n_k e_k$ when occupied by n_k bosons, and $n_k = 0, 1, 2, \ldots$. The partition function is then

$$Z = Z_0[1 + \exp(-\beta e_k) + \exp(-2\beta e_k) + \exp(-3\beta e_k) + \cdots]$$
$$= Z_0[1 - \exp(-\beta e_k)]^{-1} \tag{7.95}$$

for $e_k > 0$. Hence, Z_0 is the partition function for the remaining states $k' \neq k$, so that, by induction,

$$Z = \prod_k [1 - \exp(-\beta e_k)]^{-1} \tag{7.96}$$

the product being over all distinct one-particle states, each labeled by a different value of the label k. Note that if any e_k is negative, the geometric series diverges (as does Z). In such cases, alternative methods of statistical mechanics must be explored.

The analogue of the Fermi function (7.94) is the Bose–Einstein distribution function, obtainable from (7.95):

$$\langle n_k \rangle = [\exp(\beta e_k) - 1]^{-1} . \tag{7.97}$$

Problem 7.8.   Derive the above by relating $\langle n_k \rangle$ to the logarithmic derivative of Z.

Gaussian

There are many examples (in acoustics, classical electromagnetism, etc.) in which the energy depends on the amplitude X_k of a normal mode, assumed to be real, in the manner: $e_k X_k^2$. The partition function then takes on an aspect:

$$Z = Z_0 \int_{-\infty}^{+\infty} dX_k \exp(-\beta e_k X_k^2) = Z_0(\pi kT/e_k)^{1/2} \tag{7.98}$$

with Z_0 refering to all normal modes other than the kth which is singled out. The average energy stored in the kth normal mode is thus,

$$e_k \langle X_k^2 \rangle = e_k \int dX_k X_k^2 \exp(-\beta e_k X_k^2) \Big/ \int dX_k \exp(-\beta e_k X_k^2) = \frac{1}{2} kT . \tag{7.99}$$

Problem 7.9. Show that odd moments $\langle X_k^{2n+1} \rangle$ all vanish. Obtain the (nonvanishing) even moments by evaluating

$$\langle X_k^{2n} \rangle = Z^{-1}(-1)^n \partial^n Z/\partial(\beta e_k)^n$$

in closed form.

Again, by induction, Z can be written explicitly in terms of all the normal modes,

$$Z = \prod_k (\pi kT/e_k)^{1/2} . \tag{7.100}$$

If two normal modes are degenerate, i.e., share the same e_k, we can rewrite $e_k(X_{1k}^1 + X_{2k}^2)$ compactly as $e_k|X_k|^2$, with $X_k = X_{1k} + iX_{2k}$, and $X_k^* = X_{1k} - iX_{2k}$. In the evaluation of the partition function one should then replace integrations

$$\int dX_{1k} \int dX_{2k} \exp[-\beta e_k(X_{1k}^2 + X_{2k}^2)] \quad \text{by} \tag{7.101a}$$

$$\int dX_k \int dX_k^* \exp(-\beta e_k|X_k|^2) \quad \text{(symbolically)} \tag{7.101b}$$

treating X_k, X_k^* as independent variables. However, in evaluating integrals of type (7.101b) explicitly one uses (7.101a). And, just as in the case of bosons, the parameters e_k are required to be positive in order that these expressions make any sense.

Legendre Transformations

By modifying the Boltzmann factors one can often cure such major or minor deficiencies as $e_k < 0$ or evaluate certain correlations more conveniently. These shifts are the equivalent of the well-known Legendre transformations in thermodynamics. As an illustration consider what happens to the thermodynamic relations (7.24)–(7.31) if the energy (7.75) is modified in the following manner:

$$E \rightarrow E - \mu(M - \mathcal{M}) . \tag{7.102}$$

Let us define the entity on the rhs as $\hat{E}$, and calculate

$$\hat{Z} = e^{-\beta \hat{F}} = \text{Tr}\{e^{-\beta \hat{E}}\} . \tag{7.103}$$

The parameter μ is to be chosen such that $\langle M \rangle \equiv \mathscr{M}$, the latter being a prescribed quantity. With this device we can deal with an ensemble in which the magnetization is fixed — rather than the usual in which the magnetic field B is given. μ is not an independent parameter; given our assumptions, it must be chosen such that $\partial \hat{F}/\partial \mu = 0$. Replacing (7.26) we have $\partial \hat{F}/\partial M = \mu$. [The usual free energy F is related to $\hat{F}$ by $F = \hat{F} - \mu \mathscr{M}$. In F, μ may be taken as the independent variable, hence μ is seen to correspond to the usual magnetic field, and $\partial F/\partial \mu = -\mathscr{M}$ as in (7.26).]

Similarly, if we wished to calculate the free energy corresponding to (7.88), we might have to deal with some negative values of $-W_k/2$. This is not permitted in the Gaussian integrals, so one defines

$$\hat{E} = E + \mu \left(\sum_k |X_k|^2 - N \right), \tag{7.104}$$

choosing μ in this instance such that $(\mu - W_k/2) > 0$ for all k. The quantity $(\sum_k |X_k|^2 - N)$ *should* vanish — as may be seen by squaring both sides of (7.87) and summing over all N sites using the identity $S_n^2 = 1$. By using the "^" ensemble, we guarantee that this identity is satisfied *on average* — and avoid divergent integrals as well. (This is the basis of the "spherical" model that is presented in the next section.)

Such transformations are useful for fermions as well. If the total number of particles $\sum n_k$ is fixed, say at $\mathscr{N}$, then the modification:

$$\sum_k e_k n_k \rightarrow \sum_k e_k n_k - \mu \left(\sum_k n_k - \mathscr{N} \right) \tag{7.105}$$

may be advantageous. A condition $\partial \hat{F}/\partial \mu = 0$ establishes the conservation law, while for the actual calculations, it suffices to replace e_k by $e_k - \mu$. In this instance, μ is known as the "chemical potential", or the Fermi energy (and is often written ε_F).

7.10.   Gaussian and Spherical Models

The preceding can be used directly to construct exact solutions of the well-known *Gaussian* and *spherical* models of Berlin and Kac.[17] These two related and physically well motivated models of short-ranged spin-spin interactions

---

[17] T. Berlin and M. Kac, *Phys. Rev.* **86**, 821 (1952).

demonstrate many interesting features of a phase transition at T_c, the Curie temperature. They are soluble with or without an external magnetic field, and can be probed for the properties of cooperative phase transitions: the type of decay laws of correlation functions, the temperature-dependence of thermodynamic functions near T_c (i.e., the critical exponents), the nature of low-temperature excitations, etc. Alas, each of these models suffers from some shortcoming rendering it unreliable in the neighborhood of T_c, thereby vitiating the calculated results at the critical point. But before we discuss the shortcomings let us first search for the positive aspects.

Gaussian Model

This model assumes that the weight $P(S)dS$ for finding the nth spin S_n in the ensemble, between S and $S + dS$, is $P(S) = (2\pi)^{-1/2} \exp(-S^2/2)dS$, normalized such that

$$\int_{-\infty}^{+\infty} dS P(S) = \int_{-\infty}^{+\infty} dS S^2 P(S) = 1 \,,$$

in which respects the Gaussian model resembles the Ising model. Unlike the latter, the distribution is continuous, peaking at $S = 0$ (the Ising model is characterized by $P(S) = \delta(S^2 - 1)$).

Assuming a ferromagnetic coupling J between nearest-neighbor atoms, the Gaussian partition function Z and free energy F in the absence of an external field take the form

$$Z_G = \exp(-F_G/kT) = (2\pi)^{-1/2N} \prod_{n=1}^{N} \int dS_n \exp\left(-\frac{1}{2}S_n^2\right)$$

$$\times \left\{ \exp\left[(J/kT) \sum_{(n,m)} S_n S_m\right] \right\} \tag{7.106}$$

the integrations being over all N spins, the sum being over all nearest-neighbor pairs (n, m). The coordination of nearest-neighbors depends on the number of dimensions and on the type of lattice; we expect the answers to depend on both.

Writing the N spins as a vector array

$$\mathbf{S} = (S_1, S_2, \ldots, S_N)$$

we can express the total quadratic form in the exponent of (7.106) as

$$-\frac{1}{2}S \cdot \mathbf{A} \cdot S$$

with $\mathbf{A}$ a matrix array,

$$A_{n,m} = \delta_{n,m} - (J/kT)\varepsilon_{n,m} \tag{7.107}$$

where $\delta_{n,m}$ is the Kronecker delta (zero unless the lattice sites R_n and R_m are equal, in which case it is 1) and $\varepsilon_{n,m}$ is 1 when R_n and R_m are nearest-neighbor sites, and zero otherwise.

With periodic boundary conditions, $A_{n,m}$ is a cyclic matrix. Cyclic matrices have plane-wave eigenvectors, so their eigenvalues are easy to obtain as the Fourier transform of a typical row or column. Explicitly, if we write the eigenvalue equation as

$$\mathbf{A} \cdot \mathbf{v} = \lambda \mathbf{v}, \quad \mathbf{v} = (v_1, v_2, \ldots, v_N),$$

we know that setting $v_n = \cos \mathbf{k} \cdot \mathbf{R}_n$ or $\sin \mathbf{k} \cdot \mathbf{R}_n$ yields the kth eigenvalue. The λ's are easily calculated, and for a d-dimensional simple-cubic lattice are simply:

$$\lambda(\mathbf{k}) = \lambda(k_1, k_2, \ldots, k_d) = 1 - (2J/kT)(\cos k_1 + \cdots + \cos k_d). \tag{7.108}$$

The lattice geometry appears in this expression. In the body-centered cubic lattice (bcc), the terms $(\cos k_1 + \cos k_2 + \cos k_3)$ are replaced by $(\cos k_1 \cdot \cos k_2 \cdot \cos k_3)$ In most lattices, an expansion of these trigonometric terms about $k = 0$ yields $a_0 - a_2\mathbf{k}^2 + 0(\mathbf{k}^4)$, and for ferromagnets, it only the leading powers of $\mathbf{k}$ that determine the interesting properties. As is well known, this feature permits a universal description of critical phenomena near T_c for many different models. At present, we shall remain with the rather illuminating simple-cubic lattices, but shall consider various dimensionalities.

In terms of normal modes, the prototype integral is the Gaussian already evaluated in the preceding section. The partition function thus reduces to

$$Z_G = \prod_k [\lambda(\mathbf{k})]^{-1/2} = \exp\left[-\frac{1}{2}\sum_k \ln \lambda(\mathbf{k})\right] = [\text{Det}|\mathbf{A}|]^{-1/2} \tag{7.109}$$

while the free energy *per* spin is

$$F_G/N = \frac{1}{2}kT\frac{1}{N}\sum_k \ln \lambda(\mathbf{k}). \tag{7.110}$$

All λ's must be positive, or the theory fails; Eq. (7.108) shows that at or below $T_G \equiv 2Jd/k$ one or more λ's change sign. Proceeding to the thermodynamic limit, and assuming $T > T_G$, we convert the expression for F_G into an ordinary integral.

In 1D it is

$$F_G/N = \frac{1}{2}kT(2\pi)^{-1} \int_{-\pi}^{+\pi} dq \ln[1 - (2J/kT)\cos q], \qquad (7.111a)$$

and in 2D it is

$$F_G/N = \frac{1}{2}kT(2\pi)^{-2} \int\int_{-\pi}^{+\pi} dq_1 dq_2$$

$$\times \ln[1 - (2J/kT)(\cos q_1 + \cos q_2)] \qquad (7.111b)$$

and similarly in higher dimensions:

$$F_G/N = \frac{1}{2}kT(2\pi)^{-d} \prod_{i=1}^{d} \int_{-\pi}^{+\pi} dq_i \ln\left[1 - (2J/kT)\sum_{j=1}^{d}\cos q_i\right]. \qquad (7.111c)$$

The two-dimensional integral can be reduced to an elliptic function and the one-dimensional integral (7.111a) directly evaluated by the following integral identity:

$$(2\pi)^{-1} \int_{-\pi}^{+\pi} dq \ln(2\cosh x - 2\cos q) = |x| \qquad (7.112)$$

often attributed to L. Onsager.

The internal energy in d dimensions is

$$u = \frac{\partial}{\partial\beta}(\beta F_G/N) = \frac{1}{2}(2\pi)^{-d} \prod_{i=1}^{d} \int_{-\pi}^{+\pi} dq_i \frac{-2J\sum_{j=1}^{d}\cos q_j}{1 - 2J\beta\sum_{j=1}^{d}\cos q_i}. \qquad (7.113)$$

Adding and subtracting kT to the numerator, we can cast this integral into a somewhat more compact form:

$$u = \frac{1}{2}kT - \frac{1}{2}kT\,W\left(d, \frac{kT}{2Jd}\right) \qquad (7.114)$$

where W is a generalized Watson's integral:

$$W(d, \tau) \equiv (2\pi)^{-d} \prod_{i=1}^{d} \int_{-\pi}^{+\pi} dq_i \left(1 - \frac{1}{\tau d}\sum_{j=1}^{d}\cos q_j\right)^{-1}. \qquad (7.115)$$

For $T \geqslant T_G$ we only need this function in the range $\tau \geqslant 1$. We first encountered this integral for the special value $\tau = 1$ in 3D in connection with bound pairs of spinwaves [Sec. 5.5]. In $d = 1, 2, 3$ one can evaluate $W(d, \tau)$ exactly (as shown at the conclusion of this section). Certain limiting results are easy to estimate: in the limit $\tau \to 1$, $W(1, \tau) \propto (\tau - 1)^{-1/2}$, $W(2, \tau) \propto \ln 1/(\tau - 1)$, and for $d > 2$, $W(d, 1)$ finite, decreasing with d, arriving at a final value $W(\infty, 1) = 1$. For all d, $W(d, \infty) = 1$.

The Gaussian model shows a surprising agreement with the MFT. If we approximate the bond Hamiltonian

$$\mathscr{H}_0 = -JS_0(S_1 + S_2 + \cdots + S_{2d})$$

(which describes the interactions of a given spin S_0 with its $2d$ nearest-neighbors) by its mean-field average,

$$\langle \mathscr{H}_0 \rangle_{\mathrm{MFT}} = -JS_0 \bar{S} 2d$$

the thermal average over S_0 yields the self-consistency equation

$$\bar{S} = \tanh(\beta J 2d)\bar{S}, \quad \text{with} \quad T_c = 2dJ/k \equiv T_G,$$

which has only the trivial solution for $T > T_G$, but develops a nontrivial order parameter below T_G. Of course, the Gaussian model yields a nontrivial internal energy (7.114) above T_G and fails below T_G; whereas MFT yields insignificant results above T_G and nontrivial results below T_G. The two methods are thus complementary and agree on the value of the critical temperature. Soon, we shall see that they predict the same Curie–Weiss law for the zero-field magnetic susceptibility.

Spherical Model

Imposing the strict condition that the partition function be evaluated on the surface of an N-sphere

$$\sum_{i=1}^{N} S_i^2 = N \tag{7.116}$$

can cure some of the unphysical divergence that we noted in the Gaussian model at the critical temperature, and serves to define the spherical model. This constraint will be satisfied, on the average, by the artifice of a Lagrange multiplier μ. Incorporating the terms with the Lagrange multiplier into the Hamiltonian, we define the latter as

$$\mathscr{H} = -J \sum_{(n,m)} S_n S_m + \mu \left(\sum_n S_n^2 - N \right). \tag{7.117}$$

The value of μ is determined by requiring that (7.66) be satisfied on average or what is equivalent, $\langle S_n^2 \rangle = 1$ (because all the spins are equivalent by translational invariance and isotropy of the stated model).

With this particular μ (to be determined), the partition function is

$$Z_{\text{sph}} = \exp(\mu N/kT) \prod_n \int dS_n \exp[-(\mu/kT)S_n^2]$$

$$\times \left\{ \exp \left[(J/kT) \sum_{n,m} S_n S_m \right] \right\}. \tag{7.118}$$

It is evaluated just as in the Gaussian model, the eigenvalues λ now being functions of μ as well as T and $\mathbf{k}$:

$$\lambda(\mu, \mathbf{k}) = \frac{2}{kT} \left(\mu - J \sum_{j=1}^d \cos k_j \right). \tag{7.119}$$

Hence, the free energy is

$$F_{\text{sph}}/N = -\frac{1}{2}kT \ln(2\pi) - \mu + \frac{1}{2}kT \frac{1}{N} \sum_{\mathbf{k}} \ln \lambda(\mu, \mathbf{k}), \tag{7.120a}$$

but is more conveniently expressed in terms of $\tau \equiv \mu/Jd$, a scaling which introduces the appropriate units into the Lagrange multiplier. The free energy is now

$$F_{\text{sph}}/N = -\frac{1}{2}kT \ln(2\pi) + \frac{1}{2}kT \ln(2Jd/kT) - Jd\tau$$

$$+ \frac{1}{2}kT \frac{1}{N} \sum_{\mathbf{k}} \ln \left(\tau - \frac{1}{d} \sum_{j=1}^d \cos k_j \right). \tag{7.120b}$$

Differentiating it with respect to τ and using the spherical condition $\partial F_{\text{sph}}/\partial \tau = 0$ we obtain

$$\frac{kT}{2\tau} \frac{1}{N} \sum_{\mathbf{k}} \frac{1}{1 - (1/\tau d) \sum_{j=1}^d \cos k_j} = Jd. \tag{7.121}$$

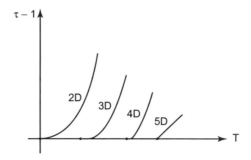

Fig. 7.11. $\tau - 1$ vs. T for various dimensionalities. Note that for 4D or less, the curve $\tau - 1$ intersects the horizontal axis with zero slope. At higher than 4D, the intersection is with finite slope.

If $\tau > 1$ the summand is nonsingular; therefore in the thermodynamic limit the sum can be replaced by an integral.

$$\frac{1}{2} kT\tau^{-1} W(d,\tau) = Jd \tag{7.122a}$$

or better,

$$\frac{kT}{2Jd} = \frac{\tau}{W(d,\tau)}. \tag{7.122b}$$

$W(d,\tau)$ is the integral (7.115) introduced in connection with the Gaussian model. Whereas (7.122a) is a transcendental equation for $\tau(T)$ which ordinarily one solves by numerical iteration, in the form (7.122b) one calculates $T(\tau)$ in a straightforward manner, obtaining the curves shown (schematically) in Fig. 7.11.

At the critical point $\tau = 1$, the above leads to a formula for the critical temperature,

$$kT_c(\text{sph.}) = 2Jd/W(d,1) \tag{7.123}$$

displayed in Table 7.1.

Table 7.1. Critical temperature in spherical model in d dimensions.

d	1	2	3	4	5	6	7	...[a]	∞
$kT_c/2Jd$	0	0	0.6595	0.8068	0.871	0.900	0.916	...[a]	1

[a] For $d \gg 1$, $kT_c = 2J(d - \frac{1}{2}) + O(1/d)$.

For any dimension $d > 2$, the spherical model yields a finite T_c, offering us a nontrivial low-temperature phase to investigate. We cannot merely set $\tau = 1$ in (7.121), as the summand now becomes singular at $k = 0$. *The integral now differs from the sum.* This deficiency is cured if we write τ in the form $1 + O(1/N)$ *before* proceeding to the thermodynamic limit. We split the sum into two contributions: the term at $k = 0$, which we treat separately, and the rest of the sum, which we now take in the thermodynamic limit, evaluating it by the corresponding integral. Thus, replacing (7.122a) over the interval $T \leqslant T_c$ we have the following:

$$\frac{kT}{2N(\tau - 1)} + \frac{1}{2}kT\tau^{-1}W(d, \tau) = Jd. \tag{7.124}$$

There is no error now in replacing τ by 1 in the second term, but the first term must be treated more carefully and requires interpretation. We define a parameter σ by

$$\frac{kT}{2Jd(\tau - 1)} \equiv N\sigma^2 \tag{7.125}$$

and inserting this into (7.74), after eliminating $W(d, 1)$ from this equation by use of (7.73), we obtain

$$|\sigma| = \left(1 - \frac{T}{T_c}\right)^{1/2} \tag{7.126}$$

for the temperature-dependence of σ, which we identify as the order parameter in the spherical model. Near T_c, the critical exponent $1/2$ agrees with MFT, as computed, e.g., in Fig. 7.2. The constants in the definition (7.75) are chosen so as to make the order parameter equal to 1 at $T = 0$.

Problem 7.10.   Justify the identification (7.125) in detail, making use of (7.67)–(7.70) and (7.99) for the specific case $\mathbf{k} = 0$. [*Hint*: see (7.127)ff.]

Problem 7.11.   With F_{sph} in (7.120a) incorporating an explicitly temperature-dependent parameter μ, prove that the thermodynamic relation $U = \langle H \rangle = \partial(\beta F)/\partial\beta$ remains valid. [*Hint*: use (7.121).]

The appearance of spontaneous ordering at $\mathbf{k} = 0$ may be understood qualitatively as a condensation phenomenon akin to Bose–Einstein condensation of He atoms. Let us write the thermal averaged spherical condition in the form

$$\langle X_0^2 \rangle + \sum_{\mathbf{k} \neq 0} \langle X_{\mathbf{k}}^2 \rangle = N \,. \tag{7.127}$$

Above T_c this equation is satisfied, with each $\langle X_{\mathbf{k}}^2 \rangle$ being $O(1)$ (including $\mathbf{k} = 0$), by adjusting τ. At or below T_c, τ sticks at its minimum value, $\tau = 1$, but the sum over terms $\mathbf{k} \neq 0$ does not add up to N. To make up the deficiency in the spherical condition, one allows $\langle X_0 \rangle$ to grow according to the above:

$$\langle X_0^2 \rangle = N - \sum_{\mathbf{k} \neq 0} \langle X_{\mathbf{k}}^2 \rangle \,,$$

T being the only variable. Setting $\langle X_0^2 \rangle = N\sigma^2$ yields (7.125) and (7.126). The reader can fill in the missing details by solving Problem 7.10.

In 2D or lower dimensions, the integrals continue to approximate the sums accurately down to $T = 0$, and there is therefore no anomalous behavior, no T_c, and no long-range order parameter.

In calculating the thermodynamic functions such as specific heat, one must take the temperature-dependence of τ into account. Making use of Problem 7.11, we obtain

$$u = \frac{1}{2}kT - Jd\tau \tag{7.128}$$

resulting in

$$c = \frac{1}{2}k \quad (T < T_c), \quad c = \frac{1}{2}k - Jd\frac{d\tau}{dT} \quad (T > T_c)\,. \tag{7.129}$$

Let us recall the famous Dulong–Petit law of classical thermodynamics: $c = k/2$ per degree of freedom. For gases, this law is approximately valid at high temperatures, although it fails at low temperatures when inter-particle interactions and quantum effects become important. In the present context, by contrast, the Dulong–Petit law is valid only at *low* temperatures where the spherical constraint affects only the $k = 0$ mode. Above T_c the correction in $d\tau/dT$ comes into play.

For $d > 4$, $d\tau/dT|_{T_c^+}$ is a finite quantity, therefore there will be a discontinuity in specific heat at T_c. For $d \leqslant 4$, $d\tau/dT$ vanishes at T_c and grows with increasing temperature, therefore there is no discontinuity in specific heat at T_c. In *all* cases, $Jd \, d\tau/dT$ approaches $k/2$ at high temperature, so that c properly vanishes at high temperature. The results are exhibited in Fig. 7.12.

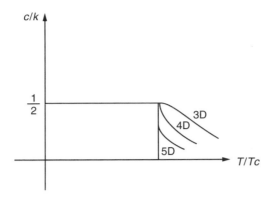

Fig. 7.12. Specific heat in spherical model theory, (7.129). The Dulong–Petit classical law $c = k/2$ is satisfied only at $T < T_c$. At T_c the specific heat is continuous (3D), has infinite slope (4D), or is discontinuous ($d > 4$), then drops from its value at T_c^+ as $(T - T_c)^{-2}$ at very high temperatures. The limiting curve ($d = \infty$), not drawn, is qualitatively similar to the elementary MFT, Fig. 7.3: finite for all $T < T_c$, zero for $T > T_c$.

We have already noted in Sec. 7.4 that the failure of $c(T)$ to vanish at absolute zero is tantamount to a failure of the third law of thermodynamics. In the present case, this results in an apparently flawed entropy. From its original definition, we know the entropy is a positive quantity. Using (7.120b) to calculate the entropy in the spherical model, we obtain

$$\mathscr{S}/N = -\partial f_{\text{sph}}/\partial T = \mathscr{S}_c/N - \frac{1}{2}k\ln(T_c/T) \qquad (7.130)$$

at $T < T_c$. The change in sign in the entropy at low temperature, and its limit $-\infty$ at $T = 0$ are unphysical consequences of the model. As these defects do not occur in the Ising model, one can infer that they are related to the failure of the spins to maintain the value $S_n^2 = 1$ (except on the average).

Watson's Integrals Generalized

The function $W(d, \tau)$ occurs in many contexts in solid-state physics. It is a special case of the lattice Green function, which in the 3D simple cubic (s.c.) lattice is given by

$$G(\mathbf{R}, \tau) = (2\pi)^{-3} \iiint_{-\pi}^{+\pi} dk_1 dk_2 dk_3 \frac{\exp(i\mathbf{k} \cdot \mathbf{R})}{1 - \frac{1}{3\tau}(\cos k_1 + \cos k_2 + \cos k_3)}$$

$$(7.131)$$

in which $\mathbf{R} = (n_1, n_2, n_3)$, $n_i = 0, 1, \ldots$. The analysis of the critical pheno-
mena in the Gaussian and spherical models depends on the value at $\mathbf{R} = 0$, $G(\mathbf{O}, \tau)$, which we examine here. We call it $W(d, \tau)$. In 1D, the integral is
trivial:

$$W(1, \tau) = \frac{1}{2\pi} \int_{-\pi}^{+\pi} dk \frac{1}{1 - \frac{1}{\tau} \cos k} = \frac{\tau}{(\tau^2 - 1)^{1/2}} \,. \tag{7.132}$$

The two-dimensional integral can be evaluated as follows:

$$W(2, \tau) = \frac{1}{2\pi} \int_{-\pi}^{+\pi} dk \frac{1}{1 - \frac{1}{2\tau} \cos k} W(1, 2\tau - \cos k)$$

$$= \frac{2}{\pi} K \left(\frac{1}{\tau} \right) \tag{7.133}$$

where K is the complete elliptic integral of the first kind. [The properties of
K are also crucial to the 2D Ising model and to the 1D Ising model in 1D;
appropriate expansions and references are found at (8.6.29)–(8.6.31).] The
three-dimensional integral $W(3, \tau)$ was first obtained by Joyce,[18] who found

$$W(3, \tau) = \left(1 - \frac{3}{4} x_1 \right)^{1/2} (1 - x_1)^{-1} (2/\pi)^2 K(k_+) K(k_-) \tag{7.134}$$

where

$$k_{\pm}^2 = \frac{1}{2} \pm \frac{1}{4} x_2 (4 - x_2)^{1/2} - \frac{1}{4} (2 - x_2)(1 - x_2)^{1/2} \quad \text{with}$$

$$x_2 = x_1 / (x_1 - 1) \quad \text{and}$$

$$x_1 = \frac{1}{2} + (1/6\tau^2) - \frac{1}{2} \left(1 - \frac{1}{\tau^2} \right)^{1/2} [1 - (1/9\tau^2)]^{1/2}$$

in the region of interest, $\tau \geqslant 1$. The value at $\tau = 1$,

$$W \equiv W(3, 1) = 1.516\ 386\ 059\ 151\ \ldots \tag{7.135}$$

recovers Watson's original result. An expansion in powers of $\tau - 1$ which is
most useful near the critical point was given by Joyce as follows

$$W(3, \tau) = W - \frac{3\sqrt{3}}{2\pi} \varepsilon + \frac{9}{32} \left(W + \frac{6}{\pi^2 W} \right) \varepsilon^2 - \frac{3\sqrt{3}}{4\pi} \varepsilon^3 + \cdots \tag{7.136}$$

---

[18] G.S. Joyce, *J. Phys.* **A5**, L65 (1972); M.L. Glasser, *J. Math. Phys.* **13**, 1145 (1972);
M.L. Glasser and I.J. Zucker, *Proc. Natl. Acad. Sci. USA* **74**, 1800 (1977). [NB: following
their Eq. (8b), $I_3(3)$ should be divided by 384π.]

with $\varepsilon^2 \equiv 1 - 1/\tau^2$.

We use it to solve (7.122) near T_c,

$$\tau - 1 \approx 1.68(1 - T_c/T)^2, \quad \text{for } T \gtrsim T_c, \tag{7.137}$$

in 3D.

In general, one can see that $W(d, \tau)$ becomes less singular as d becomes larger, by an iteration similar to (7.133):

$$W(d, \tau) = \frac{1}{2\pi} \int_{-\pi}^{+\pi} dk \frac{1}{1 - \frac{1}{d\tau} \cos k} W\left(d - 1, \frac{\tau d - \cos k}{d - 1}\right) \tag{7.138}$$

with the ultimate limit $W(\infty, \tau) = 1$ in the range $1 < \tau < \infty$. In principle, one can use the power series $(7.136)^{18}$ in the above to compute $W(4, \tau)$, then use the resulting power series to compute $W(5, \tau)$, etc., but there are probably better ways to obtain these generalized Watson's integrals.

For more information on lattice Green functions, the reader is referred to Katsura *et al.*[19]

Finally, in the analysis of critical properties ($T \sim T_c$, $\tau \gtrsim 1$) the derivative of (7.122) provides a helpful equation:

$$\partial \tau / \partial T = \frac{k/2Jd}{W(d, \tau)^{-1} + W(d, \tau)^{-2}|\partial W(d, \tau)/\partial \tau|_\tau}. \tag{7.139}$$

This is used to show that $\partial \tau / \partial T|_{T_c} = 0$ for $d \leqslant 4$.

7.11.   Magnetic Susceptibility in Gaussian and Spherical Models

We factor into the partition function Z_G of the Gaussian model an interaction with an external field B:

$$Z_G(T, B) = (2\pi)^{-1/2N} \prod_{n=1}^{N} \int dS_n \exp\left(-\frac{1}{2}S_n^2\right)$$

$$\times \left\{ \exp\left[(J/kT) \sum S_n S_m\right] \exp\left[(b/kT)B \sum S_n\right] \right\}. \tag{7.140}$$

[19] S. Katsura, T. Morita, S. Inawashiro, T. Horiguchi and Y. Abe, *J. Math. Phys.* **12**, 892 (1971); S. Katsura, S. Inawashiro and Y. Abe, *J. Math. Phys.* **12**, 895 (1971). Earlier, the extended Watson integrals and their derivatives were tabulated numerically: I. Maunari and L. Kawabata, Research Notes of Department of Physics, Okoyama University, No. 15, Dec. 21, 1964 (unpublished).

A uniform shift in the N variables of integration: $S_n = S'_n + \sigma$ eliminates the term linear in the spins provided we choose

$$-\sigma + (J/kT)2d\sigma + (bB/kT) = 0. \qquad (7.141)$$

Because $\langle S'_n \rangle = 0$, this yields the familiar Curie–Weiss law

$$\langle S_n \rangle = \sigma = B\frac{b}{k(T - T_{\mathrm{G}})} \qquad (7.142)$$

with $T_{\mathrm{G}} = 2Jd$ as before. The magnetization $m = \sigma b$ is then precisely Curie's law (7.69); but the latter was only valid at weak fields $B \to 0$, whereas the present result (7.142) is for any B, however large! This lack of saturation is a defect of the Gaussian model, a consequence of the lack of constraint on the length of the individual spins. Physically, $|\sigma|$ should not be allowed to exceed 1 however large the external field. Hopefully, the spherical model provides the cure.

For the spherical model, we similarly modify (7.118):

$$Z_{\mathrm{sph}}(T, B) = \exp(\mu N/kT) \prod_{n=1}^{N} \int dS_n \exp[-(\mu/kT)S_n^2]$$

$$\times \left[\exp\left(\frac{J}{kT}\sum S_n S_m\right) \exp\left(\frac{b}{kT}\sum S_n B\right) \right]. \qquad (7.143)$$

Again, a uniform shift $S_n = S'_n + \sigma$, $dS_n = dS'_n$ eliminates terms linear in the spins, provided we choose

$$-2\mu\sigma + J2d\sigma + bB = 0 \qquad (7.144)$$

and therefore,

$$\langle S_n \rangle = \sigma = B\frac{b}{2\mu - 2Jd} = B\frac{b}{2Jd(\tau - 1)} \qquad (7.145)$$

using $\mu = Jd\tau$. The value of τ, via the spherical condition, is now somewhat modified because the external field affects $\langle S_n^2 \rangle$.

$$\langle S_n^2 \rangle = \langle S'^2_n \rangle + \sigma^2 \qquad (7.146)$$

as the cross-terms $\langle S'_n \sigma \rangle$ vanish by symmetry. Thus, (7.121) and (7.122) must be modified. Simply,

$$\tau^{-1} W(d, \tau) = (1 - \sigma^2)2Jd/kT \qquad (7.147)$$

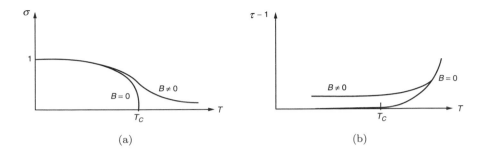

Fig. 7.13.   Order parameters (a) σ and (b) $\tau - 1$ as functions of T in spherical model. *Lower curve* of each pair is in zero field, and shows the characteristic discontinuity at T_c. *Upper curve* shows the order parameters are smooth functions of T in finite field $B \neq 0$.

replaces (7.122). Equations (7.145) for $\sigma(B, \tau)$ and (7.147) for $\tau(B, \sigma, T)$ must be solved simultaneously — numerically of course — to yield the desired $\sigma(B, T)$ and $\tau(B, T)$.

Solving for $\tau(B, \tau)$ and substituting for σ by use of (7.145) we obtain the explicit formula:

$$kT = 2Jd\left[1 - \frac{(Bb/2Jd)^2}{(\tau - 1)^2}\right]\frac{\tau}{W(d, \tau)}. \qquad (7.148)$$

At any finite B, this yields $T(\tau)$ for all positive T, with $\tau \geq 1 + |Bd/2Jd|$. T is analytic in τ in these intervals, thus the phase transition at T_c is eradicated in the presence of a finite external field, as illustrated in Fig. 7.13. The physical explanation for this applies to *all* model ferromagnets. The external field sets up long-range order even at high T, thus the order-disorder phase transition that would otherwise occur spontaneously at T_c is pushed up to $T = \infty$.

Nevertheless, we can obtain the *zero-field* susceptibility by differentiating σ with respect to B in zero field, using the value of $\tau(T)$ calculated in zero field, (7.122). We have just an estimate of this in 3D, (7.137), which shows $\tau - 1 \propto (T - T_c)^2$, hence by (7.145) $\chi_0 \propto (T - T_c)^{-2}$.

Near T_c it is customary in all theories of magnetism — not just the spherical model — to try to fit χ_0 to a power law,

$$\chi_0 = \frac{C}{(T - T_c)^\gamma}, \qquad T \gtrsim T_c, \qquad (7.149)$$

where C is a constant, and γ is the *susceptibility critical exponent*, a quantity which depends on the symmetry of the model as well as on the number d of dimensions. For example, $\gamma = 1$ in the Gaussian model for all d, whereas

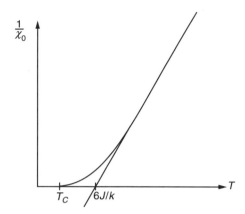

Fig. 7.14.   In 3D, $1/\chi_0$ for special model ferromagnet is parabolic near its T_c ($6J/k$ is MFT value of T_c, shown by linear extrapolation to lie above $T_{csph} \approx 4J/k$).

$\gamma = 2$ in the spherical model for $d = 3$, and $\gamma = 1$ for the same model in $d \geqslant 5$. The susceptibility for the spherical model in 3D is plotted in Fig. 7.14.

Problem 7.12.   Obtain in the special case $d = 4$, in the spherical model approximation.

7.12.   Spherical Antiferromagnet

The nearest-neighbor antiferromagnet on a bipartite lattice has precisely the same free energy as the corresponding ferromagnet in zero external field. A bipartite lattice is any lattice, in any number of dimensions, that can be decomposed into two sublattices such that the sites on one sublattice (say the A sublattice) interact only with those on the other (B) sublattice. The s.c. lattice examined in the preceding sections is a special case. The transformation of the antiferromagnet proceeds explicitly, by re-defining up and down on the B-sublattice but not on the A, thus effectively changing the sign of the coupling constant J. Although this transformation is not operative in quantum Heisenberg systems (it is impermissible to invert three spin components, as shown in Chap. 3), it applies to Ising models, Gaussian and spherical models, and to quantum XY models. Where applicable, it implies that in the absence of external fields the thermodynamic functions are even functions of J; thus, changing the sign of J has no effect.

The important difference now comes in the magnetic properties. Changing the sign of the spin direction on alternating sites results in

$$\exp\left[\frac{b}{kT}\sum_{n} S_n(-1)^n B\right] \tag{7.150}$$

as the extra factor in the computation of the partition function in an external, homogeneous, magnetic field B. This factor differs by the alternating $(-1)^n$ from the corresponding factor in (7.143).

Again, a shift in the origin of the spin coordinates eliminates terms linear in the S_n in the partition function, but the shift is no longer given by (7.144). Rather, it is

$$S_n = S_n' + (-1)^n \sigma \quad \text{with} \tag{7.151}$$

$$\sigma = \frac{bB/2Jd}{\tau + 1}. \tag{7.152}$$

Note that the net magnetization is in the direction of the applied field on *either* sublattice. The zero-field paramagnetic susceptibility is therefore,

$$\lim_{B \to 0} (\sigma/B) = \chi_0 = \frac{D}{\tau + 1} \quad (D, \text{ constant}; \ T \geqslant T_N) \tag{7.153}$$

with τ the solution of (7.122), the same as for a ferromagnet. Because of the $+1$ in the denominator, the susceptibility does not now diverge at T_c (which, properly, must be relabeled T_N in the present case). Morever, it is constant below T_N, having a value

$$\chi_0(T) = \chi_0(T_N) = \frac{D}{2}, \quad T \leqslant T_N. \tag{7.154}$$

Proceeding to finite external fields, we compute the magnitude of the spin-flop field, at which the external field is sufficiently large to force spins into parallelism despite the tendency to antiparallelism due to antiferromagnetic coupling. Equation (7.147) is unmodified, but with $(\tau + 1)$ replacing $(\tau - 1)$ in the evaluation of σ, we obtain a new equation of state replacing (7.148):

$$kT = 2Jd\left(1 - \frac{(Bb/2Jd)^2}{(\tau + 1)^2}\right)\frac{\tau}{W(d,\tau)}. \tag{7.155}$$

If $|B|$ exceeds $B_0 \equiv 4Jd/b$ there will be only a single phase, the paramagnetic. For any $|B|$ less than this, the critical temperature is finite and is calculated to be

$$kT_N(B) = 2Jd \left[1 - \frac{1}{4}(Bd/2Jd)^2\right] \bigg/ W(d,1)$$

$$= kT_N(0)[1 - (B/B_0)^2] \tag{7.156}$$

by setting $\tau = 1$ in the preceding. An antiferromagnetic (AF) phase thus persists at any given $T < T_N(B)$, or conversely, for $|B| < B_c(T)$, where $B_c(T)$ the spin-flop field can be obtained from (7.156). The phase diagram and magnetic properties are sketched in Fig. 7.15.

Physically, it is simple to see why the behavior differs from the ferromagnet. The homogeneous external field does not couple directly with the Néel mode which is the ground state of the AF. Therefore, AF correlations can appear or disappear at T_N even in the presence of an external field. The presence of this field does, however, affect the magnitude of T_N.

An interesting detail concerning the spherical model antiferromagnet is discussed in the following problem.

Problem 7.13.   The susceptibility in (7.153) is the longitudinal susceptibility, as previously defined in connection with Néel's theory (Sec. 7.7). Calculate the perpendicular susceptibility in the spherical model, and compare with the longitudinal susceptibility and with MFT.

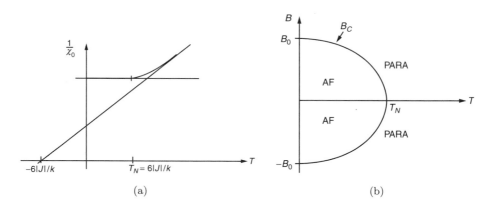

(a)   (b)

Fig. 7.15.   (a) Inverse susceptibility in spherical model antiferromagnet, showing constant portion for $T < T_N$, the linearly extrapolated MFT estimate of T_N, and the asymptotic Curie–Weiss–Néel behavior at high T. (b) B_c, the spin-flop field, as a function of T. This curve marks a thermodynamic phase boundary; all thermodynamic functions and order parameters are discontinuous or have discontinuous derivatives on crossing this curve. Within the boundary the behavior is that of an AF; outside, it is paramagnetic.

7.13. Spherical Model Spin Glass

For recent years, a great deal of interest has centered about the spin-glass phase. Physically, it concerns magnetic ions having random interactions with one another, as in moderately dilute solutions of magnetic ions such as Mn or Fe in nonmagnetic host metals such as Cu. The RKKY interactions[12] are of the form $-J_{ij}S_iS_j$ between every pair of spins, with

$$J_{ij} = J_0|R_{ij}|^{-3}\cos 2k_F|R_{ij}| \qquad (7.157)$$

and as the position of the individual spins is a random variable, this long-ranged interaction (in the sense (b) of Sec. 7.7) is oscillatory in sign and random in magnitude. In other prototype spin glasses, the interactions are short-ranged but the presence or absence of spins on the prescribed sites introduces a random variable into the analysis.

In any given model, one must know the particular geometry of the spin arrangement (e.g., randomly distributed spins, or clustered, or on a regular array) and of the bonds (e.g., nearest-neighbor ferromagnetic, or n.n. AF, or long ranged as in (7.157) and of the external fields (e.g., random external field, or homogeneous of given magnitude, etc.). The free energy is calculated as $-kT\ln Z$ for the given configuration of random variables. (We define a configuration of random variables as the statement about spin lattice geometry, the sign and magnitude of the individual J_{ij} and the external field, all subject to given probability distributions.) As the free energy is extensive, in the limit $N \to \infty$ all possible configurations will be experienced. Thus, the desired object is the free energy per spin in the thermodynamic limit:

$$f = F/N = -\frac{kT}{N}\langle\log Z\rangle \qquad (7.158)$$

where Z is the trace over spins, for the given random configuration, and $\langle\ \rangle$ signifies average over all random configurations.

If the random configurations were annealed instead of quenched, we could follow the *much* simpler procedure of averaging Z over random configurations. Unfortunately, the activation energy to move magnetic ions in the metal, or to vary any of the other random variables, is generally so large that in the range of temperatures of interest (0 K–10^3 K) these random variables have to be considered frozen-in, i.e., quenched.

The last step, the average over random configurations in (7.158), is the hardest. Fortunately there exist examples in which all random configurations (excepting a set of measure zero) yield the same Z in the thermodynamic limit. Because of this feature, they are easily solved. The one we study in this section is a version of the spherical model spin glass first solved by Kosterlitz *et al.*[20] But before specializing to their version of long-ranged forces, let us develop the formalism for arbitrary spherical models. Incorporating the constraint explicitly, one writes an Hamiltonian:

$$H = -\frac{1}{2}\sum_i\sum_{j\neq i} J_{ij}S_iS_j + \mu\left(\sum_i S_i^2 - N\right). \tag{7.159}$$

It must be noted that, in general, H depends on the particular configuration or set of $\{J_{ij}\}$ which is quenched.

Proceeding, we diagonalize H so as to evaluate Z in convenient form. Label the eigenvalues and orthonormal eigenvectors of the quadratic form by a set of N "quantum numbers" $\{\alpha\}$; thus a typical eigenvalue equation is

$$-\sum_j J_{ij}\phi_\alpha(j) = 2E_\alpha\phi_\alpha(i) \tag{7.160}$$

taking the eigenvalue to be $2E_\alpha$ for future convenience. Now H takes on the following form

$$H = \sum_\alpha (E_\alpha + \mu)X_\alpha^2 - N\mu \tag{7.161}$$

in terms of the normal mode amplitude X_α. The individual spins are now given by

$$S_i = \sum_\alpha X_\alpha\phi_\alpha(i) \tag{7.162}$$

and as the $\phi_\alpha(i)$ are an orthonormal set, the conditions $\langle\sum S_i^2\rangle = N$ becomes

$$\left\langle\sum_\alpha X_\alpha^2\right\rangle = N, \tag{7.163}$$

exactly equivalent to $\partial f/\partial\mu = 0$, after traces and configurational averages are performed.

Along with the set of eigenvalues $\{E_\alpha\}$ one introduces the concept of a density of states $\rho(E)$,

$$\rho(E) \equiv \sum_\alpha \delta(E - E_\alpha) \tag{7.164}$$

---

[20] J. Kosterlitz, D. Thouless and R.C. Jones, *Phys. Rev. Lett.* **36**, 1217 (1976).

and of a *local* density of states at the site of the spin S_i,

$$\rho_i(E) = \sum_\alpha \phi_\alpha^2(i)\delta(E - E_\alpha), \qquad (7.165)$$

clearly,

$$\sum_i \rho_i(E) = \rho(E). \quad (\text{Reader : proof?})$$

The free energy in the spin glass is obtained in a manner similar to (7.120),

$$F_{\text{SG}}/N = -\frac{1}{2}kT \ln 2\pi - \mu + \frac{1}{2}kT\frac{1}{N}\sum_\alpha \ln \frac{2}{kT}(\mu + E_\alpha)$$

$$= -\frac{1}{2}kT \ln 2\pi - \mu + \frac{1}{2}kT\frac{1}{N}\int dE\rho(E) \ln \frac{2}{kT}(\mu + E),$$

but must now be averaged over random configurations:

$$f = -\frac{1}{2}kT \ln 2\pi - \langle\mu\rangle + \frac{1}{2}kT\frac{1}{N}\int dE\langle\rho(E)\rangle \ln \frac{2}{kT}(\mu + E). \qquad (7.166)$$

This serves to define $\langle\rho(E)\rangle$, the configuration-averaged density of states, and shows that a knowledge of the thermodynamic functions requires only a knowledge of the statistical distribution of the eigenvalues of J_{ij}.

At this point, let us specialize to the spin glass of Kosterlitz *et al.*[20] The bonds J_{ij} ($i \neq j$) are random, $\langle J_{ij}\rangle = 0$ and $\langle J_{ij}^2\rangle = J^2/N$, J being the strength of the random bonds.

The distribution of eigenvalues of such a matrix is known precisely in the thermodynamic limit[21,22]; it is Wigner's semicircular density of states. For each spin S_i and configuration,

$$\rho_i(E) = \frac{2}{\pi J}[1 - (E/J)^2]^{1/2}, \quad \text{each } \mu = \langle\mu\rangle, \qquad (7.167)$$

and, as this result is explicitly independent of i, the configuration-averaged density of states is simply

$$\langle\rho(E)\rangle = \frac{2N}{\pi J}[1 - (E/J)^2]^{1/2}. \qquad (7.168)$$

Substitution of this formula into (7.166) yields the free energy in this model. The spherical constraint equation obtained by differentiating f, $\partial f/\partial\mu = 0$, has the following aspect:

[21] D. Mattis and R. Raghavan, *Phys. Lett.* **75A**, 313 (1980).
[22] M.L. Mehta, *Random Matrices*, Academic Press, New York, 1967, Appendix A29.

$$\frac{kT}{\mu J} \int_{-J}^{+J} dE[1 - (E/J)^2]^{1/2} \frac{1}{\mu + E}$$

$$= \frac{kT}{J}[\tau - (\tau^2 - 1)^{1/2}] = 1 \tag{7.169}$$

in terms of the natural variable $\tau \equiv \mu/J$. Defining $kT_c = J$, this equation yields

$$\tau = \frac{T^2 + T_c^2}{2TT_c} \quad \text{for } T \geqslant T_c. \tag{7.170}$$

with τ sticking to its value $\tau_c = 1$ in the range $T \leqslant T_c$. Comparison with the ferromagnetic spherical model, (7.128) and (7.129), shows how to obtain the specific heat. One can easily express it in closed form:

$$c = \begin{cases} \dfrac{1}{2}k & (T \leqslant T_c) \\[2mm] \dfrac{1}{2}kT_c^2/T^2 & (T > T_c). \end{cases} \tag{7.171}$$

While the specific heat is continuous at T_c, its derivative is not. According to the conventional classification scheme, the phase transition at T_c is of *third* order, as the leading discontinuity occurs in a third derivative of the free energy. But this point may well be academic, for it is virtually impossible experimentally to detect discontinuities in third order or higher, therefore the existence or nonexistence of a phase transition in this model may not

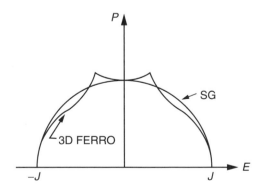

Fig. 7.16. Wigner's semicircular density of states $\rho(E)$ of a perfectly random matrix J_{ij}, as derived in Sec. 7.13. It is marked SG, and contrasted to the density of states for a nearest-neighbor sc lattice (Sec. 7.9).

be of any consequence. Indeed, we observe in the following section that *any* external field will eliminate the phase transition altogether.

In many important respects, the properties of this long-ranged (all spins interact with one another) random model mimic those of the three-dimensional ferromagnet studied previously. This is not a coincidence. If we identify J here with $3J$ in the ferromagnet, we find that the respective *dos* functions have similar features, shown in Fig. 7.16 (some details are left to the reader in Problem 7.14). As the free energy is determined entirely by the *dos*, the thermodynamic resemblence follows directly. Unfortunately, such nice mappings do not carry over into Ising and quantum-mechanical models of magnetism.

Problem 7.14. Calculate the *dos* function for the 3D n.-n. ferromagnet:

$$\rho(E) = \sum_{\mathbf{k}} \delta[E + (J/3)(\cos k_x + \cos k_y + \cos k_z)], \qquad (7.172)$$

and show that near the band edges, $E = \pm J$, it shares the 1/2 power law behavior with Wigner's semicircular *dos*, (7.168). Show the integrals over (7.168) and (7.172) are both equal to N. Discuss (7.172) at $E = \pm J/3$, the sites of Van Hove singularities (see any book on solid-state theory).

Magnetic Properties of Spin Glass

As the τ-parameter sticks to its value $\tau_c = 1$ below T_c, there must occur a condensation into the lowest mode at $E = -J$. The similarity of the random model with the 3D ferromagnet ceases if we examine the nature of the condensed phase. In the latter, it is the macroscopic magnetization that grows as the temperature is lowered. In the case of the random interactions, the corresponding normal mode varies from configuration to configuration and is a random variable. So, although there exists an internal order variable that describes the freezing out of the spin glass at low temperature — the amplitude X_0 of the condensed phase — satisfying an equation of the type (7.127),

$$\frac{T}{T_c} + (X_0^2/N) = 1 \quad (T \leqslant T_c), \qquad (7.173)$$

the corresponding normal mode is unknowable! It is therefore that much more surprising that applying *any* external magnetic field to this system eliminates the phase transition, and along with it, the macroscopic condensation.

In studying the magnetic response of the random model, we can even generalize it a bit. Let each J_{ij} be a random variable, such that

$$\langle J_{ij} \rangle = J_0/N \quad \text{and} \quad \langle J_{ij}^2 \rangle = J^2/N \,. \tag{7.174}$$

Positive J_0 signifies a tendency to ferromagnetism, negative J_0 antiferromagnetism. This model combines the long-ranged systematic interactions examined in Sec. 7.7 with the random ones. We recall that the example in Sec. 7.7 involved only the $k = 0$ plane wave. There, plane waves were normal modes. Here, we can use a procedure known as tridiagonalization[21] to obtain the spectral decomposition of each plane wave.

First, we write the Hamiltonian, including the interaction with the external field, and the spherical condition in terms of plane wave states — which we call spin wave states henceforth.

$$\mathcal{H} = \mathcal{H}_0 + \mathcal{H}' \,, \tag{7.175}$$

with

$$\mathcal{H}_0 = -\frac{1}{2} \sum_{kk'} W_{kk'} \sigma_k \sigma_{k'} + \left(\sum_k |\sigma_k|^2 - N \right) \mu \tag{7.176}$$

and

$$\mathcal{H}' = -BN^{1/2} \sigma_0 \tag{7.177}$$

using the definition of a spin wave:

$$\sigma_k = N^{-1/2} \sum_i \exp(ik \cdot R_i) S_i \,. \tag{7.178}$$

The spin wave states have an expansion in normal modes $\psi_\alpha(k)$:

$$\sigma_k = \sum_\alpha X_\alpha \psi_\alpha(k) \,. \tag{7.179}$$

In terms of the amplitudes X, the various quantities above take on the following aspect when reduced to normal modes:

$$\mathcal{H}_0 = \sum_\alpha (E_\alpha + \mu) X_\alpha^2 - N\mu \tag{7.180}$$

and

$$\mathscr{H}' = -BN^{1/2} \sum_\alpha X_\alpha \psi_\alpha(0) \,. \tag{7.181}$$

The spherical condition (7.163) remains unaffected.

We now carry out the decomposition of each spin wave into normal modes to implement (7.179). First, consider the $k = 0$ wave in (7.178):

$$\sigma_0 = N^{-1/2} \begin{bmatrix} 1 \\ 1 \\ \vdots \\ 1 \end{bmatrix} \,. \tag{7.182}$$

We will denote it $\mathbf{v}_0$, and study the effects of the J_{ij} matrix on it.

$$-\mathbf{J} \cdot \mathbf{v}_0 \equiv -\sum_j J_{ij} v_0(j) = m_{00} v_0(i) + m_{01} v_1(i) \,. \tag{7.183}$$

The individual J_{ij}'s $(i \neq j)$ are random, subject to the constraints (7.174) on their average and on their square. The various quantities are obtained as follows:

$$m_{00} = -\sum_{i,j} v_0(i) J_{ij} v_0(j) \tag{7.184}$$

assuming $\mathbf{v}_1$ is a new vector, orthogonal to $\mathbf{v}_0$ and like it, of unit length. It follows that

$$m_{01}^2 = \mathbf{v}_0 \cdot (\mathbf{J} + m_{00}\mathbf{1})^2 \cdot \mathbf{v}_0 \tag{7.185}$$

and

$$\mathbf{v}_1 = m_{01}^{-1} = (-\mathbf{J} - m_{00}\mathbf{1}) \cdot \mathbf{v}_0 \,. \tag{7.186}$$

Proceeding, one allows $\mathbf{J}$ to act on $\mathbf{v}_1$, generating a new vectors $\mathbf{v}_2$ in addition to the known $\mathbf{v}_1$ and $\mathbf{v}_0$. Continuing this procedure, one obtains a matrix $m_{n,n}$, which vanishes unless $n = n'$ or $n = n' \pm 1$ (a tridiagonal matrix). In the particular case at hand, even though the original J_{ij} are random, the $m_{n,n}$, tridiagonal matrix becomes sharp in the thermodynamic limit, having the values[21]:

$$m_{0,0} = -J_0 \,, \quad m_{n,n} = 0 \quad (n \neq 0) \,, \quad \text{and}$$

$$m_{n,n+1} = -J = m_{n+1,n} \,, \quad \text{all } n \geqslant 0 \,. \tag{7.187}$$

We have previously analyzed precisely such a matrix [see Sec. 6.13]. The conclusions were: there is a continuous spectrum of eigenvalues $2E$ ranging from $-J < E < +J$ and, in addition, a bound state *below* the continuum if $J_0 > |J|$, and above the continuum if $J_0 < -|J|$. The density of states which describes the continuum is:

$$\rho_0(E) = \frac{2}{\pi}[1 - (E/J)^2]^{1/2}\frac{J}{J_0^2 + J_0 + 2J_0E} \qquad (7.188)$$

labeling it with subscript 0 to indicate the $k = 0$ sector, and assuming $|J_0| < |J|$ so there are no bound states. Turning to the $k \neq 0$ sectors, we start with σ_k as $\mathbf{v}_0$ replacing (7.182), and follow the same procedure. The results are similar, except that J_0 does not appear in m_{00}. Thus, each sector $k \neq 0$ has a *dos*:

$$\rho_k(E) = \frac{2}{\pi J}[1 - (E/J)^2]^{1/2} . \qquad (7.189)$$

The *combined* weight of the $k \neq 0$ sectors is $N - 1$. Thus, unless something special occurs, the $k = 0$ *dos* has zero statistical weight in the thermodynamic limit and can be ignored.

The special cases where the $k = 0$ sector contributes are of two distinct types. First, if $J_0 > |J|$, a bound state develops at eigenvalue $E_0 < -|J|$. As the temperature is lowered and μ becomes smaller, there will come a temperature where $\mu - |E_0|$ will vanish. This condensation into the $k = 0$ sector occurs before any of the integrals become singular, and thus dominates the thermodynamics even though the statistical weight of the bound state is negligible. As the condensed state is not orthogonal to σ_0, there is partial ferromagnetism — a form of "mictomagnetism" distinct from the spin glass phase (insofar as *spontaneous* magnetization will be found at sufficiently low temperature). The critical temperature will be a function of J_0. Some properties of this phase were treated in Kosterlitz *et al.*;[20] we ignore it henceforth.

The second special case just involves the application of an external magnetic field. The spin glass is quite sensitive to an applied field even though its paramagnetic susceptibility is finite. We shall return to this shortly. But first, we may inquire as to the behavior of the model when $J_0 < -|J|$, i.e., when a bound state appears in the $k = 0$ sector *above* the continuum. In that case the bound state factor $(\mu + E_0)$ never vanishes, and the bound state never contributes due to the negligible statistical factor $1/N$. This lack of symmetry between negative and positive bonds was previously noted in the long-range model (Sec. 7.7). The *spin glass phase* is the term given to the stable thermodynamic condensed phase which is found for J_0 in the range $J_0 < |J|$; for example, T_c will be independent of J_0 *for any J_0 in this range*, and so is the zero-field specific heat, which continues to be given by (7.171) precisely!

We now examine the effects of an applied external field B.

The magnetization in the spin glass phase (*no* bound state) is:

$$m = B \frac{K_1}{\pi J_0} \int_{-J}^{+J} dE[1 - (E/J)^2]^{1/2} \frac{1}{\gamma_0 + (E/J)} \frac{1}{\mu + E} \qquad (7.190)$$

where $\gamma_0 \equiv (J^2 + J_0^2)/(2JJ_0)$. K_1 is just a constant of proportionality permitting us to introduce convenient units — say, express m in units of the saturation magnetization, and B in such units that we recover $\chi = \mathbb{C}/T$ at high temperature.

The spherical condition is easily expressed in terms of the spin wave states. Only the states connecting to $k = 0$ are affected by the external field, and so we obtain a formula only slightly more complex than (7.169):

$$\frac{kT}{J}[\tau - (\tau^2 - 1)^{1/2}] + (BK_2)^2 \frac{1}{\pi J_0} \int_{-J}^{+J} dE[1 - (E/J)^2]^{1/2}$$

$$\times \frac{1}{\gamma_0 + (E/J)} \frac{1}{(\mu + E)^2} = 1 \qquad (7.191)$$

with K_2 determined by the choice of units. Note that the bias interaction J_0 enters the spherical condition *only* when there is an external field. Performing the integrations and adjusting the constants K_1 and K_2 suitably, we obtain for the magnetization,

$$m = (B\mathbb{C}/2T_c)(J/J_0) \left[\frac{\tau - (\tau^2 - 1)^{1/2} - \gamma_0 + (\gamma_0^2 - 1)^{1/2}}{\gamma_0 - \tau} \right] \qquad (7.192)$$

with $kT_c = J$ as before. This spherical condition is

$$\frac{kT}{J}[\tau - (\tau^2 - 1)^{1/2}] + (B\mathbb{C}/2T_c)^2(J/J_0)$$

$$\times \left\{ (\gamma_0 - \tau)^{-2}[\gamma_0 - (\gamma_0^2 - 1)^{1/2} - \tau + (\tau^2 - 1)^{1/2}] + (\gamma_0 - \tau)^{-1} \right.$$

$$\left. \times \left[\frac{\tau}{(\tau^2 - 1)^{1/2}} - 1 \right] \right\} = 1. \qquad (7.193)$$

These equations simplify considerably if we introduce two quantities with the units of temperature, $T_F = 2J_0/k$ and $\theta = T_c[\tau - (\tau^2 - 1)^{1/2}]^{-1}$. The restriction $J_0 \leqslant |J|$ is equivalent to the inequalities,

$$\theta \geqslant T_c \geqslant T_F . \tag{7.194}$$

If J_0 is negative (AF), so is T_F. The following relations are easily demonstrated:

$$\gamma_0 = \frac{T_F^2 + T_c^2}{2T_F T_c} \quad \text{and} \quad \tau = \frac{\theta^2 + T_c^2}{2\theta T_c} . \tag{7.195}$$

It is then a matter of some algebraic manipulations to simplify the formula for m to the familiar form,

$$m = \frac{B\mathbb{C}}{\theta - T_F} \tag{7.196}$$

and the spherical condition to polynomial form,

$$\frac{T}{\theta} + \frac{(B\mathbb{C}\theta)^2}{(\theta - T_F)^2(\theta^2 - T_c^2)} = 1 . \tag{7.197}$$

Specifying the parameters T_c and T_F, and fixing B, one can vary θ in the two equations above, obtaining $T(\theta)$ and $m(\theta)$, and thereby $m(T)$. For *any* *finite* B (however small) *there is no phase transition*, as the solutions are analytic in θ. As $B \to 0$, however, a cusp develops in the function $\theta(T)$, i.e., a discontinuity in $T(\theta)$. Thus, in the $B = 0$ limit there is the previously mentioned third-order phase transition with the specific heat function still given precisely by (7.171), independent of T_F. The zero-field susceptibility *is* sensitive to T_F. Calculating the ratio $(m/B\mathbb{C})$ in the $B = 0$ limit we obtain

$$\chi_0 = \frac{\mathbb{C}}{T - T_F} \quad \text{for } T > T_c , \quad \text{and} \quad \chi_0 = \frac{\mathbb{C}}{T_c - T_F} , \quad T \leqslant T_c . \tag{7.198}$$

Combining (7.196) and (7.197) yields

$$m = \{[1 - (T_c/\theta)^2][1 - (T/\theta)]\}^{1/2} \tag{7.199}$$

for arbitrary B, T. Because θ increases monotonically with B, this expression proves that the *maximum* m is 1, the saturation magnetization. This, and the Curie–Weiss result (7.198), allow convenient comparison with experiment and with other theories. Experiments by Symko and his students on the spin-glass phase are displayed in Fig. 7.17.

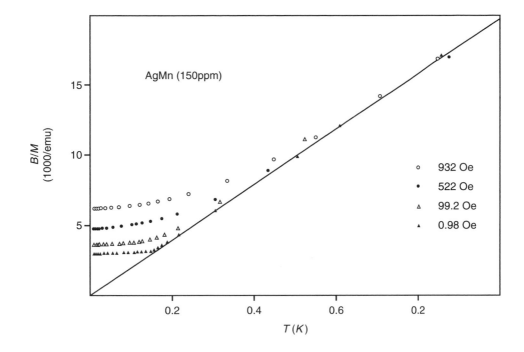

Fig. 7.17. B/M as function of T for various B in a spin glass consisting of dilute solid solution of Mn in Ag, as measured by M. Novak, O.G. Symko, D. Zheng: Private communication (1984). The dependence on B tracks the theoretical solution of (7.196) and (7.197), with $T_F =$ order of 10^{-3} K ≈ 0, and it indicates $T_c \approx 0.15$ K. These experiments show discrepancy with the spherical model theory only at low temperatures, in *strong* magnetic fields.

Other interesting or nontrivial applications of the spherical model are listed below.[23]

In concluding this section on the spherical model spin glass, it may be appropriate to note the very great difficulties that were faced in the more realistic Ising or vector spin glasses. The conditions $S_i^2 = 1$ for each of N spins are difficult to preserve alongside the stochastic features of the problem, and no satisfactory resolution exists at the date of writing. The impetus for much of the present research came in a paper by Sherrington and

---

[23] The nearest-neighbor random bond model is treated in the spherical model approximation by L. deMenezes, A. Rauh and S.R. Salinas, *Phys. Rev.* **B15**, 3485 (1977). A quantum version of the spherical spin glass is in P. Shukla and S. Singh, *Phys. Lett.* **81A**, 477 (1981). A peculiar failure of the spherical model (the spurious phase transition of a single spin!) has been discovered by E.H. Lieb and C.J. Thompson, *J. Math. Phys.* **10**, 1403 (1969). The AF spherical model was first examined by R. Mazo, *J. Chem. Phys.* **39**, 2196 (1963).

Kirkpatrick,[24] who introduced the long-range model defined by (7.174), treating the spins in the Ising manner, $S_i = 1$. Other popular realistic models include the nearest-neighbor random bond model of Edwards and Anderson[25] and an analysis of the realistic indirect exchange interactions in metals by Walker and Walstedt,[26] perforce numerical. But the mean-field aspect of the Sherrington–Kirkpatrick model made it the most attractive candidate for a universal theory, in the sense of the Weiss–Néel theories, and it has therefore been tested and studied from many different points of view.

The results of such studies have been most confusing, leading Kosterlitz *et al.* to their study[20] of the spherical version of the Sherrington–Kirkpatrick (S–K) model, which we have outlined in these sections. The lack of ergodicity of the S–K model has been found[27] to be related to *relaxation times* $O(N^{1/2})$ instead of the usual $O(1)$.

Parisi has introduced a functional order parameter,[28] while Sompolinsky[29] has a time-dependent variant thereof, and Jonsson[30a] presented an order parameter which is a function of two variables, following the suggestion of de Dominicis *et al.*[30b]

The difference between the simple spherical model solution we have reproduced here, and the more realistic models in the literature, lies in the activation barriers that prevent individual spins in the latter from flipping (from up to down or vice versa) as the temperature is changed or the magnetic field is applied. The failure of linear response theory which results, i.e., the irreversibility, was first noted by Parisi,[28] and by Bray and Moore[31a] who were studying the distribution of relaxation times. A numerical experiment conducted by Bantilan and Palmer[31b] casts some light on this situation. Studying the ground state ($T = 0$) of 200 Ising spins averaged over 50 different sample distributions of random J_{ij}'s, they found *two* distinct curves for $m(B)$. The lower one exhibits $\chi = \partial m / \partial B = 0$ at small B; this is a consequence of varying the ground state *continuously* with applied field. The

[24] D. Sherrington and S. Kirkpatrick, *Phys. Rev. Lett.* **35**, 1792 (1975), and *Phys. Rev.* **B17**, 4384 (1978).

[25] S.F. Edwards and P.W. Anderson, *J. Phys.* **F5**, 965 (1975).

[26] L.R. Walker and R.E. Walstedt, *Phys. Rev.* **B22**, 3816 (1980).

[27] N.D. Mackenzie and A.P. Young, *Phys. Rev. Lett.* **49**, 301 (1982).

[28] G. Parisi, *J. Phys.* **A13**, 1101, 1887, L115 (1980) and *Phil. Mag.* **B41**, 677 (1980).

[29] H. Sompolinsky, *Phys. Rev. Lett.* **47**, 935 (1981).

[30a] T. Jonsson, *Phys. Lett.* **91A**, 185 (1982).

[30b] C. deDominicis, M. Gabay and C. Orland, *J. Physique Lett.* **42**, L523 (1981).

[31a] A.J. Bray and M.A. Moore, *J. Phys.* **C12**, L441 (1979).

[31b] F. Bantilan and R.G. Palmer, *J. Phys.* **F11**, 261 (1981).

upper curve, displaying finite susceptibility, is the result of optimizing the ground state *ab initio* at each value of the applied field. The latter is what we calculate in the spherical model approximation, but the former may simulate the experimental situation more accurately (see the preceding figure).

Due to the many uncertainties at present, we urge the interested reader to consult the most recent literature prior to pursuing this rapidly evolving subject.[32]

Problem 7.15.   Prove that only the *maximum* calculated susceptibility is thermodynamically stable. As a consequence, the lower of the curves obtained by Bantilan and Palmer must be thermodynamically unstable, although the relaxation time might be ∞ in the thermodynamic limit.

7.14.   Thermodynamics of Magnons

Magnons are the quantized elementary excitations in magnetic systems with continuous symmetry, including both the Heisenberg and the itinerant electron models. In the classical limit they are the spin waves. But at low temperatures, where quantum effects are important, it is the study of the magnons that reveals the correct thermodynamic properties.

We have previously observed the mapping of magnon states onto harmonic oscillators (Chap. 5) and observed that at low density, their commutation relations are those of photons, phonons, ..., i.e., that of generic *bosons*. Starting with a ferromagnetic state of all spins up we introduce a small number of magnons, say $\mathcal{N}$. As each decreases the total magnetization by one unit of spin angular momentum ($\hbar = 1$), the total magnetization drops to,

$$\mathcal{M}_z = Ns - \mathcal{N} = \mathcal{M}_0 - \mathcal{N} \tag{7.200}$$

---

[32] Some 700 recent titles on the topic of mean-field spin glasses are reviewed by D. Chowdhury and A. Mookerjee, *Phys. Reports* **114**, 1–98 (1984). Additionally: M. Mezard, G. Parisi and M. Virasoro, *Spin Glass Theory and Beyond*, World Scientific, 1987, and K. Binder and W. Kob, *Glassy Materials and Disordered Solids*, World Scientific, 2005, pp. 221–309. "Ultrametricity", a branch of mathematics relating to the random walk, plays a nontrivial role in this topic. See P. Boldi and E. Baum, *Phys. Rev. Lett.* **56**, 1598 (1986) and the review by R. Rammal, G. Toulouse and M. Virasoro, *Rev. Mod. Phys.* **58**, 765 (1986).

where the product Ns represents the saturation magnetization of N spins of length s, also denoted $\mathscr{M}_0$. If the magnons are created by thermal fluctuations appropriate to temperature T, we can use (7.97) to obtain $\mathscr{M}_z$ as a function of T. We denote each normal mode by $\mathbf{q}$:

$$\mathscr{M}_z(T) = \mathscr{M}_0 - \sum_{\mathbf{q}} [e^{\omega(\mathbf{q})/kT} - 1]^{-1} . \tag{7.201}$$

Such sums are best expressed in terms of the normal mode *dos*, $\rho(\omega)$, which incorporates both the dispersion (dependence of ω on $\mathbf{q}$) and dimensionality. With the substitution

$$\sum_{\mathbf{q}} \rightarrow N \int d\omega \rho(\omega) ,$$

we find that the magnetization is

$$\mathscr{M}_z(T) = \mathscr{M}_0 \left[1 - \frac{1}{s} \int d\omega \rho(\omega)(e^{\beta\omega} - 1)^{-1} \right] . \tag{7.202}$$

For an ordinary ferromagnet in 3D, $\omega \propto \mathbf{q}^2$ and therefore $\rho(\omega) \propto \omega^{1/2}$. Assuming that the exponential factor cuts off the integral sufficiently that we may take the upper limit to ∞, we make the substitution of variables $\beta\omega = x$ and with

$$\int d\omega \rho(\omega)(e^{\beta\omega} - 1)^{-1} = (kT)^{3/2} \int dx \rho(x)(e^x - 1)^{-1}$$

factor the temperature dependence from the integral. The result is

$$\mathscr{M}_z(T) = \mathscr{M}_0[1 - (T/T_0)^{3/2}] \tag{7.203}$$

the celebrated "$T^{3/2}$ law" of F. Bloch, with T_0 a lumped constant involving the integral and s. If we define the Curie temperature by the vanishing of spontaneous long-range order (LRO), $\mathscr{M}_z$ in this case, then it is T_0. But there is reason to believe that the $T^{3/2}$ law fails near T_c owing to our neglect of a number of mathematically and physically important concerns. Figure 7.18 shows in the case of some Gd alloys that this law is substantially satisfied until T reaches approximately 2/3 of the critical temperature, i.e., it is quite satisfactory at low temperatures. At finite temperatures there are two sources of error: a change in dispersion relations at finite q (including Brillouin zone edge effects), and the interactions and scattering of magnons with one another denoted generically as nonlinear effects. The higher the temperature, the larger the q's which are excited, so the greater the first

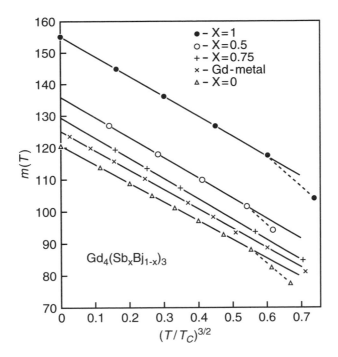

Fig. 7.18. Verification of Bloch's $T^{3/2}$ law in several gadolinium $(s = 7/2)$ alloy ferromagnets.[33] Slight downward deviations near T_c may be due to magnon interactions: cf. Fig. 7.19.

correction, but the greater the number of magnons, the greater the second as well.

Consider an Heisenberg ferromagnet with nearest-neighbor bonds on a sc lattice, spin magnitudes $s \gg 1$, so that expansions in powers of $1/s$ are justified (see Chap. 5). The magnetization, in units of $\mathcal{M}_0$, is

$$m(T) = 1 - s^{-1}(2\pi)^{-3} \iiint_{-\pi}^{+\pi} dk_x dk_y dk_z (e^{\beta\omega(\mathbf{k})} - 1)^{-1} \qquad (7.204)$$

where

$$\omega(\mathbf{k}) = 2sJ(3 - \cos k_x - \cos k_y - \cos k_z) \qquad (7.205)$$

reflects the lattice structure; it is approximated by sJk^2 only for $k^2 \ll 1$. We can make use of the properties of Bessel functions of imaginary argument I_p.

[33] F. Holtzberg *et al.*, *J. Appl. Phys.* **35/2**, 1033 (1964).

$$I_p(z) = \frac{1}{2\pi} \int_{-\pi}^{\pi} d\theta e^{z\cos\theta} \cos p\theta$$

$$= \left(\frac{z}{2}\right)^p \sum_{m=0}^{\infty} \frac{(z/2)^{2m}}{m!(|p|+m)!} . \tag{7.206}$$

Expanding the denominator of (7.204) in a geometric series in the exponential, we see that only I_0 enters into the expansion, which takes the following aspect:

$$m(T) = 1 - \frac{1}{s} \sum_{n=1}^{\infty} [e^{-ng} I_0(ng)]^3 \tag{7.207}$$

where $g = 2sJ/kT$. The bessel functions have asymptotic expansions which are convenient at large arguments, low temperature in this instance, i.e.,

$$I_0(z) \sim \frac{e^z}{(2\pi z)^{1/2}} \left[1 + \sum_{r=1}^{\infty} \frac{1^2 \times 3^2 \cdots (2r-1)^2}{r! 2^{3r} z^r}\right]$$

$$= \frac{e^z}{(2\pi z)^{1/2}} \sum_{r=0}^{\infty} \frac{[\Gamma(r+\frac{1}{2})]^2}{\pi r! (2z)^r} . \tag{7.208}$$

Inserted into the preceding, this yields both the series:

$$m(T) = 1 - B_{3/2}(T/T_0)^{3/2} - B_{5/2}(T/T_0)^{5/2} - B_{7/2}(T/T_0)^{7/2} - \cdots \tag{7.209}$$

and the values of the coefficients $B_{n/2}$, which we omit (in an external field the B's are additionally functions of the applied field; for details see Keffer[34]).

The asymptotic expansion is no longer accurate when kT is of the order of sJ, but at these temperatures the distribution becomes quasi-classical. Expanding the Bose–Einstein function in leading powers of sJ/kT, we obtain,

$$(e^{\omega/kT} - 1)^{-1} = \frac{kT}{\omega} - \frac{1}{2} + O(\omega/kT), \tag{7.210}$$

retaining only leading terms. Over the range of temperatures $T \gtrsim T_c s^{-1}$ we then calculate

---

[34] F. Keffer, Spin Waves, in *Handbuch der Physik*, Vol. 18 Part 2, H.J.P. Wijn, ed., Springer, Berlin, 1966; see pp. 1–273, and his Eq. (9.21) ff.

$$m(T) \cong 1 + \frac{1}{2s} - \frac{kT}{2Js^2}\left(\frac{1}{2\pi}\right)^3 \int_{-\pi}^{\pi} \frac{dk_x dk_y dk_z}{3 - \cos k_x - \cos k_y - \cos k_z}$$

$$\cong 1 + \frac{1}{2s} - \frac{kT}{6Js^2}W \qquad\qquad (7.211)$$

where the Watson integral $W = 1.516 \ldots$, as given in Sec. 7.10. Calculating the Curie temperature T_c by setting $m(T_c) = 0$, we obtain

$$kT_c = 3.96Js^2\left(1 + \frac{1}{2s}\right), \quad s \gg 1. \qquad\qquad (7.212)$$

For comparison, the same lattice may be treated in the molecular field approximation, where each spin interacts with a molecular field B_m,

$$\mathcal{H} = -S_i \cdot B_m$$

made up of the average force exerted by its neighbors at $R_i + \delta$,

$$B_m = 2J\left\langle \sum_\delta S_\delta \right\rangle_{\mathrm{TA}} .$$

One may calculate $\langle S_i \rangle_{\mathrm{TA}}$ in the standard way, and then solve the molecular field equation,

$$\langle S_i \rangle_{\mathrm{TA}} = \langle S_j \rangle_{\mathrm{TA}} \quad \text{for all } i, j .$$

The temperature at which this constitutive equation ceases to have only a nontrivial solution is the Curie–Weiss temperature $T_c = \theta$, the temperature at which $\chi = \mathbb{C}/(T - \theta) \to \infty$,

$$k\theta = 4Js^2\left(1 + \frac{1}{s}\right). \qquad\qquad (7.213)$$

The near-agreement of this high-temperature theory with the low-temperature result is truly astounding. But very accurate high-temperature series extrapolation methods have shown *both* estimates to be some 30% too high!

Taking into account the *real* shifts in magnon energies due to magnon-magnon interactions overcorrects this situation near T_c, as we shall now observe. We note that the energy of a magnon depends on the occupancy of all the other magnons, approximately according to the formula:

$$\varepsilon_{\mathbf{k}} = \hbar\omega_{\mathbf{k}} - \frac{1}{Ns}\sum_{\mathbf{k}'}(\hbar\omega_{\mathbf{k}} + \hbar\omega_{\mathbf{k}'} - \hbar\omega_{\mathbf{k}-\mathbf{k}'} - \hbar\omega_0)\langle \mathbf{n}_{\mathbf{k}'} \rangle. \qquad (7.214)$$

The occupation-numbers $\langle n_k \rangle$ are, on thermal average,

$$\langle \mathbf{n_k} \rangle = [\exp(\varepsilon_{\mathbf{k}}/kT) - 1]^{-1}. \tag{7.215}$$

These equations are coupled, nonlinear equations for the self-consistent determination of the nonlinear magnon spectrum. Equation (7.214) can be simplified somewhat if we take advantage of the cubic symmetry, whereupon it reduces to

$$\varepsilon_{\mathbf{k}} = \omega(\mathbf{k})[1 - b(T)] \quad \text{with} \tag{7.216}$$

$$b(T) = (2Js^2)^{-1}\frac{1}{N}\sum_{\mathbf{k}'}\omega(\mathbf{k}')\langle \mathbf{n_{k'}} \rangle \tag{7.217}$$

a function that has to be calculated self-consistently. Solutions of these equations lead to several interesting results.[35] At low temperature, the first correction to the expansion in (7.209) occurs via a term of the form

$$-C(T/T_0)^{8/2} \quad \text{(as first proved by F.J. Dyson, 1956 — cf. Chap. 5)} \tag{7.218}$$

and is practically negligible. Near T_c, the corrections are qualitatively similar to the experimental data (Fig. 7.18), but even more severe, as shown in Fig. 7.19. The consequence is that the theory predicts a first-order phase transition with a small jump in $m(T)$ at T_c. The calculated value of T_c is now within, a few percent of the exact value, but the jump in $m(T)$ at T_c is incorrect. This unphysical jump is probably the result of our total neglect of lifetime broadening effects, i.e., of magnon-magnon scattering. A theory which dealt properly with this problem in a simple manner is desirable but requires the use of Green functions, introduced in the next chapter.

So far we have concentrated only on the magnetization, but the internal energy and its derivative, the heat capacity, are of equal interest. Having determined $b(T)$ self-consistently, we can immediately write

$$U = \sum \omega(\mathbf{k})\left[1 - \frac{1}{2}b(T)\right]\langle \mathbf{n_k} \rangle \tag{7.219}$$

for the internal energy, taking care with the factor 1/2 not to double-count the interactions. The evaluation of such expressions proceeds along the model of the calculation of $m(T)$ above.

---

[35] M. Bloch, *Phys. Rev. Lett.* **9**, 286 (1962); *J. Appl. Phys.* **34**, 1151 (1963); I. Goldhirsch and V. Yakhot, *Phys. Rev.* **B21**, 2833 (1980).

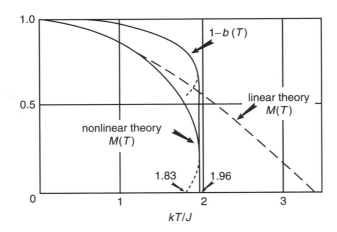

Fig. 7.19. Modifying magnon dispersion to include nonlinearities (self-energy modifications due to magnon-magnon interactions), (7.216) and (7.217), leads to improved agreement with experiment (Fig. 7.18) but also to an unphysical jump in $m(T)$ at T_c.[35]

Had we attempted to perform analogous calculations appropriate to one- and two-dimensional ferromagnets, we would have immediately discovered that integrals such as in (7.204) and (7.211) are divergent in 2D or in any lower dimension. As we shall see in the following section, this is related to the absence of long-range order for models with continuous symmetry in low-dimensional arrays.[36] This also entails the breakdown of spin-wave theory and its quantized magnon version, as these are the dynamical excitations from an ordered ground state that cannot be sustained in low dimensions. The lack of a simple framework may explain why theoreticians have been studying low-dimensional systems with such great fascination in recent years, with much new mathematical and physical understanding resulting therefrom.

Problem 7.16. Develop the low-temperature expansion of U (7.219) in powers of $T^{1/2}$, indicating the contribution of $b(T)$.

---

[36] N.D. Mermin and H. Wagner, *Phys. Rev. Lett.* **17**, 1133, 1307 (1966).

7.15.   Magnetism in Two Dimensions

In this section, we seek to reconcile two facts that are in apparent conflict: (i) there can be no long-range order in magnetic systems with continuous symmetry in 2D at any finite temperature, (ii) such systems *can* have a phase transition at finite T, with such discontinuities as infinite susceptibility, etc.

Of these two observations, (i) has the greater force, being the consequence of a rigorous theorem due to Mermin and Wagner.[36] The second was explained by Stanley and Kaplan[37] in the following terms: if $\chi(T)$ diverges at, and below, T_c, the implication is that the correlation function does not decay exponentially, but rather as some power law:

$$\Gamma(R) \equiv \langle S_0 \cdot S_R \rangle \sim 1/R^{\eta(T)} \tag{7.220}$$

with $\eta < 2$ for $T \leqslant T_c$. Then, the susceptibility diverges:

$$\chi \sim \sum \Gamma(R) = \infty$$

for $T \leqslant T_c$, while if $\Gamma(R)$ decays exponentially above T_c, the correlations are finite-ranged, hence the susceptibility will be finite above that temperature. It is interesting that this can occur without $\Gamma(\infty)$ being finite — i.e., without LRO. Such features must be very much model-dependent. They are not seen in the 2D spherical model. Stanley and Kaplan illustrate by numerical calculations on the Heisenberg ferromagnet, finding near T_c:

$$\chi \sim (T - T_c)^{-\gamma_c} \tag{7.221}$$

with $\gamma_c \cong 5/2 + 2/(3s^2)$ and $kT_c \cong 1/5(z-1)[2s(s+1)-1]J$ in lattices with coordination number $z > 3$. It seemed for a while that these results (based on some dozen terms in a high-temperature series expansion) would meet the test of time. They were experimentally confirmed by observations on a number of 2-dimensional antiferromagnets and ferromagnets.[38] However, it is difficult to tell whether 3-dimensional effects were not intruding into the experiments, and numerical "experiments" appear more reliable. The

[37] H.E. Stanley and T. Kaplan, *Phys. Rev. Lett.* **17**, 913 (1966). The proof for the existence of such a phase transition, for the plane rotator (classical XY model) is in J. Fröhlich and T. Spencer, *Phys. Rev. Lett.* **46**, 1006 (1981) and is based on the general method given in J. Fröhlich and E.H. Lieb, *Commun. Math. Phys.* **60**, 233 (1978).

[38] L.J. deJongh and A.R. Miedema, *Adv. Phys.* **23**, 1 (1974) treated Heisenberg antiferromagnets in 2D, pp. 64ff.

most recent such studies[39] have apparently demonstrated that $T_c = 0$ for the classical limit. The proximate cause appears to be a topological low-energy excitation called a "skyrmion," an hedgehog sort of vortex, that is more efficient at destroying long-range order than are the spin waves. The anisotropic version of the Heisenberg model (the plane-rotator model) *does* have a phase transition of the type predicted by Stanley and Kaplan! Its interesting properties are treated in Secs. 7.16–7.18 immediately following.

We now demonstrate the theorem of Mermin and Wagner and extend it to itinerant ferromagnets as well. The statement is that *no magnetic LRO* (spontaneous magnetization or, for antiferromagnets, sublattice magnetization) *can be sustained at finite T, except in finite external fields B*. It is implemented by an inequality:

$$|m(T)| < \frac{\text{conts.}}{T^{1/2}} |\ln|B||^{-1/2}, \quad (2\text{D}). \tag{7.222}$$

The same method establishes a somewhat stronger inequality in one dimension

$$|m(T)| < \frac{\text{conts.}}{T^{2/3}} |B|^{1/3}. \tag{7.223}$$

The inequalities are compatible both with divergent zero-field susceptibilities found by Stanley and Kaplan, and with the properties of a spherical model in external field. They concern long-range order but are unrelated to the questions concerning the existence of phase transitions in one or two dimensions.

The proof of these inequalities starts with Bogoliubov's inequality[40,41]

$$\frac{1}{2}\langle A, A^+\rangle_{\text{TA}} \langle [[C,H],C^+]\rangle_{\text{TA}} \geq kT|\langle [C,A]\rangle_{\text{TA}}|^2 \tag{7.224}$$

in which A and B are arbitrary operators, H is the Hamiltonian, and C is picked to satisfy

$$B = [C^+, H]. \tag{7.225}$$

$[,]$ is the usual commutator bracket, $\{,\}$ is the usual anticommutator bracket, and we define yet another bracket:

---

[39] A Monte-Carlo renormalization-group analysis by S. Shenker and J. Tobochnik, *Phys. Rev.* **B22**, 4462 (1980).

[40] TA = Thermal Average. Note: the reader who is not interested in tedious mathematical "details" may skip to the end of this section, starting after Eq. (7.240).

[41] N.N. Bogoliubov, *Phys. Abh. Sowj.* **6**, 1, 113, 229 (1962); H. Wagner, *Z. Phys.* **195**, 273 (1966).

$$(A, B) = \sum_i \sum_{j \neq i} (i|A|j)^* (i|B|j) \frac{W_i - W_j}{E_j - E_i} \tag{7.226}$$

with $W_i = Z^{-1} \exp(-\beta E_i)$ the normalized Boltzmann factor. By the property of the exponential function, the following is always obeyed:

$$\frac{1}{2} \beta (W_i + W_j) > \frac{W_i - W_j}{E_j - E_i} > 0 \tag{7.227}$$

and combining results,

$$(A, A) \leqslant \frac{1}{2} \beta \langle \{A, A^+\} \rangle . \tag{7.228}$$

Here, Schwartz' inequality (i.e. $1 > \cos^2 \theta$) applies in the form:

$$(A, A)(B, B) \geqslant |(A, B)|^2 \tag{7.229}$$

and with (7.225) defining C, we can obtain

$$(A, B) = \langle [C^+, A^+] \rangle \quad \text{and} \tag{7.230}$$

$$(B, B) = \langle [C^+, [H, C]] \rangle , \tag{7.231}$$

thus establishing Bogoliubov's inequality (7.224), QED.
     For the Heisenberg Hamiltonian we take

$$\mathcal{H} = -\sum_{i,j} J(R_i - R_j) S_i \cdot S_j - |B| \sum_i S_i^z \exp(iK \cdot R_i) \tag{7.232}$$

picking $K = 0$ to rule out ferromagnetism, $K = (\pi, \pi, \pi)/a$ to rule out antiferromagnetism or arbitrary K to rule out LRO with any wavevector whatever. We define Fourier transforms of the various quantities:

$$S^x(k) = \sum_i \exp(-ik \cdot R_i) S_i^x , \quad \text{etc.}, \tag{7.233}$$

and take

$$C = S^x(k) + iS^y(k) = S^+(k), \quad A = S^-(-k - K) \tag{7.234}$$

and form B from (7.225). Inequality (7.224) now takes the appearance

$$\frac{1}{2} \langle \{S^+(k+K), S^-(-k-K)\} \rangle$$

$$\geqslant N^2 kT\sigma^2 \div \left\{ \frac{1}{N} \sum_{k'} [J(k') - J(k' - k)] \left\langle S^z(-k') S^z(k') + \frac{1}{4} \{S^+(k'), S^-(-k')\} \right\rangle + N|B|\sigma/2 \right\}$$

$$\tag{7.235}$$

with $\sigma \equiv 1/N \sum \exp(iK \cdot R_i)\langle S_i^z \rangle$ the order parameter. The denominator in curly brackets on the right-hand side of this relation is positive, and is always less than

$$\frac{1}{2}N \sum_i R_i^2 |J(R_i)|s(s+1)k^2 + \frac{1}{2}N|B\sigma|. \tag{7.236}$$

Replacing the denominator by this upper bound and summing both sides over wavevectors, we conclude

$$s(s+1) > 2kT\sigma^2 \int_{BZ} \frac{d^dk}{(2\pi)^d} \left[s(s+1) \sum_i R_i^2 |J(R_i)|k^2 + |B\sigma| \right]^{-1} \tag{7.237}$$

and, replacing the BZ by the smaller d-dimensional sphere of radius k_0 which it contains, we perform the integrations and bound the order parameter as follows:

$$\sigma^2 < 2\pi s(s+1)k_0^{-2}\omega_0/[kT \ln(1 + \omega_0/|B\sigma|)] \tag{7.238}$$

where

$$\omega_0 \equiv \sum_i s(s+1)k_0^2 R_i^2 |J(R_i)|. \tag{7.239}$$

In the limit of vanishing field B, this reduces to the result quoted in (7.222). In 1D, the integration results in

$$\sigma^3 < |B|\omega_0[s(s+1)/2kT \tan^{-1}(\omega_0|B\sigma|)^{1/2}]^2 \tag{7.240}$$

instead; which in the same limit, reduces to the result quoted in (7.223).

The vanishing of the LRO parameter cannot be simply ascribed to the softness of the magnon dispersion $w(k) \propto k^2$, as it applies also to antiferromagnets in which $w(k) \propto |k|$. Nor is it limited to Heisenberg models. In the case of magnetic metals, the use of spin operators such as

$$S^+(q) = \sum_k c_{k+q\uparrow}^* c_{k\downarrow}$$

$$= \sum_i b_{i\uparrow}^* b_{i\downarrow} \exp(iq \cdot R_i) \tag{7.241}$$

and

$$S^z(K) = \frac{1}{2} \sum_i (b_{i\uparrow}^* b_{i\uparrow} - b_{i\downarrow}^* b_{i\downarrow}) \exp(iK \cdot R_i) \tag{7.242}$$

constructed out of fermion operators ($c_{k\uparrow}$ destroys an electron in Bloch state
k, spin↑, whereas $b_{i\uparrow}$ destroys an electron on site R_i with the same spin
index) permits virtually the same proof to go through. The fact that the
spin operators commute with the one- and two-body potentials ensures that
the order parameter will be the same as for a non-interacting fermion gas, i.e.,
zero in vanishing field. Thus, regardless of whether the electronic interactions
favor parallel spin alignments or antiparallel, they can only affect the short-
ranged characteristics of the material.

The literature contains several less obvious extensions of these original
discoveries. Fisher and Jasnow[42] have extended the no LRO rule to *finite*
thickness slabs (eliminating the constraint to monatomic thin films, a
serious experimental limitation if it existed). Fröhlich and Lieb[43] and
Dyson *et al.*[44] *proved* the existence of a phase transition at finite T_c in a
variety of models with continuous symmetry in 2D[43] and 3D.[44] The decay
of correlation functions at large distances has been bounded by inequalities due to Griffiths, Simon,[45] and others. It appears that 2D is "delicate":
seemingly innocuous changes in model parameters can cause profound
changes in thermodynamics.

7.16.   The XY Model: 1D

The peculiarities of low dimensions are nowhere better exemplified than in
the XY model, the Hamiltonian of which is

$$\mathscr{H} = \frac{1}{2}J\sum_{(ij)}(S_i^+ S_j^- + \text{H.c.}) - B \cdot \sum_i S_i. \qquad (7.243)$$

In 1D it is exactly solvable in the extreme quantum limit of spins $s = 1/2$
(with the external field in the z-direction) by means of the Jordan–Wigner
transformation of the spin operators to fermions. We have discussed this
extensively in Chap. 5, and now return to it. In brief, introduce a set of
fermion operators a_i^* and a_i, and their Fourier transforms c_k^* and c_k, such
that the spins are expressible in terms of the a's:

[42] M.E. Fisher and D. Jasnow, *Phys. Rev.* **B3**, 907 (1971).
[43] J. Fröhlich and E.H. Lieb, *Phys. Rev. Lett.* **38**, 440 (1977); *Commun. Math. Phys.* **60**, 233 (1978).
[44] F.J. Dyson, E.H. Lieb and B. Simon, *J. Stat. Phys.* **18**, 335 (1978).
[45] B. Simon, *Phys. Rev. Lett.* **44**, 547 (1980).

$$S_i^+ = a_i^*(-1)^{Q_i}, \quad \text{where} \tag{7.244}$$

$$Q_i = \sum_{j<i} a_j^* a_j = \sum_{j<i}\left(S_j^z + \frac{1}{2}\right),$$

and similarly for the Hermitean conjugate S_i^-, with

$$S_i^z = a_i^* a_i - \frac{1}{2}. \tag{7.245}$$

The phase factors cancel if (i, j) are nearest neighbors, so for an external field in the z-direction,

$$\mathscr{H} = -\frac{1}{2}J\sum_n (a_n^* a_{n+1} + \text{H.c.}) - B\sum_n \left(a_n^* a_n - \frac{1}{2}\right). \tag{7.246}$$

This quadratic form in fermion operators is diagonalized by a transformation to running waves:

$$\mathscr{H} = \sum_k e(k)c_k^* c_k + \frac{1}{2}BN, \tag{7.247}$$

where k runs over N equally spaced points in the interval $-\pi, +\pi$, and

$$e(k) = -(J\cos k + B) \tag{7.248}$$

is the energy of the individual fermion states. Fadeev and Takhtajan[46] have remarked, in connection with the $s = 1/2$ Heisenberg antiferromagnet in 1D, that the appropriate excitations are fermions $(\hbar/2)$ rather than magnons $(\hbar)$.[47] If we consider the XY model as an extreme limiting case of the anisotropic Heisenberg model within the range $|J_z| < |J|$, then the above is a confirmation of their remark. (Undoubtedly the fermions form *bound pairs* of total spin $\hbar$ for $|J_z| \geq |J|$, in which case the elementary excitations can no longer be trivially pulled apart into fermionic "half-excitations.") Even for a classical $(s \to \infty)$ isotropic Heisenberg chain, Nakamura and Sasada[48] again claim that the elementary excitations are *fermionic* kinks, i.e. solitons.

---

[46] L. Fadeev and L. Takhtajan, *Phys. Lett.* **85A**, 375 (1981).
[47] They point out that "elementary excitations" consist of 2 independent quasi-particles.
[48] K. Nakamura and T. Sasada, *J. Phys.* **C15**, L1013 (1982).

The thermodynamics of the Hamiltonian (7.247) is obtained by using the fermion rules (7.92)–(7.94). The internal energy is

$$U = \sum_k e(k)\langle \mathbf{n}_k \rangle, \quad \text{where } \langle \mathbf{n}_k \rangle = 1/(e^{\beta e(k)} + 1) \tag{7.249}$$

and dU/dT yields the heat capacity. For small J_z we can solve the thermal Hartree–Fock equations of the anisotropic model. To $\mathscr{H}$ we add the perturbation

$$\mathscr{H}' = -J_z \sum_n \left(a_n^* a_n - \frac{1}{2} \right) \left(a_{n+1}^* a_{n+1} - \frac{1}{2} \right) \tag{7.250}$$

and pair the operators in the manner preferred by the XY model. In zero external field, this is

$$\mathscr{H}'_{H-F} = +J_z \sum_n (a_n^* a_{n+1} \langle a_{n+1}^* a_n \rangle + \text{H.C.}) - J_z \sum_n |\langle a_n^* a_{n+1} \rangle|^2 \tag{7.251}$$

with the change in apparent sign of J_z due to the anticommutation relations, fermion statistics. (In an external field, pairings such as $\langle a_n^* a_n - 1/2 \rangle$ must also be retained.) It is easy to see that this problem involves some self-consistency requirement, which are spelled out and solved in the work of Caliri and Mattis.[49a] We show the specific heat calculated in zero field for various J_z in Fig. 7.20 taken from this work. The formulation of the correct Hartree–Fock H' when the external field is present in (7.246) is given in Glauss *et al.*[49b]

Amusingly, the spin $1/2$ XY antiferromagnet ($J < 0$) can even be solved exactly for in-plane orientations of the external field. This discovery by Thomas and coworkers[50] remains insufficiently understood. Briefly, starting wiith (7.243) and $J < 0$ with $\mathbf{B} = (B, 0, 0)$ in the x-direction, they found $\langle S^x \rangle$ to be discontinuous in B at discrete values of B, with the last discontinuity occurring at $B = \sqrt{2}J$, clearly a critical value of sorts. Although there is no known systematic solution to the XY Hamiltonian when the field is not in the z-direction, these researchers *guessed* a product state of the form,

$$\psi = |\phi_1\rangle \otimes |\phi_2\rangle \otimes \cdots \tag{7.252}$$

[49a] A. Caliri and D.C. Mattis, *Rev. Brasileira de Fisica* **13**, 322 (1983).
[49b] V. Glauss, T. Schneider and E. Stoll, *Phys. Rev.* **827**, 6770 (1983).
[50] J. Groen, T. Klaasen, N. Poulis, G. Müller, H. Thomas and H. Beck, *Phys. Rev.* **B22**, 5369 (1980). This work was extended by J. Kurmann, H. Thomas and G. Müller, *Physica* **112A**, 235 (1982).

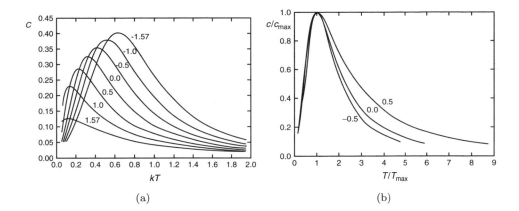

(a)   (b)

Fig. 7.20.   (a) Specific heat vs. T for various ratios J_z/J_x in Hartree–Fock approximation to Heisenberg chain with $J_x = J_y \pm J_z$.[49a] (b) Combining these results in terms of $c/c_{\max}$ vs. $T/T_{\max}$ yields a universal curve only for $T_{\max} \leqslant 1$, where it is invariably linear in T. This accords with the phase diagram (next figure).

and found that it is the *ground state* of their Hamiltonian at $B = \sqrt{2}J$. In this expression, the individual spin states are

$$|\phi_n\rangle = 2^{-1/2}\{\exp[-(-1)^n(i\pi/8)]S_n^+ + \exp[(-1)^n(i\pi/8)]\}|\phi\rangle \qquad (7.253)$$

taking $|\phi\rangle$ to be the state of spin down. It is not known whether any of these ground state features persist to finite T, although it is an intriguing possibility.

Thermal properties of the *classical models* are even easier to study in 1D than was the extreme quantum $s = 1/2$ limit. In almost all cases of interest, one can obtain the thermodynamic functions exactly using the method of *transfer matrices*, a procedure peculiarly suited to 1D. For purposes of illustration let us specialize to the *planar model* (also called plane rotator model; the spins are constrained to the XY plane.) Temporarily set $B = 0$. The effective Hamiltonian is then,

$$\mathcal{H} = -J\sum_{n} \cos(\theta_n - \theta_{n+1}) \qquad (7.254)$$

where θ_n is the angle of the nth spin with respect to some arbitrary (fixed) axis in the XY plane.

As the total Hamiltonian consists of pieces $H_{n,n+1}$ which commute with one another, we write the partition function as an ordered sequence of integrations

$$Z = \text{Tr}\{e^{-\beta H}\}$$

$$= \cdots \frac{1}{\pi} \int_{-\pi}^{\pi} d\theta_n \exp(-\beta H_{n,n+1}) \frac{1}{\pi} \int_{-\pi}^{\pi} d\theta_{n+1} \exp(-\beta H_{n+1,n+2}) \cdots$$

$$(7.255)$$

with the factors $1/\pi$ chosen so that Z is, arbitrarily, normalized to 2^N in the high-temperature limit. We view the evaluation of this repeated integration as an eigenvalue problem. For suppose we knew the eigenfunctions $f_j(\theta)$ and eigenvalues z_j in the integral eigenvalue equation,

$$\frac{1}{\pi} \int_{-\pi}^{\pi} d\theta' \exp[-\beta H(\theta, \theta')] f_j(\theta') = z_j f_j(\theta) \qquad (7.256)$$

then by iteration in Z, we obtain

$$Z = \sum_j (z_j)^N \quad (\text{PBC}) \qquad (7.257)$$

if the Nth spin is tied to the first (periodic boundary conditions). Or,

$$Z = 2z_0^{N-1} \quad (\text{free ends}) \qquad (7.258)$$

if the chain merely ends at the Nth spin (so there are only $N - 1$ bonds). Here z_0 is the largest eigenvalue of (7.256), which is the only attainable one when we start with $f(\theta_1) = 1$.

These two results are actually the same in the thermodynamic limit, for the free energy with periodic boundary conditions, is:

$$\ln Z_{\text{PBC}} = N \ln z_0 + \ln[1 + (z_1/z_0)^N + (z_2/z_0)^N + \cdots]$$

and may be compared to that with free ends:

$$\ln Z_{\text{f.e.}} = N \ln z_0 + \ln(2/z_0).$$

The extensive part is the same in both cases, whereas the correction term, which is $O(1)$, expresses the thermodynamics of the particular end conditions.

With $H(\theta, \theta') = -J\cos(\theta - \theta')$, the integrand is well known as the generating function for Bessel functions I_n of (7.206):

$$\exp(\beta J \cos(\theta - \theta')) = \sum_{n=-\infty}^{+\infty} I_n(\beta J) \exp[in(\theta - \theta')]. \qquad (7.259)$$

It is of the form

$$\exp(\beta J \cos(\theta - \theta')) = V(\theta, \theta') = \sum_n w_n \psi_n^*(\theta') \psi_n(\theta), \quad \text{i.e.,}$$

it is a diagonal matrix in the representation of its eigenfunctions $\exp(in\theta)$, with eigenvalues $w_n = I_n(\beta J)$. We call $V(\theta, \theta')$ the transfer matrix. The eigenvalues I_n are the Bessel functions of imaginary argument,

$$I_n(K) = \frac{1}{2\pi} \oint d\theta \exp(K \cos\theta + in\theta) = I_{-n}(K)$$

$$= i^n J_n(-iK) \qquad (7.260)$$

satisfying usual Bessel function identities $I_1 = dI_0(K)/dK$, etc. Their magnitudes are in natural order: $I_0 > I_1 > I_2 > \cdots$ at all values of $K \equiv \beta J$.

Consequently, the largest eigenvalue z_0 is always $2I_0(\beta J)$ and

$$Z = [2I_0(\beta J)]^N, \quad F = -NkT \ln 2I_0(\beta J) \qquad (7.261)$$

yields the thermodynamics in zero external field. In finite field, or in general if we change the form of $H_{n,n+1}$, we can expand the eigenfunctions of the new operators in terms of the complete set of ψ_n's, as in perturbation theory. Take the example of the external field in the plane:

$$H'_{n,n+1} = -\frac{1}{2}B(\cos\theta_n + \cos\theta_{n+1}). \qquad (7.262)$$

The new transfer matrix can be written as the sum of three matrices: the unperturbed matrix, a matrix linear in B and one quadratic in B. For the calculation of the paramagnetic susceptibility all we require is the free energy to second order in B, thus we treat the linear perturbation in second-order perturbation theory, and the quadratic perturbation in first-order perturbation theory.

Explicitly,

$$\delta_1 V = 2 \exp[\beta J \cos(\theta - \theta')] \left[\left(\frac{1}{2}\beta B\right)(\cos\theta + \cos\theta') \right] \qquad (7.263)$$

for which second-order perturbation theory yields

$$\delta z = \sum_{n \neq 0} \frac{|(\delta_1 V)_{n,0}|^2}{2(I_0 - I_n)}, \qquad (7.264a)$$

while the term second-order in B is

$$\delta_2 V = 2 \exp[\beta J \cos(\theta - \theta')] \left[\frac{1}{8}(\beta B)^2(\cos\theta + \cos\theta')^2 \right] \tag{7.265}$$

which, in first-order, yields

$$\delta z = (\delta_2 V)_{0,0} = \frac{1}{4}(\beta B)^2(I_0 + I_1). \tag{7.264b}$$

The evaluation of (7.264a) yields $\frac{1}{4}(\beta B)^2(I_0 + I_1)^2/(I_0 - I_1)$, so combining it with (7.264b) we find for the new eigenvalue,

$$z = 2I_0 \left[1 + \frac{1}{4}(\beta B)^2 \frac{I_0 + I_1}{I_0 - I_1} \right]. \tag{7.266}$$

We recognize $I_1/I_0 = d(\ln I_0)/dK$ as $\langle \cos(\theta_n - \theta_{n+1}) \rangle \equiv \mu$, i.e., the nearest-neighbor correlation function in zero field (recall: $K = \beta J$). Taking logarithms of z, and identifying the susceptibility through the expansion $f = f_0 - \frac{1}{2}\chi_0 B^2$, we arrive at the expression

$$\chi_0 = \frac{\mathbb{C}}{T} \left(\frac{I_0 + I_1}{I_0 - I_1} \right) = \frac{\mathbb{C}}{T} \left(\frac{1 + \mu}{1 - \mu} \right). \tag{7.267}$$

We should note that μ is an odd function of J, i.e., $\mu(-J) = -\mu(J)$. In the following chapter we derive an identical expression for the linear-chain Ising model if we take $\mu = \tanh(\beta J)$, the appropriate correlation function in that case. For the isotropic classical Heisenberg model where $\mu = \coth(\beta J) - 1/(\beta J)$ (Problem 7.17), Fisher[51] also finds a susceptibility $\chi_0 = (\mathbb{C}/T)(1 + \mu)/(1 - \mu)$. These three examples suggest an interesting relationship between the susceptibility of classical ferromagnets ($J > 0$) and antiferromagnets ($J < 0$) in 1D. If we define $X(\beta J) \equiv T\chi_0/\mathbb{C}$, they *all* satisfy:

$$X(\beta J)X(-\beta J) = 1. \tag{7.268}$$

Computational or experimental deviations from this symmetry will be caused by quantum fluctuations, longer-ranged forces, or a combination of such factors.

The well-informed reader will already have noted that the sequence of integrations (convolution) (7.255) is a Feynman path-integral or functional integral (with only the simple modification that n labels *physical* points and not an arbitrary sequence). In view of the relationship between such

---

[51] M.E. Fisher, *Am. J. Phys.* **32**, 343 (1964).

integrals and the Schrödinger equation, we should not be surprised at the relationship between the partition function and the transfer-matrix equation (7.256), one which has been particularly successful in the resolution of the two-dimensional Ising model treated in the next chapter.

Problem 7.17. Identify the transfer matrix for the classical Heisenberg linear chain, and derive its expansion in terms of spherical harmonics eigenfunctions $Y_{\ell,m}(\theta,\phi)$ and eigenvalues $j_\ell(x) = (\pi/2x)^{1/2}J_{\ell+1/2}(x)$ (spherical Bessel functions):

$$V(\theta,\phi\,|\,\theta',\phi') = \sum_{\ell=0}^{\infty}\sum_{m=-\ell}^{+\ell} i^\ell j_\ell(i\beta J)Y^*_{\ell,m}(\theta,\phi)Y_{\ell,m}(\theta',\phi')\,.$$

Calculate $\mu \equiv \langle S_n \cdot S_{n+1}\rangle = -\partial \ln z_0/\partial(\beta J)$ and compare with Fisher's result.

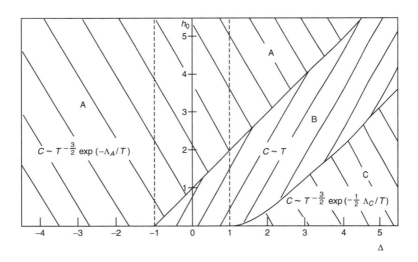

Fig. 7.21. Parameter space of XXZ anisotropic Heisenberg model.[52] $c \propto T$ (shown as region B) indicates where the spectrum of elementary excitations is gapless. In regions A and C the specific heat is activated, with energy gaps dependent on the two parameters (h_0 and Δ). The quantum XY model (at $\Delta = 0$) is the only trivially solvable model in which *all* the eigenstates and eigenvalues are known. The *antiferromagnet* ($\Delta > 0$) turns into a *ferromagnet* for $\Delta < 0$; there is little or no symmetry in Δ.

The literature on exact calculations of thermal effects at finite T in *quantum* linear chains is sparse. However, the spectrum of elementary excitations, whether obtained by Bethe's ansatz or by some other means, allows asymptotically exact formulations of low-T specific heat and other low-T thermodynamic properties. Johnson[52] has summarized much of the available information on the spin-1/2 XXZ model (Heisenberg anisotropic model in an external field h_0.) It is governed by the following Hamiltonian:

$$\mathscr{H} = \sum_n \left(\frac{1}{2} S_n^+ S_{n+1}^- + \text{H.C.} + \Delta S_n^z S_{n+1}^z - h_0 S_n^z \right). \tag{7.269}$$

A simplified version of his phase diagram is reproduced in Fig. 7.21. For more details, discussion, critical points, etc., the reader should consult the original paper with its detailed explanations.

7.17. The XY Model: 2D

The XY model is an interesting example of cooperative phenomena in 2D systems with continuous symmetry: there is no LRO, yet a phase transition occurs at finite T_c.

The information on the extreme quantum limit $s = 1/2$ originates in numerical studies by Betts and his collaborators.[53] For indications concerning the low-temperature phase, they investigated the ground state susceptibility tensor χ_{ij}^0 (χ_{zz}^0 refers to magnetization in the z-direction owing to an applied field in the z-direction, χ_{xz}^0 to magnetization in the x-direction owing to an applied field in the z-direction, etc.), finding $\chi_{zz}^0 = 0$ while $\chi_{xx}^0 = \chi_{yy}^0$ both *diverge* for $J > 0$. For $J < 0$, the latter are finite.

The proof that the ground state χ_{zz}^0 vanishes is relatively simple, and can be extended by inspection to *all* components of the susceptibility tensor for the isotropic Heisenberg antiferromagnet. It is also valid in any dimension d. We write the ground-state susceptibility in terms of the matrix elements

---

[52] J.D. Johnson, *J. Appl. Phys.* **52**, 1991 (1981).

[53] J. Oitmaa, D.D. Betts and L.G. Marland, *Phys. Lett.* **79A**, 193 (1980) (ground state); D.D. Betts, F. Salevsky and J. Rogiers, *J. Phys.* **A14**, 531 (1981) (vortex operators); J. Rogiers, T. Lookman, D.D. Betts and C.J. Elliot, *Canad. J. Phys.* **56**, 409 (1978) (high T_c series expansions). The quantum spin 1/2 XY model is studied further as a function of T in M. Makivic and H.-Q. Ding, *Phys. Rev.* **B43**, 3562 (1991). The phase transition with classical (plane rotator) model is the subject of R. Gupta and C. Baillie, *Phys. Rev.* **B45**, 2883 (1992). The literature on this topic continues, unabated, to this day.

of the total spin operator $S_{\text{tot}}^z = \sum_i S_i^z$ connecting the ground state (0) to the excited spectrum (α):

$$\chi_{zz}^0 = \sum_{\alpha \neq 0} |(S_{\text{tot}}^z)_{\alpha,0}|^2/(E_\alpha - E_0). \tag{7.270}$$

Because the ground state is an eigenfunction of S_{tot}^z with eigenvalue 0 (see Mattis[54] for a proof), this expression vanishes identically! While this does *not* prove that $\lim T \to 0 \chi_{zz}(T) = 0$, it *does* indicate the absence of long-ranged $\langle S_i^z S_j^z \rangle$ correlation in the ground state.

We briefly touched upon the need to consider *vortices* in the XY model in Chap. 5. A plausible vortex *operator* for 4 spins at the corners of an elementary square is[53]:

$$V_Q = \frac{1}{4}(S_1^x S_2^y - S_2^y S_3^x + S_3^x S_4^y - S_4^y S_1^x). \tag{7.271}$$

Of the 16 eigenstates of these 4 spins, V_Q has zero eigenvalue on 12, $+1$ on two and -1 on the remaining two. The expectation value of V_Q is zero in general, by symmetry, but its square V_Q^2 indicates the vortex-antivortex density. Numerical studies show this to be relatively small and increasing slowly with temperature, until a finite T_c is reached. Thereafter, their number increases rapidly to the theoretical maximum.[53]

A rather more roundabout approach to the $s = 1/2$ XY model in 2D has been taken by Lagendijk and De Raedt,[55] who by use of Trotter's formula $e^{A+B} = \lim_{m \to \infty}(e^{A/m} e^{B/m})^m$, useful where A and B do not commute, disentangle all the even-numbered sites from the odd-numbered sites in the evaluation of the partition function, which they perform in the $m = 1$ approximation to the above limit. Their calculation is nontrivial, as it requires a mapping onto the 8-vertex model, for which solutions are known only in certain cases. They conclude that the $m = 1$, 2D XY model has a logarithmically divergent specific heat at T_c, with T_c, given by $\sinh(2\beta_c J) = 1$, i.e., $kT_c = (2.27\cdots)J$, just as in the corresponding Ising model. Obviously, there is need for additional work in this area. The $s = 1/2$ XY model is the prototype of the hard sphere Bose gas. As such, our increased understanding

[54] D.C. Mattis, *Phys. Rev. Lett.* **42**, 1503 (1979) (proof that ground state in XY model has $S_{\text{tot}}^z = 0$.) Related work on other models was carried out by D. Mattis, *Phys. Rev.* **130**, 76 (1963); E. Lieb and D. Mattis, *J. Math. Phys.* **3**, 749 (1962).
[55] A. Lagendijk and H. De Raedt, *Phys. Rev. Lett.* **49**, 602 (1982).

of its thermodynamic properties in 2D and 3D will be directly relevant to continued progress in the related theory of superfluidity.

But undoubtedly it is the *planar* (*alias* plane rotator) *classical* model,

$$H = -J \sum_{(i,j)} \cos(\phi_i - \phi_j), \qquad (7.272)$$

that has generated the greatest activity following the discoveries by Berezinskii and Kosterlitz and Thouless[56] of its remarkable properties. As it is not yet clear which of several approach will ultimately prove the most fruitful, we indicate several paths to the resolution of this model. In the original semiphenomenological approach, bolstered by corrections due to scaling and renormalization group arguments,[57] one separates the vortex (topological) excitation degrees of freedom from the magnon (small deviations about topological configurations) normal modes. It is the former which are responsible for the anomalies, including the phase transition, while the latter provide an analytic, essentially uninteresting background (as is evident if we recall the two-dimensional spherical model).

Then, instead of the total H above, one considers primarily the vortex Hamiltonian consisting of two parts: the energy of each vortex, and their interaction energy. At low temperature, pairs of vortices of opposite sign are created at small distance from each other at a small cost in energy. At higher temperatures, they interpenetrate and form a fluid formally analogous to an electrolyte. Finally, above T_c, free vortices are created in ever increasing numbers with increasing temperature. The energy to create a single vortex was estimated[56,57] as,

$$\Delta E = 2\pi J \ln(R/a)$$

where R is the radial dimension of the sample and a the lattice parameter. The entropy gained in this process is the logarithm of the number of ways we can position the center of the vortex,

$$\Delta \mathscr{S} = k \ln(R/a)^2 \qquad (7.273)$$

and thus the excess free energy is,

$$\Delta F \approx 2(\pi J - kT) \ln(R/a). \qquad (7.274)$$

---

[56] J.M. Kosterlitz and D.J. Thouless, *J. Phys.* **C6**, 1181 (1973); and similarly, V.L. Berezinskii, *Sov. Phys. JEPT* **32**, 493 (1971).
[57] J.M. Kosterlitz, *J. Phys.* **C7**, 1046 (1974).

The phase transition occurs when vortices are spontaneously generated, i.e., at approximately

$$kT_c \approx \pi J. \tag{7.275}$$

The important corrections come from vortex interactions:

$$\Delta E_{ij} = -2\pi J q_i q_j \ln R_{ij}/a \tag{7.276}$$

and interactions with background magnon terms. Their result is to change J in (7.272) to $\frac{1}{2}J_{\text{eff}}(T)$.[57]

The straightforward evaluation of this partition function is more difficult than one would believe. A significant attempt was made by Villain,[58] whose modifications are referred to as the Villain model. Recognizing that the effective interactions will be logarithmic (so that the Boltzmann factors will become temperature-dependent *powers* of the vortex separations) he starts with an analysis of the Boltzmann factors through formal expansion. Consider the expansion of an exponential periodic in ϕ:

$$\exp(2\beta J \cos \phi) \cong \text{const.} \times \sum_{n=-\infty}^{+\infty} \exp[-\beta A(\phi - 2\pi n)^2] \tag{7.277}$$

with the factor 2 in $2\beta J$ to allow for double summation over pairs in subsequent calculations. The parameter A is a function of T, which together with the constant multiplying the sum above, is adjusted for a best fit. Both sides are formally periodic in ϕ, but in evaluating the partition function ϕ is now taken over the entire range $-\infty < \phi < +\infty$ instead of the fundamental interval. The justification is that for any periodic function $F(\phi)$,

$$\int_{-\pi}^{+\pi} d\phi F(\phi) = \lim_{\varepsilon \to 0} 2(\beta \pi \varepsilon)^{1/2} \int_{-\infty}^{+\infty} d\phi \exp(-\beta \varepsilon \phi^2) F(\phi).$$

The simplification is that the partition function will contain only Gaussian-type integrations, calculable in closed form. The effective Hamiltonian for Villain's model is thus,

$$H_{\text{eff}} = \frac{1}{2} \sum_{i,j} A(\phi_i - \phi_j - 2\pi n_{ij})^2 + B \sum_i (\phi_i - 2\pi \nu_i)^2 \tag{7.278}$$

in which the $n_{ij} = -n_{ji}$ and the ν_i are discrete which supplement the ϕ_i. Pursuing this, Villain obtained the properties of the planar model with a

---

[58] J. Villain, *J. Physique* **36**, 581 (1975).

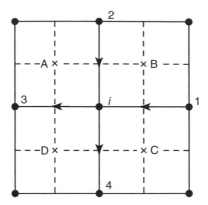

Fig. 7.22.   Construction of a dual lattice $(A, B, C, \ldots)$ to satisfy curl-free condition on lattice $(i, 1, 2, \ldots)$.

critical temperature half that in (7.275), more in line with series estimates but still too high.

This approach was taken an important step further by José *et al.*,[59] who found a neat trick for performing sums over integers n_{ij} in the partition function, and were then enabled to derive and rederive all known results by reasonably straightforward calculations. Their slight modification of Villain's expansion results in complex exponentials and reduces the evaluation of Z in zero external field to

$$Z = \int_{-\infty}^{+\infty} \prod d\phi_i \prod_{ij} \sum_{n_{ij}=-\infty}^{+\infty}$$

$$\times \exp\left\{\sum_{i,j}[F(n_{ij}) + in_{ij}(\phi_i - \phi_j)]\right\}. \tag{7.279}$$

The ϕ_i's are now eliminated. One obtains a partition function purely in terms of integer quantum numbers, which we can then identify with the vortices. Each integration over a ϕ_i forces the integers with which it interacts to add to zero, a form of momentum conservation. Let us examine this with the aid of Fig. 7.22, in which we label the neighbors of a given site i by 1, 2, 3, 4.

Integrating over ϕ_i alone, we obtain zero unless $n_{1i} + n_{2i} - n_{i3} - n_{i4} = 0$. We represent these integers by arrows, and vertify that 2 arrows go in, and

---

[59] J. José, L. Kadanoff, S. Kirkpatrick and D. Nelson, *Phys. Rev.* **B16**, 1217 (1977).

two go out of the vertex at i. This discrete version of a zero divergence condition can be satisfied by the choice of a dual field of integers n_i, located within each plaquette of the original lattice as shown in the figure. Then setting

$$n_{i1} = n_B - n_C, \quad n_{i2} = n_A - n_B,$$
$$n_{i3} = n_A - n_D \quad \text{and} \quad n_{i4} = n_C - n_D,$$
(7.280)

around i and making similar identifications throughout, the desired conditions are met identically for every point in the original lattice, and one obtains a partition function over the integers of the dual lattice. Finally, one makes use of the Poisson summation formula for each integer n_A:

$$\sum_{n=-\infty}^{+\infty} g(n) = \sum_{m=-\infty}^{+\infty} \int_{-\infty}^{+\infty} d\phi\, g(\phi) \exp(-2\pi i m\phi)$$
(7.281)

transforming Z into the form:

$$Z = \prod_i \sum_{m_i=-\infty}^{+\infty} \int_{-\infty}^{+\infty} d\phi_i \left\{ \exp\left[\sum_{i,j} F(\phi_i - \phi_j) + \sum_j 2\pi i m_j \phi_j \right] \right\},$$
(7.282)

summing over the dual lattice points. In this representation, the ϕ_j's describe the magnon modes, and the m_j's the vortex quantum numbers. Especially at low temperatures, the sums converge quickly and the integrals (F is not a very complicated function) can be performed by a variety of methods. The results of Villain are obtained, but with a renormalized J_{eff}. We summarize them as follows.

The spin-spin correlation function decreases non-exponentially, as a power law

$$\langle \cos(\theta_0 - \theta_r) \rangle = \left(\frac{a}{r} \right)^{\eta(T)} \quad \text{with}$$
(7.283)

$$0 \leqslant \eta(T) = kT/2\pi J_{\text{eff}} \leqslant \frac{1}{4},$$
(7.284)

the phase transition occurring at $kT_c = \frac{1}{2}\pi J_{\text{eff}}(T_c)$. Above T_c the decay is exponential with a finite correlation length $\xi(T) \approx C \exp[b/(T - T_c)^\nu]$ (with $\nu \approx 0.76$[60] or $\nu = 0.5$,[57] C and b being constants; the magnetic susceptibility χ is proportional to ξ at $T \gtrsim T_c$ and both are essentially infinite for $T \leqslant T_c$.[60]

---

[60] J. Tobochnick and G.V. Chester, *Phys. Rev.* **B20**, 3761 (1979).

This behavior differs considerably from models with LRO below T_c. Because exponents are continuously varying up to T_c, one describes the phase transition as a "line of critical points from 0 to T_c," rather than as a unique phase transition at T_c. The main surprise is that $\eta_c = 1/4$, far from the theoretical maximum $\eta = 2$ derived in (7.220), but the same as for the Ising model in 2D. The power-law behavior is precisely what one expects from free magnons with dispersion $\omega(q) = \frac{1}{2}J_{\text{eff}}q^2$, as we now show.

$$\langle\cos(\theta_0 - \theta_r)\rangle = 1 - \frac{1}{2}\langle(\theta_0 - \theta_r)^2\rangle + \cdots$$

$$\approx \exp\left[-\frac{1}{2}\langle(\theta_0 - \theta_r)^2\rangle\right]$$

$$= \exp[-G(r)] \qquad (7.285)$$

with

$$G(r) = \frac{kT}{J_{\text{eff}}N}\sum_q(1 - e^{iq\cdot r})q^{-2} \approx \frac{kT}{2\pi J_{\text{eff}}}\ln(r/a) \qquad (7.286)$$

from which (7.283) follows.

These properties and the specific heat were examined numerically by Monte Carlo simulation (which generally confirms the theoretical picture). We reproduce the numerical experiments on the specific heat by Tobochnik

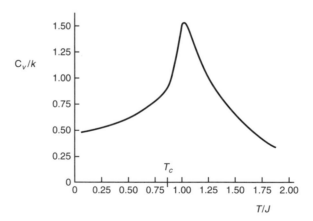

Fig. 7.23. Specific heat in plane rotator model in 2D, as obtained by numerical (Monte Carlo) experiments. (—) replaces experimental points published by Tobochnick and Chester[60] to better show some features: maximum near $kT = 1.02$ J, smooth phase transition at $kT_c = 0.89$ J.

and Chester[60] in Fig. 7.23. The interesting features are a specific heat maximum at $kT = 1.02$ J, with the phase transition occurring at the lower temperature $kT_c = 0.89$ J. There appear to be no specific heat anomalies at T_c, (defined as the temperature at which the susceptibility diverges).

Is it possible that, finally, a simple theory can yield this plethora of interesting results? In the following section we prepare for this contingency, by deriving the transfer matrix in a rather transparent form.[61]

7.18.   Transfer Matrix of Plane Rotator Model

The evaluation of the partition function of the 2D or 3D plane rotator model is reducible to an eigenvalue problem, that of obtaining the largest eigenvalue of the transfer matrix. We derive this matrix in simplest possible form,[61] but its complete resolution is another matter![62]

Let us illustrate with two connected chains, to see what needs be generalized in the earlier formulation of the one-dimensional problem. Label the coordinates on the first chain θ_n, and on the second, ϕ_n. Then,

$$H = \sum_n^N H_{n,n+1} \quad \text{where} \tag{7.287}$$

$$H_{n,n+1} = -J_1[\cos(\theta_n - \theta_{n+1}) + \cos(\phi_n - \phi_{n+1})] - J_2 \cos(\theta_{n+1} - \phi_{n+1})$$
$$\qquad\quad Chain\ 1 \qquad\qquad\quad Chain\ 2 \qquad\qquad\quad Interchain$$

$$\tag{7.288}$$

in which we distinguish horizontal (intrachain) bonds J_1 from vertical (interchain) bonds J_2 for greater generality. The coupling constants J_1 and J_2 can be set equal at the end of the calculation, if desired.

Writing Z as an ordered sequence of integrations — similar to (7.255) — find it is given by $Z = z_0^N$, with z_0 the maximal eigenvalue of the integral equation:

---

[61] The first attempt at a transfer matrix in this problem seems to be that of A. Luther and D. Scalapino, *Phys. Rev.* **B16**, 1153 (1977), who truncate it into the form of a $s = 1$ chain in an external field: $\Delta(S_i^z)^2$. Such chains have been most recently studied by S.T. Chui and K.B. Ma, *Phys. Rev.* **B29**, 1287 (1984), and it seems almost certain they *can* yield the general features of the plane rotator model. However, the derivation of the transfer matrix in the present book seems to be the first rigorous one.

[62] The procedure: mapping of the transfer matrix onto the $s = 1/2$ anisotropic Heisenberg antiferromagnetic linear chain, was originally published by D. Mattis, *Phys. Lett.* **104**, 357 (1984).

$$\frac{1}{\pi^2} \iint_{-\pi}^{+\pi} d\theta_n d\phi_n \exp(-\beta H_{n,n+1}) \chi(\theta_n, \phi_n)$$

$$= z_0 \chi(\theta_{n+1}, \phi_{n+1}). \tag{7.289}$$

The transfer matrix $V_{n,n+1} = 4\exp(-\beta H_{n,n+1})$ is a product $V^{(1)}V^{(2)}$, where

$$V^{(1)}_{n,n+1} = 4I_0^2(\beta J_1) \sum_{p,q=-\infty}^{+\infty} \exp\{-\beta[w(p) + w(q)]\} \exp[ip(\theta_n - \theta_{n+1})]$$

$$\times \exp[iq(\phi_n - \phi_{n+1})] \tag{7.290}$$

represents the two individual chains, according to (7.259), with

$$w(p) \equiv kT \ln[I_0(\beta J_1)/I_p(\beta J_1)] \tag{7.291}$$

defined for convenience; note s is the subscript on rhs. $V^{(2)}$ contains the interchain links:

$$V^{(2)}_{n,n+1} = \exp[\beta J_2 \cos(\theta_{n+1} - \phi_{n+1})]. \tag{7.292}$$

Performing the integration indicated in (7.289) we obtain the eigenvalue equation in the form of an exponentiated differential equation:

$$4I_0^2(\beta J_1) \exp\{-\beta[w(p_\theta) + w(p_\phi)]\} \exp[\beta J_2 \cos(\theta - \phi)]\chi = z_0 \chi \tag{7.293}$$

with $p_\theta = -i\partial/\partial\theta$ and $p_\phi = -i\partial/\partial\phi$. This can be simplified further, writing $z_0 = 4I_0^2(\beta J_1) \exp(-\beta f)$, with f (the "interaction free energy") the new eigenvalue. As usual, we want only the *lowest* f. The above equation now reads:

$$\exp\{-\beta[w(p_\theta) + w(p_\phi)]\} \exp[\beta J_2 \cos(\theta - \phi)]\chi(\theta, \phi) = \exp(-\beta f)\chi(\theta, \phi). \tag{7.294}$$

Note. To fix ideas, we discuss the behavior of $w(p)$. From series and asymptotic expansions we can analyze $w(p)$ at small $\beta J_1 = K_1$ and also at large K_1. In the vicinity of the 2D phase transition, K tends to be $O(1)$ so there the exact definition of the I_p must be used in (7.295).

High temperature $(K_1 \to 0)$: $I_p(K_1) \approx (\frac{1}{2}K_1)^p/\Gamma(p+1)$, thus

$$w(p) = kT|p|[\ln(2kT|p|/eJ_1)]. \tag{7.295a}$$

This formula is also applicable at large $|p| \gg 1$, for arbitrary K_1.

Low temperature ($K_1 \to \infty$): the asymptotic expansions[63] again lead to (7.295a) for large $|p|$, but at small $|p|$ we find:

$$w(p) = \frac{1}{2}(kT)^2 p^2 / J_1 \quad \text{instead}. \tag{7.295b}$$

In general, $w(p)$ is a smoothly increasing function of $|p|$, with $w(0) = 0$.

We would like now to solve the eigenvalue problem. First, for some insight into the nature of the solutions, we ignore the fact that the exponents of $\mathbf{V}^{(1)}$ and $\mathbf{V}^{(2)}$ do not commute, and simply go ahead combine them, and solve:

$$[w(p_\theta) + w(p_\phi) - J_2 \cos(\theta - \phi)]\chi = f\chi \tag{7.296}$$

for its ground-state eigenvalue. There is no proof that this leads to an accurate solution, but the equivalent approximation in the Ising model transfer matrix fortuitously yields excellent results. Because the interaction conserves angular momentum, we may write $\chi(\theta, \phi)$ in the form:

$$\chi(\theta, \phi) = \exp[iQ(\theta + \phi)] \sum_k A_k \exp[ik(\theta - \phi)] \tag{7.297}$$

and set $Q = 0$ in the ground state. The eigenvalue equation now takes the form of a difference equation

$$[2w(k) - J_2 - f]A_k = \frac{1}{2}J_2(A_{k+1} + A_{k-1} - 2A_k), \tag{7.298}$$

reasonably simple equation to analyze (it is the equation of a particle hopping on a linear chain labeled by integers k, subjected to a potential $2w(|k|) - J_2$ which has only bound states). Certainly, f and its derivatives are continuous in the parameters J_1, J_2 and T, and there can be no phase transition. This is expected as there are only *two* chains.

The solution by means of (7.297) of the *exact* equation (7.295) takes on a more formidable mien. It is convenient to left-multiply both sides of the equation by $\exp \beta[w(p_\theta) + w(p_\phi)]$, and express $\exp(\beta J_2)$ in a series expansion involving the Bessel functions. We obtain

$$I_0(\beta J_2) \sum_{q=-\infty}^{+\infty} \exp[-\beta \hat{w}(q)]A_{k-q} = \exp\{-\beta[f - 2\hat{w}(k)]\}A_k \tag{7.299}$$

---

[63] M. Abramowitz and I. Stegun (eds.), *Handbook of Math. Functions* (National Bureau of Standards, Washington, 1964), Secs. 9.6, 9.7.

with

$$\hat{w}(q) = \hat{w}(-q) = kT\ln[I_0(\beta J_2)/I_q(\beta J_2)] \tag{7.300}$$

by analogy with (7.291). This equation is of a less familiar type, but analysis of it indicates that the solution is even smoother function of the parameters and that f agrees with the solution of the approximate equation (7.298) both at low and at high temperatures.

Without entering the domain of numerical analysis of the above, we now turn to the *full* 2D transfer matrix. Its formulation is a simple extension of the preceding. We define the nth line (2D) [or nth plane (3D)], to contain all the spins at the nth position of the linear chains. They are labeled θ_i (in the example of the two chains, $\theta_1 = \theta$ and $\theta_2 = \phi$) with $i = 1, 2, \ldots, N'$. θ_i interacts only with $\theta_{i\pm1}$, its nearest-neighbors on this line. (In 3D, *two* subscripts label a spin in the plane of the transfer matrix, and again, only nearest-neighbors interact.) As $N' \to \infty$, a phase transition is allowed. Now, we factor $2^{N'} I_0^{N'}(\beta J_1)$ from the largest eigenvalue,

$$Z = z_0^N, \quad \text{and} \quad z_0 = 2^{N'} I_0^{N'}(\beta J_1)\exp(-\beta f N') \tag{7.301}$$

in which form f is the interaction free energy *per* spin. The eigenvalue equation is now:

$$\mathbf{V}\chi(\theta_1, \theta_2, \ldots, \theta_{N'}) = \exp(-\beta f N')\chi(\theta_1, \theta_2, \ldots, \theta_{N'}) \tag{7.302}$$

where $\mathbf{V} = \mathbf{V}^{(1)}\mathbf{V}^{(2)}$, and

$$\mathbf{V}^{(1)} = \exp\left[-\beta\sum_{n=1}^{N'} w\left(\frac{1}{i}\partial/\partial\theta_n\right)\right] \equiv \exp[-\beta H^{(1)}] \tag{7.303a}$$

and

$$\mathbf{V}^{(2)} = \exp\left[\beta J_2\sum_{n=1}^{N'}\cos(\theta_n - \theta_{n+1})\right] \equiv \exp[-\beta H^{(2)}] \tag{7.303b}$$

in 2D. [In 3D the sum in (a) ranges over the 2D plane of the transfer matrix, and in (b) over the nearest-neighbor pairs in that plane.] The solution of (7.302) yields the *exact* thermodynamic properties of the model. Once again, it is instructive to examine the simpler model eigenvalue problem first:

$$[H^{(1)} + H^{(2)}]\chi = (fN')\chi \quad \text{(see definition of the H's above)} \tag{7.304}$$

which is recognizable as a linear chain of coupled pendulums, subject to zero-point motion induced by the kinetic energy $w(p_i)$. The kinetic energy is the expression in $w(p)$; the higher the physical parameter T, the higher the "kinetic energy" parameter in the present equation. Thus, as temperature is raised, the equivalent of the uncertainty principle causes the pendulums to become misaligned. At high T each pendulum is approximately homogeneously distributed over the interval $0 - 2\pi$, and is essentially uncorrelated to the position of its neighbors. At T_c the correlation to nearest-neighbor begins, until at $T = 0$ all the pendulums are at the same angle $\theta_1 = \theta_2 = \cdots = \theta_{N'}$ which we can take to be $\theta = 0$. The value of T_c found by solving (7.304)[62] satisfies:

$$(2kT_c/J)\ln(2kT_c/J) = 1 \qquad (7.305)$$

which yields: $kT_c = 0.88$ J, in good agreement with the numerical "experiments"[61] for 2D.

The actual manipulations that result in the formula above, involve mapping the operators in the exponents onto those of an anisotropic one-dimensional Heisenberg model. The ground state of this particular anisotropic model has been studied extensively, and is by now well understood. It yields almost all the thermodynamic information one normally requires: correlation functions, heat capacity, etc., once the connection between anisotropy parameter and T is made. The details are too cumbersome for the present text but can be found in the author's original publication.[62]

This is but one example among many, where the calculation of thermodynamic properties of a classical system in 2D have been transformed into the calculation of ground state properties of a quantum model in 1D thanks to the transfer-matrix approach. Because of the enormous attention directed at one-dimensional quantum models in recent years, an equal number of two-dimensional models of statistical mechanics have thereby been reduced to quadrature. The correspondence works both ways. For example, where conformal invariance has resulted in understanding some phase transition in 2D it has (by extension) been used to solve the corresponding *one*-dimensional quantum field theories (also called *two*-dimensional if one considers the added dimension of time.) The curious reader is directed to an excellent compendium on the uses of the conformal group in 2D,[64] which includes, but also transcends, the study of correlation functions in magnetic systems and of magnetic phenomena.

---

[64] *Conformal Invariance and Applications to Statistical Mechanics*, C. Itzykson, H. Saleur and J.-B. Zuber, eds., World Scientific, 1988.

Chapter 8

The Ising Model

Ising's model may be one of the simplest, yet most thoroughly investigated topics in all of theoretical physics. Its applications are much wider than one might at first suspect: fire-fighting, politics, etc. In this prototype theory of ferromagnetism — and of many other physical phenomena as well — a spin $S_i = \pm 1$ is assigned to each of N sites on a fixed lattice. The spins, which live on the vertices of the lattice, interact with one another by means of bonds (the links of the lattice). These have strengths J_{ij} in energy units. In addition, the spins can interact with external fields B_i of arbitrary strengths. The total energy is then given by:

$$H = - \sum J_{ij} S_i S_j - \sum B_i S_i$$

and can be directly evaluated in any of the 2^N spin configurations. In the most familiar version of the Ising model, the interactions are limited to nearest-neighbors on the lattice and the magnetic field is homogeneous, $B_i = \text{constant}$.

A historical review of this model is of some interest, as it illustrates the frequently devious paths along which research proceeds in such matters. If the reader's appetite for history is whetted, he will find a rather more thorough account in Brush's review.[1]

While the idea of a microscopic theory in which elementary spin dipoles are constrained to quantized directions dates back to Lenz in 1920,[2] the calculational burden fell upon his student Ising some years later. But Ising's

---

[1] S. G. Brush, The history of the Lenz-Ising model, *Rev. Mod. Phys.* **39**, 883 (1967).
[2] W. Lenz, *Phys. Z.* **21**, 613 (1920).

Ph.D. thesis and later publication[3] was not very encouraging. Based as it was on a one-dimensional analysis, it revealed no ferromagnetic phase above the absolute zero of temperature. The ground state was ordered: all spins parallel (all up or all down), yet the long-range order disappeared at any finite positive temperature. Thus, despite the nearest-neighbor bonds, this model displayed only paramagnetic behavior at finite T. Concerning his thesis, Ising has written:

> At the time...Stern and Gerlach were working in the same institute (Hamburg) on their famous experiment on space quantization...I discussed the results of my paper widely with Prof. Lenz and Dr. Wolfgang Pauli, who at that time was teaching in Hamburg. There was some disappointment that the linear model did not show the expected ferromagnetic properties.[1]

As a consequence, interest in Lenz' model died out, while other avenues of inquiry were opened, such as Heisenberg's fully quantum-mechanical model and the itinerant electron theories of Bloch and, later, Stoner. For it was commonly held that Lenz' semiclassical model was incapable of explaining the known laws of ferromagnetism, including the phase transition at the Curie temperature.

Ultimately, the study of what came to be known as Ising's model was revived by a parallel development in alloy physics. The Bragg-Williams molecular-field theory of the thermodynamics of alloys, patterned after Weiss, was justified by cluster calculations by Bethe[4] on a model which was mathematically, if not physically, identical to that of Lenz: If in the theory of an AB alloy, one allows spin up to refer to an atom of type A and spin down to an atom B, arranging the magnitudes of the bonds J_{ij} and of the external field B so as to correspond to realistic AA, AB and BB bonding energies and chemical potentials, one finds a perfect correspondence to the Ising model in an external field. The solution of one entails the other. This approach to the structure of alloys is incorporated into metallurgy theory, appearing already in the 1939 text *Statistical Thermodynamics* of Fowler and Guggenheim.[5] In an alternative application, the B's can represent vacancies in a solid of

---

[3] E. Ising, *Z. Physik* **31**, 253 (1925).

[4] H. Bethe, *Proc. Roy. Soc.* (*London*) **A150**, 552 (1935); *J. Appl. Phys.* **9**, 244 (1938). Also, F. Cernuschi and H. Eyring, *J. Chem. Phys.* **7**, 547 (1939).

[5] R. Fowler and E. Guggenheim, *Statistical Thermodynamics*, Cambridge Univ. Press, Cambridge, 1939, Chap. 13.

A's. When sufficiently dilute, the latter form a lattice gas, which may condense at sufficiently high density or low temperature. Thus, the study of the Ising model contains the seeds of a first-principles theory of the phases of nonmagnetic matter: solid, liquid and gas.

The exact phase diagram of the lattice gas was first obtained by Lee and Yang in two remarkable papers,[6] based on the large amount of information then known about the Ising model. The equation of state they obtained gave the mixed phase region correctly without invoking Maxwell's construction — which had previously been thought to be unavoidable. Lee and Yang also drew attention to the importance of extending the model parameters, such as the external field, into the complex plane. Their study of the zeros of the partition function in complex fields created a new tool for the study of phase transitions.

However, Lee and Yang were not the first to treat the Ising model of magnetism (and the lattice-gas problem) within a unified formalism. Right after Bethe's work,[4] Peierls recognized that ferromagnetism, order-disorder transformations in alloys, and the very existence of phase transitions in microscopic theory could all be reduced to the examination of the Ising model in two, or three dimensions — but *not* in one![7] Peierls contributed an important theorem that in two dimensions or higher, long-range order (LRO) persists at sufficiently low temperature. Conversely, LRO is known to disappear above some critical temperature, when there is sufficient thermal disorder. Thus, Peierls established the relevance of the Ising model to the Curie problem. In the technicalities of his proof of this theorem, Peierls erred and was ultimately corrected by Griffiths[8] some 28 years later.

The exact solution of the thermodynamic properties of the two-dimensional Ising model was the consequence of innovative research by several individuals, working out ideas that culminated in the exact

[6] C. N. Yang and T. D. Lee, *Phys. Rev.* **87**, 404 (1952); T. D. Lee and C. N. Yang, *Phys. Rev.* **87**, 410 (1952). Recent work on the complex zeros includes: E. Marinari, *Nucl. Phys.* **B235** (FS11), 123 (1984), 3D Ising model; K. De'Bell and M. L. Glasser, *Phys. Lett.* **104A**, 255 (1984), Cayley tree; W. Saarloos and D. Kurtze, *J. Phys.* **A17**, 1301 (1984), Ising model; A. Caliri and D. Mattis, *Phys. Lett.* **106A**, 74 (1984), long-range model of (2.7.2) with $J_0 \gtrless 0$.

[7] R. Peierls, *Proc. Camb. Phil. Soc.* **32**, 477 (1936).

[8] R. Griffiths, *Phys. Rev.* **136**, A437 (1964). See further corrections and extension in C.-Y. Weng, R. Griffiths and M. Fisher, *Phys. Rev.* **162**, 475 (1967).

calculation by Onsager[9] of the partition function of this model (in the absence of an external field). In 1942, Kramers and Wannier,[10] assuming the existence of a unique phase transition, located T_c *exactly* by means of a duality transformation they discovered. Such a transformation maps the high- and low-temperature properties of the partition function into one another; the fixed point of this mapping is then T_c. These authors, invoking what is now termed "finite-size scaling" arguments, derived a logarithmically divergent specific heat anomaly at T_c, quite unlike anything which has been obtained heretofore in statistical mechanics. But it was Onsager[9] who, in 1944, provided the exact calculations. The tale continues in the following pages. We terminate these prefatory remarks by noting that the prospects of solving the two-dimensional Ising model in an external magnetic field, or of solving the three-dimensional Ising model, remain as tantalizingly elusive today as they were in 1925 despite an impressive array of methods and theorems dedicated to these tasks: a clear challenge to the present reader!

8.1. High Temperature Expansions

In general, high-temperature series can be generated for *any* spin Hamiltonian by straightforward expansion of the exponential in a Taylor series:

$$Z_N(\beta) = Z_N(0) \left\langle 1 - \beta \mathscr{H} + \frac{1}{2} \beta^2 \mathscr{H}^2 - \cdots \right\rangle_0 \qquad (8.1)$$

where $\langle \ \rangle_0$ stands for an unweighted average over all spin parameters, or equivalently, for a thermal average at $\beta = 0$. But regardless of how small β may become (or, however, high the temperature) each succeeding term is one order of N greater than the preceding so that the series clearly diverges in the thermodynamic limit! This difficulty is resolved by noting that Z_N must be of the form $\exp(-\beta F)$, where $F = fN$ and f is independent of N. Thus the right-hand side must *also* be precisely exponential in N. (Contributions which do not lead to an extensive F but are, say, higher-order in N would destroy the existence of a thermodynamic limit, while terms lower-order in N are surface or edge terms, that do not survive the thermodynamic limit.)

---

[9] L. Onsager, *Phys. Rev.* **65**, 117 (1944), algebraic formulation; B. Kaufman, *Phys. Rev.* **76**, 1232 (1949), spinor reformulation; L. Onsager, *Nuovo Cimento (Suppl.)* **6**, 261 (1949), spontaneous magnetization; C. N. Yang, *Phys. Rev.* **85**, 809 (1952) gave the first derivation of Onsager's formula for magnetization *in the literature*; Onsager had previously announced it!
[10] H. Kramers and G. Wannier, *Phys. Rev.* **60**, 252, 263 (1941).

Let us test this by expanding $\log Z_N$, suitably regrouped into contributions for each power of β [NB: in our notation $\log \equiv \ln$ is *always* base e]:

$$\log Z_N(\beta) = \log Z_N(0) - \beta[\langle\mathscr{H}\rangle_0] + \frac{1}{2}\beta^2[\langle\mathscr{H}^2\rangle_0 - \langle\mathscr{H}\rangle_0^2]$$

$$-\frac{1}{3!}\beta^3[\langle\mathscr{H}^3\rangle_0 - 3\langle\mathscr{H}^2\rangle_0\langle\mathscr{H}\rangle_0 + 2\langle\mathscr{H}\rangle_0^3]$$

$$+\frac{1}{4!}\beta^4[\langle\mathscr{H}^4\rangle_0 - 4\langle\mathscr{H}^3\rangle_0\langle\mathscr{H}\rangle_0 - 3\langle\mathscr{H}^2\rangle_0^2 + 12\langle\mathscr{H}^2\rangle_0\langle\mathscr{H}\rangle_0^2$$

$$-6\langle\mathscr{H}\rangle_0^4] - \cdots + \frac{1}{m!}(-\beta)^m[\ ]_m + \cdots \qquad (8.2)$$

The brackets $[\ ]_m$ are called *mth-order cumulants*. For finite-ranged interactions J_{ij}, each cumulant may be verified to be $O(N)$, so that the summation yields a free energy which *is* extensive, term by term. The proof is simple. The powers of β are linearly independent, hence their coefficients must each be *individually extensive*. Through comparison with the above series, one can develop a formula for the cumulants:

$$[\ ]_m = m! \sum_{\{n\}_m} (-1)^{M-1}(M-1)! \prod_{i=1}^{m} \frac{1}{n_i!}\left(\frac{1}{i!}\langle\mathscr{H}^i\rangle_0\right)^{n_i} \qquad (8.3)$$

in which $\{n\}_m$ stands for a distinct set of non-negative integers n_i satisfying the condition $\sum_i i n_i = m$; within each set, $M \equiv \sum_i n_i$. The sum in (8.3) is over all such sets.

Problem 8.1.   Use the above formula to generate the given cumulants, (8.2), and a new one: $[\ ]_5$.

Exercise: As a simple application of the above, assume each $S_i = \pm1$ (spins one-half) and let the magnetic field be B, a constant. We calculate F to $O(T^{-2})$ for arbitrary J_{ij}. Because $\langle S_i\rangle_0 = 0$, terms involving single spins vanish in all orders, whereas $S_i^2 = 1$. With these simplifications, we obtain:

$$F = -kTN\ln 2 - \frac{1}{2}\beta\left(\frac{1}{2}\sum_{ij}J_{ij}^2 + NB^2\right) - \frac{1}{6}\beta^2\left(\sum_{ijk}J_{ij}J_{jk}J_{ki} + 6B^2\sum_{ij}J_{ij}\right)$$

$$+\cdots \qquad (8.4)$$

The calculation of the induced magnetization yields some plausible results. Differentiating the above series term by term, with the Boltzmann constant

k and the Curie constant $\mathbb{C}_{1/2}$

$$\mathscr{M} = -k\mathbb{C}_{1/2}\partial F/\partial B = N\mathbb{C}_{1/2}B\frac{1}{T}\left(1 + \frac{2}{kT}\frac{1}{N}\sum_{ij}J_{ij} + \cdots\right)$$

which, to the given order, is the same as

$$= N\mathbb{C}_{1/2}B\frac{1}{T}\left(1 - \frac{2}{kT}\frac{1}{N}\sum_{ij}J_{ij} + \cdots\right)^{-1}$$

$$= \frac{N\mathbb{C}_{1/2}B}{T - \theta}. \tag{8.5}$$

Here, $\theta = k^{-1}\frac{2}{N}\sum J_{ij}$ is the same as T_c in the MFT if the bonds J_{ij} are ferromagnetic (positive) on the average; but it is $-T_N$, the negative of the Néel temperature, if they are antiferromagnetic on average. It is amusing to recover this simple formula from the leading terms in the high-temperature expansion. Unfortunately, it is the asymptotic terms in the expansion that determine the behavior near T_c or θ, so the above is valid only for high temperatures, $T \gg \theta$.

Problem 8.2.   Decomposing an arbitrary classical Hamiltonian into two parts:

$\mathscr{H} = \mathscr{H}_0 + \delta\mathscr{H}$, where $\mathscr{H}_0$ and $\delta\mathscr{H}$ commute, define $Z_0 = \mathrm{Tr}\{\exp(-\beta\mathscr{H}_0)\}$ and $\langle\delta\mathscr{H}\rangle_0 \equiv Z_0^{-1}\mathrm{Tr}\{\delta\mathscr{H}\exp(-\beta\mathscr{H}_0)\}$.

Show that the total free energy $F = -kT\log\mathrm{Tr}\{\exp(-\beta\mathscr{H})\}$ is given by a series similar to (8.2):

$$F = -kT\log Z_0 + \langle\delta\mathscr{H}\rangle_0 - \frac{kT}{2!}\beta^2[\langle(\delta\mathscr{H})^2\rangle_0 - \langle\delta\mathscr{H}\rangle_0^2] + \cdots$$

$$- \frac{kT}{m!}(-\beta)^m[\ ]_m \cdots$$

where $[\ ]_m$ is given by (8.3) with H^i replaced by $(\delta\mathscr{H})^i$ using weighted averages using $\frac{1}{Z_0}\exp(-\beta\mathscr{H}_0)$. This perturbative expansion may be useful in cases where $\delta\mathscr{H}$ is complicated but small by comparison with $\mathscr{H}_0$, provided Z_0 and $\langle\ \rangle_0$ are readily calculable.

Problem 8.3.   Use the above to derive the following identity for the zero-field susceptibility of an arbitrary system in an external field $B \to 0$:

$$\chi_0 = \frac{1}{kT}(\langle(\partial\mathcal{H}/\partial B)^2\rangle_0 - \langle\partial\mathcal{H}/\partial B\rangle_0^2).$$

Compare with formulas given earlier in the book.

Although the equations above apply to arbitrary Hamiltonians in which the various parts commute, and for arbitrary spin magnitudes, there is considerable advantage and simplification in specializing to an Ising model, spins $\frac{1}{2}(S_i = \pm 1$; thus $S_i^2 = 1, S_i^3 = S_i$, etc.). The exponentials in Z_N are easily resummed:

$$\exp(\beta J_{ij} S_i S_j) = \cosh\beta J_{ij} + S_i S_j \sinh\beta J_{ij}$$

and

$$\exp(\beta B_i S_i) = \cosh\beta B_i + S_i \sinh\beta B_i. \tag{8.6}$$

Z_N is seen to be a polynomial in quantities denoted w:

$$w_{ij} = \tanh\beta J_{ij} \quad \text{and} \quad w_i = \tanh\beta B_i, \tag{8.7}$$

after factorization of the related $\cosh\beta J_{ij}$ and $\cosh\beta B_i$ which, by themselves, are the leading contribution to the high-temperature partition function.
Explicitly:

$$Z_N = \left(\prod_{ij}\cosh\beta J_{ij}\right)\left(\prod_i\cosh\beta B_i\right)$$

$$\times \text{Tr}\left\{\prod_{ij}(1 + S_i S_j w_{ij})\prod_n(1 + S_n w_n)\right\}. \tag{8.8}$$

The trace in this expression yields a power series in the w's:

$$\text{Tr}\{\ \} = 2^N\left(1 + \sum_{ij}w_{ij}w_iw_j + \cdots\right) \tag{8.9}$$

which converges better than the original series (8.1). Even when βJ_{ij} or βB_i are large the w's can never exceed magnitude unity. We note that at finite N, the series (8.9) has a finite number of terms, unlike (8.1) which is always an infinite series. Being a polynomial in the w's, Z_N is *analytic* in β, hence cannot exhibit a phase transition. To study the expected phase transition at or near T_c, it will be necessary to *first* take the limit $N \to \infty$, then to calculate the contributions of asymptotically (arbitrarily high) powers of the w's.

The evaluation of Z_N appears at first sight to be as complicated in 1D as it is in 2D or 3D. However, a plethora of tricks enable us to solve the 1D problem trivially. Another variety of methods can be brought to bear on the substantially more difficult 2D problem, but can solve it only in vanishingly small external field. For the 2D model in external field, or in 3D, our only known recourse is to numerical analysis and approximate methods. For these, graph theory has provided a powerful and suggestive procedure, to which we turn briefly.

8.2. Graph Theory

If the bonds J_{ij} connecting the N spins S_i are reasonably long-ranged the nature of the space lattice on which the spins are disposed is of secondary importance although the dimensionality will, in general, matter. For short-ranged forces the topology of the lattice can play a role, especially for antiferromagnetic couplings. For this reason, we distinguish two-dimensional structures: sq (short-hand for simple quadratic or square), triangular and honeycomb (denoted T and H), and the three-dimensional structures: sc (simple cubic), bcc (body-centered), fcc (face-centered), hexagonal, and other frequently encountered lattice types.

For present purposes let us extinguish the external field and allow *only* *nearest-neighbor* bonds to differ from zero. Assigning to them a strength J, which can be positive (i.e., ferromagnetic) or negative (AF), $w = \tanh \beta J$ becomes the sole parameter in the series expansion (8.9). The trace eliminates any single spins. To appear twice, each spin must belong to two distinct, neighboring, bonds. Thus, among the objects contributing to Z_N, there can appear only closed figures — polygons of various sorts, or combinations thereof.

If all B_i are set to zero, Z_N takes the form

$$Z_N = 2^N (\cosh \beta J)^{1/2Nz} \left[1 + \sum_{r=1}^{1/2Nz} p(r) w^r \right]. \tag{8.10}$$

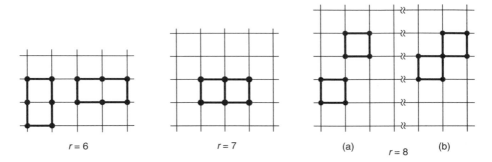

$r = 6$ $r = 7$ (a)   (b)
                                          $r = 8$

Fig. 8.1. Typical diagrams with $r = 6$ and 8 links which contribute to high-temperature expansion (8.10). The contributions of disconnected loops (a) differs from that of connected loops (b). A "connection" such as illustrated for $r = 7$ does not contribute, as shown in Problem 8.4.

In this expression, z denotes the coordination number of the lattice ($z = 2d$ for the simple lattices in d dimensions such as linear chain, sq, sc, etc.). The total number of links is $\frac{1}{2}Nz$. Also, $p(r)$ is the number of independent closed figures that can be drawn on the given lattice using r links once each.

In 1D, there are no closed circuits except if we adopt periodic boundary conditions (PBC). For free ends, all $p(r) = 0$. With PBC, all $p(r)$ vanish except for $r = N$, where $p(N) = 1$. Thus the partition function depends on boundary conditions as follows:

$$1\mathrm{D}: \quad Z_N = 2(2 \cosh \beta J)^{N-1} \quad \text{(free ends)}$$

$$= (2 \cosh \beta J)^N + (2 \sinh \beta J)^N \quad \text{(PBC)}. \tag{8.11}$$

On the two-dimensional sq lattice, $p(1) = p(2) = p(3) = 0$. The lowest order nonvanishing polygon is the square with 4 sides of minimum length. As there are N such polygons, $p(4) = N$. The next nonvanishing order is $r = 6$, of which the $2N$ contributions are illustrated in Fig. 8.1. The vanishing contribution of $r = 7$ polygons is the subject of the following problem, thus the next nontrivial contribution comes at $r = 8$ for which two distinct diagrams must be considered: disconnected squares, which can be placed at $\frac{1}{2}N(N-9)$ locations, and connected squares, of which there are $2N$ distinct contributions.

Problem 8.4.   A 7-sided polygon is illustrated in Fig. 8.1. Prove that $p(7) = 0$, using the observation that in $p(r)$ each bond occurs only once but each spin must appear twice.

Thus, the expansion for the sq lattice is:

2D, sq : $Z_N = (2 \cosh \beta J)^{2N}$

$$\times \left[1 + Nw^4 + 2Nw^6 + 2Nw^8 + \frac{1}{2}N(N-9)w^8 + \cdots \right]$$

(8.12)

a mixed series in powers of w and N. As Z_N must be of the form $(Z_1)^N$, we obtain the w-dependence of $Z_1 = \exp(-\beta f)$, by extracting the Nth root of the above.

In 1D,

$$1D : \quad Z_1 = 2 \cosh \beta J, \tag{8.13}$$

with both boundary conditions yielding precisely the same result in the thermodynamic limit. In 2D,

$$2D, \text{ sq} : \quad Z_1 = 2 \cosh^2 \beta J (1 + w^4 + 2w^6 - 2w^8 + \cdots) \tag{8.14}$$

where the coefficients $(1, 1, 2, -2, \ldots)$ are picked so as to fit Z_1^N to Z_N (8.12), to leading orders in N. The same exercise in 3D yields[11]:

sc : $Z_1 = 2 \cosh^3 \beta J (1 + 3w^4 + 22w^6 + 192w^8$

$\qquad + 2046w^{10} + 24853w^{12} + \cdots)$

bcc : $Z_1 = 2 \cosh^4 \beta J (1 + 12w^4 + 148w^6 + 2568w^8 + 53944w^{10} + \cdots)$

fcc : $Z_1 = 2 \cosh^6 \beta J (1 + 8w^3 + 33w^4 + 168w^5 + 962w^6$

$\qquad + 5928w^7 + 38907w^8 + 268056w^9 + \cdots).$

(8.15)

The series are derived for ferromagnets, J (hence w) > 0. For antiferromagnets, one reverses the sign of J, hence of w, finding that only the fcc in (8.15) is affected. (The 2D triangular lattice similarly contains odd powers of w, and is also affected by a change in sign.) Quite generally, it is found that on *bipartite* lattices (and only on bipartite lattices) the partition function of the ferromagnet (F) and antiferromagnet (AF) are identical in zero external field. We have already encountered this property in connection with the spherical model AF (Sec. 7.11). If the lattice can be decomposed into two interpenetrating nets, such that the spins on one sublattice interact only with those on the other, a change in the definition of up and down on one

---

[11] See C. Domb, "On the theory of cooperative phenomena in crystals," *Adv. Phys.* **9**, 149–361 (1960). The fit of $T_c(d)$ on hypercubic lattices to 2 straight lines was performed by G. Cocho, G. Martinez-Mekler and R. Martinez-Enriquez, *Phys. Rev.* **B26**, 2666 (1982).

sublattice (but not on the other) effectively changes the sign of J. While the Ising model on a bipartite lattice has a free energy which is an even function of J, the odd powers of J enter in more general cases, including the fcc example above and the triangular lattice illustrated in Fig. 8.3. In the figure, it is clear that some of the neighbors of any given spin are themselves neighbors. The consequences for the AF on the triangular and fcc lattices are rather drastic, as we shall ultimately determine. As a preparation, the reader will want to examine the triangular lattice in Problem 8.5.

Problem 8.5.   Obtain Z_1 to sixth order in w for the triangular lattice, illustrated in Fig. 8.3. To this order, plot $\partial^2 Z_1/\partial w^2$ for $0 < w < 1$ and $0 > w > -1$.

We have encountered the lack of symmetry in the sign of the bonds in Sec. 7.7 on long-ranged interactions, where the phenomenon of "frustration" was first examined. As in the long-ranged model, frustration in the triangular AF Ising model destroys the phase transition that exists in the corresponding ferromagnet. The behavior of a quantity analogous to the specific heat is examined in Problem 8.5, serving to illustrate the asymmetrical behavior.

8.3.   Low Temperature Expansions and the Duality Relations

Consider the ferromagnetic state, all spins up, as contributing the leading term in an expansion of Z in the number of overturned spins. Overturning a single spin changes the interaction energy with z neighbors from $-zJ$ to $+zJ$ — i.e., each bond contributes $+2J$. The partition function is a power series in $\exp(-\beta 2J)$ but is not simple, especially if overturned spins are neighbors.

In 2D, there exists a geometric relation between any given lattice and a second lattice, denoted the dual lattice. Given the first, the second is obtained by placing points in each elementary cell, connecting them by lines such that each line in the old grid is crossed by a new one. Figure 8.2 shows the sq lattice, which is *self-dual*, and Fig. 8.3 the triangular and hexagonal nets, duals to one another. We shall see that the low temperature expansion on a given lattice is precisely related to the w expansion on its dual.

Including a factor 2 for the degeneracy (the spins in the ferromagnetic state can equally be all down as up), the low-temperature partition function

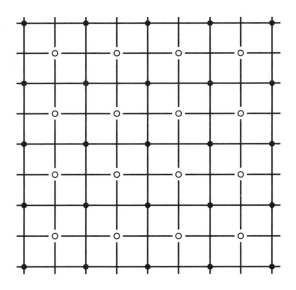

Fig. 8.2.   Two-dimensional sq net is self-dual.

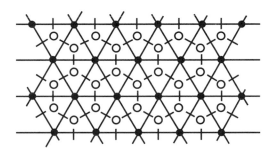

Fig. 8.3.   Triangular (T) lattice (• • •, —) is dual of hexagonal (H) lattice (o o o, —) and vice versa.

takes the form

$$Z_N = 2 \, \exp\left(\frac{1}{2}\beta J z N\right) \left[1 + \sum_{r \geq 1} \nu(r) \exp(-2r\beta J)\right] \qquad (8.16)$$

where $\nu(r)$ is the number of configurations having r broken bonds. (A broken bond has its 2 spins antiparallel.)

There is a geometric relation, illustrated in Fig. 8.4 for 1, 2, and 3 over-turned spins, i.e.: *for any r broken bonds one can draw a closed path of length*

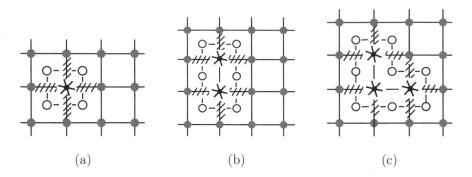

(a)   (b)   (c)

Fig. 8.4.   Effects of overturning spins. In (a) one overturned spin (*) "breaks" 4 bonds (cross-hatching). A closed path of length $r = 4$ encloses this spin on the dual lattice. (b) 2 overturned spins break 6 bonds, the shortest path on the dual lattice which crosses these bonds and encloses the 2 spins has length $r = 6$. (c) Similar for 3 spins, $r = 8$.

r *on the dual lattice, which crosses every broken bond.* If we denote parameters of the dual lattice by *, we have just proved: $\nu(r) = p^*(r)$ and vice versa, $p(r) = \nu^*(r)$. Thus,

$$Z_N^* = 2^{N^*}(\cosh \beta^* J)^{1/2N^* z^*}\left[1 + \sum_{r \geqslant 1} p^*(r) w^{*r}\right] \tag{8.17}$$

with $w^* = \tanh \beta^* J$. If we pick β and β^* to satisfy the relation,

$$\tanh \beta^* J = \exp(-2\beta J) \tag{8.18}$$

we find (a) this relation to be reciprocal, i.e.,

$$\tanh \beta J = \exp(-2\beta^* J) \tag{8.19}$$

and (b) the square brackets in (8.16) and (8.17) are equal. Therefore,

$$\frac{Z_N^*(\beta^*)}{2N^*(\cosh \beta^* J)^{1/2N^* z^*}} = \frac{Z_N(\beta)}{2 \exp(\frac{1}{2}\beta J z N)}. \tag{8.20}$$

To relate the coordination numbers z and z^* and the lattice numbers N and N^* we need look again at the construction of the dual nets, as in Figs. 8.2

and 8.3, from which one deduces

$$Nz = N^* z^* \quad \text{and} \quad N + N^* = \frac{1}{2} Nz + 2 \,. \tag{8.21}$$

Using these relations together with (8.18)–(8.20), one proves another useful correspondence:

$$Z_N(\beta) 2^{-1/2N} (\cosh\ 2\beta J)^{-1/4Nz} = Z_N^* (\beta^*) 2^{-1/2N^*} (\cosh\ 2\beta^* J)^{-1/4N^* z^*} \tag{8.22}$$

which is equivalent to, but more symmetric than (8.20).

Equations (8.18) and (8.19) have the common solution:

$$\sinh\ 2\beta^* J\ \sinh\ 2\beta J = 1 \,. \tag{8.23}$$

In the case of the sq lattice which is self-dual, the critical point — if such exists, and if it is unique — must occur at $\beta^* = \beta$. The corresponding T_c ($\sinh\ 2J/kT_c = 1$) is one of the values given in Table 8.1. In this table, we present T_c on other lattices, by way of comparison. These are obtained by the exact "duality" and "star-triangle" transformations discussed in this chapter for 2D, and by approximate series-expansion methods for 3D and higher.

Table 8.1. T_c for Ising model ferromagnet on various lattices.[11]

Lattice type	Coordination number (z)	kT_c/zJ
Linear chain	2	0
Simple quadratic (sq)	4	0.567
Simple cubic (sc)	6	0.752
Hypercubic	$2d$	$\left(1 - \frac{1}{d}\right) \Big/ \ln(1 + \sqrt{2})$
(generalization of the above to d dimensions)	$1 \leqslant d \leqslant 4$	
	$2d$ $d \geqslant 4$	$(1 - 0.5962/d)$
Honeycomb	3	0.506
Triangular	6	0.607
Body-centered cubic (bcc)	8	0.794
Face-centered cubic (fcc)	12	0.816

This table for the Ising ferromagnet may be compared with Table 7.1 for T_c in the spherical model on the hypercubic lattices. Despite a lack of agreement in two dimensions ($d = 2$) and a discrepancy of some 15% for $d = 3$, both models ultimately agree with each other and with MFT as $d \to \infty$. The disagreement at $d = 2$ (the spherical model predicts $T_c = 0$ while the duality relation yields a finite value for the Ising model) might occasion some worry, were it not for an independent proof of the existence of a phase transition which we present next.

8.4. Peierls' Proof of Long Range Order

Here, we summarize the arguments by Peierls[7] subsequently perfected by Griffiths,[8] arguments that can be adapted to all dimensionalities $d \geqslant 2$. They concern the persistence of long-range order (LRO) at finite temperature. Calculations based on this procedure yield a lower bound to the critical temperature T_c, just as MFT always provides an upper bound.[12]

As a preliminary illustration of the line of reasoning, consider first a 1D chain of N spins initially all "up", at $T = 0$. The energetic cost to create a single domain of N_- spins "down" is $2J$ at each boundary, for a total of $4J$. The size of the domain can extend from 1 to $N - 2$ without any change in energy, thus the length of the domain averages $\frac{1}{2}N$ and the average magnetization per spin, the measure of LRO, is zero. Now, if a single domain destroys LRO, the formation of smaller domains (e.g., regions of spins up within the initial domain, regions of spins down within *these*, etc.) will further compound the disorder. At any finite temperature the Boltzmann factor $\exp(-2\beta J)$ for the formation of an individual domain wall is finite, thus the number of such domain walls must be proportional to this Boltzmann factor and to N, the length of the chain. We conclude LRO is surely destroyed at any finite T.

In two dimensions or higher, starting from a situation of all spins up at $T = 0$, we wish to prove that N_- is a small fraction of N at low temperature, and that the average magnetization per spin $m = 1 - 2N_-/N$ is nonzero. The merging of overturned spins into small *domains*, and the length of domain walls, are essential considerations. For if the overturned spins were isolated, their free energy would be the minimum of the function:

$$F = 8JN_- + kT[N_- \ln N_- + (N - N_-)\ln(N - N_-)] \tag{8.24}$$

---

[12] M. E. Fisher, *Phys. Rev.* **162**, 480 (1967).

on a sq lattice, i.e.,

$$N = N(1 + e^{8J/kT})^{-1}, \quad \text{or } m = \tanh 4J/kT \tag{8.25}$$

indicating that the magnetization persists to infinite T. This result is seriously in error, for once the number of overturned spins N_- becomes appreciable, they can no longer be isolated. The energetic cost of adding a spin down to a cluster of overturned spins is significantly less than the energy $8J$ to overturn a single spin, and must in fact become precisely zero at T_c. [If two neighbors of a given spin are up, and two are down, it no longer has any tendency to be up (or down).]

Peierl's procedure takes into account the variety of shapes of clusters of everturned spins, their energy, and their contribution to the partition function. Because a border of length b encloses at most $b^2/16$ overturned spins, and because we can find an upper bound to the number of such borders $n(b)$ and a lower bound to the energy of each, we can establish a bound on the total number N_- of spins reversed.

First, the total number of possible borders of length b, $n(b)$, satisfies the inequality

$$n(b) \leqslant 4N3^b/3b. \tag{8.26}$$

We *over*estimate the corresponding Boltzmann factor by *under*estimating the cost in energy as $+2Jb$. Hence, the total number of spins reversed is bounded:

$$N_- \leqslant \sum_{b=4}^{\infty} (b^2/16)(4N3^b/3b)e^{-2Jb/kT}$$

$$\leqslant \frac{Nq^4}{6} \frac{2 - q^2}{(1 - q^2)^2} \tag{8.27}$$

where $q = 3 \exp(-2J/kT)$ must be < 1. Solving this inequality for the value of q that makes $N_- = \frac{1}{2}N$, we can obtain a lower bound on the true T_c at which the spontaneous magnetization disappears. In this way, we find

$$kT_c/4J \geqslant 0.37 \tag{8.28}$$

or about two-thirds the exact value (Table 8.1, sq. lattice).

When applied to other lattices in 2D and 3D, this procedure again yields reasonable lower bounds on the Curie temperature. Fisher has shown that MFT provides upper bounds,[12] thus $kT_c \leqslant zJ$ with $z = 4$ here.

To complete the proof of the existence of a phase transition, one must show that the spins are *uncorrelated* at sufficiently high temperature. (One simple argument states that as all thermodynamic functions depend on the ratio J/kT only, the limit $T = \infty$ is the same as $J = 0$; there we know that all spins are uncorrelated!) Quantitatively, to evaluate $\langle S_n S_m \rangle_{\mathrm{TA}}$ we can use (8.8):

$$\langle S_n S_m \rangle_{\mathrm{TA}} = \frac{1}{Z_N} \partial^2 Z_N / \partial w_n \partial w_m \Big|_{\substack{\text{all} \\ w_j = 0}} . \tag{8.29}$$

The so-called magnetic graphs which contribute in the above start at n and terminate at m, thus involve L_{nm} products of the nearest-neighbor functions w_{ij}, where L_{nm} is the number of links separating n from m. At high temperatures, $w_{ij} \simeq J/kT \to 0$, so we expect the shortest path to dominate the sum, hence *an exponential decrease in the correlation function* (such as one finds in 1D.) Thus,

$$\langle S_n S_m \rangle_{\mathrm{TA}} \sim \exp(-L_{nm}/\xi) \tag{8.30}$$

with the correlation length ξ a function of T. Evidently, $\xi \to \infty$ at T_c with the onset of long-range order (LRO); assuming a power-law behavior for the correlation length, one defines a new critical index ν:

$$\xi = \left(\frac{T - T_c}{T_c} \right)^{-\nu} . \tag{8.31}$$

The divergence of ξ is closely related to the divergence in the susceptibility. (The left-hand side of (8.30) is just the summand in the high-temperature susceptibility). The assumption above, of a simple power law, is just that: an assumption. It fits the boundary conditions (at T_c and ∞), but then so do a lot of other functions! It is, however, key to a simple theory of critical phenomena.

8.5.   1D Ising Model in Longitudinal Fields

In this section we derive the exact solution of the 1D Ising model with nearest-neighbor interactions, including such parallel "longitudinal" interactions as an external magnetic field parallel to the axis of quantization, and interactions with the underlying normal modes (phonons) of the molecule or solid to which the spins are pinned. We leave to the following section the study of transverse fields — those that involve operators such as S_i^x or S_i^y

that do not commute with the S_i^z of the Ising model. (The *combination* of transverse *and* longitudinal external fields has not been solved but appears to be solvable.) The longitudinal case is classical — all the operators commute, all spatial configurations can be specified without uncertainty — but the study of its associated statistical mechanics quickly involves us in the algebra of operators that — perversely — do *not* commute with one another.

A reasonably general starting point is provided by the Hamiltonian,

$$\mathcal{H} = -\sum_{n=1}^{N} J_n S_n S_{n+1} - \sum_{n=1}^{N} B_n S_n . \tag{8.32}$$

We write this as $\sum \mathcal{H}_n$ and associate with each $\mathcal{H}_n$ a Boltzmann factor V_n:

$$V_n = \exp\left[\beta J_n S_n S_{n+1} + \frac{1}{2}\beta(B_n S_n + B_{n+1} S_{n+1})\right], \tag{8.33}$$

with some allowance at $n = 1$ and N for desired boundary conditions.

The partition function is then

$$Z_N = \text{Tr}\{\cdots V_n V_{n+1} \cdots\}. \tag{8.34}$$

There is some advantage in considering the V_n's to be matrices of the following form:

$$V_n \to \mathbf{V}(S_n; S_{n+1}) \tag{8.35}$$

i.e., 2×2 matrices in the indices $S_n = \pm 1$, and $S_{n+1} = \pm 1$. Matrix multiplication takes the form

$$\mathbf{V}(S_{n-1}; S_n){\cdot}\mathbf{V}(S_n; S_{n+1}) \equiv \sum_{S_n=-1}^{+1} \mathbf{V}(S_{n-1}; S_n)\mathbf{V}(S_n; S_{n+1})$$

$$\equiv \mathbf{W}(S_{n-1}; S_{n+1}) \tag{8.36}$$

where $\mathbf{W}$ is the resultant product matrix, also 2×2.

The advantage lies in the fact that *the operation "Tr" is identical to* $\Pi_n(\sum_{S_n})$, *i.e., to repeated matrix multiplication.*

So, according to (8.36),

$$Z_N = \mathrm{Tr}\{\mathbf{V}_1 \cdot \mathbf{V}_2 \cdots \mathbf{V}_n \cdot \mathbf{V}_{n+1} \cdots \mathbf{V}_N\}$$

$$= \mathrm{tr}\{\mathbf{U}(S_1; S_N)\} \qquad (8.37)$$

where the last operation, tr, is the sum of the diagonal elements of the final 2×2 matrix

$$\mathbf{U} = \mathbf{V}_1 \cdots \mathbf{V}_n \cdot \mathbf{V}_{n+1} \cdots \mathbf{V}_N . \qquad (8.38)$$

Explicitly, each $\mathbf{V}_n$ matrix takes the form:

S_n		S_{n+1}	
		$+1$	-1
$\mathbf{V}_n =$	$+1$	$\exp\left[\beta J_n + \frac{1}{2}\beta(B_n + B_{n+1})\right]$	$\exp\left[-\beta J_n + \frac{1}{2}\beta(B_n + B_{n+1})\right]$
	-1	$\exp\left[-\beta J_n - \frac{1}{2}\beta(B_n - B_{n+1})\right]$	$\exp\left[\beta J_n - \frac{1}{2}\beta(B_n + B_{n+1})\right]$

$$(8.39)$$

Expanding in the Pauli spin matrices, $\sigma_y = \begin{pmatrix} 0 & -i \\ i & 0 \end{pmatrix}$

$$\mathbf{1} = \begin{bmatrix} 1 & 0 \\ 0 & 1 \end{bmatrix}, \quad \sigma_x = \begin{bmatrix} 0 & 1 \\ 1 & 0 \end{bmatrix}, \quad \sigma_y = \begin{bmatrix} 0 & -i \\ i & 0 \end{bmatrix}, \quad \text{and} \quad \sigma_z = \begin{bmatrix} 1 & 0 \\ 0 & -1 \end{bmatrix}$$

(8.39) can be expressed as follows:

$$\mathbf{V}_n = \exp(\beta J_n) \cosh \frac{1}{2}\beta(B_n + B_{n+1})\mathbf{1}$$

$$+ \exp(\beta J_n) \sinh \frac{1}{2}\beta(B_n + B_{n+1})\boldsymbol{\sigma}_z$$

$$+ \exp(-\beta J_n) \cosh \frac{1}{2}\beta(B_n - B_{n+1})\boldsymbol{\sigma}_x$$

$$+ i\exp(-\beta J_n) \sinh \frac{1}{2}\beta(B_n - B_{n+1})\boldsymbol{\sigma}_y . \qquad (8.40)$$

As we have stated before, $\mathbf{V}_1$ and $\mathbf{V}_N$ are special cases. The manner in which we treat them does not influence extensive quantities, but is important in the evaluation of boundary phenomena such as surface tension.

In the case of constant nearest-neighbor interactions $J_n = 1$, periodic boundary conditions and homogeneous field $B_n = B$, the $\mathbf{V}$ matrices are all identical. $\mathbf{U}$ in (8.37) becomes

$$\mathbf{U} = \mathbf{V}^N , \tag{8.41}$$

and therefore

$$Z_N = \lambda_+^N + \lambda_-^N \tag{8.42}$$

where $\lambda_\pm$ are the two eigenvalues of $\mathbf{V}$. We solve for them in the usual way:

$$\det \left\| \begin{matrix} K_1 K_2 - \lambda & K_1^{-1} \\ K_1^{-1} & K_1 K_2^{-1} - \lambda \end{matrix} \right\| = 0 . \tag{8.43}$$

with $K_1 \equiv \exp(\beta J)$ and $K_2 \equiv \exp(\beta B)$. The solution is

$$\lambda_\pm = \frac{1}{2} K_1 (K_2 + K_2^{-1}) \pm \left[\frac{1}{4} K_1^2 (K_2 + K_2^{-1})^2 - (K_1^2 - K_1^{-2}) \right]^{1/2}$$

$$= e^{\beta J} \cosh \beta B \pm (e^{2\beta J} \sinh^2 \beta B + e^{-2\beta J})^{1/2} . \tag{8.44}$$

Evidently, λ_+ exceeds λ_-. When each is raised to the Nth power and N becomes large, we find

$$Z_N = \lambda_+^N [1 + (\lambda_-/\lambda_+)^N] \to \lambda_+^N \tag{8.45}$$

as the ratio $(\lambda_-/\lambda_+)^N$ becomes exponentially small in the limit of large N.

The above illustrates a general principle in calculating Z_N. $\mathbf{V}$ is denoted the *transfer matrix* and λ_+ is its largest or *optimal* eigenvalue. *We need obtain only the largest eigenvalue of the transfer matrix*, the contribution of the other(s) vanishing at large N. This is a consequence of the criterion developed in Chap. 7, the minimization of the free energy.

The Nth root of (8.45) is trivially

$$Z_1 = \lambda_+ \tag{8.46}$$

with λ_+ given in (8.44). For zero external field, this yields precise agreement with (8.11) and (8.13) found by series expansion in zero field. In vanishing

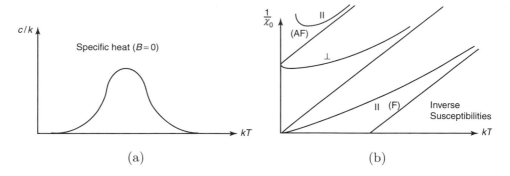

Fig. 8.5. Properties of 1D Ising model: Specific heat in zero field, and various zero-field susceptibilities. $1/\chi_0$ is plotted to show asymptotic approach to Curie-Weiss laws for antiferromagnet (AF) and ferromagnet (F), and are labelled $\parallel$ to indicate magnetic field is applied parallel to spins. The perpendicular susceptibility ($\perp$) is the same for F and AF, see (8.91).

field, we now obtain the specific heat and paramagnetic susceptibility:

$$c = k(\beta J)^2 / \cosh^2 \beta J \qquad (8.47)$$

and

$$\chi_0 = \frac{\mathbb{C}_{1/2}}{T} e^{2\beta J} . \qquad (8.48)$$

The latter *diverges* at $T = 0$ for ferromagnetic coupling ($J > 0$), and *vanishes* exponentially for antiferromagnetic coupling ($J < 0$) at low temperature, a common occurrence in 1D, as previously discussed [see Chap. 7, where ferromagnetic and antiferromagnetic susceptibility are compared in 1D]. At high T,

$$\chi_0^{-1} \propto T[1 - 2\beta J + O(\beta J)^2] = T - 2J/k + O(1/T). \qquad (8.49)$$

While this is precisely the Curie-Weiss law, as usual this expansion is misleading near the critical point, $T = 0$ in this instance. Equations (8.47)–(8.49) are illustrated in Fig. 8.5, which indicates the high-temperature extrapolation.

In zero external field the transfer matrices all commute and can be diagonalized simultaneously. Consequently, we obtain Z for an *arbitrary* distribution of J_n

$$Z_N = \prod_n 2 \cosh \beta J_n \, ,$$

or better

$$f = -\frac{kT}{N} \sum_n \ln(2 \cosh \beta J_n) = -kT \langle \ln(2 \cosh \beta J) \rangle$$

$$= -kT \int dJ P(J) \ln 2 \cosh \beta J \tag{8.50}$$

where the average is over the given distribution of J_n's. We can treat the case of "random bonds" similarly, that is, the case when the bond strengths are given only by a statistical distribution.

Given arbitrary bonds *and* external field, the evaluation of the partition function becomes more difficult as, in general, the $\mathbf{V}_n$ matrices (8.40) no longer commute with one another. Let us demonstrate the procedure when first one transfer matrix differs from the rest and later, when two such matrices are distinguished. Let the pth matrix in (8.37) be $\mathbf{V}' \neq \mathbf{V}$. Only the eigenvector u_+ associated with λ_+ contributes, therefore we only need calculate:

$$Z_N = Z_N^0 \frac{u_+^T \cdot \mathbf{V}^{p-1} \cdot \mathbf{V}' \cdot \mathbf{V}^{N-p} \cdot u_+}{u_+^T \cdot \mathbf{V}^{p-1} \cdot \mathbf{V} \cdot \mathbf{V}^{N-p} \cdot u_+} \, . \tag{8.51}$$

Here, Z_N^0 is the partition function when all the matrices are equal, and the eigenvectors (and transposes) are the vectors that satisfy:

$$\mathbf{V} \cdot u_+ = \lambda_+ u_+ \, , \quad u_+^T \cdot \mathbf{V} = \lambda_+ u_+^T \tag{8.52}$$

with similar expressions for u_-, the second eigenvector (associated with λ_-). Using the eigenvalue relation above, we reduce (8.51) to a ground state expectation value

$$Z_N = Z_N^0 \frac{1}{\lambda_+} u_+^T \cdot \mathbf{V}' \cdot u_+ \, . \tag{8.53}$$

In terms of free energy

$$F = F^0 + kT [\ln \lambda_+ - \ln(u_+^T \cdot \mathbf{V}' \cdot u_+)] \, . \tag{8.54}$$

The case of *two* different **V**'s is more illuminating. Assume they are separated by n host (i.e., unperturbed) transfer matrices. We then find

$$Z_N = Z_N^0 \frac{u_+^T \cdot \mathbf{V}' \cdot \mathbf{V}^n \cdot \mathbf{V}'' \cdot u_+}{u_+^T \cdot \mathbf{V}^{n+2} \cdot u_+} = Z_N^0 \lambda_+^{-2} [(u_+^T \cdot \mathbf{V}' \cdot u_+)(u_+^T \cdot \mathbf{V}'' \cdot u_+)$$

$$+ (\lambda_-/\lambda_+)^n (u_+^T \cdot \mathbf{V}' \cdot u_-)(u_-^T \cdot \mathbf{V}'' \cdot u_+)] . \tag{8.55}$$

This is obtained by using *completeness* $u_+ u_+^T + u_- u_-^T = 1$ twice: once following $\mathbf{V}'$, and once preceding $\mathbf{V}''$. Taking logarithms, we find that this expression contains the free energy of the individual defects as in (8.54) plus a two-defect interaction term which involves u_- and λ_-.

This expression also suggests how one is to obtain *correlation functions* in the homogeneous chain (all $J_n = J$, all $B_n = B$). Suppose we wished to obtain the thermal averaged correlation function of S_p and S_{p+n+1}, i.e., the two spins separated by n intervening sites. We define two matrices $\mathbf{V}'$ and $\mathbf{V}''$ as follows:

$$\mathbf{V}' = V(S_{p-1}; S_p)S_p \quad \text{and} \quad \mathbf{V}'' = V(S_{p+n}; S_{p+n+1})S_{p+n+1} . \tag{8.56}$$

Problem 8.6.   Using the above, derive the 2×2 matrix σ in the following relations:

$$m \equiv \langle S_p \rangle_{\text{TA}} = u_+^T \cdot \sigma \cdot u_+ \quad \text{and} \tag{8.57}$$

$$\langle (S_p - m)(S_{p+n+1} - m) \rangle_{\text{TA}} = \left(\frac{\lambda_-}{\lambda_+} \right)^{n+1} (u_+^T \cdot \sigma \cdot u_-)(u_-^T \cdot \sigma' \cdot u_+), \tag{8.58}$$

and show that the first agrees with $-\partial F/\partial B$ calculated from (8.45). Calculate (8.58) in zero field $(B = 0)$ and verify that it yields the bond energy at $n = 0$, as well as χ_0, (8.48), summing all n.

The result (8.58) leads to two significant observations. First, that the excited spectrum of the transfer matrix (λ_- in this instance) is needed to calculate the correlation functions. Second, in 1D correlations decay exponentially with a "correlation length" $\xi \sim 1/\ln(\lambda_+/\lambda_-)$. For thermodynamic functions $(Z, m, \text{etc.})$ only λ_+ is needed. But for evaluating the properties of a system, e.g., by neutron scattering where the correlation functions are invaluable, we should have the complete spectrum λ's and u's of $\mathbf{V}$.

In practice, J_n could exhibit some variability due to *magnetostriction*. This rather general phenomenon comes about as the result of coupling between the magnetic and mechanical degrees of freedom of the lattice, and is particularly easy to analyze in one dimension. Suppose the *ideal* inter-atomic distances are a, and that the exchange bonds J_n vary with *actual* interatomic distance. Then, expand the interactions in powers of the small excursions:

$$J(x_{n+1} - x_n) = J(\mathbf{a}) \left[1 + \eta_1 \frac{(x_{n+1} - x_n)}{a} + \eta_2 \frac{(x_{n+1} - x_n)^2}{a^2} + \cdots \right]$$

$$(8.59)$$

where η_1 and η_2 characterize the leading and sub-leading corrections. The total Hamiltonian includes the unperturbed Hamiltonian,

$$\mathscr{H}_{\text{phonon}} = \sum_n \left[p_n^2/2M + \frac{1}{2} M \omega_0^2 (x_{n+1} - x_n)^2 \right]$$

$$\mathscr{H}_{\text{Ising}} = -J(\mathbf{a}) \sum_n S_n S_{n+1} - \sum_n B_n S_n .$$

$$(8.60)$$

The first-order coupling is

$$\mathscr{H}_1 = -\frac{J(\mathbf{a})}{a} \eta_1 \sum_n (x_{n+1} - x_n) S_n S_{n+1} ,$$

$$(8.61)$$

and the second-order coupling is

$$\mathscr{H}_2 = -\frac{J(\mathbf{a})}{a^2} \eta_2 \sum_n (x_{n+1} - x_n)^2 S_n S_{n+1} .$$

$$(8.62)$$

The effects of the first-order coupling can be derived without further approximation.

The method is straightforward. We merely shift the origin of each atomic coordinate by an amount which depends on the magnetic configuration. Thus, with

$$x_n \to x_n + \frac{J(\mathbf{a}) \eta_1}{M \omega_0^2 a} \sum_{j<n} S_j S_{j+1}$$

$$(8.63)$$

we eliminate $\mathcal{H}_1$ and obtain

$$\mathcal{H} = \mathcal{H}_{\text{ph}} + H_{\text{I}} - \frac{1}{2}N\frac{J^2(\mathbf{a})\eta_1^2}{M\omega_0^2 a^2} + \mathcal{H}_2\,. \tag{8.64}$$

The effect of $\mathcal{H}_2$ can be estimated (Problem 8.7). According to (8.63), the main contribution of η_1 is to the length of the chain — hence the term magnetostriction. If $\langle x_N - x_0 \rangle + Na$ was initially L, we have now (for $B = 0$):

$$L \to L - (\eta_1/M\omega_0^2 a)U(T) \tag{8.65}$$

where $U(T) = \langle H_{\text{Ising}} \rangle_{\text{TA}}$ is the internal energy of the Ising Hamiltonian with $B = 0$. (In finite field, the expression is somewhat more cumbersome.) The sign of the magnetostrictive effect depends on the sign of n_1. The result, $\partial L/\partial T \propto -\eta_1 c(T)$ agrees with experiment even in three dimensions — even in MFT.[13]

Problem 8.7. Suppose $\eta_2 = O(\eta_1^2)$, and treat $\mathcal{H}_2$ in leading order, obtaining quantities such as new speed of sound, new J, etc.[14]

8.6.   1D Ising Model in Transverse Fields

Until now, the spins and the external and internal forces acting on them have all lain along a preferred axis. Here consider situations in which the two-body forces involve components of the spins along one axis, while the external field interacts with spin components along a perpendicular axis. We observed that in MFT, the susceptibility of the antiferromagnet is quite different when the field is transverse rather than longitudinal, the discrepancies being highlighted in Fig. 7.8. In this section we corroborate this in the context of a microscopic theory. The calculations themselves are good practice for the statistical mechanics of the *two*-dimensional Ising model considered later in this chapter.

---

[13] H. R. Ott *et al.*, *Phys. Rev.* **B25**, 477 (1982); Z. Chen and M. Kardar, *Phys. Rev.* **B30**, 4113 (1984).
[14] M. E. Lines, *Phys. Reports* **55**, 133 (1979).

The Hamiltonian now contains non-commuting operators S_i^z and S_i^x, for example:

$$\mathcal{H} = -J \sum_n S_n^z S_{n+1}^z - B \sum_n S_n^x . \tag{8.66}$$

This describes a simple, homogeneous case: constant nearest-neighbor bonds and a constant external field. Nevertheless, we can no longer write

$$\exp(-\beta\mathcal{H}) = \exp(-\beta\mathcal{H}_1)\exp(-\beta\mathcal{H}_2)\cdots$$

as in (8.33)–(8.37) because the $\mathcal{H}_n$'s do not commute with their neighbors $\mathcal{H}_{n\pm1}$. We are thus forced to seek an alternative to the transfer matrix approach, a complication typical of *quantum* statistical mechanics, generally requiring one to diagonalize the complete Hamiltonian before attempting to evaluate Z_N duality.

In this instance we possess an additional tool, a *"duality"* transformation that maps J and B into each other. The details of this nonlinear, nonlocal transformation now follow:

Let us write S_n^z and S_n^x as products over a second set of Pauli operators T_m^z and T_m^x:

$$S_n^z = \prod_{m=1}^{n} T_m^x \quad \text{and} \quad S_n^x = T_n^z T_{n+1}^z . \tag{8.67}$$

Problem 8.8.   Show that

$$S_n^y = -T_n^y T_{n+1}^z \prod_{m=1}^{n-1} T_m^x . \tag{8.68}$$

Prove that the Pauli matrices, as defined in the two equations above in terms of the T's, satisfy the correct algebra, i.e., commute if on different sites and anticommute on the same site.

When the operators in the new representation are inserted in (8.66),

$$\mathcal{H} = -B \sum_n T_n^z T_{n+1}^z - J \sum_n T_n^x \tag{8.69}$$

the only visible effect is that J, B are interchanged. We can infer that if there are only two phases — one corresponding to large $|J/B|$ and the other to small $|J/B|$, the *critical* point must lie at $|J/B| = 1$. Quantities such as the ground state energy will be symmetric about that *critical* point.

For definiteness, the following calculations are based on (8.66) although they could just as easily apply to (8.69).

Equation (8.66) under consideration contains both quadratic and linear expressions in the spins. Unlike similar expressions in the Gaussian and spherical models, these cannot be reduced to a diagonal, quadratic, form in the spin operators in which the various normal modes are decoupled. The principal reason lies with the commutation relations. Whereas Pauli matrices on a given site satisfy anticommutation relations, matrices associated with two different spins commute with one another. Linear combinations of spin operators neither commute nor anticommute with one another — consequently, their algebraic properties are highly entangled.

As a remedy we shall seek to construct a set of operators that anticommute everywhere. The transformation from Pauli operators to fermions by means of the Jordan–Wigner transformation previously examined in Chap. 3 is entirely suitable for this purpose.

We introduce a set of anticommuting operators c_j and $c_j^\dagger$, and a "number operator" $\mathbf{n}_j = c_j^\dagger c_j$ the eigenvalues of which are 0, 1. The spins can be expressed entirely with the aid of these operators, in the form

$$S_j^x = S_j^+ + S_j^- , \quad \text{where} \quad S_j^- = c_j \exp\left(i\pi \sum_{r<j} \mathbf{n}_r \right) \quad \text{and} \quad S_j^+ = (S_j^-)^+$$

$$\text{and} \quad S_j^z = 2\mathbf{n}_j - 1 . \tag{8.70}$$

On a given site, the anticommutation relations of c_j and $c_j^\dagger$ are the same as for Pauli matrices; the exponential phase factors cancel out. On two different sites, these phase factors change a commutator into an *anti*commutator. In this sense this transformation is *not* unitary, as it does not preserve commutation relations. Nevertheless, it is well defined and leads to an ordinary eigenvalue problem for spins 1/2 with nearest-neighbor interactions on the linear chain. Consider the quadratic form $S_j^x S_{j+p}^x$:

$$(S_j^+ S_{j+p}^+ + S_j^+ S_{j+p}^-) + \text{H.c.} \tag{8.71}$$

where H.c. stands for Hermitean conjugate also indicated by $^+$ as in (8.70).

In terms of Fermi operators:

$$S_j^+ S_{j+p}^+ = c_j^\dagger c_{j+p}^\dagger \exp\left(i\pi \sum_{j\leqslant r<j+p} \mathbf{n}_r\right)$$

$$S_j^+ S_{j+p}^- = c_j^\dagger c_{j+p} \exp\left(i\pi \sum_{j\leqslant r<j+p} \mathbf{n}_r\right).$$

(8.72)

It is useful to note the identity,

$$\exp(i\pi\mathbf{n}_r) \equiv 1 - 2\mathbf{n}_r \equiv (c_r^\dagger + c_r)(c_r^\dagger - c_r)$$

(8.73)

and that the square of each side is 1. Further,

$$c_r^\dagger \exp(i\pi\mathbf{n}_r) = c_r^\dagger, \quad c_r \exp(i\pi\mathbf{n}_r) = -c_r.$$

(8.74)

Thus, for separations $p = 1$ in (8.72) we obtain simply

$$S_j^+ S_{j+1}^+ = c_j^\dagger c_{j+1}^\dagger \quad \text{and}$$

$$S_j^+ S_{j+1}^- = c_j^\dagger c_{j+1}.$$

(8.75)

For separations $p = 2$, the expressions are

$$S_j^+ S_{j+2}^+ = c_j^\dagger c_{j+2}^\dagger (1 - 2\mathbf{n}_{j+1})$$

$$S_j^+ S_{j+2}^- = c_j^\dagger c_{j+2}(1 - 2\mathbf{n}_{j+1}).$$

(8.76)

As the number operator $\mathbf{n}_{j+1} = c_{j+1}^\dagger c_{j+1}$ is itself bilinear in fermion field operators it follows that for $p = 1$ the expressions are quadratic, whereas for $p = 2$ they are quartic, etc.

Fortunately, because the starting Hamiltonian (8.66) involves *only* nearest-neighbor bonds, $p = 1$. As it is written, with the bonds in terms of S_j^z operators, the transformation is not of much use. But an initial rotation of 90° about the y-axis in spin space: $S_j^x \to -S_j^z$, $S_j^z \to +S_j^x$ for all j, transforms $\mathscr{H}$ into the form:

$$\mathscr{H} \to -J\sum_j S_j^x S_{j+1}^x + B\sum_j S_j^z.$$

(8.77)

This can be rendered uniformly quadratic in fermions by (8.70):

$$\mathcal{H} \rightarrow -J \sum_j (c_j^\dagger c_{j+1}^\dagger + c_j^\dagger c_{j+1} + c_{j+1}^\dagger c_j + c_{j+1} c_j)$$

$$+ B \sum_j (2 c_j^\dagger c_j - 1), \tag{8.78}$$

in which form it can be easily diagonalized after Fourier transformation. In the duality transformation and in the Jordan–Wigner transformation to fermions we have neglected end effects (at $j = 1$ or N). We continue to do this, for simplicity in obtaining the normal modes. Assuming translational invariance (PBC in the present form of H) one starts by Fourier transforming, introducing a new complete set of fermion operators a_k and $a_k^\dagger$ [$k = (2\pi/N) \cdot$ integer] by what amounts to a unitary transformation from the set of c's to the set of a's.

$$c_n = N^{-1/2} \sum_k a_k e^{ikn} \quad \text{and}$$

$$c_n^\dagger = N^{-1/2} \sum_k a_k^\dagger e^{-ikn} \tag{8.79}$$

where sums over k range over the N equally spaced points in the interval $-\pi < k \leqslant +\pi$. Like the c_n's, the a_k's also satisfy fermion anticommutation relations:

$$a_k^\dagger a_{k'}^\dagger + a_{k'}^\dagger a_k^\dagger \equiv \{a_k^\dagger, a_{k'}^\dagger\} = 0 = \{a_{k'}, a_k\} \quad \text{and}$$

$$\{a_k^\dagger, a_{k'}\} = \delta_{k,k'}. \tag{8.80}$$

That *such* algebraic bilinear relations are preserved by a unitary transformation (8.79), is verified in Problem 8.9.

Problem 8.9. Using (8.79) and (8.80) above to *define* the c_n's and $c_n^\dagger$'s, verify

$$c_j^2 = c_j^{\dagger 2} = 0, \quad \{c_j^\dagger, c_r\} = \delta_{jr}$$

and finally, prove (8.74).

We insert (8.79) into $\mathcal{H}$, (8.78), obtaining:

$$\mathcal{H} = -JN^{-1}\sum_j\sum_k\sum_{k'} e^{i(k-k')j}$$

$$\times [a_k^\dagger a_{-k'}^\dagger e^{ik'} + a_k^\dagger a_{k'}(e^{ik'} + e^{-ik}) + a_{-k}a_{k'}e^{-ik}]$$

$$+ 2BN^{-1}\sum_j\sum_k\sum_{k'} e^{i(k'-k)j}a_k^\dagger a_{k'} - NB. \tag{8.81}$$

In this context

$$\frac{1}{N}\sum_j e^{i(k'-k)j} = \delta_{k,k'} \tag{8.82}$$

is tantamount to conservation of momentum. It serves to eliminate the sums over j and k' and bring $\mathcal{H}$ to the form $\sum \mathcal{H}_k$,

$$\mathcal{H} = -J\sum_{\pi>k>0}[(a_k^\dagger a_{-k}^\dagger - a_{-k}a_k)(2i\sin k) + (a_k^\dagger a_k + a_{-k}^\dagger a_{-k})(2\cos k)]$$

$$+ 2B\sum_{\pi>k>0}(a_k^\dagger a_k + a_{-k}^\dagger a_{-k}) + \mathcal{H}_0 + \mathcal{H}_\pi - NB. \tag{8.83}$$

The individual $\mathcal{H}_k$ now commute with one another and can be individually diagonalized. At $k=0$ or π, they are *already* diagonal:

$$\mathcal{H}_0 = -2(J-B)a_0^\dagger a_0 \quad \text{and}$$
$$\mathcal{H}_\pi = 2(J+B)a_\pi^\dagger a_\pi. \tag{8.84}$$

The $k=0$ mode is significant for ferromagnetism ($J>0$) whereas $k=\pi$ is important for antiferromagnetism ($J<0$). In either case we see confirmation of the critical point at $|J/B|=1$, with the occupancy of the relevant mode changing from 0 to 1.

But, whatever the behavior of these two particular modes, this can have no bearing on the *total* energies which are $O(N)$. We must diagonalize *all* the $\mathcal{H}_k$, then sum over their ground state energies (at $T=0$) or over their free energies to obtain the system properties. We write $\mathcal{H}_k$ in the form

$$\mathcal{H}_k = -J(a_k^\dagger a_{-k}^\dagger - \text{H.c.})2i\sin k$$

$$+ 2(B - J\cos k)(a_k^\dagger a_k + a_{-k}^\dagger a_{-k} - 1) \tag{8.85}$$

after adding and subtracting $-2J\sum_{\pi>k>0}\cos k = 0$ in (8.83) for convenience.

The operators $a^\dagger$ occur only in pairs $a^\dagger_k a^\dagger_{-k}$, so $\mathcal{H}_k$ connects only even-occupancy states with each other, or odd-occupancy states with each other. Thus, $|0\rangle$ is connected by $\mathcal{H}_k$ to $a^\dagger_k a^\dagger_{-k}|0\rangle$ but not to $a^\dagger_k|0\rangle$ or $a^\dagger_{-k}|0\rangle$. In fact, the last two are — by inspection — individual eignstates of $\mathcal{H}_k$ whose eigenvalues are zero. Within the even-occupancy subspace, $\mathcal{H}_k$ takes the form:

$$\mathcal{H}_k = -2J \sin k\, S^y_k + 2(B - J \cos k) S^z_k \qquad (8.86)$$

where the S^y and S^z matrices are the usual Pauli matrices in the space of:

$$|0\rangle = \begin{bmatrix} 0 \\ 1 \end{bmatrix} \quad \text{and} \quad a^\dagger_k a^\dagger_{-k}|0\rangle = \begin{bmatrix} 1 \\ 0 \end{bmatrix}. \qquad (8.87)$$

This is the Hamiltonian of a single "spin" in an external field tilted in the y, z plane. The eigenvalues are $\pm\Delta_k$, the magnitude of this pseudo-external field:

$$\Delta_k \equiv 2[(J \sin k)^2 + (B - J \cos k)^2]^{1/2}. \qquad (8.88)$$

The ground state is always $-\Delta_k$; the remaining eigenvalues are 0, 0 and $+\Delta_k$; however, all four must be taken into account in evaluating the free energy. For F_k we find $-kT \ln(2 + 2 \cosh \beta\Delta_k)$, i.e.

$$F_k = -kT \ln\{4 \cosh^2 \beta[(J \sin k)^2 + (B - J \cos k)^2]^{1/2}\} \qquad (8.89)$$

after use of a half-angle trigonometric identity. The total free energy is thus

$$F = -kTN \frac{1}{2\pi} \int_{-\pi}^{+\pi} dk \ln\{2 \cosh \beta\Delta_k/2\}. \qquad (8.90)$$

Similarly, the ground state energy is

$$E_0 = -\frac{N}{2\pi} \int_{-\pi}^{+\pi} dk\,\Delta_k/2. \qquad (8.91)$$

Without further analysis, we obtain the transverse magnetic susceptibility $\chi_\perp$ in vanishing field. Differentiate F twice, then set $B = 0$. This results in trivial integrations, with the result:

$$\chi_\perp = -\mathbb{C}_{1/2} \frac{\partial^2 F}{\partial B^2}\bigg|_{B=0} = \frac{1}{2} N \mathbb{C}_{1/2} [J^{-1} \tanh \beta J + \beta \cosh^{-2} \beta J]. \qquad (8.92)$$

This quantity differs from the longitudinal susceptibility, calculated earlier, in being symmetric in J. It thus extrapolates to a pure Curie law at high T, not to a Curie-Weiss law. It also is finite at $T = 0$, unlike the longitudinal susceptibility which is either 0 (AF) or ∞ (ferromagnetic couplings). These various results are compared graphically in Fig. 8.5 of the preceding section.

At finite B the results are considerably more involved. F must be evaluated numerically and its derivatives are expressible only in terms of elliptic functions. These are tabulated and their properties near critical points are well known, but we can obtain a qualitative understanding more simply by analyzing the spectrum Δ_k of elementary excitations.

This spectrum is given by Δ_k (the lowest excitation energy is from the ground state of any H_k to either of the odd-occupation states having energy zero). For $J > 0$ and $B = 0$ these excitations are easily interpreted as the energy $2J$ to create one additional *domain wall* in the system (taking any two parallel spins, and overturning the second member of this pair together with all the spins to the right of it). As B is increased, the geometric interpretation becomes more difficult. But we note that with increasing B, the gap against the elementary excitations decreases until, at $B = J$ it vanishes altogether. Indeed, the symmetry between B and J which was revealed through the duality transformation (8.67) is also reflected in the spectrum of Δ_k, which we now rewrite in symmetric form:

$$\Delta_k = 2[J^2 + B^2]\left(1 - \frac{2BJ}{B^2 + J^2}\cos k\right)^{1/2}$$

$$= 2(|J| + |B|)\left(1 - \frac{4|BJ|}{(|J| + |B|)^2}g_k\right)^{1/2} \tag{8.93}$$

where $g_k = \cos^2 \frac{1}{2}k$ for $BJ > 0$ and $\sin^2 \frac{1}{2}k$ for $BJ < 0$.

This spectrum is plotted in Fig. 8.6 for the general cases $B \neq J$ as well as in the singular cases $|B| = |J|$ for which the gap $||J| - B|$ vanishes.

To evaluate (8.91) and its derivatives, we invoke the complete elliptic integrals[15]:

$$E(k) = \int_0^{1/2\pi} d\phi(1 - k^2\sin^2\phi)^{1/2} \quad \text{and}$$

$$K(k) = \int_0^{1/2\pi} d\phi(1 - k^2\sin^2\phi)^{-1/2}. \tag{8.94}$$

---

[15] E. Jahnke and F. Emde, *Tables of Functions*, Dover, New York, 1945.

$$c_j^\dagger c_{j+1} \rightarrow \frac{1}{N} \sum_{k,k'} \exp(-ikj)\exp[ik'(j+1)]a_k^\dagger a_{k'}$$

$$\rightarrow \frac{1}{N} \sum_{k,k'} \exp(-ikj)\exp[ik'(j+1)](a_k^\dagger \cos\theta_k + a_{-k}\sin\theta_k)$$

$$\times (a_{k'}\cos\theta_{k'} + a_{-k'}^\dagger \sin\theta_{k'}) \qquad (8.99)$$

with the angles determined in (8.97). The thermal average is taken in the occupation-number representation, in which

$$\langle a_k^\dagger a_{k'}\rangle_{\mathrm{TA}} = \delta_{k,k'}(1+e^{\beta\Delta_k})^{-1}. \qquad (8.100)$$

After some algebra, we find

$$\langle S_j^+ S_{j+1}\rangle_{\mathrm{TA}} = -\frac{1}{2\pi}\int_{-\pi}^{+\pi} dk(\cos k)(J\cos k - B)\frac{\tanh(\beta\Delta_k/2)}{\Delta_k}. \qquad (8.101)$$

The generalization to non-nearest-neighbor spins can be made on the following pattern. $\langle S_0^+ S_p^-\rangle_{\mathrm{TA}}$ is first written in terms of the fermions, by (8.72):

$$\langle S_0^+ S_p^-\rangle_{\mathrm{TA}} = \left\langle c_0^\dagger \exp\left(i\pi \sum_{0\leqslant r<p}\mathbf{n}_r\right)c_p\right\rangle_{\mathrm{TA}}$$

$$= \langle c_0^\dagger (c_1^\dagger + c_1)(c_1^\dagger - c_1)\cdots(c_{p-1}^\dagger - c_{p-1})c_p\rangle_{\mathrm{TA}}.$$

The c_j's are expanded in plane wave operators a_k which are then transformed by (8.97). The average over resultant multinomials in the a_k's is now performed in the representation in which H is diagonal. Thus, only pairings of type (8.100) and their products survive. An explicit example of this type of calculation occurs in connection with the spontaneous magnetization in the 2D Ising model, and is detailed later in this chapter.

Let us now examine time-dependent correlations. Because the time-developed state $|t\rangle$ satisfies Schrödinger's equation,

$$\mathscr{H}|t\rangle = \hbar i \frac{\partial}{\partial t}|t\rangle \qquad (8.102)$$

it may be given the formal solution

$$|t\rangle = e^{-i\mathscr{H}t}|0\rangle \qquad (8.103)$$

the only normalized state satisfying the initial condition. It follows that the probability of decaying into a state $|\alpha\rangle$ at time t is:

$$P_\alpha(t) \equiv |\langle\alpha|e^{-i\mathscr{H}t}|0\rangle|^2 . \tag{8.104}$$

In particular, we are interested in the probability that the system stays in the initial state.

The probability of remaining in $|0\rangle$ is:

$$P_0(t) = |\langle 0|e^{-i\mathscr{H}t}|0\rangle|^2 \tag{8.105}$$

while the decay is measured by $1 - P_0(t)$. The $\mathscr{H}$ that appears above is given in (8.77). As we have seen, there is a sequence of unitary transformations that bring it into diagonal form (H_D). Denote these by a single compound operator Ω, so symbolically:

$$e^{-i\Omega}\mathscr{H}(8.77)e^{i\Omega} = H_D \tag{8.106}$$

and therefore,

$$P_0(t) = |\langle 0|e^{i\Omega}e^{-i\mathscr{H}_D t}e^{-i\Omega}e^{i\mathscr{H}_D t}e^{-i\mathscr{H}_D t}|0\rangle|^2$$

$$= |\langle 0|e^{i\Omega(0)}e^{-i\Omega(t)}e^{-i\mathscr{H}_D t}|0\rangle|^2 , \tag{8.107}$$

in which

$$\Omega(t) = e^{-i\mathscr{H}_D t}\Omega e^{+i\mathscr{H}_D t} \tag{8.108}$$

is defined in a natural way. At this point, the Bogolubor transformation (8.97) *defines* Ω, and we find

$$P_0(t) = \prod_{k>0}|\langle 0|\exp[\theta_k(a_k^\dagger a_{-k}^\dagger - \text{H.c.})]\exp[-\theta_k(a_k^\dagger a_{-k}^\dagger e^{-i\Delta_k t} - \text{H.c.})]|0\rangle|^2$$

$$= \exp\left[-\sum_{k>0}\ln\left(1 - \sin^2 2\theta_k \sin^2 \frac{1}{2}\Delta_k t\right)^{-1}\right]$$

$$= \exp\left[-\frac{N}{2\pi}\int_0^\pi dk \ln\left(1 - \sin^2 2\theta_k \sin^2 \frac{1}{2}\Delta_k t\right)^{-1}\right]. \tag{8.109}$$

When $J = 0$, the state $|0\rangle$ as defined above is the ground state, the parameters θ_k are all zero, and $P(t) \equiv 1$. When $J \neq 0$, the probability $P(t)$ becomes zero instantly [in time $O(1/N)$] because of the factor N in the exponent; there are no recursion times (Poincaré cycles) when the sum is replaced by an integral as in the equation above, as is proper in the thermodynamic limit. The above is an introduction to relaxation phenomena.

The decay of states closer to the eigenstates of $\mathscr{H}$ is less dramatic. The reader interested in magnetic relaxation will find a substantial literature.[16]

We next turn to some purely mathematical considerations generalizing these results. The casual reader will wish to skip the next section, and turn instead to the theory of the two-dimensional Ising model which follows it.

8.7.   Concerning Quadratic Forms of Fermion Operators

In the preceding, we ignored boundary conditions, variable J_n's and B_n's and yet other complications, in order to acquaint the reader with the salient features of what turned out to be, after all, a rather involved problem. This section is devoted to the "loose ends" and to some esoterica.

First, concerning boundary conditions, if we really have *free ends* we cannot use plane waves and the simplifying transformation (8.79) that permits the decoupling into normal modes, is inapplicable. If we have *periodic boundary conditions*, then the fermion Hamiltonian possesses a subtly more complicated structure than we had supposed and there may exist *two* ground states. Let us re-examine the Jordan–Wigner transformation of (8.70), assuming the Nth spin is bonded to the first spin. We need to know operators such as $S_N^{\pm} S_1^{\pm}$ in Fermi operator language. By using the identities $\exp(i\pi \mathbf{n}_N) c_N = c_N$ and $\exp(i\pi \mathbf{n}_N) c_N^{\dagger} = -c_N^{\dagger}$ we can incorporate *all* the occupation number operators into the phase factors:

$$
\begin{aligned}
S_N^- S_1^+ &\to e^{i\pi \mathbf{n}} c_N c_1^{\dagger} \quad \text{and} \\
S_N^+ S_1^+ &\to -e^{i\pi \mathbf{n}} c_N^{\dagger} c_1^{\dagger},
\end{aligned}
\tag{8.110}
$$

($S_N^- S_1^-$ and $S_N^+ S_1^-$ are obtainable by Hermitean conjugation). The operator $\mathbf{N}$ is the total occupation-number operator. Although it does not commute with the Hamiltonian [which creates and destroys particles in pairs, cf. (8.83)–(8.85)] parity *is* conserved (it is even or it is odd) as H does not connect even occupation numbers to odd ones, and vice versa. Therefore the phase factors in (8.110) are both $+1$ for *even* occupancies and both -1 for *odd* occupancies.

---

[16] T. A. Tjon, *Phys. Rev.* **B2**, 2411 (1970); B. McCoy, J. Perk and R. Schrock, *Nucl. Phys.* **B220**, 35, 269 (1983) and references therein.

In general, the two ground states will not be degenerate, but we do know that when $B = 0$ the Ising ground state is two-fold degenerate. It is this degeneracy which persists, in the thermodynamic limit of long chains $(N \to \infty)$, out to the critical point $|B/J| = 1$. For $|B| > |J|$, the ground state becomes unique and increasingly oriented in the direction of B as this quantity is further increased.

It is important to note that although we *start* with a unique, given, spin Hamiltonian, it transforms into *two* distinct quadratic forms in fermions. The one corresponding to even-occupancy states has a different operator representing the dynamics of the $(N, 1)$ bond than does the quadratic form representing the odd-occupancy states. *Both* differ from the typical bond $(n, n + 1)$. The even-occupancy fermion Hamiltonian should not be used to study odd-occupancy states, nor should the odd-occupancy fermion Hamiltonian be used in the calculation of even-occupancy states. If the total Hilbert space contains 2^N configurations, then each of the two distinct fermion Hamiltonians is applicable only in a subspace of $1/2 \times 2^N$ states. Certainly, the differences are small (1 bond out of N) and negligible in many cases, so that using only the periodic fermion Hamiltonian appropriate to the even-occupancy states, as in the preceding section, is adequate for many purposes.

But this entire situation is interesting in that it forces us to examine the case of fermion quadratic forms that are not explicitly homogeneous. In general, we may have arbitrary J_n's and B_n's, and we require a systematic method — not based on plane waves — to calculate the eigenvalue spectrum. The procedure we now outline is taken almost *verbatim* from the LSM paper.[17]

We wish to diagonalize the quadratic form

$$\mathcal{H} = \sum_{ij} \left[c_i^\dagger A_{ij} c_j + \frac{1}{2} (c_i^\dagger B_{ij} c_j^\dagger + \text{H.c.}) \right] \tag{8.111}$$

where the c's and $c^\dagger$'s are fermion annihilation and creation operators, A_{ij} is a real symmetric matrix and B_{ij} is a real antisymmetric matrix. (If they are not in this form, trivial phase transformations of the type $c_j, c_j^\dagger \to \exp[i\phi(j)]c_j, \exp[-i\phi(j)]c_j^\dagger$ can make them so, provided H is Hermitean.)

---

[17] E. Lieb, T. Schultz and D. Mattis, *Ann. Phys.* (*NY*) **16**, 407 (1961), Appendix A.

Problem 8.11. Verify that $\mathscr{H}$ in (8.78) is in the canonical form defined above in (8.111). Also, with k and $-k$ replacing i and j, show how (8.85) may be put in canonical form by a phase transformation.

We seek a linear transformation of the form,

$$b_k = \sum_i (g_{ki} c_i + h_{ki} c_i^\dagger) \quad \text{and}$$

$$b_k^\dagger = \sum (g_{ki} c_i^\dagger + h_{ki} c_i), \tag{8.112}$$

with real g_{ki} and h_{ki} coefficients, which is canonical so that the b_k's and $b_k^\dagger$'s satisfy the fermion anticommutation relations (8.80). It must also diagonalize $\mathscr{H}$:

$$\mathscr{H} = \sum_k E_k b_k^\dagger b_k + \text{const.} \tag{8.113}$$

If this is possible, b_k is a *lowering operator* of $\mathscr{H}$ and satisfies:

$$[b_k, \mathscr{H}] = E_k b_k. \tag{8.114}$$

Substituting (8.112) into (8.114) and equating coefficients of each operator c_i and $c_i^\dagger$ we obtain sets of equations for the g's, h's, and E_k's:

$$\sum_j (g_{kj} A_{ji} - h_{kj} B_{ji}) = E_k g_{ki} \quad \text{and}$$

$$\sum_j (g_{kj} B_{ji} - h_{kj} A_{ji}) = E_k h_{ki}. \tag{8.115}$$

These are simplified by means of the linear combinations

$$\phi_{ki} = g_{ki} + h_{ki} \quad \text{and} \quad \psi_{ki} = g_{ki} - h_{ki} \tag{8.116}$$

in terms of which the coupled equations are:

$$\phi_k(\mathbf{A} - \mathbf{B}) = E_k \psi_k \quad \text{and}$$

$$\psi_k(\mathbf{A} + \mathbf{B}) = E_k \phi_k \tag{8.117}$$

in obvious matrix notation. Eliminating ϕ or ψ, we have either

$$\phi_k(\mathbf{A} - \mathbf{B})(\mathbf{A} + \mathbf{B}) = E_k^2 \phi_k \quad \text{or}$$

$$\psi_k(\mathbf{A} + \mathbf{B})(\mathbf{A} - \mathbf{B}) = E_k^2 \psi_k. \tag{8.118}$$

For $E_k = 0$, (8.117) provides the solutions. For $E_k \neq 0$, one may solve the eigenvalue equations (8.118) and use (8.117) to relate the phase of the two solutions.

Because $\mathbf{A}$ is symmetric and $\mathbf{B}$ antisymmetric, $(\mathbf{A} + \mathbf{B})^{\mathrm{T}} = \mathbf{A} - \mathbf{B}$ so that the product matrices in (8.118) are symmetric and non-negative. With all the eigenvalues E_k real, one can choose the eigenvectors to be real as well as orthogonal. If the ϕ_k's are normalized $(\sum_i \phi_{ki}^2 = 1)$ then the ψ_k's are also, or can be chosen to be so. This ensures that

$$\sum_i (g_{ki}g_{k'i} + h_{ki}h_{k'i}) = \delta_{kk'} \tag{8.119a}$$

and

$$\sum_i (g_{ki}h_{k'i} - g_{k'i}h_{ki}) = 0, \tag{8.119b}$$

the necessary and sufficient condition for the $b_k, b_k^\dagger$ to satisfy the fermion anticommutation relations.

The energies E_k are thus the eigenvalues that emerge from the equations above. The constant in $\mathcal{H}$, (8.113), is obtained by the invariance of the trace of $\mathcal{H}$, and thus we have for (8.113):

$$\mathcal{H} = \sum_k E_k b_k^\dagger b_k + \frac{1}{2}\left(\sum_i A_{ii} - \sum_k E_k\right). \tag{8.120}$$

Now consider the case of arbitrary bonds connecting nearest-neighbor spins J_n and arbitrary fields B_n, as solved by Pfeuty.[18] The $\mathbf{A} - \mathbf{B}$ matrix takes the form

$$\mathbf{A} - \mathbf{B} = \begin{bmatrix} B_1 & \cdot & \cdot & J_N \\ J_1 & B_2 & 0 & \vdots \\ & J_2 & \cdot & \\ 0 & \cdot & \cdot & B_N \\ & & \cdot & \end{bmatrix} \tag{8.121}$$

All the B_n's and J_n's can be made $\geqslant 0$ by appropriate rotations of each of the spins through $180°$ where necessary.

The secular equations (8.119) must, in general, be solved numerically. We may, however, inquire when is there a switch-over from J_n to B_n domination,

---

[18] P. Pfeuty, *Phys. Lett.* **72A**, 245 (1979).

i.e., where is the critical point analogous to $|J/B| = 1$ of the homogeneous case. Evidently, this occurs when an eigenvalue is zero, i.e., when the determinant of the product matrix vanishes:

$$\mathrm{Det} \|(\mathbf{A} - \mathbf{B})(\mathbf{A} + \mathbf{B})\| = 0$$

but because $\mathbf{A} - \mathbf{B}$ is the transpose of $\mathbf{A} + \mathbf{B}$ it suffices to determine the point at which

$$\mathrm{Det}\|\mathbf{A} - \mathbf{B}\| = \prod_n |B_n| - \prod_n |J_n| = 0 \qquad (8.122)$$

vanishes. Taking logarithms, we establish the *critical point* for *any* given ensemble of *arbitrary or random* B_n's and J_n's *as*

$$\langle \ln |B_n| \rangle = \langle \ln |J_n| \rangle , \qquad (8.123)$$

where the averages are taken over the given ensemble.

After diagonalization, one often wishes to calculate ground-state or thermal-averaged correlation functions. This can be done directly with the matrices introduced in this section, but the algebra becomes rather involved and the interested reader is referred to the original LSM paper.[17]

Finally, from the title of this section it is clear that we can do nothing for next-nearest neighbor interactions (quartic in fermions), spins greater than spins 1/2 (fermions are altogether inappropriate there), or the full Heisenberg model. However, two generalizations of the Ising model: the *XY* and the alternating Heisenberg-Ising chain *can* be reduced to quadratic forms in fermions, and were solved in LSM.

8.8.   Two-Dimensional Ising Model: The Transfer Matrix

The manner in which one builds up a square lattice starting from linear chains dictates the construction of the transfer matrix in the two-dimensional Ising model. Figure 8.7 illustrates the process of laying out the chains horizontally, one below the next. The bonds *within* the jth chain are $J^x_{j,1}$, $J^x_{j,2}, \ldots, J^x_{j,n}, \ldots, J^x_{j,N-1}$ with the superscript x referring to the horizontal direction of the bond, the first subscript to the chain and the second subscript to the two spins $(n, n+1)$ which are connected by it. (In treating the *single* chain of Sec. 8.5 we could dispense with both superscript x and chain index j, and wrote such bonds merely as J_n.)

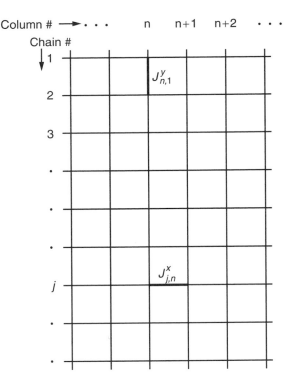

Fig. 8.7. Arbitrary bonds J^x (horizontal) and J^y (vertical) on sq lattice. In the homogeneous case, these are labeled J_1 and J_2.

The feature which distinguishes the two-dimensional model from an array of independent chains is the presence of vertical bonds similarly labeled $J^y_{n,j}$. Here y indicates the direction of the bond (now vertical), n the column, and j the pair of spins in the jth and $j + 1$st chains which are connected by it. In addition to the horizontal bonds, the transfer matrix must incorporate these inter-chain connections.

Let us start with the horizontal contributions. The transfer matrix of (8.40) $\mathbf{V}_n$ must now be relabeled, $\mathbf{V}^x_{j,n}$ and the total transfer operator is the product over all chains,

$$\prod_{j=1}^{N} \mathbf{V}^x_{j,n} \equiv \mathbf{V}^x_n. \tag{8.124}$$

The vertical contributions, i.e., the bonds internal to (within) the nth column, carry a Boltzmann factor denoted $\mathbf{V}^y_{j,n}$, each being

$$\mathbf{V}^y_{j,n} \equiv \exp(\beta J^y_{n,j}, S^z_{n,j} S^z_{n,j+1})$$

and the total being

$$\mathbf{V}_n^y = \prod_{j=1}^{N} \mathbf{V}_{n,j}^y \,. \tag{8.125}$$

Defining the transfer matrix as the combination

$$\mathbf{V}_n \equiv \mathbf{V}_n^x \mathbf{V}_n^y \,, \tag{8.126}$$

we can evaluate the partition function

$$Z = \mathrm{Tr}\{\mathbf{V}_1 \mathbf{V}_2 \cdots \mathbf{V}_{N-1}\} \tag{8.127}$$

for an array of N columns. If the last column is connected to the first (PBC, i.e., periodic boundary conditions) we label the additional transfer operator $\mathbf{V}_N$, and include it also.

It would be desirable to be able to evaluate Z for arbitrary bonds connecting (n, j) to $(n \pm 1, j)$ and $(n, j \pm 1)$, for, in addition to the ferromagnet, that would include as special cases (a) the Edwards-Anderson (E.-A.) spin glass,[19] (b) arbitrary frustration,[20] (c) the extended Ising model[21] and most importantly, (d) the three-dimensional Ising model.[22] This last has stood notoriously unsolved to this very day, while the preceding examples are also all instances of acknowledged difficulty. It seems we can only evaluate Z in the simplest instances, some of which are demonstrated in this chapter.

Our first choice is for a sq lattice (2D), all horizontal n.-n. bonds being the same J_1 and all vertical n.-n. bonds being the same J_2, actually no more complicated than the isotropic limit $J_1 = J_2 = J$. With $B_{n,j} = B$, a constant and N_x the number of columns, the trace in (8.127) can be evaluated as just λ^{N_x}, where λ is the largest eigenvalue of the horizontal *transfer matrix* $\mathbf{V}$ that propagates the nth column to the $n + 1$st,

$$V_n \cdot u = \lambda u \,. \tag{8.128}$$

---

[19] EA spin glass; here, the bonds are all $\pm J$ with $\pm$ assigned at random in proportion p, $1 - p$, with p being the adjustable parameter of the theory.

[20] Consider the product of the bonds forming the perimeter of any given plaquette: if the product is negative (AF) we say the plaquette is *frustrated*. As an example, the E-A spin glass on the sq lattice is 50% frustrated for $p = 0.5$.

[21] By setting some bonds $J = \infty$ or 0 one effectively alters the topology.

[22] Of various approaches the "Grassmann algebra" formulation offers the beat hope for success. Grassmann variables are defined at each site; they are "classical" in every way, except they *anticommute*. For more information see K. Nojima, "Review: the Grassmann representation of Ising Model," *Int. J. Mod. Phys.* **B12**, 1995 (1998).

The dimensions of $\mathbf{V}$ are $2^{N_y} \times 2^{N_y}$, N_y being the height of each column. By extensivity, $\lambda = \exp(AN_y)$. Then $Z = [\exp(AN_y)]^{N_x} = \exp(AN)$, proving that the free energy $(F = AkT)$ is also properly extensive — no matter what the aspect ratio of the rectangle might be.

Although at this point the eigenvalue problem is well posed and its solution will yield Z, hence $f = F/N$ and all the thermodynamic functions that are calculated as derivatives of f, the form of V^x leaves something to be desired. This operator includes B and is complicated even in homogeneous fields $(B_n = \text{const.})$. It so happens that we can simplify the problem considerably by including the external field terms with the vertical bonds, in V^y. We therefore reformulate, starting with $\mathcal{H}$. Let

$$\mathcal{H} = \sum_n \mathcal{H}_n \qquad (8.129)$$

where each $\mathcal{H}_n$ contains all the new bonds introduced when the $n + 1$st column is added, and includes the interactions of spins in that column with any external field. Thus, $\mathcal{H}_n = \mathcal{H}_n^x + \mathcal{H}_n^y + \mathcal{H}_n^B$, where the horizontal bonds are contained in $\mathcal{H}_n^x$,

$$\mathcal{H}_n^x = -J_1 \sum_{j=1}^{N} S_{n,j}^z S_{n+1,j}^z , \qquad (8.130a)$$

the vertical bonds being

$$\mathcal{H}_n^y = -J_2 \sum_{j=1}^{N-1} S_{n+1,j}^z S_{n+1,j+1}^z . \qquad (8.130b)$$

Finally, the interactions which the spins in the new column have with an external field are included in

$$\mathcal{H}_n^B = -B \sum_{j=1}^{N} S_{n+1,j}^z . \qquad (8.130c)$$

The reader will verify that a sum of the above operators over all n yields the total Hamiltonian for the model, with the possible exception of boundary effects irrelevant to the computation of f, the free energy per spin.

Exponentiating the $\mathscr{H}^x$ operator, we can use (8.40) with $B = 0$ to obtain for the individual chains,

$$\mathbf{V}^x_{j,n} = \exp(\beta J_1)\mathbf{1} + \exp(-\beta J_1)\boldsymbol{\sigma}^x_j$$

$$= [2 \, \sinh(2\beta J_1)]^{1/2} \exp(K^*_1 \boldsymbol{\sigma}^x_j) \,. \tag{8.131}$$

This serves to define a new real parameter K^*_1, using the identity $\exp(K^* \boldsymbol{\sigma}^x) = \cosh K^* + \boldsymbol{\sigma}^x \sinh K^*$, and the results of Problem 8.12.

Problem 8.12.   Verify that (8.131) requires

$$\tanh K^*_1 = \exp(-2K_1) \quad (\text{with } K_1 \equiv \beta J_1) \,, \tag{8.132}$$

and also check the duality relations,

$$\tanh K_1 = \exp(-2K^*_1) \quad \text{and} \quad \sinh 2K_1 \sinh 2K^*_1 = 1 \,. \tag{8.133}$$

With V^x_j in the form of the rhs of (8.131) it is an easy matter to include the contributions of all horizontal chains $j = 1, 2, \ldots$:

$$\mathbf{V}^x = [2 \, \sinh(2K_1)]^{1/2N} \exp\left(K^*_1 \sum_j \sigma^x_j \right) \,. \tag{8.134a}$$

The interchain contributions are summed in

$$\mathbf{V}^y = \exp\left(K_2 \sum_j \sigma^z_j \sigma^z_{j+1} \right) \,, \tag{8.134b}$$

and the external field appears in

$$\mathbf{V}^B = \exp\left(h \sum_j \sigma^z_j \right) \tag{8.134c}$$

with $K_2 = \beta J_2$ and $h = \beta B$, in an obvious notation, and $\sigma^z_j = S^z_j$.

Now, the eigenvalue equation (8.128) has $V = V^x V^y V^B$, which suffers from the usual inconveniences in dealing with the exponentials (8.134a)–(8.134c) of noncommuting operators.

But even if we ignored the lack of commutation and combined the three exponents, the combined term takes on the appearance of the 1D Ising model in *both* transverse and longitudinal fields (K^*_1 being the effective transverse field, and h the longitudinal). In Sec. 8.5 we analyzed the latter, in Sec. 8.6

the former. Some difficulties in obtaining the eigenstates analytically in the presence of *both* fields were apparent. For this reason it is most convenient to set $B = 0$, and to calculate the magnetic properties with the aid of zero-field correlation functions. In finite B, the appropriate analysis has to be numerical. In Sec. 8.11, in which *complex* fields are analyzed, the transfer matrix formalism is abandoned altogether.

In selecting the form of V to use in the eigenvalue equation (8.128), we have several options that differ from each other by similarity transformations. The *obvious* choice, $V^x V^y V^B$ is not self-adjoint. This is an inconvenience if we seek to calculate correlation functions, as u^T will not be the left-hand eigenvector. There *are* a number of self-adjoint combinations leading to the same optimum eigenvalue;

$$(\mathbf{V}^y \mathbf{V}^B)^{1/2} \mathbf{V}^x (\mathbf{V}^y \mathbf{V}^B)^{1/2} \quad \text{and} \quad \mathbf{V}^{x1/2}(\mathbf{V}^y \mathbf{V}^B) \mathbf{V}^{x1/2} \qquad (8.135)$$

are just two of them. (Because V^y and V^B involve the same Pauli matrices, they commute and can be written in any order, or combined into a single exponential form.)

Although the derivation of the transfer matrix and the resulting discussions have centered about the 2D case, the generalization to 3D or higher, or to non-nearest neighbor bonds is obvious. It suffices to include in V^y *all* the bonds which go into growing the lattice layer by layer. In 3D, the column becomes a plane; each spin must be labeled by *two* indices — its position in the plane. In 4D it is a volume and 3 indices are required. In d dimensions, the transfer matrix describes a $d - 1$-dimensional Ising model subject to the transverse operators in V^x; not surprisingly, these are called "disorder operators" and are dominant at high temperature (K^* increases with T, while K decreases). The operators in V^y and V^B are the "order operators", dominating at low temperature.

Where it is solvable, the transfer matrix formalism simplifies the statistical mechanics. We need only the largest eigenvalue of a model in one fewer dimension, complicated only by the presence of the new disorder operators.

There exists an alternative derivation of the transfer matrix formalism, somewhat more natural than the preceding as it clarifies the meaning of the eigenvector $\mathbf{u}$ in (8.128), the eigenvector belonging to the optimum eigenvalue λ, henceforth denoted the optimal eigenvector.

Frobenius' theorem concerning the optimum eigenvector of a matrix, all elements of which are positive, states that all the components of the optimum eigenvector are themselves positive. The proof is intuitive: any change in sign can only lower the eigenvalue.

It follows that we can consider the various elements u_j of $\mathbf{u}$ as probabilities, provided the eigenvector is normalized according to the rule,

$$\sum_j u_j = 1. \tag{8.136}$$

It is also apparent that any vector with all positive elements is not orthogonal to $\mathbf{u}$. Suppose we start the first column in a 2D Ising model in *total disorder* — corresponding to each element in $\mathbf{u}_1$ being $u_{1j} = 2^{-N}$. With this choice, all configurations appear with equal probability. If we expand $\mathbf{u}_1$ in eigenvectors of $\mathbf{V}$, the optimum eigenvector $\mathbf{u}$ appears with a coefficient a_0:

$$\mathbf{u}_1 = a_0\mathbf{u} + a_1\mathbf{u}' + a_2\mathbf{u}'' + \cdots \tag{8.137}$$

indicating the other eigenvectors by $\mathbf{u}'$, $\mathbf{u}''$, etc. Applying the transfer matrix n times brings us to the $n+$1st column and yields

$$\mathbf{V}^n \cdot \mathbf{u}_1 = a_0\lambda^n\mathbf{u} + a_1\lambda'^n\mathbf{u}' + a_2\lambda''^n\mathbf{u}'' + \cdots . \tag{8.138}$$

Regardless of the original values of a_1, a_2, etc., the resulting coefficients $a_1\lambda'^n$, $a_2\lambda''^n$, etc., all tend exponentially to zero by comparison with the optimum value $a_0\lambda^n$, which alone survives in the asymptotic limit $n \to \infty$. (The possibility $a_0 = 0$ is eliminated, by noting that two vectors $\mathbf{u}$ and $\mathbf{u}_1$, both of which have all positive elements, cannot be orthogonal.)

Thus, even if we started with disorder, $u_{1j} = 2^{-N}$, repeated application of the transfer matrix brings the nth vector into the form of the optimum eigenvector, reflecting the probabilities of having the various configurations at the given temperature T. The optimal eigenvector is the "reduced density matrix" for the problem.

Repeated applications of V are tantamount to the approach to thermal equilibrium. This approach is fastest, in fact exponential, when the leading eigenvalue is separated from the others by a finite gap: $\lambda'/\lambda < 1$. This is generally the case, and convergence by this method is far more efficient than the usual solution of a secular determinant. The exception: at T_c, in the absence of any applied field, the gap disappears and the approach to thermal equilibrium is "critically damped." With this exception, the iterative approach to the largest eigenvalue is the method of choice when numerical calculations are required. Of course, it is not required to start with a perfectly random configuration. Once $\mathbf{u}$ is known for a given set of parameters (such as T or B) *it* can serve as the initial $\mathbf{u}_1$ for different parameters ($T + dT$, $B + dB$), and convergence is enhanced even if just low-order perturbation theory is used to get the answers.

8.9. Solution of Two-Dimensional Ising Model in Zero Field

It is particularly convenient to rotate the spin operators by $90°$, so that $\sigma^x \to -\sigma^z$ and $\sigma^z \to +\sigma^x$ just as in (8.77), and obtain

$$\mathbf{V}^x = [2 \sinh(2K_1)]^{1/2N} \exp\left(-K_1^* \sum_j \sigma_j^z\right), \qquad (8.139a)$$

$$\mathbf{V}^y = \exp\left(K_2 \sum_j \sigma_j^x \sigma_{j+1}^x\right) \qquad (8.139b)$$

and

$$\mathbf{V}^B = \exp\left(h \sum_j \sigma_j^x\right). \qquad (8.139c)$$

In zero field, $h = 0$, and $\mathbf{V}^B = 1$. The exponents in (8.139a) and (8.139b) are precisely the two terms in (8.77) — with K_1^* replacing B in the earlier equation, and K_2 replacing J. Our method of solution will be similar in its outline and follows the paper of SML.[23]

First, writing $\sigma_j^x = S_j^+ + S_j^-$ and $\sigma_j^z = 2S_j^+ S_j^- - 1$, then invoking the Jordan–Wigner transformation to fermion operators c_j and $c_j^\dagger$, we obtain:

$$\mathbf{V}^x = [2 \sinh(2K_1)]^{1/2N} \exp\left[-2K_1^* \sum_j \left(c_j^\dagger c_j - \frac{1}{2}\right)\right] \qquad (8.140a)$$

and

$$\mathbf{V}^y = \exp\left[K_2 \sum_j (c_j^\dagger - c_j)(c_{j+1}^\dagger + c_{j+1})\right]. \qquad (8.140b)$$

In Sec. 8.7 we already noted that the boundary conditions may affect the Jordan–Wigner transformation. Supposing that we want the *spin*

---

[23] T. Schultz, D. Mattis and E. Lieb, *Rev. Mod. Phys.* **36**, 856 (1964).

Hamiltonian to be periodic in the y direction (forming a cylinder of circumference N and length N) then mathematically we can retain the form of the exponentials in fermions in (8.140) without modification, the bond connecting $N \to N+1$ being the same as the others, provided we identify c_{N+1} with c_1 in the correct way. This requires:

$$c_{N+1} = -c_1 \quad \text{and} \quad c_{N+1}^\dagger = -c_1^\dagger \tag{8.141a}$$

for even occupancy states, whereas

$$c_{N+1} = c_1 \quad \text{and} \quad c_{N+1}^\dagger = c_1^\dagger \tag{8.141b}$$

for odd-occupancy states.

Both are achievable by an expansion in plane waves; the first by the use of anti-periodic boundary conditions, the second with the more conventional periodic boundary conditions(PBC). Label the transfer matrix for the even occupancy case V^+, for the odd-occupancy case, V^-. The odd eigenstates of V^+ should be discarded, and likewise, the even eigenstates of V^- are also to be discarded.

The second possible boundary condition on the spins, free ends, is conceptually simpler: the N–to–1 bond is absent; the natural waveforms are sin kj rather than the plane waves. We shall leave the analysis of this case (which should parallel closely the following derivations for the periodic/antiperiodic cases) as an exercise for the reader, with Sec. 8.7 as a guide.

Because of translational symmetry, we again try a plane-wave expansion, writing the c_j's in the form

$$c_j = \frac{1}{N^{1/2}} e^{-1/4i\pi} \sum_q e^{iqj} a_q \tag{8.142}$$

and similarly for the $c_j^\dagger$'s. The factor $\exp(-i\pi/4)$ is incorporated in anticipation of a transformation similar to (8.97). The a_q's are a set of fermion operators, with values of q chosen to satisfy anticyclic boundary conditions for even occupancy, or PBC for odd occupancy.

That is, from (8.141a) above, we see $\exp(iqN) = -1$, and

$$q = \pm\pi/N, \pm 3\pi/N, \ldots, \pm(N-1)\pi/N, \tag{8.143a}$$

whereas from (8.141b) with $\exp(iqN) = +1$, we obtain the more conventional

$$q = 0, \pm 2\pi/N, \pm 4\pi/N, \ldots, \pm(N-2)\pi/N, +\pi. \tag{8.143b}$$

Each of V^x and V^y is now of the form $\prod V_q^x$ and $\prod V_q^y$ with $0 \leqslant q \leqslant \pi$. Factoring out the ubiquitous $[2 \sinh(2K_1)]^{1/2N}$, what remains for each normal mode is

$$V_q^x = \exp[-2K_1^*(a_q^\dagger a_q + a_{-q}^\dagger a_{-q} - 1)] \quad \text{and} \tag{8.144}$$

$$V_q^y = \exp\{2K_2[\cos q(a_q^\dagger a_q + a_{-q}^\dagger a_{-q} - 1) + \sin q(a_q a_{-q} + a_{-q}^\dagger a_q^\dagger)]\} \tag{8.145}$$

for $q \neq 0$ or π. (In those special cases, for which there is no $-q$, we have

$$V_0 = \exp\left[-2(K_1^* - K_2)\left(a_0^\dagger a_0 - \frac{1}{2}\right)\right],$$

$$V_\pi = \exp\left[-2(K_1^* + K_2)\left(a_\pi^\dagger a_\pi - \frac{1}{2}\right)\right], \tag{8.146}$$

combining V^x and V^y at these points where, exceptionally, they commute.)

It is also possible to combine V_q^x and V_q^y at arbitrary q by some sort of Baker-Hausdorff expansion. Many algorithms have been devised to combine exponential operators into a single exponential or, conversely, to disentangle an exponential of noncommuting operators into products of exponentials. One such relation, the Zassenhaus formula, is

$$\exp[g(A + B)] = \exp(gA)\exp(gB)$$
$$\times \exp(g^2 C_2)\exp(g^3 C_3)\cdots\exp(g^n C_n)\cdots \tag{8.147}$$

where

$$C_2 = \frac{1}{2}[B, A], \quad C_3 = \frac{1}{6}[C_2, A + 2B], \ldots$$

$$C_n = \frac{1}{n!}\left[\frac{d^n}{dg^n}\left(\exp(-g^{n-1}C_{n-1})\cdots\exp(-g^2 C_2)\right.\right.$$

$$\left.\left.\cdot \exp(-gB)\cdot\exp(-gA)\exp[-g(A + B)]\right)\right]_{g=0}. \tag{8.148}$$

Read from left to right, (8.147) is used to disentangle A and B; from right to left, to combine them. This formula has many uses beyond its present application.

For our purposes, it is sufficient to note that if A and B are bilinear in fermion operators, so are C_2, C_3, and all the other C_n's. In fact, they are *all* linear combinations of what amounts to Pauli spin-matrices in occupation-number space, as explicitly demonstrated in Problem 8.13.

Problem 8.13.   With A and B any linear combinations of the operators:

$$X \equiv a^{\dagger}_{-q}a^{\dagger}_{q} + a_{q}a_{-q}, \quad Y \equiv -i(a^{\dagger}_{-q}a^{\dagger}_{q} - a_{q}a_{-q}), \quad Z \equiv n_{q} + n_{-q} - 1$$

show that C_2, C_3, ... will all be of this form also. Show that these three operators are the Pauli matrices in the space $|0\rangle$, $a^{\dagger}_{-q}a^{\dagger}_{q}|0\rangle$, but all *vanish* in the space of $a^{\dagger}_{q}|0\rangle$ and $a^{\dagger}_{-q}|0\rangle$.

Thus, with V^{x}_{q} of the form $\exp(-2K^{*}_{1} Z)$ and $V^{y}_{q} = \exp\{2K_{2}[(\cos q) Z + (\sin q)] X]\}$, it is clear that we can combine them into a single exponential of the form

$$V^{y1/2}_{q}V^{x}_{q}V^{y1/2}_{q} = c \, \exp(a X + b Z) \tag{8.149}$$

(in which a, b, c are constants), and then brought into diagonal form — defined as V_q — by a rotation in spin space. Since the lhs of this equation yields 1 for the odd-occupancy states, as does the rhs, the equality would extend to all 4 states in the $q, -q$ subspace. Or, alternatively we use an elegant trick to obtain the same result. Given the diagonal matrix

$$V_q = c_q \exp\{\varepsilon_q(n_q + n_{-q} - 1)\}, \tag{8.150}$$

in which

$$b_q = a_q \cos\theta_q + a^{\dagger}_{-q}\sin\theta_q, \quad \text{and} \quad n_q = b^{\dagger}_q b_q; \tag{8.151}$$

then by inspection: $V^{-1}_q b_q V_q = b_q e^{\varepsilon_q}$, a sort of generalization of equations of motion. By the same token, $((V^{y}_{q})^{1/2}V^{x}_{q}(V^{y}_{q})^{1/2})^{-1}(a_q \cos\theta_q + a^{\dagger}_{-q}\sin\theta_q)(V^{y}_{q})^{1/2}V^{x}_{q}(V^{y}_{q})^{1/2}$ is *exactly the same* and should therefore equal $(a_q \cos\theta_q + a^{\dagger}_{-q}\sin\theta_q)e^{\varepsilon_q}$. This can be solved for $\vartheta_q, \varepsilon_q$. Implementation is left to our diligent reader, with the following result[23]:

$$\tan(2\theta_q) = \frac{2C(q)}{B(q) - A(q)} \tag{8.152}$$

in which $\text{sgn}(2\theta_q) = \text{sgn}(q)$, and

$$A(q) = \exp(-2K_1^*)(\cosh K_2 + \sinh K_2 \cos q)^2$$
$$+ \exp(2K_1^*)(\sinh K_2 \sin q)^2 \tag{8.153a}$$

$$B(q) = \exp(-2K_1^*)(\sinh K_2 \sin q)^2$$
$$+ \exp(2K_1^*)(\cosh K_2 - \sinh K_2 \cos q)^2 \tag{8.153b}$$

$$C(q) = (2 \sinh K_2 \sin q)$$
$$\times (\cosh 2K_1^* \cosh K_2 - \sinh 2K_1^* \sinh K_2 \cos q). \tag{8.153c}$$

Finally, ε_q is the positive root of

$$\cosh \varepsilon_q = \cosh 2K_2 \cosh 2K_1^* - \sinh 2K_2 \sinh 2K_1^* \cos q. \tag{8.154}$$

In this diagonal representation,

$$\mathbf{V} = (2 \sinh 2K_1)^N \exp\left[-\sum_{\text{all } q} |\varepsilon_q|\left(b_q^* b_q - \frac{1}{2}\right)\right]. \tag{8.155}$$

For *even* total occupancy, the set (8.143a) of q's is to be used in the sum; for *odd* total occupancy, the set (b). Precisely at $K_1^* = K_2$ the eigenvalue of the $q = 0$ mode vanishes, so *it can be occupied or not*, without changing Z. At this temperature — and, below it as well, the vacua for even occupancy and for odd occupancy are degenerate in the large N limit. It is a simple algebraic exercise to verify: (i) only at this critical point $K_1^* = K_2$ is the spectrum gapless and linear at small q ($\lim q \to 0\ \varepsilon_q \sim |q|$), (ii) at all other temperatures there is an energy gap against excitations (of magnitude $|K_1^* - K_2|$), and (iii) that this yields the same critical temperature as the more familiar condition

$$\sinh(2J_1/kT_c)\ \sinh(2J_2/kT_c) = 1, \tag{8.156}$$

i.e., $kT_c/J = 2.269185\ldots$ for $J_1 = J_2 = J$. It is almost self-evident that Z is given by setting all $b_k^\dagger b_n = 0$ in $\mathbf{V}$,

$$Z = (2 \sinh 2K_1)^N \exp\left(\frac{1}{2}\sum_q |\varepsilon_q|\right) \tag{8.157}$$

and the free energy per spin, f, by

$$f = -kT \left[\ln(2 \sinh 2K_1)^{1/2} + \frac{1}{4\pi} \int_{-\pi}^{+\pi} dq |\varepsilon_q| \right]. \tag{8.158}$$

It is possible to solve for ε_q ((8.154) is for $\cosh \varepsilon_q$) and evaluate f. Although this must be done numerically, the derivatives of f can be expressed in terms of elliptic functions, and one finds that the second derivative — the specific heat — involves K [given in (8.94) and (8.95)] and therefore diverges logarithmically at T_c. And while (8.158) is not manifestly symmetric in the two parameters J_1 and J_2, a simple integral identity transforms it into a two-dimensional integral, proving F to be fully symmetric under a 90° rotation of the plane.

With ε_q replacing x in Onsager's identity we have for the integral above

$$\frac{1}{4\pi} \int_{-\pi}^{+\pi} dq \frac{1}{2\pi} \int_{-\pi}^{+\pi} dq' \ln(2 \cosh \varepsilon_q - 2 \cos q')$$

into which we may insert the solution (8.154) for $\cosh \varepsilon_q$,

$$\frac{1}{4\pi} \int dq \frac{1}{2\pi} \int dq'$$

$$\times \ln(2 \cosh 2K_2 \cosh 2K_1^* - 2 \sinh 2K_2 \sinh 2K_1^* \cos q - 2 \cos q').$$

Factoring out $2 \sinh 2K_1^* = (\frac{1}{2} \sinh 2K_1)^{-1}$ (cf. (8.133) in Problem 8.12) we obtain

$$-\frac{(2\pi)^2}{8\pi^2} \ln \left(\frac{1}{2} \sinh 2K_1 \right) + \frac{1}{8\pi^2} \int dq_1 \int dq_2$$

$$\times \ln(\cosh 2K_2 \coth 2K_1^* - \sinh 2K_1 \cos q_1 - \sinh 2K_2 \cos q_2)$$

so that finally, with the aid of the duality relation (8.132) and standard trigonometric identities we eliminate $\coth 2K_1^*$

$$\coth 2K_1^* = \frac{1 + \exp(-4K_1^*)}{1 - \exp(-4K_1^*)} = \frac{1 + \tanh^2 K_1}{1 - \tanh^2 K_1} = \cosh 2K_1,$$

to arrive at Onsager's expression:

$$f = -kT\left[\ln 2 + \frac{1}{2\pi^2}\int\!\!\int_0^\pi dq_1 dq_2 \ln(\cosh 2K_1 \cosh 2K_2\right.$$

$$\left. - \sinh 2K_1 \cos q_1 - \sinh 2K_2 \cos q_2)\right]. \tag{8.159}$$

Despite a surprising resemblance to a similar expression in the 2-dimensional Gaussian model, Eqs. (7.111) *this* free energy is well-behaved both above and below T_c. A new feature — the hyperbolic functions, trademarks of the Ising model — makes for well-behaved thermodynamic functions.

To compare with the Gaussian result, consider the isotropic case $J_1 = J_2$ for which T_c is given by $\sinh(2J/kT_c) = 1$.

If we write

$$\sinh 2K = 1 - t, \tag{8.160}$$

we may consider t a small parameter in the critical region, and f can now be written

$$f_I \cong -kT\left\{\ln[2(1-t)^{1/2}] + \frac{1}{2\pi^2}\int\!\!\int_0^\pi dq_1 dq_2\right.$$

$$\left.\times \ln\left(\frac{t^2}{1-t} + 2 - \cos q_1 - \cos q_2\right)\right\}. \tag{8.161}$$

By contrast, the Gaussian model on the sq lattice, yields

$$f_G \cong +\frac{1}{2}kT\left[\ln(2J/kT) + \frac{1}{2\pi^2}\int\!\!\int_0^\pi dq_1 dq_2 \ln(\hat{t} + 2 - \cos q_1 - \cos q_2)\right]$$

with $\hat{t} = \frac{k}{2J}(T - T_G)$, where $T_G = 4J/k$ on this lattice. $\tag{8.162}$

(The spherical model can be written in similar form, with $\hat{t}$ involving μ (i.e., τ) as well, but as $T_c = 0$ for the spherical model in 2D it is not interesting to pursue the analogy.)

So the main difference between the Ising and the Gaussian free energies is the overall $\pm$sign. Secondarily, the singularity at t (or $\hat{t}$) $= 0$ is approached as t^2 in the former, and linearly as $\hat{t}$ in the latter. So while the internal energy diverges (unphysically!) in the Gaussian approximation, it is the specific heat

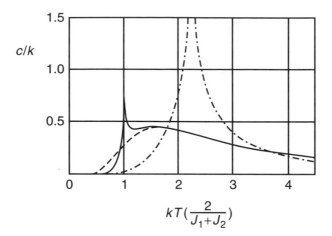

Fig. 8.8.   Specific heat for anisotropic 2D Ising model on sq lattice: $J_2 = J_1/100$ (———), limits of independent chains $J_2 = 0$ (- - - -), as compared to the isotropic result, (8.164) (-·-·-·-).[9]

— a second derivative of f with respect to t — that is weakly, logarithmically, singular at T_c in the present case.

The specific heat for this case and in the anisotropic $(J_1 \neq J_2)$ limits is sketched in Fig. 8.8.

The exact formulas for the internal energy and specific heat of the Ising ferromagnet on an isotropic sq lattice are:

$$U/N = -J \coth(2K) \left[1 + \frac{2}{\pi} k_1'' K(k_1) \right] \quad \text{with} \quad K = \beta J,$$

$$k_1 = 2 \sinh(2K)/\cosh^2 2K \quad \text{and} \quad k_1'' = 2 \tanh^2 2K - 1.$$

$$(8.163)$$

The specific heat is logarithmically singular at T_c.

$$c/k_B = \frac{2}{\pi}(K \coth K)^2$$

$$\times \left\{ 2K(k_1) - 2E(k_1) - (1 - k_1'') \left[\frac{1}{2}\pi + k_1'' K(k_1) \right] \right\} \qquad (8.164)$$

the elliptic integrals are defined in (8.94).

It is not surprising that the same integrals occur in this study as in the ground-state properties of the one-dimensional Ising model in a transverse field. For if in the present section we had merely combined the exponentials

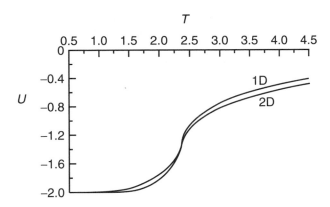

Fig. 8.9.   Plot of internal energy $U(T)$ of isotropic Ising model on 2D sq lattice versus equivalent quantity obtained from ground state of 1D Ising model in transverse magnetic field. The agreement is exact at the three critical points: $T = 0$, T_c and ∞.[29]

in (8.139a) and (8.139b) without regard to the commutators (8.147) which go into the exact expression, the present problem would have reduced to that of Sec. 8.6. Indeed, the correspondences are:

$$\text{Zero-Field 2D Ising Model} \longleftrightarrow \text{1D Ising in Transverse Field}$$

$$-f/kT \qquad\qquad\qquad E_0/N$$
$$K_1^* \qquad\qquad\qquad B$$
$$K_2 \qquad\qquad\qquad J$$

The internal energy as calculated with the method of Sec. 8.6 is compared with the exact results, (8.163), in Fig. 8.9. The agreement is excellent at low T, at high T (and at T_c which is the same in the approximate and exact analyses).

In Table 8.2 below, we list some of the calculated properties of the Ising ferromagnet on three of the important two-dimensional lattices: sq, triangular and honeycomb (hexagonal). Although *anti*ferromagnets can also be studied by these same methods, their solutions yield surprises as we shall see subsequently.

In the following section, we turn to the magnetic properties: spontaneous magnetization of the ferromagnet below T_c, and magnetic susceptibility above T_c.

Table 8.2.   Critical properties of Ising ferromagnets on the three principal two-dimensional lattices[a] [Domb[11], pp. 235ff].

	Sq	Triangular	Honeycomb (Hex.)
$\exp(-2J/kT_c)$[a]	$2^{1/2}-1$	$3^{-1/2}$	$2-3^{1/2}$
kT_c/J	2.269185	3.640957	1.518652
U_c/NJ (Internal energy at T_c)	$-2^{1/2}$	-2	$-2/3^{1/2}$
S_c/Nk (Entropy at T_c)[b]	0.30647	0.33028	0.26471

[a]L. Onsager has noted that for these 3 lattices, arctan $(\sinh 2K_c) = \pi/z$, with $z =$ coordination number (4, 6, 3).
[b]Note: $S_\infty/Nk = \ln 2 = 0.69315$.

8.10.   Spontaneous Magnetization and Magnetic Susceptibility

There are compelling reasons for identifying the magnetization with long-range order (LRO). In molecular-field theory, they are indistinguishable. In *any* theory based on spin variables,

$$M_z^2(T) = \frac{1}{N}\left\langle\left(\sum_i S_i^z\right)^2\right\rangle_{\text{TA}} = \lim_{R_{ji}\to\infty}\langle S_i^z S_j^z\rangle_{\text{TA}}. \tag{8.165}$$

Divison by the total number of spins N in the thermodynamic limit elimi-nates all but the asymptotic correlations.

The susceptibility, on the other hand, is related to the manner in which the correlations decay to the asymptotic value. Specifically,

$$\chi_{zz}(T) = \frac{\beta}{N}\left\langle\left[\sum_i S_i^z - M_z(T)\right]^2\right\rangle_{\text{TA}}$$

$$= \beta\sum_j\langle S_i^z[S_j^z - M_z(T)]\rangle_{\text{TA}} \tag{8.166}$$

indicating by the subscript zz that this expression stands for the second derivative, $-\partial^2 f/\partial B_z^2|_{B=0}$. One derives similar expressions for mixed deriva-tives, e.g., $-\partial^2 f/\partial B_z\partial B_x|_{B=0}$ and labels them analogously. The transverse susceptibility discussed in Sec. 8.6 is χ_{xx}.

By calculating correlation functions such as the above, one dispenses with the magnetic field in the transfer matrix to make use of the results obtained so far. Actually, in *numerical* calculations of the largest eigenvalue of the transfer matrix, or in certain series calculations, an external field speeds up the convergence. Unfortunately, it introduces great complexity into the Jordan–Wigner transformation. We choose distant spins in (8.165) along the same column, avoiding sandwiching transfer matrices between the spins, cf. (8.56) and (8.58). We must follow the transformations through which we diagonalized $\mathbf{V}$, starting with the substitution of S_i^z by the Pauli matrix $\sigma_i^z \to \sigma_i^x$, thence to fermions and finally to normal modes (eigenstates of $\mathbf{V}$). It is convenient to use an identity, (8.73),

$$\exp(i\pi \mathbf{n}_m) = (c_m^\dagger + c_m)(c_m^\dagger - c_m) \tag{8.167}$$

to obtain a compact expression for the correlations among the S_i^z of (8.130):

$$S_i^z S_j^z \Rightarrow (c_i^\dagger - c_i)(c_{i+1}^\dagger + c_{i+1})(c_{i+1}^\dagger - c_{i+1}) \cdots (c_{j-1}^\dagger - c_{j-1})(c_j^\dagger + c_j)$$

and

$$\langle S_i^z S_j^z \rangle_{\text{TA}} \Rightarrow (0|(c_i^\dagger - c_j) \cdots (c_j^\dagger + c_j)|0) . \tag{8.168}$$

The thermal average is just the ground-state expectation value of this expression, taking care that there may be two ground states! It is easily calculable by Wick's theorem, which states: "associate the operators in pairs in their original order, replace each pair by its ground state expectation, multiply the product of these pairs by $(-)^p$ (where $(-)^p$ is the parity of the permutation required to bring the paired operators back into the original ordering) then sum over all possible distinct such pairings". We distinguish three possible pairings, but the only *nonvanishing-type pairing* takes the form:

$$a_{ij} = (0|(c_i^\dagger - c_i)(c_{j+1}^\dagger + c_{j+1})|0) . \tag{8.169}$$

With the inverse of the Bogoliubov transformation (8.151) being

$$a_q = b_q \cos \theta_q - b_{-q}^\dagger \sin \theta_q \tag{8.170}$$

(with θ_q given in (8.152) odd in q), and (8.142) giving the c_i's in terms of the a_q's, we express (8.169) in terms of the b_q and $b_q^\dagger$, the *exact* lowering and raising operators of the transfer matrix. That is,

$$c_n^\dagger \pm c_n = \frac{1}{N^{1/2}} \sum_q e^{iqn}$$

$$\times [b_q(\sin\theta_q \pm \cos\theta_q) + b_{-q}^\dagger(\cos\theta_q \mp \sin\theta_q)] \tag{8.171}$$

and by virtue of orthogonality,

$$(0|b_q b_{q'}^\dagger|0) = \delta_{q,q'}, \tag{8.172}$$

it becomes possible to evaluate the factors a_{ij},

$$a_{ij} = a(j-i) = -\frac{1}{N} \sum_q \exp[-iq(j-1)] \exp[-i(2\theta_q + q)]. \tag{8.173}$$

Problem 8.14.   (a) Fill in the details leading to (8.173) above.
(b) Prove that the contractions such as the following, vanish:

$$(0|(c_n^\dagger \pm c_n)(c_m^\dagger \pm c_m)|0) = 0 \quad \text{for } n \neq m \tag{8.174}$$

by using (8.171) and (8.172), then verify using anticommutation relations directly.
(c) Obtain their value for $n = m$ by the same methods, then confirm the results by transforming back to Pauli spin operators. (*Hint:* $\sigma_x^2 = 1$)

Using Wick's theorem to obtain the ground-state expectation of (8.168), we take a product of the a_{ij}'s plus all the signed permutations $\mathbb{P}$:

$$\langle S_i^z S_j^z \rangle_{\text{TA}} \Rightarrow a_{ii}a_{i+1,i+1}\cdots a_{j-1,j-1} + \sum_p (-)^\mathbb{P}\mathbb{P}(a_{ii}a_{i+1,i+1}\cdots a_{j-1,j-1}).$$

The result is, by inspection, a determinant:

$$\langle S_i^z S_j^z \rangle_{\text{TA}} = \begin{Vmatrix} a_{ii} & a_{i,i+1} & \cdots & & a_{i,j-1} \\ & a_{i+1,i+1} & \cdots & & \\ & & \cdot & \cdot & \cdot \\ & & & \cdot & \cdot \\ a_{j-2,i} & & & \cdot & \\ a_{j-1,i} & & & & a_{j-1,j-1} \end{Vmatrix} \tag{8.175}$$

the elements a_{ik} of which depend only on the distance from the main diagonal, $i-k$. These are known as *Toeplitz* determinants and differ only slightly from the corresponding *cyclic* determinants, although this difference can be significant as we see in the example below. Only the contractions of type (8.169) contribute, the other possible contractions vanishing according to (8.174).

Example:

$$
\text{Toeplitz}: \quad
\begin{Vmatrix}
010000\ldots \\
0010000.. \\
00010000. \\
\cdot\cdot\cdot\cdot\cdot \\
\cdot\cdot\cdot\cdot\cdot\cdot\cdot \\
000000000
\end{Vmatrix}
= 0,
$$

whereas

$$
\text{Cyclic}: \quad
\begin{Vmatrix}
010000\ldots \\
0010000.. \\
\cdot\cdot\cdot\cdot\cdot\cdot\cdot\cdot\cdot \\
\cdot\cdot\cdot\cdot\cdot\cdot\cdot\cdot \\
\cdot\cdot\cdot\cdot\cdot\cdot\cdot\cdot\cdot \\
100000000
\end{Vmatrix}
= 1
$$

for all dimensionality.

Happily, there exists a relevant theorem by Szegö and Kac (which might even have originated with Onsager[24] relating the Toeplitz form to the easily calculated cyclic determinant, with an accuracy $0(1/|j-i|)$. (This accuracy is just sufficient for the calculation of the magnetization but insufficient for the calculation of the susceptibility, which is perhaps one reason for the ongoing research into Toeplitz forms.) Briefly, the Kac-Szegö theorem states that the

---

[24] A review of Toeplitz matrices, and various improvements and applications thereof to statistical mechanics has been published by M. Fisher and R. Hartwig, *Adv. Chem. Phys.* **15**, 333–354 (1968). The original application to the Ising model in the familiar literature seems to be E. Montroll, R. Potts and J. Ward, *J. Math. Phys.* **4**, 308 (1963) in the Onsager anniversary issue of that journal. But Montroll *et al.* disclaim first use, and credit Onsager: "... this is one of the methods used by Onsager himself. Mark Kac alerted the authors to a limit formula for the calculation of large Toeplitz determinants which appear naturally in the theory of spin correlations in a two-dimensional Ising lattice. This formula was first discussed by Szegö (Commun. Semin. Math. Univ. Lund, Tome Suppl. (1952) dédié à M. Riesz, p. 228). Perusal of the Szegö paper shows that the problem was proposed to Szegö by the Yale mathematician S. Kakutani, who apparently heard it from Onsager. ..."

Toeplitz determinant $\mathscr{T}$ is proportional to the analogous cyclic determinant $\mathscr{C}$ as follows

$$\mathscr{T} = \mathscr{C} \exp\left(\sum_{n=1}^{\infty} n k_n k_{-n}\right). \tag{8.176}$$

The k's are obtained from the eigenvalues $F(q)$ of the cyclic matrix,

$$F(q) \equiv \sum_m e^{iqm} a(m) \tag{8.177}$$

by the relation

$$\ln F(q) = \sum_{n=-\infty}^{+\infty} k_n e^{inq}. \tag{8.178}$$

Finally, the cyclic determinant is just the product,

$$\mathscr{C} = \prod_q F(q)$$

$$= \exp \int \left[|j - i| \int_{-\pi}^{+\pi} \frac{dq}{2\pi} \ln F(q) \right]. \tag{8.179}$$

The second expression being valid in the limit $|j - i| \to \infty$.

With the $a(m)$ given in (8.173), we find F by inspection:

$$F(q) = \exp[i(2\theta_q + q)] \quad \text{and} \quad \ln F(q) = i(2\theta_q + q). \tag{8.180}$$

As q and θ_q are both odd functions, *the exponent in $\mathscr{C}$ (8.179) vanishes* and $\mathscr{C} = 1$. To obtain the k_n's, some elementary algebra is required. Solve first for $\exp(-2i\theta_q)$ in (8.152) and (8.153),

$$\exp(-2i\theta_q) = e^{iq} \left[\frac{(1 - x_1^{-1}e^{iq})(1 - x_2 e^{-iq})}{(1 - x_1^{-1}e^{-iq})(1 - x_2 e^{iq})}\right]^{1/2} \quad (T < T_c) \tag{8.181a}$$

and

$$\exp(-2i\theta_q) = -\left[\frac{(1 - x_1^{-1}e^{iq})(1 - x_2^{-1}e^{iq})}{(1 - x_1^{-1}e^{-iq})(1 - x_2^{-1}e^{-iq})}\right]^{1/2} \quad (T > T_c) \tag{8.181b}$$

in which

$$x_1 = \coth K_1^* \coth K_2 > 1 \quad \text{and}$$
$$x_2 = \coth K_2 \tanh K_1^* \gtrless 1 \quad \text{for } T \gtrless T_c. \tag{8.182}$$

The phase angles in all four factors in (8.181a) or (8.181b) are taken in the interval $-\frac{1}{2}\pi, +\frac{1}{2}\pi$. K_1 and K_2 retain the meanings, $J_1\beta$ and $J_2\beta$.

Next, solve (8.178) for k_n, using (8.181a):

$$
k_n = \frac{i}{2\pi} \int_{-\pi}^{+\pi} dq\, e^{-inq}(2\theta_q + q)
$$

$$
= \frac{1}{4\pi} \int_{-\pi}^{+\pi} dq\, e^{-inq}[\ln(1 - x_1^{-1} e^{-iq}) - \ln(1 - x_1^{-1} e^{iq})
$$

$$
+ \ln(1 - x_2 e^{+iq}) - \ln(1 - x_2 e^{-iq})]. \tag{8.183}
$$

Because of the inequalities (8.182) the logarithms are expandable in a Taylor series:

$$
\ln(1 - x) = -\sum_{n=1}^{+\infty} \frac{x^n}{n},
$$

so that the above integrations are easily performed. Obtain,

$$
k_n = \frac{1}{2n}(x_1^{-|n|} - x_2^{|n|}), \quad \text{hence} \tag{8.184}
$$

$$
\sum_{n=1}^{\infty} n k_n k_{-n} = -\frac{1}{4} \sum_{n=1}^{\infty} \frac{1}{n}[x_1^{-2n} + x_2^{2n} - 2(x_2 x_1^{-1})^n]
$$

$$
= \frac{1}{4} \ln[(1 - x_1^{-2})(1 - x_2^2)(1 - x_2 x_1^{-1})^{-2}]. \tag{8.185}
$$

When this expression is inserted into the exponential in (8.176) it yields the correlation function,

$$
\lim_{|j-i| \to \infty} \langle S_i^z S_j^z \rangle_{\mathrm{TA}} = \mathscr{T} = \left(1 - \frac{1}{\sinh^2 2K_1 \sinh^2 2K_2}\right)^{1/4} \tag{8.186}
$$

for $T \leqslant T_c$ [note: T_c is thus given by $\sinh(2K_{1c})\sinh(2K_{2c}) = 1$]. As an obvious consequence, the spontaneous magnetization $m(T)$ at $T \leqslant T_c$ is

$$
m(T) = |M_z(T)| = \left(1 - \frac{1}{\sinh^2 2K_1 \sinh^2 2K_2}\right)^{1/8} \tag{8.187}
$$

which can be interpreted crudely as $m \propto (T - T_c)^{1/8}$ (in sharp contrast with MFT 1/2 power dependence).

None of these calculations depended on whether the vacuum had even or odd occupation, therefore the conclusions are general.

Problem 8.15.   Using the appropriate form (8.181b), show $m = 0$ *above* T_c, in the manner of (8.183)–(8.187).

The isotropy of the above results is somewhat amazing. We calculated correlations along a column, finding a function symmetric in J_1 and J_2. Obviously, we would find the same along a row (by rotating the direction of propagation of the transfer matrix 90°). This implies that long-range order is truly isotropic (independent both of distance and of angle relative to the crystal axes in the large distance limit.) Being a scalar quantity (square root of M^2,) the magnetization is thus isotropic and uniform.

The magnetic susceptibility, a tensor quantity, is a more complicated matter. Oguchi[25] showed that the zero-field susceptibility was a power series in the $w_{ij} = \tanh \beta J_{ij}$, and obtained for the nearest-neighbor model ($J_{ij} = J$ if i and j are nearest neighbors, $= 0$ otherwise),

$$\chi = \frac{\mathbb{C}_{1/2}}{T} \left(\sum_r a_r w^r \right) \equiv \frac{\mathbb{C}_{1/2}}{T} [X] \tag{8.188}$$

where $a_0 = 1$ and $a_r =$ twice the term linear in N in the total number of ways of placing a graph of r lines on the lattice of N sites, such that all but two of the vertices are the meeting points of an even number of lines. This first theory of magnetic graphs can be derived on the basis of (8.8), but we shall not dwell on it. There have been a number of early attempts to calculate X, mainly by numerical techniques and lately, by scaling or renormalization-group methodology.

M. F. Sykes has used bond-counting techniques to obtain a large number of terms in this expansion, for various lattices in 2D and 3D. His calculations reveal the first 15 terms for the sq lattice[26]:

$$X = 1 + 4w + 12w^2 + 36w^3 + 100w^4 + 276w^5 + 740w^6 + 1972w^7$$

$$+ 5172w^8 + 13492w^9 + 34876w^{10} + 89764w^{11} + 229628w^{12}$$

$$+ 585508w^{13} + 1486308w^{14} + 3763460w^{15} \cdots . \tag{8.189}$$

---

[25] T. Oguchi, *J. Phys. Soc. Jpn.* **6**, 31 (1951).
[26] M. F. Sykes, *J. Math. Phys.* **2**, 52 (1961).

Unlike our primitive series in (8.5) which contained just the first two terms in this expansion, there is sufficient information in this series to unambiguously fix an exponent γ in the expression,

$$X \sim \left(1 - \frac{T_c}{T}\right)^{-\gamma} \tag{8.190}$$

by curve-fitting and numerical analysis. The result is

$$\gamma = 7/4 \quad (\pm 10^{-2} \text{ or better}). \tag{8.191}$$

This critical exponent is found to fit the Ising ferromagnet on *all* principal lattices in two dimensions. In 3D, a similar calculation yields $\gamma = 5/4$ for the Ising model on all type lattices. (We expect that as the number d of dimensions is increased further, the MFT result $\gamma = 1$ is approached at $d \to \infty$. In fact, there is good reason to believe that the MFT limit is achieved already at $d = 4$ or 5, as for the spherical model!) These inferences were confirmed subsequently by the powerful Padé approximant techniques of Baker[27] and, *in 2D*, by a theoretical analysis of the asymptotic behavior of the correlation function. (We describe this work by Barouch *et al.*[28] in the briefest manner. After establishing that the correlation function in (8.166) decays as

$$R^{-1/4} F_1(t) + R^{-5/4} F_2(t) + O(R^{-9/4+0^+})$$

with R a temperature-dependent length related to the actual distance $|j - i|$, and replacing sums by integrals (in the critical region where this is permissible) they obtain the coefficients in the expression, valid *near* T_c:

$$X = C_{0\pm}|1 - T_c/T|^{-7/4} + C_{1\pm}|1 - T_c/T|^{-3/4} + O(1)$$

as $C_{0-} = 0.0255,\ldots,$ $C_{0+} = 0.9625,\ldots,$ $C_{1-} = -0.00199,\ldots$ and $C_{1+} = 0.07499,\ldots,$ where $\pm$ is for $T \lessgtr T_c$. With the corrections $C_{1\pm}$ small, it certainly seems that the critical phenomena are well represented by the leading singularity in this instance.)

At the *precise* point T_c where χ diverges but the spontaneous magnetization still vanishes, it is reasonable to expect that $m(T_c, B)$ depends

[27] G. A. Baker, Jr., *Phys. Rev.* **124**, 768 (1961).
[28] E. Barouch, B. McCoy and T. T. Wu, *Phys. Rev. Lett.* **31**, 1409 (1973).

sublinearly on B, e.g.,

$$|m| = A|B|^{1/\delta}, \quad \text{with} \quad \delta > 1. \tag{8.192}$$

The exponent δ is an important critical exponent. Although it is impossible to diagonalize the infinite transfer matrix in the presence of an external field by the methods developed in this chapter, the numerical evaluation of the largest eigenvalue for a finite two-dimensional strip converges quickly and yields accurate prediction. In particular, $\delta = 15$ for the 2D Ising model is obtained from a fit of the data to the formula[29]:

$$|m|^{15} = |\tanh B/kT_{\mathrm{c}}|. \tag{8.193}$$

It describes the approach to saturation qualitatively well, and the small-field behavior precisely.

8.11.   Zeros of the Partition Function

Lee and Yang discovered a remarkable property of the Ising model, and of the analogous lattice gas as well.[6] Briefly, by allowing the quantity $-2B/kT$ to be complex, and determining the values of this quantity at which Z vanishes they were able to infer the essential analytic properties of the free energy. Defining the complex variable ζ,

$$\zeta = e^{-2B/kT} \tag{8.194}$$

they showed the partition function for the N spins of an Ising ferromagnet to be a polynomial in ζ of degree N, the zeros of which all lie on the unit circle $|\zeta| = 1$. In the thermodynamic limit the distribution of zeros is therefore a continuous T-dependent function $g(\theta)$, where θ is the angle of ζ in the complex plane. From this observation follow several interesting results. First, the spontaneous magnetization, $m(T)$, is

$$m = 2\pi g(0). \tag{8.195}$$

---

[29] M. Plischke and D. Mattis, *Phys. Rev.* **B2**, 2660 (1970).

Generally, the magnetization in finite (complex) field is given by an integral,

$$M(\zeta) = 1 - 4\zeta \int_0^\pi g(\theta) \frac{\zeta - \cos\theta}{\zeta^2 - 2\zeta\cos\theta + 1} d\theta$$

$$= 1 - 2\zeta \int_{-\pi}^\pi g(\theta)(\zeta - e^{i\theta})^{-1} d\theta . \tag{8.196}$$

The free energy per spin is, with E_0 the ground state energy,

$$f = \frac{E_0}{N} - B + kT \int_0^\pi g(\theta) \ln(\zeta^2 - 2\zeta\cos\theta + 1) d\theta . \tag{8.197}$$

The function $g(\theta) = -g(-\theta)$ is symmetric, positive, and is normalized such that the integral around the unit circle is 1, or

$$\int_0^\pi g(\theta) d\theta = \frac{1}{2} . \tag{8.198}$$

Thus, it is the temperature-dependence of $g(\theta)$ which yields the interesting temperature dependence of f and M, and $g(T, \theta)$ can itself be considered a thermodynamic function of considerable importance. With it, one can compute the properties of an Ising ferromagnet in finite field, in the thermodynamic limit ($N \to \infty$) where all previous non-extrapolatory methods had failed. Unfortunately, to this date $g(T, \theta)$ is known only in 1D. (We know $f(T)$, so if it were possible to invert (8.197) we would know $g(T, \theta)$ in 2D! But this does not seem feasible...)

We now derive the above formulas and their applications as simply as possible. We consider an Ising ferromagnet in which all nonzero bonds (not necessarily nearest-neighbor) are ferromagnetic in sign. Denote the corresponding Hamiltonian H_0, the interaction with an external field being $-B \sum_i S_i$. As usual, the $S_i = \pm 1$. With the trivial identity

$$\exp[B(S_i - 1)/kT] = \frac{1}{2}(1 + \zeta) + \frac{1}{2}(1 - \zeta)S_i \tag{8.199}$$

we write the partition function in the form

$$Z = \exp(NB/kT) \exp(-E_0/kT)Q \tag{8.200}$$

where E_0 is the energy of H_0 when all spins are up (all $S_i = +1$) and

$$Q = \mathrm{Tr} \left\{ \exp[-(H_0 - E_0)/kT] \prod_i \exp[B(S_i - 1)/kT] \right\}$$

$$= \sum_{n=0}^{N} P_n \zeta^n \,. \tag{8.201}$$

It is easily verified that the coefficients $P_n = P_{N-n}$ are real, positive, and $P_0 = P_N = 1$. Clearly, knowledge of the above polynomial in ζ is tantamount to calculation of the partition function.

The Lee-Yang theorem that all the zeros of Z (hence of Q) lie on the unit circle of complex ζ is relatively lengthy to prove, although it is quite plausible. In the absence of H_0, Q is easily seen to be just $(1 + \zeta)^N$, and the P_n are the ordinary binomial coefficients. The zeros all lie at $\zeta = -1$. Introducing an interaction between 2 spins S_1 and S_2 and no others, Q becomes $(1 + \zeta)^{N-2}(1 + 2\zeta e^{-2K} + \zeta^2)$, where $K = J/kT$ as usual. Now, $N - 2$ zeros lie at -1 and 2 lie at conjugate points:

$$\zeta_\pm = -e^{-2K} \pm i(1 - e^{-4K}) \,,$$

again on the unit circle.

Problem 8.16. With 4 spins interacting: $H_0 - E_0 = -\frac{1}{2}J(S_1 + S_2 + S_3 + S_4)^2 + 8J$ locate the zeros of the partition function on the unit circle and their angles, showing they are symmetrically distributed about $\theta = 0$.

There is no spontaneous magnetization and no phase transition unless the zeros pinch the real axis near $\theta = 0$. Thus it is interesting and important to know the distribution of zeros. The Lee-Yang theorem[6] requires a number of lemmas and a proof by induction, too complex to be reproduced here. An important observation, which limits the applications to ferromagnets, is that the more interactions we introduce, the smaller the coefficients P_n become compared to the binomial coefficients; and the closer n is to $\frac{1}{2}N$, the greater is this depression. In the limit of very low temperatures, where the factors $\exp(-2K)$ are essentially zero,

$$\lim_{T \to 0} Q = 1 + \zeta^N + 0(e^{-2K}) \,,$$

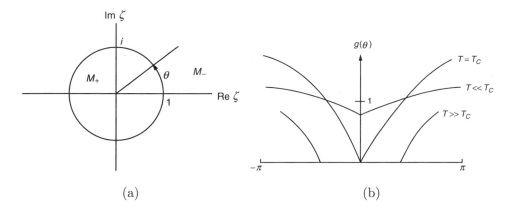

Fig. 8.10. (a) Complex $\zeta = \exp(-2B/kT)$ plane. M is analytic outside unit circle, M_+ within it. The discontinuity $M_+ - M_-$ is related to $g(\theta)$ by (8.122). The behavior of $g(\theta)$ with θ is shown in (b) at various temperatures.

the point $\exp(i\theta_r)$ with $\theta_r = (2r+1)\pi/N$, r an integer in the interval $-\frac{1}{2}N$, $+\frac{1}{2}N$, mark the vanishing of Q. In this example, g would be a constant $1/2\pi$, and the spontaneous magnetiztion $m = 1$ according to (8.195). In the opposite limit of no interactions, we have seen all the zeros are at -1. Raising the temperature *generally* moves the distribution from the first and fourth quadrants into the second and third quandrants of the unit circle, as indicated in the Fig. 8.10. Thus, spontaneous magnetization always disappears at high T.

Once we know the N zeros are on the unit circle, we can denote them $\exp(i\theta_k)$, and $Q = \prod_k [\zeta - \exp(i\theta_k)]$ by definition. Taking logarithms and using the symmetry of $g(\theta)$ establishes (8.197). The formula for M then follows by differentiation with respect to B. While these are straightforward, the simple result (8.195) for the spontaneous magnetization requires an explanation. For ζ on the positive real axis outside the unit circle M is real but negative. Inside the unit circle, it is real and positive, for ζ on the real axis in the interval $0, 1$. The discontinuity at $\zeta = 1$ is obtained from the branch cut in (8.196).

The function in (8.196) has a discontinuity everywhere on the unit circle, not merely on the real axis. If we define M_- to be this function outside the unit circle and M_+ the same function inside of it, and also define the discontinuity as $2M_0(\theta) = M_+(\theta) - M_-(\theta)$ at $r = 1$, we then calculate it to have the value,

$$M_0(\theta) = 2\pi\, e^{i\theta}\, g(\theta)\,. \tag{8.202}$$

This relation generalizes (8.195) and identifies $g(\theta)$ with the thermodynamic function $M_0(\theta)$. But it should be noted that M_0 is the usual magnetization $m(T)$ only at $\theta = 0$, elsewhere it can even exceed 1. In general, there has been no identification of $g(\theta)$ or $M_0(\theta)$ with a suitable correlation function, although this in itself would mark useful progress.

We proceed with an example, and find the zeros of the one-dimensional Ising chain analyzed in Sec. 8.5. From (8.44),

$$\lambda_\pm = e^{(B+J)/kT} \left\{ \frac{1}{2}(1+\zeta) \pm \left[\frac{1}{4}(1-\zeta)^2 + \zeta e^{-4K} \right]^{1/2} \right\} \tag{8.203}$$

are the eigenvalues of the transfer matrix, in the new language using ζ.

Z will vanish if

$$\lambda_+^N + \lambda_-^N = 0, \tag{8.204}$$

therefore we solve for $\lambda_- = \lambda_+ \exp(i\pi p/N)$, with p any odd integer in the interval $-N, +N$. After some more elementary algebra, we obtain the roots of Z at $z_p \equiv \exp(i\theta_p)$:

$$z_p = -e^{-4K} + (1 - e^{-4K}) \cos \frac{\pi p}{N}$$

$$\pm i \left\{ 1 - \left[-e^{-4K} + (1 - e^{-4K}) \cos \frac{p}{N} \right]^2 \right\}^{1/2}. \tag{8.205}$$

This shows $g(\theta)$ is symmetric, and allows us to compute it in the limit $N \to \infty$:

$$g(\theta) = \frac{1}{2\pi} \frac{\sin \frac{1}{2}\theta}{(\sin^2 \frac{1}{2}\theta - e^{-4K})^{1/2}} \tag{8.206}$$

for θ in the interval $[-\pi, \cos^{-1}(1 - 2e^{-4K})]$ and $g = 0$ for θ in the interval $[0, \cos^{-1}(1 - 2e^{-4K}]$.

Problem 8.17.   Show that with this formula for $g(\theta)$ the integral (8.197) will yield the correct free energy $-kT \ln \lambda_+$ in 1D. Derive

$$M(\zeta) = \left(\frac{\zeta^2 - 2\zeta + 1}{\zeta^2 - 2\zeta(1 - 2e^{-4K}) + 1} \right)^{1/2}, \tag{8.207}$$

and use this to compute the *susceptibility*, verifying (8.48).

Terminating this section, it is appropriate to mention that the solution of the 2D Ising model in a magnetic field has been reformulated by Barouch,[30] who produced an exact formula for Z in terms of the — alas unknown! — distribution of prime numbers.

8.12. Miscellania, Including 2D Antiferromagnets

The dual of the triangular lattice is the honeycomb (hexagonal), and vice versa, as we saw in Fig. 8.3. This duality cannot immediately locate the critical point of either structure, nor map the high temperature properties onto the low temperature properties for a given lattice in a direct way. Onsager invented the *star-triangle* transformation to relate the partition functions on these two lattices, thus determining them both. This methodology has recently been used by Baxter and Enting in a paper amusingly entitled "399th solution of the Ising model" to connect the two abovementioned lattices to the square, and then to obtain the partition functions of all of them without use of transfer matrices, series summations, or other special

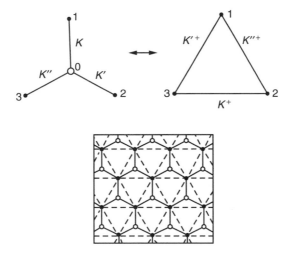

Fig. 8.11. Star-triangle transformation. Its application to stars on the H lattice (∘ ∘ ∘ and their 3 neighbors) leads to the T lattice.

[30] E. Barouch, *Physica* **1D**, 333 (1980); Generalizations of the Lee-Yang methods[6] to other models have also appeared, notably: M. Bander and C. Itzykson, *Phys. Rev.* **B30**, 6485 (1984) for $O(N)$ spin models; D. Kurtze and M. Fisher, *J. Stat. Phys.* **19**, 205 (1978) for spherical models.

tricks.[31] The star-triangle transformation is essential in the study of the triangular lattice, on which one has an antiferromagnet with very unusual properties. We now derive its properties, following pp. 182ff in Domb's 1960 exposition.[11]

Figure 8.11 illustrates the basic star and triangle in question. The star contains an extra spin, so it is the partial trace of the former that will be equal to the Boltzmann factor for the latter. We sum over S_0 in the star:

$$\sum_{S_0=\pm 1} \exp[K S_0 (S_1 + S_2 + S_3)]$$

$$= 2 \cosh^3 K + 2(\sinh^2 K)(\cosh K)(S_1 S_2 + S_2 S_3 + S_3 S_1) \qquad (8.208)$$

and denoting the parameter for the triangle as K^+, we can compare the above with

$$\exp[K^+ (S_1 S_2 + S_2 S_3 + S_3 S_1)]$$

$$= (\cosh^3 K^+ + \sinh^3 K^+) + \frac{1}{2} e^{K^+} (\sinh 2K^2)(S_1 S_2 + S_2 S_3 + S_3 S_1).$$

$$(8.209)$$

The two expressions will agree if the second is multiplied by

$$f = 2 \cosh^3 K / (\cosh^3 K^+ + \sinh^3 K^+) \qquad (8.210)$$

and if we choose K^+ to satisfy

$$\frac{\frac{1}{2} e^{K^+} \sinh(2K^+)}{\cosh^3 K^+ + \sinh^3 K^+} = \tanh^2 K. \qquad (8.211)$$

Now applying the star-triangle transformation to the open circles in Fig. 8.12 we readily transform the honeycomb (H) into the triangle (T) lattice,

$$Z_{2N}^{(H)}(K) = f^N Z_N^{(T)}(K^+). \qquad (8.212)$$

The duality relationships (8.17)–(8.22) relate properties on one lattice at K with those on the other at K^*:

$$\frac{Z_N^{(T)}(K^*)}{2^{1/2N}(\cosh 2K^*)^{3N/2}} = \frac{Z_{2N}^{(H)}(K)}{2^N(\cosh 2K)^{3N/2}}. \qquad (8.213)$$

---

[31] R. Baxter and I. Enting, *J. Phys.* **A11**, 2463 (1978).

Simplifying (8.211) we find

$$e^{4K^+} = 2\cosh 2K - 1 \tag{8.214}$$

and with the help of (8.23),

$$(e^{4K^+} - 1)(e^{4K^*} - 1) = 4. \tag{8.215}$$

The critical point on the triangular (T) lattice is at $K^+ = K^*$, i.e., at $\exp(4K_c^+) = 3$, as recorded in Table 8.2. That of the H lattice is at $K = (K^+)^*$, which, after some algebra, is located at $\exp(2K_c) = 2 + \sqrt{3}$ also given in the table. For details, see the original work of Wannier[32] or the review[11] or below in the present section, where formulas for anisotropic lattices (different J's in different directions) are derived and used.

The critical point for the sq ferromagnetic Ising model was given as $\sinh^2 2K_c = 1$ in Sec. 8.3. It is evident that for antiferromagnetic coupling on the same lattice, i.e., $J < 0$, the critical temperature now called T_N is given by *precisely the same expression*. In the absence of an external field, a symmetry operation (reversing every second spin) effectively reverses the sign of J without affecting the partition function.

This is patently untrue for the T lattice. For when J is negative, K^+ is negative and there is no K with which one can satisfy (8.214)! Indeed, there is no phase transition for the isotropic T antiferromagnet, which is in many other respects a remarkable system. Its ground state has energy $1/3$ that of the corresponding ferromagnet and is so degenerate — that is, there are so many configurations having this energy — that the high-temperature disordered phase persists down to $T = 0$. Now, it is obvious that within each elementary triangle, the optimum energy is obtained for 2 spins of one sign and the 3rd spin of the opposite sign. The energy of such a state — one-third the ground state energy of the corresponding ferromagnet — is thus a *lower* bound. But, starting with a single cell in this configuration and working outward, one may construct such configurations in every triangle of the plane. We have this as a variational ground state — an upper bound. Being an upper and lower bound, it must be the exact ground state energy. As this ground state is enormously degenerate, $0(2^{pN})$ with $0 < p < 1$, it is desired to obtain the ground state entropy. Wannier, first to investigate this remarkable antiferromagnet[31] obtained the exact

---

[32] G. Wannier, *Phys. Rev.* **79**, 357 (1950).

transfer matrix and was able to establish the formula for the ground-state entropy:

$$\mathscr{S}_0/Nk = \frac{3}{\pi} \int_0^{\pi/6} d\theta \ln(2\cos\theta) = 0.32306\cdots. \qquad (8.216)$$

This number is approximately equal to the entropy $\mathscr{S}_c$ of the corresponding ferromagnet at its (nonzero) T_c. In Table 8.2 we observe that the 2D Ising model usually has a *critical* entropy of this magnitude, approximately half the maximum value ($\ln 2$). In a few pages, we shall see that Wannier's antiferromagnet has, indeed, its phase transition at $T = 0$.

A large number of lattices have been solved in 2D, as listed in Domb's review.[11] Their study often follows the same routine which we illustrate for the sq, T and H lattices (following Wannier[32] as generalized by Utiyama[33]). A transfer matrix is derived, the solution of which (by a transformation to free fermions) yields all the thermodynamic properties.

The difficulty with the straightforward approach in the case of arbitrary lattices, is that the transfer matrix is now not one-to-one. A single spin on the nth row connects to 2 spins on the $n + 1$st row, and one cannot express this by a simple $\mathbf{V}^x$ matrix as in (8.124). But if we introduce extra spins to restore rectangular symmetry (or almost) and then freeze them in or out by setting certain of the coupling constants equal to infinity (or zero), we may achieve the desired results. There is no unique way to do this; however Fig. 8.12 contains the three principal lattices as special cases for the indicated values of the J's. The spins have to be moved somewhat (without disturbing the connectivity or crossing any of the lines) to display the desired T or H geometries explicitly.

If we wish to establish the general transfer operator for the grid illustrated in Fig. 8.12, then specialize to the indicated J's to calculate the properties of the three special lattices, then we should transfer from the nth column to the $n + 2$nd, or from the $n + 1$st to the $n + 3$rd, alternate columns (or rows) being distinct in general. As we must include two columns in the operator, the eigenvalue will be λ^2. We then solve

$$\mathbf{W} \cdot \mathbf{u} = \lambda^2 \mathbf{u} \qquad (8.217)$$

---

[33] T. Utiyama, *Progr. Theor. Phys.* **6**, 907 (1951).

where the following **W** replaces $\mathbf{V}^x$, $\mathbf{V}^y$ and $\mathbf{V}^B$ of (8.139):

$$\mathbf{W} = A\mathbf{V}_1\mathbf{V}_2\mathbf{V}_3\mathbf{V}_4\mathbf{V}_5\mathbf{V}_6 \,. \tag{8.218}$$

With

$$A = (2\sinh 2K_0)^{1/2N}(2\sinh 2K_0')^{1/2N} \text{ is a constant},$$

$$\mathbf{V}_1 = \exp\left[-\left(K_0^*\sum\sigma_{2j}^z + K_0'^*\sum\sigma_{2j+1}^z\right]\right\}$$

$$\mathbf{V}_2 = \exp\left(K\sum\sigma_{2j}^x\sigma_{2j+1}^x + K_1\sum\sigma_{2j+1}^x\sigma_{2j+2}^x\right)$$

$$\mathbf{V}_3 = \exp\left(h\sum\sigma_j^x\right)$$

$$\mathbf{V}_4 = \exp\left[-\left(K_0^*\sum\sigma_{2j+1}^z + K_0'^*\sum\sigma_{2j}^z\right)\right]$$

$$\mathbf{V}_5 = \exp\left(K\sum\sigma_{2j+1}^x\sigma_{2j+2}^x + K_1\sum\sigma_{2j}^x\sigma_{2j+1}^x\right)$$

$$\mathbf{V}_6 = \exp\left(h\sum\sigma_j^x\right).$$

A solution in closed form will be possible only for $h = 0$, $i\pi/2$ or $i\pi$. For $h = i\pi$, the operators $\mathbf{V}_3$ anf $\mathbf{V}_6$ reduce to ± 1 and can be trivially absorbed

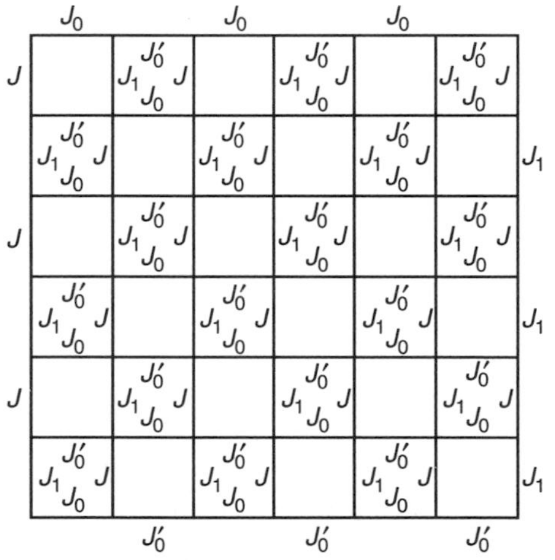

Fig. 8.12.   The checker lattice becomes the sq ($J = J_1$, $J_0 = J_0'$), the hexagonal ($J = 0$) and triangular ($J = \infty$).

into the constant. For $h = i\pi/2$, i.e., $\zeta = -1$,

$$\exp\left(\frac{1}{2}i\pi\sigma_j^x\right) = i\sigma_j^x \tag{8.219}$$

is a trivial identity that we can use twice:

$$\exp\left(\frac{1}{2}i\pi\sigma_{2j}^x\right)\exp\left(\frac{1}{2}i\pi\sigma_{2j+1}^x\right) = i\exp\left[\frac{1}{2}i\pi(\sigma_{2j}^x\sigma_{2j+1}^x)\right]. \tag{8.220}$$

The product of i's can be absorbed into the constant, and the complex exponentials into $\mathbf{V}_2$ and $\mathbf{V}_5$ by redefining

$$K \to K + \frac{1}{2}i\pi \quad \text{and} \quad K_1 \to K_1 + \frac{1}{2}i\pi. \tag{8.221}$$

Subsequently, the calculation proceeds as for real K's and one finds formulas originally given (inscrutably, without explanation,) in the papers of Yang and Lee.[6]

To calculate the optimum eigenvalue in (8.217) one proceeds almost as for the sq lattice, except that he takes 2 sites $(2j, 2j+1)$ as the unit cell. The transformation to fermions is facilitated if operators on even-numbered sites are denoted c_{2j}, those on odd-numbered sites, b_{2j+1}. For each k, this yields 4×4 matrices resulting from the b and c operators, rather than the 2×2 in the original method of Sec. 8.9. Leaving algebraic details aside, we quote the result for the free energy in the *triangular lattice* with totally anisotropic bonds, J, J' and J'' in zero field after it is brought into its most symmetric form:

$$f = -kT\left\{ \ln 2 + \frac{1}{8\pi^2}\int_{-\pi}^{+\pi} dq_1 \int_{-\pi}^{+\pi} dq_2 \right.$$

$$\times \ln[\cosh 2K \cosh 2K' \cosh 2K'' + \sinh 2K \sinh 2K' \sinh 2K''$$

$$\left. - \sinh 2K \cos q_1 - \sinh 2K' \cos q_2 - \sinh 2K'' \cos(q_1 + q_2)]\right\}. \tag{8.222}$$

Notice that setting $J'' = 0$ results in the free energy previously computed for the sq lattice, (8.159). The phase transition occurs at the value of kT for which [ ] in the above integral just vanishes. For the ferromagnet (all $J > 0$) this occurs at $q_1 = q_2 = 0$, and it is not difficult to prove that there will always be a solution.

For the *isotropic antiferromagnet* $J = J' = J'' < 0$ discussed in (8.216) and *supra*, the breakdown of the star-triangle transformation and the high degeneracy of the ground state seemed to preclude a phase transition. To analyze this point further, let $J = J' < 0$ and $J'' \leqslant 0$ be independent variables. For $|J''| < |J|$ the minimum of [ ] occurs at $q_1 = g_2 = \pi$. Thus, the conditions for the bracket to vanish at that point determine the critical temperature. The equation is

$$\cosh^2 2K_c \cosh 2K_c'' + \sinh 2K_c''(1 - \sinh^2 2K_c) - 2 \sinh 2K_c = 0 \quad (8.223)$$

where $K_c = |J|/kT_c$ and $K_c'' = |J''|/kT_c$. It has the solution

$$\exp(2K_c'') = \sinh 2K_c \quad\quad\quad (8.224)$$

from which we can observe kT_c vanishing as J'' approaches J:

$$kT_c \sim 2.89(|J| - |J''|). \quad\quad\quad (8.225)$$

For $|J''| > |J|$, T_c *stays* zero, as there is no possible way for [ ] to vanish, regardless of q_1, q_2. This will be analyzed by the reader in Problem 8.18. Thus, the isotropic AF is a singular case — the *precise* limit at which $T_c \to 0$, and we should not be surprised at the ground state entropy's equalling the typical critical entropy in 2D.

Problem 8.18.   Study the octant in parameter space $J \leqslant J' \leqslant J'' \leqslant 0$ for the triangular AF, showing all the regions where $T_c = 0$.

In an *external field* the ground state degeneracies are lifted and there is no reason not to have one or more phase transitions. This situation is interesting, although it resists exact analysis.

The development by Baxter and Enting[31] also relates the three principal lattices in 2D, with the same results as in Fig. 8.12 but by a different procedure. These authors use a sequence of star-triangle transformations to functionally transform the honeycomb into sq and T sections, in a manner reminiscent of the work *Metamorphosis* by the great Dutch artist, M. Escher. Their work, compared to his, is reproduced in Fig. 8.13, perhaps a new instance of nature imitating art!

We now return to the important topic of the sq *antiferromagnet*. We already know that in the absence of an external field, the thermal properties are identical to those of the corresponding ferromagnet. In an external field, we can take some of the expressions which are accurately known for the ferromagnet and by reversing J (or $w = \tanh J/kT$) obtain the expressions relevant to the antiferromagnet.

Consider the susceptibility as expanded to 15th order in w (8.189). Reversing the sign of J (hence of w) introduces an alternating sign in the expansion which now *vanishes* at a temperature just above T_N. This unphysical behavior due to a failure of convergence has been analyzed by Fisher and Sykes,[34] who determined that the susceptibility series should have singularities at both $+w_c$ and $-w_c$. The ferromagnetic singularity $(1-T_c/T)^{-7/4}$ becomes an innocuous $(1+T_N/T)^{-7/4}$ in the antiferromagnet. Nevertheless, comparing the expansion of this quantity with the known series leads to tangible results. Supposing X to be of the form:

$$X = (1 + w/w_c)^{-7/4} F(w/w_c) \tag{8.226}$$

they determined

$$F = C + D \left(1 - \frac{w}{w_c} \right) \ln \left(1 - \frac{w}{w_c} \right) \tag{8.227}$$

for the AF, with C, D constants to be adjusted for best fit.

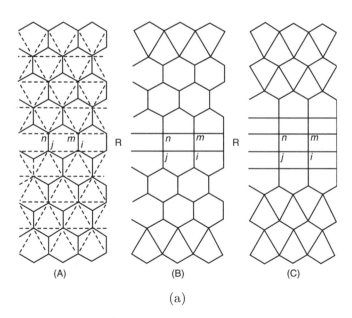

(a)

Fig. 8.13.   Metamorphosis: (a) the effects of repeated star-triangle and triangle-star transformations on H lattice[31] and (b) the work of M. Escher.

---

[34] M. Sykes and M. Fisher, *Physica* **28**, 919, 939 (1962).

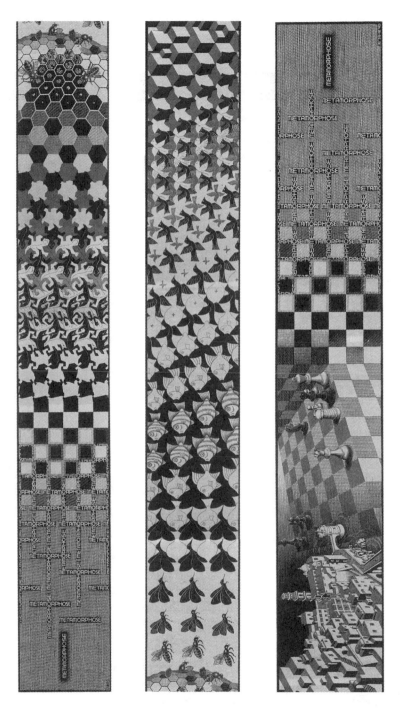

Fig. 8.13. (b)

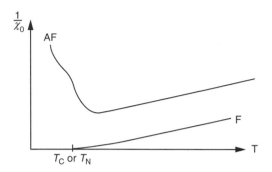

Fig. 8.14. Inverse susceptibility for Ising sq antiferromagnet in 2D, compared with that of analogous ferromagnet (lower curve).

It follows that $\chi_{\parallel}$ *peaks* just above T_N, then dropping as the temperature is lowered — with infinite slope at T_N — thence, approaching zero rapidly as $T \to 0$. This behavior differs in detail from that in the AF models studied in Chap. 7, namely the MFT in Fig. 7.8 and the spherical antiferromagnet in Fig. 7.14. To compare, in Fig. 8.14 we show $1/\chi_0$ for the Ising AF and the same quantity for the Ising ferromagnet.

There has been some study of the Ising model AF in finite, homogeneous, external field, with a view to establishing the spin-flop field $B_c(T)$ and the nature of the phase transition across this boundary. The results are not dissimilar from the properties of the simple antiferromagnets considered in Chap. 7, except in some details. Lieb and Ruelle[35] proved a theorem analogous to that of Lee and Yang, that there is no phase transition in sufficiently weak (real or complex) fields above T_N. Now, at $T = 0$ an external field $B > 2|J|$ does break up the AF ordering in the sq lattice, so an interpolation such as

$$\cosh B_c/kT = \sinh^2 2K \qquad (8.228)$$

cannot be too much in error. This is the formula proposed by Zittartz[36] and Müller-Hartmann[37] in a somewhat generalized form suitable for anisotropic couplings. Indeed, Lin and Wu[38] probing the validity of such approximations, decided they apply also to the anisotropic antiferromagnet which, they found, accommodates *several* distinct phases in an applied field. The original

[35] E. Lieb and D. Ruelle, *J. Math. Phys.* **13**, 781 (1972).
[36] J. Zittartz, *Z. Physik* **B40**, 233 (1980).
[37] E. Müller-Hartmann and J. Zittartz, *Z. Physik* **B27**, 261 (1977).
[38] K. Y. Lin and F. Y. Wu, *Z. Physik* **B33**, 181 (1979).

investigation of the AF Ising model in an external field goes back to Bienenstock,[39] followed by Plischke and Mattis[40] in connection with Plischke's 1970 doctoral dissertation. These studies, which included next-nearest neighbor interactions, concluded that the logarithmic specific heat anomaly persists at finite field.

8.13. The Three-Dimensional Ising Model

If this section on the three-dimensional Ising model were limited to exact results, it would be short indeed. As we shall show below, the agreement between "exact" (but numerical) series calculations and experiment is excellent, but theory — in the form of closed form expressions, relations between physical quantities or manageable approximations — remains sparse, except in the relatively specialized area of the calculation of critical exponents.

Two antiferromagnets Rb_3CoCl_5 and Cs_3CoCl_5 fullfil the physical requirements for the Ising model antiferromagnet on a sc lattice (the magnetic atoms, Co, are on a sc sublattice and their spins are restricted to $\pm 3/2$ due to crystal-field splittings). These substances show specific heat anomalies at their respective values of T_N, as shown in Fig. 8.15.[41] Using the series extrapolations and analytical results of Sykes *et al.*, Fisher, Domb, and their collaborators, de Jongh and Miedema collected many physically relevant properties and compared them with experiment.[42] We extract from their Table 11 some information on the critical temperatures T_c (or T_N), critical entropy $\mathscr{S}_c$, high-temperature entropy $\mathscr{S}_\infty$, ground state energy E_0 and critical internal energy U_c. These quantities are compared to experiment for the sc ($z = 6$) structure, and also to theory for bcc ($z = 8$) and fcc ($z = 12$) ferromagnets in Table 8.3. (Purely magnetic properties that differ in ferro- and antiferro-magnets are not listed.)

The principal sources of reliable information on the 3D Ising model have come from various series expansions. The low T expansion, the high T expansion (there is no known duality linking them in 3D), and critical-point theories combine into a coherent picture which we now attempt to draw,

---

[39] A. Bienenstock, *J. Appl. Phys.* **37**, 1459 (1966).
[40] M. Plischke and D. C. Mattis, *Phys. Rev.* **A3**, 2092 (1971).
[41] H. Blöte and W. Huiskamp, *Phys. Lett.* **A29**, 304 (1969).
[42] L. de Jongh and A. Miedema, Experiments on simple magnetic model systems, *Adv. Phys.* **23**, 1–260 (1974).

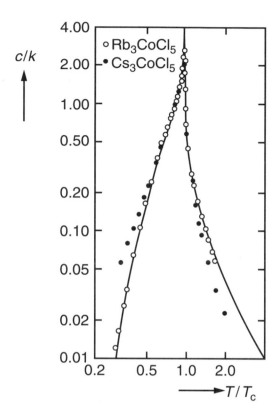

Fig. 8.15. Specific heat for CoRb$_3$Cl$_5$ and CoCs$_3$Cl$_5$ ($\cdots$) compared with theory for sc Ising model in 3D.[41] Note logarithmic scale.

Table 8.3. Critical parameters of 3D ferromagnetic or antiferromagnetic Ising lattices.[42]

Lattice	kT_c/zJ	$\mathscr{S}_c/Nk$[a]	$(\mathscr{S}_\infty - \mathscr{S}_c)/Nk$	$-E_0/NkT_c$	$-U_c/NkT_c$
Ising sc $z = 6$	0.7518	0.5579	0.1352	0.6651	0.2200
Ising bcc $z = 8$	0.7942	0.5820	0.1111	0.6296	0.1720
Ising fcc $z = 12$	0.8163	0.5902	0.1029	0.6126	0.1516
CoRb$_3$Cl$_5$ $z = 6$	0.74	0.563	0.137	0.673	0.226
CoCs$_3$Cl$_5$ $z > 6$	0.79	0.593	0.106	0.632	0.173

[a]Recall $\mathscr{S}_\infty/Nk = \ln 2 = 0.69315$.

starting with the methodology. The low-T expansions are principally the development of Sykes, Essam, Gaunt and various collaborators over the span of a decade[43] in a series of papers which the reader should consult for many important features omitted here. High-T series have existed since the dawn of statistical physics, but for present application the papers of Sykes *et al.*[44] form a complete documentation.

In the calculation of the free energy and its most sensitive derivatives (c, χ) there is a substantive question of how to extract the discontinuities at T_c and predict the functional form of thermodynamic quantities in a manner most suitable for comparison with experiment and, ultimately, with an exact theory — when such becomes available! Thanks to the efforts of a large number of dilligent researchers, among whom the names of G. A. Baker, Jr., M. F. Sykes, C. Domb, and M. E. Fisher stand out, the analysis of power series — whether in $\exp(-J/kT)$ or in $\tanh(J/kT)$ — has turned into a refined discipline. An introduction to series analysis has been given by Gaunt and Guttmann,[45] contributors to the remarkable book, Vol. 3 in the series *Phase Transitions and Critical Phenomena* ed. by C. Domb and M. S. Green.[46] In addition to the cited paper, this book contains chapters on graph theory (C. Domb), computational techniques for lattice sums (J. Martin), the linked cluster expansion (M. Wortis), the Ising model (C. Domb) and includes the study of other magnetic systems (e.g., Heisenberg and XY models) that are treated elsewhere in the present text. The information on the 3D Ising model which we now present is, in large measure, called from this sourcebook[46] to which the interested reader is referred for full details.

Consider the susceptibility series of the type given in (8.189) for the sq lattice. The calculation of the critical exponent 7/4 proceeds easily, because T_c is known exactly. From the equivalent series in 3D, we must have the

[43] M. Sykes, J. Essam and D. Gaunt, *J. Math. Phys.* **6**, 283 (1965); M. Sykes, D. Gaunt, J. Essam and D. Hunter, *J. Math. Phys.* **14**, 1060 (1973); M. Sykes, D. Gaunt, S. Mattingly, J. Essam and C. Elliott, *J. Math. Phys.* **14**, 1066 (1973); M. Sykes, D. Gaunt, J. Martin, S. Mattingly and J. Essam, *J. Math. Phys.* **14**, 1071 (1973); M. Sykes, D. Gaunt, J. Essam, B. Heap, C. Elliott and S. Mattingly, *J. Phys.* **A6**, 1498 (1973); M. Sykes, D. Gaunt, J. Essam and C. Elliott, *J. Phys.* **A6**, 1506 (1973); D. Gaunt and M. Sykes, *J. Phys.* **A6**, 1517 (1973).
[44] M. Sykes, D. Gaunt, P. Roberts and J. Wyles, *J. Phys.* **A5**, 624, 640 (1972); M. Sykes, D. Hunter, D. McKenzie and B. Heap, *J. Phys.* **A5**, 667 (1972).
[45] D. Gaunt and J. Guttmann, *Asymptotic Analysis of Coefficients*, in *Phase Transitions and Critical Phenomena*, Vol. 3, eds. C. Domb and M. Green, Academic Press, New York, 1974.
[46] C. Domb and M. Green (eds.), *Phase Transitions and Critical Phenomena*, Vol. 3 Academic Press, New York, 1974.

ability to extract *both* T_c and the relevant exponents, and more. Fortunately, the diagrammatic expansions have been carried out to a large number of terms, with published series such as that for X for the sc lattice:

$$X = 1 + 6w + 30w^2 + 150w^3 + 726w^4 + 3510w^5 + 16710w^6$$

$$+ 79494w^7 + 375174w^8 + 1769686w^9 \cdots \qquad (8.229)$$

and for the face-centered cubic (fcc) lattice:

$$X = 1 + 12w + 132w^2 + 1404w^3 + 14652w^4 + 151116w^5 + 1546332w^6$$

$$+ 15734460w^7 + 159425580w^8 + 1609987708w^9 + \cdots . \qquad (8.230)$$

Coefficients are given by Gaust and Guttmann[44] to $O(w^{17})$ for the sc, $O(w^{12})$ for the fcc, and $O(w^{15})$ for body-centered cubic (bcc).

By methods that we shall discuss shortly in Eqs. (8.234)ff, one extracts from such series the following information for ferromagnets:

$$X = A(1 - T_c/T)^{-5/4} \quad (T \to T_c^+) \qquad (8.231)$$

with A, calculated to 5 decimals, found to be approximately (but not exactly) 1 in all three cubic lattices. T_c, or $w_c = \tanh J/kT_c$, is obtained to better than 1 part in 10^4, as

$$1/w_c(\text{sc}) = 4.5844, \ (\text{bcc}) \ 6.4055 \text{ and } (\text{fcc}) \ 9.8290 . \qquad (8.232)$$

Although the singular behavior (8.231) describes the behavior of the power series (8.229) and (8.230) over a large range of temperatures above T_c, the same thing cannot be said for the specific heat anomaly. After much research, the critical specific heat has been found to be in the form[47]

$$c/k = A_\pm (1 - T_c/T)^{-1/8} - B_\pm . \qquad (8.233)$$

At first, series analysis had seemed incapable of providing the critical exponent correctly. (With A_+ and B_+ the parameters above T_c both $O(1)$, the constant term remains comparable to the divergent one in the critical region until T is within 1 part in 10^4 of T_c.) Similar considerations apply to A_- and B_- below T_c, and it is an ironic fact that the specific heat anomaly first noted by Kramers, Wannier and Onsager, sparking the intense interest in critical phenomena, remains the hardest to analyze in *all* magnetic models.

---

[47] See the analysis and references in S. Jensen and O. Mouritsen, *J. Phys.* **A15**, 2631 (1982) or in Domb and Green.[46]

The susceptibility and other correlation functions are generally much better understood than the specific heat, the critical exponent of which is small (and, in some cases, uncertain even as to sign).

Of course, given a sufficiently long series one can always hope to approach T_c so closely that the leading singularity dominates. In such cases, the critical temperature itself and the critical exponent are calculated as follows. Suppose we know the power series $F(w)$,

$$F(w) = 1 + \sum_{n=1}^{\infty} a_n w^n \tag{8.234}$$

for a quantity believed to diverge at T_c. Then, presumably,

$$\lim_{n \to \infty} |a_n|^{-1/n} = w_c \tag{8.235}$$

fixes w_c. But this rarely occurs; if dominant singularity are $w_c \exp(\pm i\theta)$, a complex pair, the a_n will oscillate asymptotically as

$$a_n \sim \frac{f(n)}{w_c^n} \cos n\theta \tag{8.236}$$

with $f(n)$ a slowly varying function. With additional singularities on the complex circle at $|w_c|$ the behavior becomes unpredictable. The *ratio method* is often helpful. It presumes

$$\lim_{n \to \infty} [a_n / f(n)\mu^n] = 1 \tag{8.237}$$

with $f(n)$ neither growing nor decaying sufficiently to violate

$$\lim[f(n)]^{1/n} = 1. \tag{8.238}$$

Ultimately, $\mu = 1/w_c$; the question is, how to determine it. It is helpful to assume an asymptotic behavior for $f(n)$:

$$f(n) \sim An^g / g!. \tag{8.239}$$

This satisfies (8.238) and provides us 2 parameters, A and g, with which to help fit the series. We try successive approximations to μ, calculating a_n / a_{n-1}:

$$a_n / a_{n-1} = \mu \left[1 + \frac{g}{n} + O(n^{-2}) \right]. \tag{8.240}$$

A plot of a_n / a_{n-1} versus $1/n$ will yield μ as the $1/n = 0$ intercept, with $(\mu g) = $ slope at that point. Now, a knowledge of g yields the critical index,

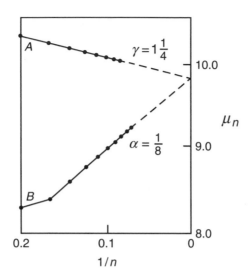

Fig. 8.16.   Ratios of successive coefficients of the susceptibility and specific heat on fcc lattice, plotted versus $1/n$. The extrapolated point yields $1/w_c = 9.8290\ldots$ while the respective slopes yield the critical exponents $\gamma = 5/4$ and $\alpha = 1/8$.[47]

for the series so defined approaches w_c (i.e., $1/\mu$) as

$$F(w) = A(1 - \mu w)^{-(1+g)} + \text{corrections}, \qquad (8.241)$$

diverging as specified for $g > -1$ and w positive $\rightarrow w_c$. In Fig. 8.16 we display Sykes' simultaneous analyses of the susceptibility and specific heat series for the fcc lattice. The susceptibility series ratios become linear already at small n, whereas the specific heat series ratios become reliable only for $n \gtrsim 10$. The use of *both* series yields a most reliable $w_c = 1/9.8290$, and critical exponents $\gamma = 5/4$ and $\alpha = 1/8$.[48]

If the correction term such as $B_\pm$ in the specific heat is an important quantity, it too can be obtained by analysis of the correction series $R(w)$, once $f(n)\,(A, g, \mu)$ is known:

$$R \equiv \sum_n (a_n - f(n)\mu)w^n \qquad (8.242)$$

which should be rapidly convergent, or have easily diagnosed singularities. When this is not the case, the method of Padé approximants becomes quite useful. Briefly, if a function has singularities such as zeros or poles, it can

---

[48] M. Sykes *et al.*, *J. Phys.* **A5**, 640 (1972), see Appendix; R. B. Griffiths, *J. Math. Phys.* **10**, 1559 (1969).

be written in the form $P(w)/Q(w)$, where P is a polynomial constructed to have its zeros in the desired points, and Q another polynomial having its zeros at the desired poles. We may *approximate* arbitrary functions $F(z)$ by

$$[L, M] = \frac{p_0 + p_1 z + \cdots + p_L z^L}{1 + q_1 z + \cdots + q_M z^M} \tag{8.243}$$

called the "L, M Padé approximant," most useful in the study of either $\ln X$, $\ln c/k$ or $X^{1/\gamma}$ or $(c/k)^{1/\alpha}$ in cases γ or α are precisely known. The applications of Padé methods are now quite widespread, and documented in books.

The types of series which have been thus subjected to analysis include high T series such as in the examples above, low T series in $\exp(-J/kT)$, and "density expansions" in which the coefficient of a given power of $\exp(-2B/kT)$ is studied. Both ferromagnets and antiferromagnets have been analyzed, for $s = \frac{1}{2}$ and spins $s > 1/2$ as well. We summarize some of the conclusions:

(i) The critical exponents are independent of spin magnitude s. In current jargon, one says that all Ising models of varying s belong to the same "universality class."

(ii) The critical exponents *do* depend strongly on d, the dimensionality. Thus, the diamond lattice ($z = 4$) shares critical exponents with the fcc ($z = 12$), rather than with the sq ($z = 4$).

(iii) Properties of 3D Ising models are closer to MFT than in 2D, while at or above 4D, the critical behavior closely approximates MFT. As examples, consider the susceptibility exponent $\gamma = 5/4$, approaching the MFT value ($\gamma = 1$), versus $\gamma = 7/4$ in 2D. The specific heat is anomalous over a smaller interval of temperature in 3D than in 2D; the principal effect of moving through T_c is analogous to the discontinuity found in Chap. 7. The critical temperature itself is closer to the MFT prediction. Antiferromagnets in bipartite lattices (sc, bcc) have a parallel suscepti-bility maximum at $\approx 1.08\ T_c$ rather than at $\approx 1.5\ T_c$ for the sq lattice, closer to the MFT limit which occurs at T_c precisely.

We illustrate with magnetic and thermal data on $DyPO_4$, an analog Ising antiferromagnet on the diamond lattice, $z = 4$.[41] Figure 8.17 illustrates the specific heat — qualitatively that of MFT except near T_c, Fig. 8.18 the parallel susceptibility, and Fig. 8.19 the sublattice magnetization. It is known that just below T_c the order parameter in 3D Ising models is

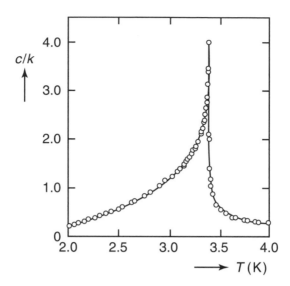

Fig. 8.17. Specific heat of Ising spins, $DyPO_4$ on diamond lattice ($z = 4$) as function of T. Data points are ($\circ \circ \circ$), (—) are from theoretical high- and low-temperature series.[42]

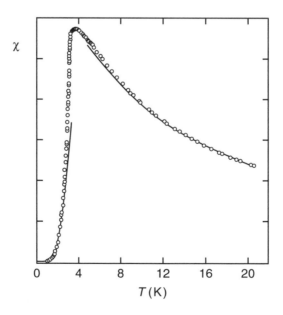

Fig. 8.18. Antiferromagnetic susceptibility $\chi T/\mathbb{C} = X$ versus T for $DyPO_4$.[42]

$$\sigma(T) = D(1 - T/T_c)^\beta \tag{8.244}$$

with the magnetization critical exponent (not to be confused with $1/kT$!) $\beta = 5/16 = 0.3125$. This should be compared with $\beta = 1/2$ for MFT and $\beta = 1/8$ in the 2D Ising model. The experiment in Fig. 8.19 is best fit by $\beta = 0.314$, in excellent agreement with theory.

With the passage of time there have been more interesting developments in the Ising model than can possibly be recounted here. We alluded to an extension to higher spin, so let us briefly discuss some properties of higher-spin Ising models before terminating this chapter.

The spin 1 Ising model has $S_i^z = 1, 0, -1$, while the spin 3/2 model has $S_i^z = 3/2, 1/2, -1/2, -3/2$, etc. In general, if we wish to consider an Ising model for spins of magnitude $p/2$, the variable $2S_i^z$ can take on the values $p, p - 2, \ldots, -p$ and can be written:

$$2S_i^z = \sigma_1 + \sigma_2 + \sigma_3 + \cdots + \sigma_p \tag{8.245}$$

with each individual $\sigma_n = \pm 1$. Thus, the generalized Ising model can be expressed in terms of the Pauli matrices. With this artifice, Griffiths[48] has extended to arbitrary p the results previously proved for $p = 1$, such as the Lee-Yang theorem[6] that the zeros of the partition function lie on the unit

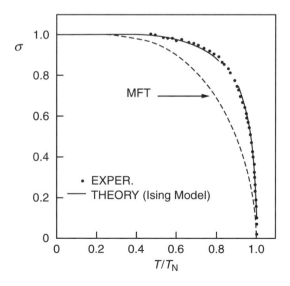

Fig. 8.19. Spontaneous sublattice magnetization (AF order parameter) σ versus T for DyPO$_4$.[42]

circle in the complex $\exp(-2B/kT)$ plane. He also proved an inequality for arbitrary p (Griffiths' inequality)[49]: If A and B are spins or products thereof,

$$\langle AB \rangle \geq \langle A \rangle \langle B \rangle \qquad (8.246)$$

in ferromagnets. For $p \neq 1$, one can also include crystal field effects via $(S_i^z)^2$ type terms, and consideration of these has led to the identification of several new phase transitions, some of first-order. We can only cite a few representative titles here.[50]

8.14.   Ising Gauge Glass

Recall from Eq. (6.42) the "separable" model of a random bond: $J_{ij} = \varepsilon_i \varepsilon_j F(R_{ij})$. In this example, the disorder — here, the random variables ε_i — can be purged by a trivial gauge transformation. In such cases, the disorder has no effect on the internal thermodynamics but does affect the interactions of the system with an external field. Here we shall work this particular model out in some detail.

The first step is to define new variables, $T_j = \varepsilon_j S_j$ where the T_j's, ε's and S's *all* are restricted to the two values ± 1. The value of each ε_j, whether $(+)$ or $(-)$, is "frozen-in." The S's or T's are "dummy" variables for the partition sum. The free energy and all other extensive thermodynamic functions (i.e., those that $\propto N$) must be computed for each distinct set of N such random variables, then averaged over the set. In the absence of an external field, the Hamiltonian lends itself to this procedure admirably:

$$\mathcal{H}_0 = -\sum_{(i,j)} F(R_{ij})\varepsilon_i \varepsilon_j S_i S_j \Rightarrow -\sum_{(i,j)} F(R_{ij}) T_i T_j . \qquad (8.247)$$

If $F = J$ when R_i and R_j are nearest-neighbors, and zero otherwise, this describes the ordinary Ising n.-n. ferromagnet $(J > 0)$ or antiferromagnet $(J < 0)$ treated earlier in this chapter, on a lattice in arbitrary dimensions. Thus the thermodynamic properties (internal energy, specific heat, etc.) of this random system are identical to those of the ordinary Ising model onto which it can be mapped. But the zero-field magnetic susceptibility and

[49] R. B. Griffiths, *J. Math. Phys.* **8**, 478, 484 (1967).

[50] M. Blume, *Phys. Rev.* **141**, 517 (1966); H. W. Capel, *Physica* **37**, 423 (1967) and references therein (Blume-Capel model); H. Chen and P. M. Levy, *Phys. Rev.* **B7**, 4267 (1973); D. Furman, S. Dattagupta and R. B. Griffiths, *Phys. Rev.* **B15**, 441 (1977); E. K. Riedel and F. J. Wegner, *Phys. Rev.* **B9**, 294 (1974); G. B. Taggart, *Phys. Rev.* **B20**, 3886 (1979).

magnetization in finite field have to be examined with more care. The perturbation due to a homogeneous magnetic field is,

$$\mathscr{H}' = -B\sum_j S_j \Rightarrow -B\sum_j \varepsilon_j T_j \tag{8.248}$$

and is here the equivalent of a random staggered field. Because the random array cannot "lock" onto any natural mode of the system, we expect χ_0 will not diverge — whatever the temperature. Thus, Z and F are:

$$Z(\{\beta B \varepsilon_j\}) = \text{Tr}\{e^{-\beta H_0(\{T_j\})}e^{-\beta B \sum_j \varepsilon_j T_j}\};$$

$$F = -kT\frac{\text{Tr}_{\{\varepsilon\}}\{\log(Z(\{\beta B \varepsilon_j\}\}}{2^N}$$

$$= -kT\langle\log(Z(\{\beta B \varepsilon_j\}))\rangle_{\text{config.}} \tag{8.249}$$

in which $\text{Tr}_{\{\varepsilon\}}$ stands for the trace over the ε variables. Because Z is not extensive and F is, it is the latter that is averaged over the random variables. Clearly there exists a *spin-flop* field $B_c(d,T)$ that overwhelms the internal forces, such that every spin is preferentially aligned with the external field. For a given lattice it depends on dimensionality d and temperature T. At large $B > B_c$ it is clearly preferable to keep the representation in terms of the S_j variables and to treat H_0 as a perturbation. (B_A might be zero in some cases.)

Mathematically and physically, the more interesting region is that of weak-field. There, we evaluate $\chi_0 \equiv -\partial^2 F/\partial B^2|_{B=0}$ explicitly as follows, as the fluctuation in the magnetization:

$$\chi_0 = \frac{-1}{kT}\left\{\left\langle\left(\sum_j \varepsilon_j T_j\right)^2\right\rangle - \left\langle\sum_j \varepsilon_j T_j\right\rangle^2\right\} \tag{8.250a}$$

where $\langle\cdots\rangle$ now stands for both *thermal* and *configurational* averages. The thermal average of $T_j = \langle T_j\rangle_{\text{TA}} = m$, the thermodynamic order parameter, the magnetization in an ordinary Ising model. Here, it refers to the ordering of the randomly transformed variables. Because only m^2 enters the expression, here *there will be no difference* between the cases of $J > 0$ and $J < 0$. Now, assuming each $\langle\varepsilon_j\rangle = 0$ and $\varepsilon_j^2 = 1$, and using the identity $T_j^2 = 1$, when we average over each and every ε_j the above reduces to,

$$\chi_0 = \frac{-1}{kT}\{N - Nm^2\} = -N\left(\frac{1-m^2}{kT}\right). \tag{8.250b}$$

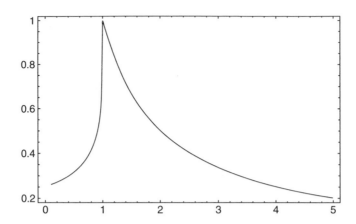

Fig. 8.20.   Zero-field longitudinal susceptibility χ_0 versus T/T_c in arbitrary units, showing the cusp in χ_0 of the separable Ising gauge glass in 2D.

In mean-field theory the quantity in parenthesis is a constant below T_c and $1/kT$ above it. Note: $m = 1$ at $T = 0$ by definition, hence, in all generality, χ has a cusp at T_c. In the accompanying figure, we plot this result for the two-dimensional Ising model, taking $m^2 = (1 - T/T_c)^{1/4}$ as a good approximation.

8.15.   Frustration

We briefly touched upon this topic earlier in this chapter and previously, in Chap. 6. Figure 6.15 showed that the sq Ising ferromagnetic lattice with a fraction $0 < p < 0.25$ antiferromagnetic (AF) bonds is a diluted ferromagnet, the energy of which rises with the concentration of "wrong" bonds, and that by symmetry, in the range $0.75 < p < 1$ it is similarly a dilute antiferromagnet. But in the intervening range $0.25 < p < 0.75$ this lattice has a stable ground state energy $\approx -1.4J$ approximately *independent* of p. This was recognized as a random, glassy phase, the Edwards-Anderson (E-A) spin-glass.[51] As we shall see, it is also a highly *frustrated* phase, with important thermodynamic consequences.

Frustration is the aptly named phenomenon that affects the triangular Wannier AF, as well as other systems in which conflicting interactions prevent the ground state from having a unique, well-defined structure

---

[51] S. F. Edwards and P. W. Anderson, *J. Phys.* **F5**, 965 (1975).

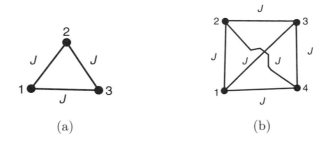

Fig. 8.21. Unit cells of two fully frustrated lattices.

in which all bonds are optimized. The following pictures of frustrated plaquettes, in which all bonds are J (AF), illustrate *geometrical frustration*. An antiferromagnetic lattice made of triangles (a) is a regular lattice in which every plaquette is frustrated. A rectangular lattice made of squares (b) with its internal AF connections is equally frustrated: these lattices have translational and rotational symmetries, but their ground states are multiply degenerate and do not exhibit the symmetry of the lattice.

Frustration occurs in antiferromagnets when nearest-neighbors of a site are *themselves* nearest-neighbors. The energy of the illustrated plaquettes can be written, $E_a = J/2[(S_1 + S_2 + S_3)^2 - 3]$ and $E_b = J/2[(S_1 + S_2 + S_3 + S_4)^2 - 4]$, so clearly the ground state of the single plaquette (a) is $E_{0,a} = -J$, 6-fold degenerate (out of only 8 possible states!) and of (b) $E_{0,b} = -2J$, also 6-fold degenerate (out of 16 states total).

Geometric frustration should be distinguished from the *random* frustration in the E-A spin glass. For example, consider the following 2 plaquettes carved out of the E-A lattice:

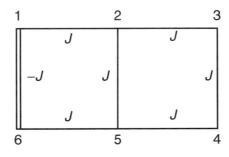

Fig. 8.22. Two connected plaquettes, the first one frustrated and the other not.

If J favors AF alignment, then a possible ground state is $\uparrow_1\downarrow_2\uparrow_3\downarrow_4\uparrow_5\downarrow_6$, the bond connecting S_1 to S_6 being "broken" because it is an odd-ball ferromagnetic bond; the energy is $E_0 = -5J$. But what is equivalent to a gauge transformation, i.e., flipping spin S_1, moves the frustration around the square. $\downarrow_1\downarrow_2\uparrow_3\downarrow_4\uparrow_5\downarrow_6$ has precisely the same energy, but the broken bond now is the one that connects S_1 to S_2. We can repeat this, finally breaking every one of 4 bonds: 1 to 2, 2 to 5, 5 to 6, and back to 6 to 1, and always recover the same energy. These are 4 distinct ground states. However attempts to move the break to the second plaquette and retain the lowest energy, $-5J$, will fail.

Our example shows that frustration is localized on the plaquette and not on any given bond. In the figure, the first plaquette is frustrated and the second one is not. Note also that the symmetry is broken: the two plaquettes are obviously inequivalent. Thus, the extended sq lattice, having some AF bonds and ferromagnetic bonds at random, exhibits *disorder frustration*, as contrasted to the geometrical effects seen earlier.

Problem 8.19.   (a) Out of a total of 64 states, how many ground states do the 2 plaquettes above have? (b) Show that if the bond connecting spin 2 to spin 5 in the above figure were *also* $-J$, it would be the *second* plaquette that is frustrated and *not* the first. How many ground states then?

The following figure on the sq lattice shows that it is possible to have *every* plaquette frustrated.

We see that *every* plaquette in a sq array *can* be frustrated. For this to occur, p need not originally be $1/2$ but can be in the range 0.25–0.75, as remarked previously. Then, by a sequence of gauge transformations the lattice can be made unitarily equivalent to the one shown (figure, right) in which every second riser is composed of "wrong" bonds. This transforms a lattice fully frustrated by *disorder* into a *geometrically* frustrated, but ordered lattice! and showing that these concepts are not that diametrically opposed.

It is now possible to set up and evaluate the transfer matrix for the fully frustrated sq lattice using the procedure of Sec. 8.12. However there is one question that has to be resolved first: do we propagate the rows vertically, by means of a single transfer matrix, or horizontally, where the unit cell consists of 2 risers, by the product of two inequivalent matrices, $V_F V_{AF}$? Each

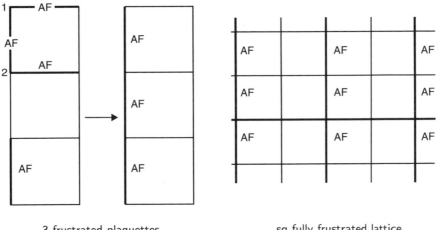

3 frustrated plaquettes   sq fully frustrated lattice

Fig. 8.23. On the left, all 3 plaquettes are frustrated (AF indicates antiferromagnetic bonds J, thin lines are ferromagnetic bonds $-J$. The top plaquette is frustrated because it has 3 odd bonds, the middle and bottom ones because they have 1 each.) By flipping the spins numbered 1 and 2 (i.e., by a gauge transformation) one obtains 3 transformed plaquettes; this shows the frustration much more clearly and exhibits the symmetry in this example. (Note that the fraction p of "wrong" bonds (marked AF) is not conserved: in the original drawing $p = 4$ (out of 10) whereas the transformed plaquettes, which are *unitarily equivalent*, only have $p = 3$. But the energies depend only the number of frustrated plaquettes and not on the fraction of "wrong" bonds.)

On the right we exhibit a *fully frustrated* sq lattice after all "bad" bonds have been transported to alternating vertical axes. In this highly symmetric representation, $p = 0.25$ (or 0.75).

presents a challenge. The vertical transfer matrix that transfers rows has regular horizontal bonds but then the vertical bonds alternate, F and AF. We refer to Problem 8.12 where we are asked to verify $\tanh K_1^* = \exp -2\beta J_1$, etc., and note that if $J_1 < 0$, K_1^* is necessarily complex. Because half the transfer operators are therefore complex, this choice of direction does require special care.

The horizontal transfer operator that transfers columns (or ladders, in the case of Fig. 8.23) has no such problem, but it has to include *two* columns by means of a product we can denote $V_F V_{AF}$. The product of these infinite matrices is possible only because, by momentum conservation, they can be decomposed into finite matrices in noninteracting, finite, sectors.

The final conclusion for this fully frustrated model: $T_c = 0$ and there is an essential singularity in the thermodynamic properties as $T \to 0$. This indicates that any attempt to obtain the thermodynamic functions c, χ, etc., by expanding around a ground state will fail. This should not be unexpected, given the huge ground state entropy $\mathscr{S}_0 = Nk_B 0.2916\ldots$ (almost half the $T \to \infty$ limit $\mathscr{S} = Nk_B \log 2 = Nk_B 0.693\ldots$) All critical exponents are shared with the triangular antiferromagnet, a better known frustrated lattice in 2D. It is also possible to decrease the degree of frustration by sequencing $n \geq 2$ unfrustrated columns following each frustrate column. The resulting plane is *ferromagnetic*! Details are given elsewhere.[52]

A simple view: the more frustration, the less neighboring plaquettes can interact. Suppose we are interested in a truly disordered system of *any* kind, in which the individual units have a set of n energies e_α, $\alpha = 0, 1, \ldots, -1$ with corresponding degeneracies g_α, believed to be weakly interacting with neighboring units. As a start we could just use the MFT to deal with their thermal properties, then take the interactions into account as a relevant perturbation. This yields for the initial free energy F and entropy $\mathscr{S}$,

$$F = -NkT \log \sum_\alpha g_\alpha e^{-\beta e_\alpha} \, ,$$

with $T \to 0$ limits : $F = -Ne_0 - NkT \log g_0$, $\mathscr{S} \to Nk \log g_0$.

$$(8.251)$$

Problem 8.20.   Take a ladder such as the one on the lhs of Fig. 8.23 above. Assume each plaquette is frustrated randomly, such that there is a 50% chance of being frustrated or not. Assuming neighboring plaquettes are independent, calculate F using the formula above and compare with an exact result.[53]

For extensive information concerning spin glasses, frustration and disparate methods of dealing with these phenomena (replica method, ultrametricity, etc.) there exist excellent specialized texts.[54]

---

[52] J. Wu and D. C. Mattis, *Phys. Rev. B* **67**, 224414 (2003) and references therein.
[53] D. C. Mattis and P. Paul, *Phys. Rev. Lett.* **83**, 3733 (1999), see their Fig. 4.
[54] M. Mézard, G. Parisi and M. Virasoro, *Spin Glass Theory and Beyond*, World Scientific, Singapore, 1987; D. Chowdhury, *Spin Glasses and Other Frustrated Systems*, World Scientific, 1986.

Chapter 9

Miscellaneous Advanced Topics

Where we introduce or reorganize some special topics to bridge the textual material and ongoing research and tie up loose ends.

9.1. Critical Phenomena

Throughout the text we alluded to critical exponents, under the assumption that all the important thermodynamic information satisfies simple power-law behavior — albeit with fractional or otherwise bizarre powers. Critical exponents were discussed in connection with the two-dimensional XY model, the Ising model, and the mean-field theory (MFT). The Table 9.1 summarizes some of the best known exponents and their values. We shall see later by what means they are obtained. The most important variable here is $t = (T - T_c)/T_c$, the temperature (in dimensionless variables) as measured from the assumed second-order phase transition at T_c. Thus, t is negative below the transition, positive above it. Similarly, B is the external field in some dimensionless units, m is the order parameter and η is the "anomalous dimension." Table 9.2 lists some values obtained for these exponents. The *scaling hypothesis* assumes any magnetic system is subject to an equation of state that, to a first approximation (small t, m, B, etc.,) takes the form $B = m^\delta \psi(t/m^{1/\beta})$. To the same approximation, the singular part of the free energy associated with the phase transition takes the form $F(t, B) = t^{2-\alpha} \phi(t/B^{1/\beta\delta})$. All quantities having dimension "length" are proportional to ξ, regardless of whether they originate in an external field $B \neq 0$ at $t = 0$ or in finite t at $B = 0$.

Table 9.1.   Definitions of magnetic thermodynamic and dynamical critical exponents: α, β, γ, δ, η, v, and z.

Magnetic Property	Power Law
Zero-Field Specific Heat c_B	$c \propto 1/|t|^{\alpha}$, t
Zero-Field Spontaneous Magnetization m	$m \propto (-t)^{\beta}$
Zero-Field Susceptibility χ_0	$\chi_0 \propto 1/|t|^{\gamma}$
Correlation Length	$\xi \propto 1/|t|^{\nu}$
Spin-Spin Correlation Function at T_c	$G(R) \propto 1/R^{d-2+\eta}$
Critical Isotherm $(t = 0)$	$|B| \propto |m|^{\delta}$
Relaxation in time τ at t	$M \sim 1/\tau^{\beta/vz}$

Table 9.2.   Some models and their critical exponents.

Name or Universality Class	Symmetry	α	β	γ	δ	ν	η	z
2D Ising	1-component scalar	0	1/8	7/4	15	1	1/4	2.167
3D Ising	//	0.1	0.33	5/4	4.8	0.63	0.04	?
3D XY	2-component vector	0.01	0.34	1.3	4.8	0.66	0.04	?
3D Heisenberg	3-component vector	−0.12	0.36	1.39	4.8	0.71	0.04	?
Mean-Field	any	0	1/2	1	3	1/2	0	?

The set of exponents in Table 9.1 is subject to a number of *scaling relations*. The idea is that equations that relate various quantities, such as c_B and c_M (specific heat in constant B or in constant magnetization) will also relate their exponents. Even though some took the form of inequalities, in most applications these have proved to be, quite accurately, equalities. The empirical "relations" are:

$$\alpha + 2\beta + \gamma = 2\,,$$

$$\alpha + \beta(1 + \delta) = 2\,,$$

$$\gamma = \beta(\delta - 1)\,,$$

$$\gamma = (2 - \eta)\nu\,,$$

$$2 - \alpha = d\nu\,, \tag{9.1}$$

in which d is the dimensionality. These are frequently supplemented by the following:

$$\alpha = 2 - d/y_1\,, \quad \beta = (d - y_2)/y_1\,,$$

$$\gamma = (2y_2 - d)/y_1 \quad \text{and} \quad \delta = y_2/(d - y_2)\,, \tag{9.2}$$

such that for any d, the set of 3 variables: y_1, y_2 and η suffice to calculate *all* other critical exponents.

The results summarized in the above tables and scaling relations confirm many of the results quoted or derived in the preceding chapters, although they transcend these specific applications. They are the result of numerous papers and collaborations analyzing correlation functions, series expansions in B, T or $1/T$, performed over a period of several decades starting in the late 1960's. More details are to be found elsewhere.[1] Insofar as they relate to "critical phenomena," they have been confirmed by much more sophisticated mathematics, including the use of conformal and renormalization group adapted for the region at or near the phase transition. Additionally there are critical exponents (e.g., z), defined for diffusive and relaxation processes at, or near, a critical point; e.g., $\mu(t) \sim 1/\tau^{\beta/\nu z}$ is the decay in time τ of the ferromagnetic state at $T = T_c$, $B = 0$. The exponents are listed in Table 9.1. We shall return to this topic shortly, but first we have to have some way to compute correlation functions. For this purpose we first turn to the topic of so-called "thermodynamic" time-dependent Green functions, also an invention of the 1960's.

---

[1] H. E. Stanley, *Introduction to Phase Transitions and Critical Phenomena*, Oxford Univ. Press, Oxford, 1971, see also J. M. Yeomans, *Statistical Mechanics of Phase Transitions*, Clarendon Press, Oxford, 1992. Scaling and "hypersealing" relations are the subjects of V. Privman and M. E. Fisher, *Phys. Rev.* **B30**, 322 (1984) and K. Binder *et al.*, *Phys. Rev.* **B31**, 1498 (1985). Critical exponents obtained by an expansion in powers of $\varepsilon = 4 - d$ are derived in K. G. Wilson and J. Kogut, *Phys. Reports* **12**, 75–200 (1974). The dynamical exponent z is the subject of C. Kalle, *J. Phys.* **A17**, L801 (1984) and D. Stauffer, *Int. J. Mod. Phys.* **C10**, 931 (1999).

9.2.   Green Functions Formalism

There is a lot of information contained in the thermal average $\langle S_n S_m \rangle_{\text{TA}}$ as a function of distance R_{nm}, temperature T (as implied in "TA") and internal and external fields J, B. At a phase transition where all natural scales of length disappear (correlations become infinite-ranged, etc.,) only an exponent can survive: $\langle S_n S_m \rangle_{\text{TA}} \propto (a/R_{nm})^\eta$. If a length scale ξ appears above (or below) the phase transition, this should not affect the power law. In that case one expects $\langle S_n S_m \rangle_{\text{TA}} \propto (a/R_{nm})^\eta \exp -R_{nm}/\xi$ with the same η, although it is now ξ that dominates the asymptotic behavior.

In generalizing to time-dependent phenomena, the technique of choice uses time-dependent Green functions. These are written by some authors as $G_{n,m}(t-t')$ and by others as $((S_m(t)|S_n(t')))$ and defined as the thermal average of the commutator at times $t > t'$, i.e., $-i\vartheta(t-t')\langle [S_m(t), S_n(t')] \rangle_{\text{TA}}$.[2,3]

Why would one define such arcane quantities? The answer lies in their Fourier transform, here denoted $\langle\langle S_m | S_n \rangle\rangle(\omega)$ somewhat awkwardly but to emphasize the dependence on the 2 operators and on a complex variable ω that has dimensions of energy (with $\hbar = 1$). Generally, it can be proven[2] that $\langle\langle A|B \rangle\rangle(\omega)$ can determine the correlation function $\langle B(t')A(t) \rangle_{\text{TA}}$ by means of the following formula:

$$\langle B(t')A(t) \rangle_{\text{TA}} = i \int_{-\infty}^{\infty} d\omega \frac{e^{-i\omega(t-t')}}{e^{\omega/kT}-1} \{ \langle\langle A|B \rangle\rangle(\omega+i\varepsilon) - \langle\langle A|B \rangle\rangle(\omega-i\varepsilon) \}_{\lim .\varepsilon \to 0}$$

$$(9.3)$$

This formula is simpler than it looks, because $\langle\langle A|B \rangle\rangle(\omega)$ is analytic in the upper half ω plane (and in the lower half also), with all singularities restricted to the real axis. *Poles* on the real axis arise from any discrete levels of the Hamiltonian $\mathscr{H}$ and *branch cuts* from any continua (e.g., the spin-wave spectrum). Discontinuities on the real ω axis are what contribute to correlation functions evaluated with the aid of (9.3).

One straightforward method to locate the singularities uses "equations of motion" of the Green functions in ω. These are:[2]

$$\omega \langle\langle A|B \rangle\rangle(\omega) = \frac{1}{2\pi} \langle [A, B] \rangle_{\text{TA}} + \langle\langle [A, \mathscr{H}]|B \rangle\rangle(\omega) \qquad (9.4)$$

---

[2] See D. N. Zubarev, *Sov. Phys. (Nauk)* **71**, 71 (1960), also G. D. Mahan, *Many-Particle Physics*, 2nd Ed., Plenum, New York, 1990, pp. 81–234.

[3] [NB. The symbol t as used in this section stands for *time*, and should not be confused with the dimensionless *temperature* in the preceding section.]

Here $[A, B]$ is the usual (equal-time) commutator of the two operators and $\mathcal{H}$ the Hamiltonian. If the commutator $[A, \mathcal{H}]$ is proportional to A, e.g., $[A, \mathcal{H}] = \omega_0 A$ (this happens if A is either a raising or a lowering operator of $\mathcal{H}$), then (9.4) is solved by inspection:

$$(\omega - \omega_0)\langle\!\langle A|B\rangle\!\rangle = \frac{\langle[A, B]\rangle_{\text{TA}}}{2\pi},$$

i.e.,

$$\langle\!\langle A|B\rangle\!\rangle(\omega \pm i\varepsilon) = \frac{\langle[A, B]\rangle_{\text{TA}}}{2\pi}\frac{1}{(\omega - \omega_0 \pm i\varepsilon)}$$

The "hair-splitting" mathematical identity $\frac{1}{x \pm i\varepsilon} = PP\frac{1}{x} \mp i\pi\delta(x)$ proves useful here (the pole on the real axis is split into two halves). The principal part PP cancels in (9.3) whereas the pole contributes $-2\pi i$. One finds: $\langle B(t')A(t)\rangle_{\text{TA}} = \langle[A, B]\rangle_{\text{TA}} \times \frac{e^{-i\omega_0(t-t')}}{e^{\omega_0/kT}-1}$, as the reader may wish to verify. This procedure works for *spins*, *bosons*, *pairs of fermions*, etc.

For individual *fermion* operators or odd products thereof, the appropriate Green functions are defined using *anticommutators* and -1 in the denominator of (9.3) is replaced by $+1$. With this minor change (9.3) can be used for *fermion correlation functions*. However, Eq. (9.4) has to be modified:

$$\omega\langle\!\langle A|B\rangle\!\rangle(\omega) = \frac{1}{2\pi}\langle\{A, B\}\rangle_{\text{TA}} + \langle\!\langle[A, \mathcal{H}]|B\rangle\!\rangle(\omega) \tag{9.5}$$

for fermion operators A and B, where $\{A, B\} = AB + BA$ is the equal-time *anticommutator*.

It is unusual for A to be an exact raising/lowering operator of $\mathcal{H}$, although in general one should be able to expand in such exact lowering and raising operators.[4] For suppose the exact (and suitably *normalized*) lowering/raising operators to be $A(\omega)$ and $A^\dagger(\omega)$; expand $A(t)$ as, $A(t) = \int_0^\infty d\omega\{e^{i\omega t}\rho_1(\omega)A(\omega) + e^{-i\omega t}\rho_2(\omega)A^\dagger(\omega)\}$, the ρ's being weight functions characterizing the operator A. Expanding B in a similar way allows to evaluate $\langle B(t')A(t)\rangle_{\text{TA}}$ in terms of their respective "spectral densities," the ρ's, in Eq. (9.3).

---

[4] The proof is simple: if $A(t) = e^{-iHt/\hbar}Ae^{iHt/\hbar}$, then $\frac{d}{dt}A(t) = \frac{i}{\hbar}[A(t), \mathcal{H}]$. Fourier transform both: $A(t) = \int d\omega e^{i\omega t}A(\omega)$ and $\frac{d}{dt}A(t) = \int d\omega e^{i\omega t}A(\omega)i\omega = \int d\omega e^{i\omega t}\frac{1}{\hbar}[A(\omega), \mathcal{H}]$. Whenever $A(\omega)$ is nonzero, it satisfies $[A(\omega), H] = \hbar\omega A(\omega)$, i.e., it is a lowering operator for $\omega > 0$ in the amount $\hbar\omega$, it is similarly a raising operator for $\omega < 0$. Hence ρ_1 in the text is nonzero only for $\omega > 0$, ρ_2 for $\omega < 0$.

A simple (albeit awkward) example illustrates the Green function approach. Suppose we wished to study a collection of spins $1/2$ with long-range Ising-type ferromagnetic interactions $-B_0/N$, such that each spin is subject to an internal and an external field, $B_{\text{int}} + B_{\text{ext}}$. Here, $B_{\text{int}} = B_0\langle S_j^z\rangle$ (for any j) defines the internal parameter B_0 and the order parameter $m = \langle S_j^z\rangle$. Consider the equations of motion:

$$\omega\langle\langle S_j^-\,|S_n^+\rangle\rangle = -\frac{1}{\pi}\delta_{j,n}\langle S_j^z\rangle_{\text{TA}}$$

$$+ \left\langle\!\left\langle\left[S_j^-\,, -\frac{B_0}{2N}\left(\sum_{i=1}^N S_i^z\right)^2 - B_{\text{ext}}\sum_{i=1}^N S_i^z\right]\middle|S_n^+\right\rangle\!\right\rangle.$$

$$(9.6)$$

The first task is evaluation of the commutators in the new Green functions generated on the rhs.

Problem 9.1.   Obtain the result in the following Eq. (9.7).

$$\left[S_j^-\,, -\frac{B_0}{2N}\left(\sum_{i=1}^N S_i^z\right)^2 - B_{\text{ext}}\sum_{i=1}^N S_i^z\right] = -(B_0 m + B_{\text{ext}})S_j^-$$

$$(9.7)$$

hence we know the Green function: $(\omega + B_0 m + B_{\text{ext}})\langle\langle S_j^-\,|S_n^+\rangle\rangle = -\frac{1}{\pi}\delta_{j,n}\langle S_j^z\rangle_{\text{TA}}$. According to Eq. (9.3) this knowledge allows to evaluate $\langle S_n^+ S_n^-\rangle_{\text{TA}}$ as follows,

$$\langle S_n^+ S_n^-\rangle_{\text{TA}} = i\int d\omega\,\frac{1}{e^{\omega/kT}-1}\{\langle\langle S_n^-\,|S_n^+\rangle\rangle_{\omega+i\varepsilon} - \langle\langle S_n^-\,|S_n^+\rangle\rangle_{\omega-i\varepsilon}\}$$

$$= -2\langle S_n^z\rangle_{\text{TA}}\frac{1}{2\pi}\int d\omega\,\frac{1}{e^{\omega/kT}-1}$$

$$\times\left\{\frac{1}{\omega + B_0 m + B_{\text{ext}} + i\varepsilon} - \frac{1}{\omega + B_0 m + B_{\text{ext}} - i\varepsilon}\right\}.$$

The integral on the rhs collapses using $\{\cdots\} = -2\pi i\delta(\omega + B_0 m + B_{\text{ext}})$. On the lhs we use an identity: $\langle S_n^+ S_n^-\rangle_{\text{TA}} = \langle S_n^z\rangle_{\text{TA}} + \frac{1}{2}$. Finally we recover a result previously derived in mean-field theory,

$$m \equiv \langle S_n^z\rangle_{\text{TA}} = \frac{1}{2}\tanh\frac{B_0 m + B_{\text{ext}}}{2kT}.$$

$$(9.8)$$

The following problem examines the quantum corrections.

Problem 9.2. Solve the same set of equations if $\mathcal{H} = -\frac{B_0}{2N}(\sum_{i=1}^{N} \mathbf{S}_i)^2 - B_{\text{ext}} \sum_{i=1}^{N} S_i^z$ is the Hamiltonian instead (i.e., if the internal interactions preserve rotational symmetry). What then is the magnon spectrum?

9.3. Nonlinear Responses and Chaos

Once a single spin is excited by an electromagnetic wave it can absorb energy — but only to a limited extent. This saturation can be perfectly analyzed using time-dependent Green functions such as were introduced above. But something even more interesting happens to a collection of interacting spins that could only be described in the language of nonlinear and chaotic phenomena. In the example of Yttrium iron garnet (YIG) stimulated by 8.86 Ghz radiation at low temperature (4 K), Yamazaki[5] at first found absorption, as might be expected by absoption of energy by magnons. But upon increasing the power level, he found subharmonic waves being generated, followed by period doubling and then chaotic behavior. At even higher power levels the oscillations became periodic again followed by *period halving*. These results are summarized in the Fig. 9.1, adapted from Yamazaki's papers.[5] From the data in the chaotic region at the power level of 3.60 db, Yamazaki was able to draw a Lorenz map as a single-humped curve. (This Lorenz map connects a maximum in the response V_n to the following maximum V_{n+1}; its shape is the signature of nonlinear, chaotic behavior.)

Magnetism abounds in nonlinearities. Looking no further than Eq. (9.8), if $B_{\text{ext}} = B_0 \cos(\omega t)$, the response $m(t)$ will display Fourier components $A_n \cos(n\omega t)$ for all integer n, i.e., a spectrum of all harmonics. The reader can easily verify this.

9.4. Kondo Phenomenon: The *s-d* Model Redux

The Kondo problem has received more attention than it deserves perhaps because, like the chaotic behavior shown in the preceding pages, it too is a

---

[5] H. Yamazaki, *J. Appl. Phys.* **64**, 5391 (1988), also Fractal properties in magnetic crystal, in *Nonlinear Phenomena and Chaos in Magnetic Materials*, P. E. Wigen, ed., World Scientific, Singapore, 1994, pp. 191ff. In more purely theoretical investigations, it has been shown that the time-dependent magnetic field can change the *order* of the phase transition (from second- to first-order) in MC simulations of an Ising model: A. Krawiecki, *Int. J. Mod. Phys.* **B19**, 4769 (2005).

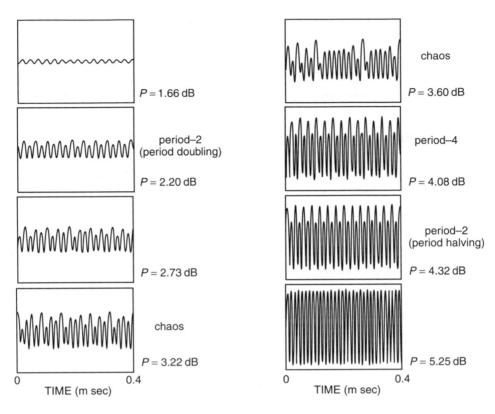

Fig. 9.1. Observed oscillations at various microwave power levels in YIG. Note chaotic oscillations at intermediate power levels and entrainment at the highest power level.[5]

severe test of the nonlinear physics associated with magnetism. In general terms, it has to do with point scatterers in metals that possess an internal degree of freedom. Because the manufacture of copper metals involves the use of steel, trace amounts of iron impurity atoms are generally present within the otherwise pure copper. Scattering of the conduction electrons by normal impurities is independent of temperature. For magnetic impurities, such as iron in this instance, the resistance apparently *increases* as the temperature is lowered, an effect that is all the more noticeable as the other sources of electrical resistance (mainly, electron-phonon interactions) are quenched at low T. The physical origins of this phenomenon were discussed at some length in Sec. 6.8, where low-order perturbation theory was found to reveal some logarithmic divergences as the discontinuity in occupation number at the Fermi level became sharper. For this, our last attack on the problem, we take advantage of the Green function formalism.

In the search for interesting applications of the renormalization group he helped invent, particle-physicist and Nobel prize-winner Ken Wilson also concentrated on the Kondo effect.[6] By dint of his numerical solution, Wilson was able to exorcise the logarithmic singularities and calculate all the interesting thermodynamic and transport properties associated with this phenomenon. For example, in comparing the paramagnetic susceptibility for a magnetic field acting on the impurity alone χ_{imp} with the excess heat capacity due to this atom with an internal degree of freedom, c_{imp}, he found that the ratio was $\lim_{T\to 0}(\frac{c_{\text{imp}}(T)}{T\chi_{\text{imp}}(T)}) = 2\pi^2/3$ ("Wilson's ratio") in weak coupling, $J \to 0$, half the value it attains in strong coupling $J \to \infty$.

About a decade later, P. B. Wiegman and N. Andrei attacked the very same problem, separately and independently,[7] assuming linearized dispersion and making use of Bethe ansatz. The reader will recall this approach, originally invented by H. Bethe for the purpose of solving the one-dimensional spin 1/2 antiferromagnet and presented earlier in this book in that context.[8] Using distinct but ultimately equivalent analyses, these two authors were able to identify the eigenstates in this model as well, especially the phase shifts that control the scattering. Their calculations yield Wilson's ratio and many more details that had remained as puzzles to be solved.

That this should be at all possible is not a given. Attempts to apply Bethe ansatz to chains of spin 1 or 3/2 have been found to give entirely the wrong answers; for example, the Bethe ansatz "solution" of the spin 1 chain does not uncover the least sign of Haldane's gap. This is due to the failure of the ansatz to satisfy the Yang-Baxter equation in those case.[9] This is an equation that establishes whether repeated scatterings among *three* particles might create destructive interference. If the final phase depends on the *order* in which particles 1, 2 and 3 scatter, then Bethe's ansatz is wrong.

---

[6] K. G. Wilson, Solution of the spin 1/2 Kondo hamiltonian, in *Collective Properties of Physical Systems*, Proc. of the 24th Nobel Symposium, B. Lundqvist and S. Lundqvist, eds., Academic Press, N.Y., 1973; also, at greater length, in *Rev. Mod. Phys.* **47**, 773 (1975).

[7] P. B. Wiegmann, *JETP Lett.* **31**, 364 (1980); N. Andrei, *Phys. Rev. Lett.* **45**, 379 (1980). For a detailed and fairly complete review, see especially, N. Andrei, K. Furuya and J. H. Lowenstein, *Rev. Mod. Phys.* **55**, 331 (1983).

[8] See Sec. 5.7.

[9] See, for example, R. J. Baxter, Exactly solved models in statistical mechanics, in *Integrable Systems in Statistical Mechanics*, G. M. D'Ariano, A. Montorsi and M. G. Rasetti, eds., World Scientific, 1985, pp. 5–63. For mathematicians: Z.-Q. Ma, *Yang-Baxter Equation and Quantum Enveloping Algebras*, World Scientific, 1993.
C. N. Yang recently recalled that he and C. P. Yang were the first to give Bethe's hypothesis a name. He also makes the connections with the Yang-Baxter Equation (YBE), "quantum groups", Jones polynomials and knot theory, in his brief historical contribution to *Hans Bethe and His Physics*, World Scientific, 2006, forthcoming.

Violation of this rule vitiates Bethe ansatz as a solution to *most* many-spin (or many-electron) problems.

Fortunately, neither the extended-Bethe-ansatz solutions found by Andrei and by Wiegmann, nor the famous Lieb-Wu solution of the one-dimensional Hubbard chain, suffers from this defect. Although the analysis of either model is beyond the scope of the mathematics we have developed so far in this book, it is good to know that such problems can be, and have been, "reduced to quadrature." But, as in the case of other seemingly insurmountable problems that were originally solved by pioneering methods deemed incomprehensible at the time,[10] it is possible to obtain similar results by simpler means. In this instance, the Green function approach introduced earlier will do nicely if applied to the *s-d* model.[11] Some fact are known exactly:[12] the impurity's spin (the "*d*"-electron) becomes increasingly screened by the conduction-band electrons with energies near the Fermi level (the "*s*"-band) in its vicinity as the temperature is lowered. For a spin 1/2, the local magnetic moment in a large sphere surrounding the impurity is entirely cancelled out at $T = 0$.[12] The *s-d* Hamiltonian takes the form,

$$\mathscr{H} = \sum_{k,\sigma} \varepsilon(k) c_{k,\sigma}^{\dagger} c_{k,\sigma} + J\left(\mathbf{S} \cdot \boldsymbol{\sigma} - \frac{1}{4}\hat{n}_0\right) - BS^z \qquad (9.9)$$

where the Bloch energy ε is measured relative to the Fermi level. The magnetic field B is assumed to act only on the impurity's spin as the paramagnetic response of the host electrons is not an issue here. The operator $\hat{n}_0 = \frac{1}{N} \sum_{k,k',\sigma} c_{k,\sigma}^{\dagger} c_{k',\sigma}$ is the occupation number operator of the conduction electrons at the origin. The origin is taken at the site of the impurity, where the spin 1/2 operator $\mathbf{S}$ is situated. The coupling J is antiferromagnetic. Subtraction of $J\hat{n}_0/4$ ensures that the coupling is purely antiferromagnetic, i.e., that *only* electrons of spin antiparallel to $\mathbf{S}$ are scattered.

Let us make use of the Fermion representation presented in Eq. (3.97), setting $S^z = \frac{1}{2}(a_{\uparrow}^{\dagger} a_{\uparrow} - a_{\downarrow}^{\dagger} a_{\downarrow})$, $S^+ = a_{\uparrow}^{\dagger} a_{\downarrow}$ and $S^- = a_{\downarrow}^{\dagger} a_{\uparrow}$, and requiring

---

[10] Take, for example, Onsager's solution of the thermodynamical properties of the two-dimensional Ising model that spurred dozens of other, equivalent or inequivalent, methods of solution. In the instance of the Kondo model, an even larger number of theories have been tested: field-theoretic, renormalization-group based, etc. Anderson has remarked that, according to the London *Times*, "Kondoism" in Swahili translates into "robbery with violence." Any violence and robbery here would be to scientific research budgets ... See, P. W. Anderson, *Rev. Mod. Phys.* **50**, 191 (1978), p. 195.

[11] G. D. Mahan, *Many-Particle Physics*, 2nd Ed., Plenum, New York, 1981, chapter 11.

[12] For example, the ground state is a singlet, D. C. Mattis, *Phys. Rev. Lett.* **19**, 1478 (1967).

$a_\uparrow^\dagger a_\uparrow + a_\downarrow^\dagger a_\downarrow = 1$, as is appropriate for spin 1/2. Insert these expressions into (9.9) along with $J\hat{n}_0/4$ replaced by $j\hat{n}_0(a_\uparrow^\dagger a_\uparrow + a_\downarrow^\dagger a_\downarrow)/4$. As in Chap. 5, the Hamiltonian with these substitutions simplifies:

$$\mathcal{H} = \sum_{k,\sigma} \varepsilon(k) c_{k,\sigma}^\dagger c_{k,\sigma} - \frac{J}{2}\Omega^\dagger\Omega - \frac{B}{2}(a_\uparrow^\dagger a_\uparrow - a_\downarrow^\dagger a_\downarrow) \tag{9.10}$$

where $\Omega \equiv \frac{1}{\sqrt{N}}\sum_k \Omega_k$, where $\Omega_k = (c_{k,\downarrow} a_\uparrow - c_{k,\uparrow} a_\downarrow)$ is bilinear in fermions, hence almost a boson operator ... but not quite.

Rather, if the Hamiltonian is linearized a fermionic Green function scheme suggests itself. For if Ω acquires a nonzero expectation value *and* if the variance $\langle(\Omega^\dagger - \langle\Omega^\dagger\rangle_{TA})(\Omega - \langle\Omega\rangle_{TA})\rangle_{TA}$ is not too great, a linearized Hamiltonian can be solved exactly and the various expectation values evaluated self-consistently. In this way it even becomes possible in principle to test the accuracy of the linearization objectively. The linearized Hamiltonian we study next,

$$\mathcal{H}_{\text{lin}} = v_F \sum_{k,\sigma}(k - k_F) c_{k,\sigma}^\dagger c_{k,\sigma} - \frac{\Delta}{\sqrt{N}}\sum_k (a_\uparrow^\dagger c_{k,\downarrow}^\dagger - a_\downarrow^\dagger c_{k,\uparrow}^\dagger)$$

$$- \frac{\Delta^*}{\sqrt{N}}\sum_k (c_{k,\downarrow} a_\uparrow - c_{k,\uparrow} a_\downarrow) + 2\frac{|\Delta|^2}{J} - \frac{1}{2}B(a_\uparrow^\dagger a_\uparrow - a_\downarrow^\dagger a_\downarrow) \tag{9.11}$$

is just one of a number of possible versions of a linearization, albeit a natural one. Here $\Delta = J/2\langle\Omega\rangle_{TA}$ defines the assumed expectation value. The quantity $2\frac{|\Delta|^2}{J}$ is added back into avoid double-counting. Note that one could additionally have paired $a_\uparrow$ with $c_{k,\uparrow}^\dagger$, etc. A linearized Hamiltonian with all possible pairings that can be formulated in this way is known as the "generalized" Hartree-Fock approximation. For the purposes of a self-consistent evaluation of such an Hamiltonian at finite T, *fermionic* Green functions of the type $\langle\langle a_\uparrow | c_{k,\downarrow}\rangle\rangle(\omega) =$ Fourier transform of $\{-i\vartheta(t-t')\langle\{a_\uparrow(t), c_{k,\downarrow}(t')\}\rangle_{TA}\}$ are required. The curly bracket $\{a, c\} = ac + ca$ indicates an *anticommutator*. For magnetic susceptibility one requires knowing the function $\langle\langle a_\sigma | a_\sigma^\dagger\rangle\rangle(\omega)$, while for scattering (transport) lifetimes it is $\langle\langle c_{k,\sigma} | c_{k,\sigma}^\dagger\rangle\rangle(\omega)$ with k at or near the Fermi surface. The Green functions serve a dual purpose: their equations of motion yield the energy spectrum for arbitrary expectation values such as Δ, Δ^* in the linearized Hamiltonian above. Then a formula similar to (9.3), suitably modified for fermions, completes the task of obtaining these expectation values self-consistently. The calculations that follow are a template of what is required under more general circumstances.

Omitting (for typographical simplicity) the explicit ω dependence in the Green functions $\langle\langle xx|yy \rangle\rangle(\omega)$ that follow, the equations of motion are:

$$(\omega + B/2)\langle\langle a_\uparrow|c_{k,\downarrow}\rangle\rangle = \frac{-\Delta}{\sqrt{N}}\sum_{k'}\langle\langle c^\dagger_{k',\downarrow}|c_{k,\downarrow}\rangle\rangle \qquad (9.12\text{A})$$

and

$$(\omega + v_F(k'-k_F))\langle\langle c^\dagger_{k',\downarrow}|c_{k,\downarrow}\rangle\rangle = \frac{\delta_{k,k'}}{2\pi} + \frac{\Delta^*}{\sqrt{N}}\langle\langle a_\uparrow|c_{k,\downarrow}\rangle\rangle. \qquad (9.12\text{B})$$

The $\delta_{k,k'}$ results from the equal-time anticommutator $\{c^\dagger_a, c_b\} = \delta_{a,b}$ that characterizes the kinematics of fermion fields, just as the equal time commutator does for the bosons. Thus, there is no inhomogeneous term in (A) because $\{a_\uparrow, c_{k,\downarrow}\} = 0$. Similar equations obtain for the opposite spins, with $B \to -B$.

To calculate the magnetic susceptibility a different set of Green functions is required, *viz.*,

$$(\omega + B/2)\langle\langle a_\uparrow|a^\dagger_\uparrow\rangle\rangle = \frac{1}{2\pi} - \frac{\Delta}{\sqrt{N}}\sum_{k'}\langle\langle c^\dagger_{k',\downarrow}|a^\dagger_\uparrow\rangle\rangle. \qquad (9.13\text{A})$$

A second equation for the quantity on the *rhs* helps achieve closure:

$$(\omega + v_F(k'-k_F))\langle\langle c^\dagger_{k',\downarrow}|a^\dagger_\uparrow\rangle\rangle = \frac{\Delta^*}{\sqrt{N}}\langle\langle a_\uparrow|a^\dagger_\uparrow\rangle\rangle. \qquad (9.13\text{B})$$

Equations (9.13) and (9.12) are, in fact, related, but here there is no need to use this information here.

Equations (9.12) are solved simultaneously by inserting (A) into (B):

$$\{\omega + v_F(k'-k_F)\}\langle\langle c^\dagger_{k',\downarrow}|c_{k,\downarrow}\rangle\rangle = \frac{\delta_{k,k'}}{2\pi} - \frac{\Delta\Delta^*}{\omega + B/2}\frac{1}{N}\sum_{k''}\langle\langle c^\dagger_{k'',\downarrow}|c_{k,\downarrow}\rangle\rangle. \qquad (9.14)$$

Division by $N\{\ldots\}$ followed by summation over k' solves this integral equation. That is,

$$\frac{1}{N}\sum_{k''}\langle\langle c^\dagger_{k'',\downarrow}|c_{k,\downarrow}\rangle\rangle = \frac{1}{2\pi N}\frac{1}{(\omega + v_F(k-k_F))} \times \frac{\omega + B/2}{\omega + B/2 + |\Delta|^2\Lambda(\omega)}, \qquad (9.15)$$

in which the function $\Lambda(\omega)$ is defined (and approximated) as follows:

$$\Lambda(\omega + i\varepsilon) \equiv \frac{1}{N} \sum_{k''} \frac{1}{\omega + i\varepsilon + v_F(k'' - k_F)}$$

$$\approx \frac{1}{v_F Q} \log\left(\frac{Q - k_F}{k_F}\right) + \frac{\omega}{v_F^2 k_F(Q - k_F)} - i\pi\rho(v_F k_F - \omega)$$

Here $\rho \geq 0$ is the density-of-states function. $\Lambda(\omega + i\varepsilon)$ is also written, $\Lambda(\omega) = R(\omega) - iI(\omega)$. With the particular choice of cutoff $Q = 2v_F k_F$, these functions simplify: $R(\omega) = \omega/(v_F k_F)^2$ and $I(\omega) = \pi/(2v_F k_F)$ for ω in the range $-v_F k_F < \omega < +v_F k_F$, and $I(\omega) = 0$ for ω outside this range. For purposes of illustration that is what will be assumed in the calculations that follow.

If the spin are flipped, the equations of motion and their solutions stay identical to the above except for trivial sign changes: $B \to -B$ and Δ, $\Delta^* \to -\Delta, -\Delta^*$. Because the calculation of Δ (unlike that of χ) is insensitive to B when $|B|$ is small it is permissible to set $B = 0$ in this expression, further simplifying the algebra that follows. Then, using the explicit solution of (9.12) and the similar Green function with flipped spins, one evaluates the correlation functions that have to be summed to obtain Δ:

$$\Delta = \frac{J}{2} \frac{1}{\sqrt{N}} \sum_k (\langle c_{k,\downarrow} a_\uparrow \rangle_{\mathrm{TA}} - \langle c_{k,\uparrow} a_\downarrow \rangle_{\mathrm{TA}})$$

$$= \frac{J}{2} i \int d\omega \frac{1}{e^{\beta\omega} + 1} \{F(\omega + i\varepsilon) - F(\omega - i\varepsilon)\}_{\varepsilon - 0} \qquad (9.16\text{A})$$

where $F(\omega) = \frac{-\Delta\Lambda(\omega)/\pi}{\omega + |\Delta|^2\Lambda(\omega)}$. Note the $(+)$ sign that replaces the $(-)$ sign in (9.3). As usual, $\beta = 1/kT$. Insertion of this expression into (9.16A) yields:

$$\Delta = -\frac{\Delta J}{\pi} \int_{-v_F k_F}^{+v_F k_F} d\omega \left(\frac{1}{e^{\beta\omega} + 1}\right) \frac{\omega I(\omega)}{(\omega + |\Delta|^2 R(\omega))^2 + |\Delta|^4 I^2(\omega)}. \qquad (9.16\text{B})$$

The trivial solution $\Delta(T) = 0$ is of little interest. At $T = 0$ (indeed, at all T,) negative values of ω contribute the most to the integral. Because of the explicit $(-)$ sign, a nontrivial solution $\Delta(T) \neq 0$ exists *only* if $J > 0$, i.e. only for antiferromagnetic coupling. That is reassuring, as the Hamiltonian excludes scattering of electrons with spins parallel to that of the impurity. Note that the ubiquitous expression $\omega + |\Delta|^2\Lambda(\omega + i\varepsilon) = \omega(1 + |\Delta/v_F k_F|^2) - i\pi|\Delta|^2/v_F k_F \equiv \alpha^2\omega - i\gamma^2$ (which *defines* α and γ) is actually quite simple. Evaluation of (9.16B) at very small J (i.e., $0 < J \ll v_F k_F$)

and $T = 0$ can proceed, and yields a nonzero Δ:

$$\Delta(0) = 0.8 v_F k_F e^{\frac{-v_F k_F}{J}} . \tag{9.17}$$

The quantity measures the condensation of a weakly-bound composite impurity-conduction band state. At finite temperature this binding energy parameter $\Delta(T)$ decreases with increasing T.

Problem 9.3.   With I and R approximated as in the text above, calculate $\Delta(T)$ as function of $kT/v_F k_F$ with $J/v_F k_F$ as the (small) control parameter. Comparison of (9.14) at $B = 0$ with a similar expression for scattering by an ordinary point-potential of strength V, determines $V = \frac{|\Delta|^2}{w\alpha^2 + B/2 - i\gamma^2} = V(T, \omega)$ here, with $\Delta(T)$, α as calculated above. The Kondo resistivity *per* impurity atom is $R(T) \propto |V(T,0)|^2$. Using your calculated $\Delta(T)$ compare $R(T)/R(0)$ to be experimental resistance, Figs. 6.10, 6.11 and estimate $T_K(J)$.

The magnetic susceptibility requires knowledge of the spin-correlation functions. These are $\langle a_\uparrow^\dagger a_\uparrow \rangle$ and $\langle a_\downarrow^\dagger a_\downarrow \rangle$ and they are found by means of the Green functions of Eqs. (9.13). Solving for these,

$$\langle\!\langle a_\uparrow | a_\uparrow^\dagger \rangle\!\rangle = \frac{1/2\pi}{w\alpha^2 + B/2 - i\gamma^2} = \text{and } \langle\!\langle a_\downarrow | a_\downarrow^\dagger \rangle\!\rangle = \frac{1/2\pi}{w\alpha^2 - B/2 - i\gamma^2} . \tag{9.18}$$

As in the calculation of $\Delta(T)$, the expectation value of the spin impurity's magnetization, $m = \frac{1}{2}(\langle a_\uparrow^\dagger a_\uparrow \rangle - \langle a_\downarrow^\dagger a_\downarrow \rangle)$, is calculated using the fermion version of Eq. (9.3),

$$m(T) = \frac{1}{2\pi} \int_{-\varepsilon_F}^{\varepsilon_F} d\omega \frac{\gamma}{e^{\beta\omega} + 1} \left\{ \frac{1}{\left(w\alpha^2 + \frac{B}{2}\right)^2 + \gamma^2} - \frac{1}{\left(w\alpha^2 - \frac{B}{2}\right)^2 + \gamma^2} \right\}$$

$$= -B\frac{\gamma}{\pi\alpha^2} \int_{-\varepsilon_F}^{\varepsilon_F} d\omega \alpha^2 \frac{1}{e^{\beta\omega} + 1} \left\{ \frac{w\alpha^2}{\left[w^2\alpha^4 + \frac{B^2}{4} + \gamma^2\right]^2 - B^2 w^2 \alpha^4} \right\}$$

$$= -B\frac{\gamma}{\pi\alpha^2} \int_{-\varepsilon_F\alpha^2}^{\varepsilon_F\alpha^2} dv \frac{1}{e^{v/(kT\alpha^2)} + 1} \left\{ \frac{v}{\left[v^2 + \frac{B^2}{4} + \gamma^2\right]^2 - B^2 v^2} \right\} \tag{9.19}$$

The zero-field susceptibility χ_0 is just m/B, setting $B^2 = 0$ in the integrand. If desired, the internal energy of the impurity can also be computed, using

$$\mathscr{E}_{\text{lin}} - \mathscr{E}_0 = \sum_{k,\sigma} \varepsilon(k) \left(\langle c_{k,\sigma}^\dagger c_{k,\sigma} \rangle - \frac{1}{e^{\beta\varepsilon(k)} + 1} \right) - \frac{|\Delta|^2}{J} - \frac{1}{2}Bm(T) \quad (9.20)$$

Its derivative w.r. to T is the specific heat.

A more detailed Green function analysis of this model can be found in more technically specialized texts such as that by Mahan.[13]

The intimate relation between bosons and fermions in $1 + 1$ dimensional quantum field theory on the one hand, and the conformal theory that allows to solve for critical phenomena in 2D on the other, is one of the topics elegantly addressed in Stone's book.[14]

9.5. Scaling, Renormalization and Information Theory

Much of what we know about magnetic systems and their phase transitions comes from clever approximations anchored to some rigorous limits. A great deal of attention has been paid to critical phenomena, because it is only in the critical regimes that the gritty lattices on which spins normally live can be replaced by a continuum and differential equations reign. It is the fluctuations that determine the critical phenomena, and they live on some highly nonlinear surfaces that are quite special in 2D. For if one adds the assumption of conformal invariance to any of the usual translational and rotational symmetries that one likes to invoke, the correlation functions become intricately intertwined and relations such as in Eqs. (9.1) and (9.2) become inevitable. All known two-dimensional phase transitions can be characterized by a single parameter c (the "central charge") in the conformal theory. But as Cardy has emphasized,[15] there exists no rigorous proof of conformal invariance at a critical point. It is just a "principle," the consequences of which can be compared to numerical results and, ultimately, to experiment.

---

[13] G. D. Mahan, *Many-Particle Physics*, 2nd Edition, Plenum, New York, 1981, Chap. 11.
[14] M. Stone in *Bosonization*, M. Stone, ed., World Scientific, 1994, especially Chapters 1 and 2.
[15] J. L. Cardy, Conformal invariance, in *Phase Transitions and Critical Phenomena*, Vol. 11, C. Domb and J. L. Lebowitz, eds., Academic Press, London, 1987, Chap. 2.

The mapping of a plane onto a cylinder is a conformal transformation. Its importance lies in the converse, the finite-size scaling and renormalization procedures, that have allowed numerical calculation on a finite strip to be mapped onto the infinite plane with exquisite precision.

In 2D the conformal group is isomorphic to the group of analytic functions, and it is much larger there than in 3D or higher dimensions. This may be the reason conformal theory has not been fruitful in 3D. However, the last word has not been said nor written. The student who wishes to proceed in this important line of study is advised to first consult the introductory book by Cardy on statistical physics in general and conformal theory in particular (his Chapter 11)[16] before consulting the more technical papers.[15]

A different, equally esoteric but equally interesting branch of research involves the *spin glasses*. The present book offered no more than an introduction to this exciting topic, which has been taken to an entirely different level in a highly original book by Nishimori.[17] This author shows how model spin glasses can be used as templates in coding, decoding and image restoration procedures.

Everything, including this chapter and this book, has to come to an end. But in looking back over the preceding pages and chapters, I cannot resist pointing out the obvious: independently of the technical importance of magnetic substances — which cannot be denied — the *theory* of magnetism continues to lie at the core of *practically all* of contemporary theoretical physics. Its study continues to be a perpetual delight and inspiration.

---

[16] John Cardy, *Scaling and Renormalization in Statistical Physics*, Cambridge Univ. Press, Cambridge, 1996.

[17] H. Nishimori, *Statistical Physics of Spin Glasses and Information Processing*, Oxford Univ. Press, Oxford, 2001.

Bibliography

Abbott, E. A., *Flatland*, Grant Dahlstrom, Pasadena, California, 1884.

Abramowitz, M. and I. Stegun (eds.), *Handbook of Math. Functions*, National Bureau of Standards, Washington, 1964, Secs. 9.6, 9.7.

Adams, H., *The Education of Henry Adams*, Random House, New York, 1931.

Affleck, I., T. Kennedy, E. H. Lieb and H. Tasaki, *Phys. Rev. Lett.* **59**, 799 (1987); *Commun. Math. Phys.* **115**, 477 (1988).

Aharoni, A., *Introduction to the Theory of Ferromagnetism*, Oxford Univ. Press, Oxford, 2000.

Anderson, P. W., *Phys. Rev.* **86**, 694 (1952).

Anderson, P. W., *Phys. Rev.* **124**, 41 (1961).

Anderson, P. W., *Phys. Rev.* **164**, 352 (1967).

Andrei, N., *Phys. Rev. Lett.* **45**, 379 (1980).

Andrei, N., K. Furuya and J. H. Lowenstein, *Rev. Mod. Phys.* **55**, 331 (1983).

Anfuso, F. and S. Eggert, cond-mat/0511001 v1, 31 Oct. 2005.

Antoulas, A., R. Schilling and W. Baltensperger, *Solid State Commun.* **18**, 1435 (1976).

Arovas, D. P. and A. Auerbach, *Phys. Rev.* **B38**, 316 (1988).

Bader, H. and R. Schilling, *Phys. Rev.* **B19**, 3556 (1979); **20**, 1977 (1979).

Baker, G. A., Jr., *Phys. Rev.* **124**, 768 (1961).

Bander, M. and C. Itzykson, *Phys. Rev.* **B30**, 6485 (1984).

Bardeen, J., *Phys. Rev.* **49**, 653 (1936).

Barouch, E., *Physica* **1D**, 333 (1980).

Barouch, E., B. McCoy and T. T. Wu, *Phys. Rev. Lett.* **31**, 1409 (1973).

Bartkowski, R., *Phys. Rev.* **B5**, 4536 (1972).

Baxter, R. J., *Ann. Phys. (N.Y.)* **70**, 323 (1972).

Baxter, R. J. and I. Enting, *J. Phys.* **A11**, 2463 (1978).

Bazylinski, D. A. and R. B. Frankel, *Nature Rev. Microbiol.* **2**, 217 (2004).

Berlin, T. and M. Kac, *Phys. Rev.* **86**, 821 (1952).

Bertaut, E. F., in *Magnetism*, Vol. 3, G. Rado and H. Suhl, eds., Academic Press, New York, 1963, Chap. 4.

Bethe, H., *Z. Phys.* **71**, 205 (1931). English language translation in D. C. Mattis, *The Many-Body Problem*, World Scientific, 1993, pp. 689ff.

Bethe, H., *Proc. Roy. Soc. (London)* **A150**, 552 (1935); *J. Appl. Phys.* **9**, 244 (1938).

Betts, D. D., F. Salevsky and J. Rogiers, *J. Phys.* **A14**, 531 (1981).

Bevis, M., A. Sievers, J. Harrison, D. Taylor and D. Thouless, *Phys. Rev. Lett.* **41**, 987 (1978).

Biegel, D., M. Karbach and G. Müller, *Europhys. Lett.* **59**, 882 (2002); *J. Phys.* **A36**, 5361 (2003); *J. Phys. Soc. Jpn.* **73**, 3008 (2004).

Bienenstock, A., *J. Appl. Phys.* **37**, 1459 (1966).

Binder, K., *et al.*, *Phys. Rev.* **B31**, 1498 (1985).

Binder, K. and W. Kob, *Glassy Materials and Disordered Solids*, World Scientific, 2005, pp. 221ff.

Binder, K. and K. Schröder, *Phys. Rev.* **B14**, 2142 (1976).

Birgeneau, R. and G. Shirane, *Phys. Today*, Dec. (1978), p. 32.

Bitter, F., *Phys. Rev.* **54**, 79 (1937).

Bloch, F., *Zeit. Phys.* **57**, 545 (1929).

Bloch, F., *Zeit. Phys.* **61**, 206 (1930).

Bloch, M., *Phys. Rev. Lett.* **9**, 286 (1962); *J. Appl. Phys.* **34**, 1151 (1963).

Bloembergen, N. and T. J. Rowland, *Phys. Rev.* **97**, 1679 (1955).

Blois, M., *J. Appl. Phys.* **26**, 975 (1955).

Blöte, H. and W. Huiskamp, *Phys. Lett.* **A29**, 304 (1969).

Blume, M., *Phys. Rev.* **141**, 517 (1966).

Bobeck, A. H., *Bell Syst. Tech. J.* **46**, 1901 (1967).

Bonner, J. C., B. Sutherland and P. Richards, *AIP Conf. Proc.* **24**, 335 (1975).

Born, M. and N. Wiener, *J. Math. Phys. (M.I.T.)* **5**, 84 (1926).

Bozorth, R., *Bell Syst. Tech. J.* **19**, 1 (1940).

Brandt, U. and A. Giesekus, *Phys. Rev. Lett.* **68**, 2648 (1992).

Brillouin, L., *J. Phys. Radium* **8**, 74 (1927).

Brush, S. G., *Rev. Mod. Phys.* **39**, 883 (1967).

Cajori, F., *A History of Physics*, Dover, New York, 1962, p. 102.

Caliri, A. and D. Mattis, *Phys. Lett.* **106A**, 74 (1984).

Callen, H. and R. M. Josephs, *J. Appl. Phys.* **42**, 1977 (1971).

Capel, H. W., *Physica* **37**, 423 (1967).

Cardy, J. L., "Conformal invariance," in *Phase Transitions and Critical Phenomena*, Vol. 11, C. Domb and J. L. Lebowitz, eds., Academic Press, London, 1987, Chap. 2.

Caux, J.-S. and J.-M. Maillet, cond-mat/0502365 v1, 15 Feb. 2005.

Cernuschi, F. and H. Eyring, *J. Chem. Phys.* **7**, 547 (1939).

Chen, H. and P. M. Levy, *Phys. Rev.* **B7**, 4267 (1973).

Chen, Z. and M. Kardar, *Phys. Rev.* **B30**, 4113 (1984).

Ching, W. Y. and D. L. Huber, *Phys. Rev.* **B20**, 4721 (1979).

Chowdhury, D., *Spin Glasses and Other Frustrated Systems*, World Scientific, 1986.

Chudnovsky, E. M. and J. Tejada, *Macroscopic Quantum Tunneling of the Magnetic Moment*, Cambridge Univ. Press, 1998.

Clogston, A., *Phys. Rev.* **125**, 541 (1962).

Cocho, G., G. Martinez-Mekler and R. Martinez-Enriquez, *Phys. Rev.* **B26**, 2666 (1982).

Cohen, M. H. and F. Keffer, *Phys. Rev.* **99**, 1128, 1135 (1955).

Compton, A. H., *J. Franklin Inst.* **192**, 144 (1921).

Condon, E. U. and G. Shortley, *The Theory of Atomic Spectra*, Cambridge Univ. Press, New York, 1935.

Cragg, D. and P. Lloyd, *J. Phys.* **C12**, L215 (1979).

Curie, P., *Ann. Chim. Phys.* **5** (7), 289 (1895); *Oeuvres*, Paris, 1908.

Davis, H. L., *Phys. Rev.* **120**, 789 (1960).

Daybell, M., "Thermal and Transport Properties," in Ref. 6.41.

Daybell, M. D. and W. Steyert, *Phys. Rev. Lett.* **18**, 398 (1967).

De'Bell, K. and M. L. Glasser, *Phys. Lett.* **104A**, 255 (1984).

de Gennes, P. G., *J. Phys.* **23**, 510 (1962).

de Jongh, L. and A. Miedema, *Adv. Phys.* **23**, 1 (1974).

Des Cloizeaux, J. and J. J. Pearson, *Phys. Rev.* **128**, 2131 (1962).

Dirac, P. A. M., *Proc. Roy. Soc.* **117A**, 610 (1928).

Dirac, P. A. M., *Proc. Camb. Phil. Soc.* **26**, 376 (1930).

Dirac, P. A. M., *Proc. Roy. Soc.* **126A**, 360 (1930).

Dirac, P. A. M., *Proc. Roy. Soc.* (*London*) **A123**, 60 (1931).

Domb, C., *Adv. Phys.* **9**, 149–361 (1960).

Domb, C. and M. Green (eds.), *Phase Transitions and Critical Phenomena*, Vol. 3, Academic Press, New York, 1974.

Duff, K. and T. Das, *Phys. Rev.* **B3**, 192, 2294 (1971).

Dunin-Borkowski, R., *et al.*, *Science* **282**, 1868 (1998).

Eckart, C., *Phys. Rev.* **28**, 711 (1926).

Edwards, S. F. and P. W. Anderson, *J. Phys.* **F5**, 965 (1975).

Einstein, A., *The Method of Theoretical Physics*, Oxford, 1933.

Elliott, R. J., *Phys. Rev.* **124**, 346 (1961).

Eschenfelder, A. H., *Magnetic Bubble Technology*, Springer Series in Solid State Sciences, Vol. 14, Springer, Berlin, Heidelberg, New York, 1980.

Essler, F. H. L., V. E. Korepin and K. Schoutens, *Nucl. Phys.* **B384**, 431 (1982).

Fadeev, L. and L. Takhtajan, *Phys. Lett.* **85A**, 375 (1981).

Fisher, M. E., *Phys. Rev.* **162**, 480 (1967).

Fogedby, H. *J. Phys.* **C11**, 4767 (1978).

Fogedby, H., "Theoretical Aspects of Mainly Low-Dimensional Magnetic Systems," notes from the Inst. Laue-Langevin, Grenoble, France (March 1979); *J. Phys.* **A13**, 1467 (1980).

Fowler, M., *Phys. Rev.* **B17**, 2989 (1978); *J. Phys.* **C11**, L977 (1978).

Fowler, M. and M. Puga, *Phys. Rev.* **B18**, 421 (1978).

Fowler, R. and E. Guggenheim, *Statistical Thermodynamics*, Cambridge Univ. Press, Cambridge, 1939.

Fradkin, E., B. Huberman and S. Shenker, *Phys. Rev.* **B18**, 4789 (1978).

Franck, J. P., F. D. Manchester and D. L. Martin, *Proc. Roy. Soc.* (*London*) **A263**, 494 (1961).

Frenkel, Y., *Z. Physik* **49**, 31 (1928).

Friedel, J., *Canad. J. Phys.* **34**, 1190 (1956).

Friedel, J., *J. Phys. Rad.* **19**, 573 (1958).

Fröhlich, J. and E.H. Lieb, *Phys. Rev. Lett.* **38**, 440 (1977); *Commun. Math. Phys.* **60**, 233 (1978).

Fukuda, N. and M. Wortis, *J. Phys. Chem. Sol.* **24**, 1675 (1963).

Fukuyama, H. and A. Sakurai, *Progr. Theor. Phys.* **62**, 595 (1979).

Furman, D., S. Dattagupta and R. B. Griffiths, *Phys. Rev.* **B15**, 441 (1977).

Gaunt, D. and J. Guttmann, "Asymptotic Analysis of Coefficients," in *Phase Transitions and Critical Phenomena*, Vol. 3, C. Domb and M. Green, eds., Academic Press, New York, 1974.

Gaunt, D. and M. Sykes, *J. Phys.* **A6**, 1517 (1973).

Gerlach, W. and O. Stern, *Z. Phys.* **9**, 349 (1922).

Gilbert, W., *De Magnete*, trans., Gilbert Club, London, 1900, rev. ed., Basic Books, New York, 1958, p. 3.

Glasser, M. L., *J. Math. Phys.* **5**, 1150 (1964); *Erratum, J. Math. Phys.* **7**, 1340 (1966).

Glasser, M. L., *J. Math. Phys.* **13**, 1145 (1972).

Gold, A. V., L. Hodges, P. Panousis and R. Stone, *Int. J. Magn.* **2**, 357 (1971).

Goldhirsch, I. and V. Yakhot, *Phys. Rev.* **B21**, 2833 (1980).

Goldstein, H., *Classical Mechanics*, Addison-Wesley, Reading, Mass., 1951.

Griffith, J. S., *The Theory of Transition-Metal Ions*, Cambridge Univ. Press, New York, 1961.

Griffiths, R. B., *Phys. Rev.* **133**, A768 (1964).

Griffiths, R. B., *Phys. Rev.* **136**, A437 (1964).

Griffiths, R. B., *J. Math. Phys.* **8**, 478, 484 (1967).

Griffiths, R. B., *Phys. Rev.* **176**, 655 (1968).

Griffiths, R. B., *J. Math. Phys.* **10**, 1559 (1969).

Groen, J., T. Klaasen, N. Poulis, G. Müller, H. Thomas and H. Beck, *Phys. Rev.* **B22**, 5369 (1980).

Gulácsi, Z. and D. Vollhardt, "Exact Ground States of the Periodic Anderson Model," arXiv:cond-mat/0504174v1 (7 April 2005), in press.

Gunnarson, O. and K. Schönhammer, *Phys. Rev.* **B31**, 4815 (1985).

Gupta, R. and C. Baillie, *Phys. Rev.* **B45**, 2883 (1992).

Gutzwiller, M. C., *Phys. Rev. Lett.* **10**, 159 (1963).

Haldane, F. D. M., *Phys. Lett.* **A93**, 454 (1983); *Phys. Rev. Lett.* **50**, 1153 (1983).

Hamann, D., *Phys. Rev.* **B2**, 1373 (1970).

Haydock, R., V. Heine and M. Kelly, *J. Phys.* **C5**, 2845 (1972).

Heeger, A., *Solid State Phys.* **23**, 283 (1969).

Heeger, A., "Localized Moments and Nonmoments in Metals: the Kondo Effect," in *Solid State Physics*, Vol. 23, F. Seitz, D. Turnbull and H. Ehrenreich, eds., Academic Press, New York, 1969, p. 284 and Fig. 28 on p. 380.

Heeger, A., A. Klein and P. Tu, *Phys. Rev. Lett.* **17**, 803 (1966).

Heisenberg, W., M. Born and P. Jordan, *Z. Phys.* **35**, 557 (1926).

Hepp, K., *Solid State Commun.* **8**, 2087 (1970).

Hirsch, J., *J. Appl. Phys.* **67**, 4549 (1990).

Hirsch, J., *Phys. Rev.* **B65**, 184502 (2002).

Höglund, K. H. and A. W. Sandvik, *Phys. Rev.* **B70**, 24406 (2004).

Hohenberg, P. and W. Brinkman, *Phys. Rev.* **B10**, 128 (1974).

Holstein, T. and H. Primakoff, *Phys. Rev.* **58**, 1048 (1940).

Holtzberg, F., *et al.*, *J. Appl. Phys.* **35/2**, 1033 (1964).

Hubbard, J., *Proc. Roy. Soc.* **A276**, 238 (1963).

Hubbard, J., *Phys. Rev.* **B19**, 2626; **20**, 4584 (1979).

Hubert, A., *Theorie der Domänewände in geordneten Medien*, Lecture Notes in Physics, Vol. 26, Springer, Berlin, Heidelberg, New York, 1974.

Hulthén, L., *Ark. Met. Astron. Fysik* **26A**, No. 11 (1938).

Ising, E., *Z. Physik* **31**, 253 (1925).

Jahnke, E. and F. Emde, *Tables of Functions*, Dover, New York, 1945.

Jimbo, M. and T. Miwa, *Algebraic Analysis of Solvable Lattice Problems*, AMS, Providence, 1995.

Johnson, J. D., *Phys. Rev.* **A9**, 1743 (1974).

Johnson, J. D. and B. McCoy, *Phys. Rev.* **A6**, 1613 (1972).

Johnson, J., S. Krinsky and B. McCoy, *Phys. Rev.* **A8**, 2526 (1973).

Jordan, P. and E. Wigner, *Z. Phys.* **47**, 631 (1928).

José, J. and P. Sahni, *Phys. Rev. Lett.* **43**, 78 (1978), *Erratum*, **43**, 1843 (1978).

José, J., L. Kadanoff, S. Kirkpatrick and D. Nelson, *Phys. Rev.* **B16**, 1217 (1977).

Joyce, G. S., *J. Phys.* **A5**, L65 (1972).

Jullien, R., *et al.*, *Phys. Rev. Lett.* **44**, 1551 (1980).

Kalle, C., *J. Phys.* **A17**, L801 (1984).

Kaplan, T. A., *Phys. Rev.* **124**, 329 (1961).

Kaplan, T. A. and D. H. Lyons, *Phys. Rev.* **129**, 2072 (1963).

Katsura, S., *Phys. Rev.* **127**, 1508 (1962).

Katsura, S., S. Inawashiro and Y. Abe, *J. Math. Phys.* **12**, 895 (1971).

Katsura, S., T. Morita, S. Inawashiro, T. Horiguchi and Y. Abe, *J. Math. Phys.* **12**, 892 (1971).

Kaufman, B., *Phys. Rev.* **76**, 1232 (1949).

Keffer, F., "Spin Waves," in *Handbuch der Physik*, Vol. 18 Part 2, H.J.P. Wijn, Springer, Berlin, 1966.

Keffer, F. and T. Oguchi, *Phys. Rev.* **117**, 718 (1960).

Keil, J., *Introductio ad Veram Physicam*, 1705 (transl., 1776).

Kirkpatrick, S., *Phys. Rev.* **B16**, 4630 (1977).

Kitanine, N., J.-M. Maillet and V. Terras, *Nucl. Phys.* **B554**, 647 (1999); **B567**, 554 (2000).

Kogut, J., *Rev. Mod. Phys.* **51**, 659 (1979).

Kojima, D. and A. Isihara, *Phys. Rev.* **B20**, 489 (1979).

Kondo, J., *Prog. Theor. Phys.* **28**, 846 (1962); **32**, 37 (1964).

Kooy, C. and U. Enz, *Philips Res. Report* **15**, 7 (1960).

Korenman, V., J. Murray and R. Prange, *Phys. Rev.* **B16**, 4032, 4048, 4058 (1977).

Korepin, V. E., *Commun. Math. Phys.* **86**, 391 (1982).

Korepin, V. E. and F. H. L. Essler, eds., *Exactly Solvable Models of Strongly Correlated Electrons*, World Scientific, 1994.

Kosterlitz, J., *J. Phys.* **C7**, 1046 (1974).

Kosterlitz, J. and D. Thouless, *J. Phys.* **C6**, 1181 (1973).

Kosterlitz, J., D. Thouless and R.C. Jones, *Phys. Rev. Lett.* **36**, 1217 (1976).

Kotliar, G. and A. E. Ruckenstein, *Phys. Rev. Lett.* **57**, 1362 (1986).

Kramers, H. A., *Physica* **1**, 182 (1934).

Krishna-Murthy, H., J. Wilkins and K. G. Wilson, *Phys. Rev.* **B21**, 1003–1083 (1980).

Kuebler, J., *Theory of Itinerant Electron Magnetism*, Clarendon Press, Oxford, 2000, p. 193.

Kurmann, J., H. Thomas and G. Müller, *Physica* **112A**, 235 (1982).

Kurtze, D. and M. Fisher, *J. Stat. Phys.* **19**, 205 (1978).

Lagendijk, A. and H. De Raedt, *Phys. Rev. Lett.* **49**, 602 (1982).

Lakshmanan, M., T. Ruijgrok and C. Thompson, *Physica* **84A**, 577 (1976).

Langevin, P., *Ann. Chim. Phys.* **5**(8), 70 (1905); *J. Phys.* **4**(4), 678 (1905).

Lee, P. A. *et al.*, *Comments Condens. Matter Phys.* **12**, 99 (1986).

Lee, T. D. and C. N. Yang, *Phys. Rev.* **87**, 410 (1952).

Lenz, W., *Phys. Z.* **21**, 613 (1920).

Leung, K., D. Hone, D. Mills, P. Riseborough and S. Trullinger, *Phys. Rev.* **B21**, 4017 (1980).

Lieb, E., *Phys. Rev. Lett.* **62**, 1201, 1927 (1989).

Lieb, E. and D. Mattis, *J. Math. Phys.* **3**, 749 (1962).

Lieb, E. and D. Mattis, *Phys. Rev.* **125**, 164 (1962).

Lieb, E. and D. Ruelle, *J. Math. Phys.* **13**, 781 (1972).

Lieb, E. and F. Wu, *Phys. Rev. Lett.* **20**, 1445 (1968).

Lieb, E., T. Schultz and D. Mattis, *Ann. Phys. (N.Y.)* **16**, 407 (1961).

Lin, K. Y. and F. Y. Wu, *Z. Physik* **B33**, 181 (1979).

Lines, M. E., *Phys. Reports* **55**, 133 (1979).

Liu, S. H., *Phys. Rev.* **B17**, 3629 (1978).

Long, M. W. and X. Zotos, *Phys. Rev.* **B48**, 317 (1993).

Luther, A. and I. Peschel, *Phys. Rev.* **B9**, 2911 (1974); **12**, 3908 (1975).

Luttinger, J. M., *Phys. Rev. Lett.* **37**, 778 (1976).

Lyons, D. and T. Kaplan, *Phys. Rev.* **120**, 1580 (1960).

Lyons, D., T. Kaplan, K. Dwight and N. Menyuk, *Phys. Rev.* **126**, 540 (1962).

Mahan, G. D., *Many-Particle Physics*, 2nd Edition, Plenum, New York, 1981, Chap. 11.

Majumdar, C. K. and D. K. Ghosh, *J. Math. Phys.* **10**, 1388, 1399 (1969).

Makivic, M. and H.-Q. Ding, *Phys. Rev.* **B43**, 3562 (1991).

Marinari, E., *Nucl. Phys.* **B235** (FS11), 123 (1984).

Marland, L. and D. Betts, *Phys. Rev. Lett.* **43**, 1618 (1979).

Marshall, W., T. Cranshaw, C. Johnson and M. Ridout, *Rev. Mod. Phys.* **36**, 399 (1964).

Matthias, B. and R. Bozorth, *Phys. Rev.* **109**, 604 (1958).

Matthias, B. T., R. Bozorth and J. H. Van Vleck, *Phys. Rev. Lett.* **7**, 160 (1961).

Mattis, D. C., *Phys. Rev.* **130**, 76 (1963).

Mattis, D. C., *Phys. Rev. Lett.* **19**, 1478 (1967).

Mattis, D. C., *Phys. Lett.* **56A**, 421 (1976); *Erratum*, **60A**, 492 (1977).

Mattis, D. C., *Phys. Rev. Lett.* **42**, 1503 (1979).

Mattis, D. C., *The Many-Body Problem: An Encyclopedia of Exactly Solved Models in 1D*, World Scientific, 1992, Chap. 4.

Mattis, D. C., *Mod. Phys. Lett.* **B12**, 655 (1998).

Mattis, D. C., *Int. J. Nanoscience* **2**, 165 (2003).

Mattis, D. C., *J. Stat. Phys.* **116**, 773 (2004).

Mattis, D. C. and S. Nam, *J. Math. Phys.* **13**, 1185 (1972).

Mattis, D. C. and P. Paul, *Phys. Rev. Lett.* **83**, 3733 (1999).

Mattis, D. C. and R. E. Peña, *Phys. Rev.* **B10**, 1006 (1974).

Mattis, D. C. and R. Raghavan, *Phys. Lett.* **75A**, 313 (1980).

Maunari, I. and L. Kawabata, Research Notes of Department of Physics, Okoyama University, No. 15, Dec. 21, 1964 (unpublished).

Maxwell, J. C., *A Treatise on Electricity and Magnetism*, 1873, reprinted by Dover, New York, 1954.

McCoy, B., J. Perk and R. Schrock, *Nucl. Phys.* **B220**, 35, 269 (1983).

Mehta, M. L., *Random Matrices*, Academic Press, New York, 1967.

Meissner, W. and G. Voigt, *Ann. Physik* **7**, 761, 892 (1930).

Mermin, N. and H. Wagner, *Phys. Rev. Lett.* **17**, 1133 (1966).

Mézard, M., G. Parisi and M. Virasoro, *Spin Glass Theory and Beyond*, World Scientific, 1987.

Mikeska, H., *J. Phys.* **C11**, L29 (1978); **13**, 2913 (1980).

Mikeska, H. and W. Pesch, *J. Phys.* **C12**, L37 (1979).

Miwa, H. and K. Yosida, *Prog. Theor. Phys. (Kyoto)* **26**, 693 (1961).

Montroll, E. W., "Lattice Statistics," in *Applied Combinatorial Mathematics*, E. F. Beckenback, ed., Wiley, New York, 1964.

Morgenstern, L. and K. Binder, *Phys. Rev. Lett.* **43**, 1615 (1979).

Müller, G., H. Beck and J. Bonner, *Phys. Rev. Lett.* **43**, 75 (1979).

Müller-Hartmann, E. and J. Zittartz, *Z. Physik* **B27**, 261 (1977).

Nagamiya, T., K. Yosida and R. Kubo, *Adv. Phys. (Philos. Mag. Suppl.)* **4** (13), 1 (1955).

Nagaoka, Y., *Phys. Rev.* **147**, 392 (1966).

Nagosa, N., Y. Hatsugai and M. Imada, *J. Phys. Soc. Jpn.* **58**, 978 (1989).

Nakamura, K. and T. Sasada, *Phys. Lett.* **A48**, 321 (1974).

Nakamura, K. and T. Sasada, *J. Phys.* **C15**, L1013 (1982).

Nakamura, T. and M. Bloch, *Phys. Rev.* **132**, 2528 (1963).

Néel, L., *Ann. Phys.* (Paris) **17**, 64 (1932); *J. Phys. Radium* **3**, 160 (1932).

Niemeijer, T., *Physica* **36**, 377 (1967).

Nishimori, H., *Statistical Physics of Spin Glasses and Information Processing*, Oxford Univ. Press, Oxford, 2001.

Oguchi, T., *J. Phys. Soc. Jpn.* **6**, 31 (1951).

Oguchi, T., *J. Phys. Chem. Sol.* **24**, 1649 (1963).

Oitmaa, J. and D. D. Betts, *Phys. Lett.* **68A**, 450 (1978).

Oitmaa, J. and D. D. Betts, *Canad. J. Phys.* **56**, 897 (1978).

Oitmaa, J., D. D. Betts and L. G. Marland, *Phys. Lett.* **79A**, 193 (1980).

Onsager, L., *Phys. Rev.* **65**, 117 (1944).

Onsager, L. *Nuovo Cimento* (*Suppl.*) **6**, 261 (1949).

Orbach, R., *Phys. Rev.* **112**, 309 (1958).

Ortiz, G., M. Harris and P. Ballone, *Phys. Rev. Lett.* **82**, 5317 (1999).

Ott, H. R., *et al.*, *Phys. Rev.* **B25**, 477 (1982).

Ovchinnikov, A., *Sov. Phys. JETP* **29**, 727 (1969).

Palanichamy, R. and K. Iyakutti, *Int. J. Mod. Phys.* **B16**, 1353 (2002).

Palmer, R. and C. Pond, *J. Phys.* **F9**, 1451 (1979).

Parisi, G., *J. Phys.* **A13**, 1101, 1887, L115 (1980); *Phil. Mag.* **B41**, 677 (1980).

Pauli, W., in *Le Magnétisme*, 6th Solvay Conf., Gauthier-Villars, Paris, 1932, p. 212.

Peierls, R. E., *Proc. Camb. Phil. Soc.* **32**, 477 (1936).

Peierls, R. E., *Quantum Theory of Solids*, Oxford Univ. Press, Oxford, 1955, p. 148.

Pfeuty, P., *Phys. Lett.* **72A**, 245 (1979).

Pickart, S., H. Alperin, G. Shirane and R. Nathans, *Phys. Rev. Lett.* **12**, 444 (1964).

Plischke, M. and D. C. Mattis, *Phys. Rev.* **B2**, 2660 (1970).

Plischke, M. and D. C. Mattis, *Phys. Rev.* **A3**, 2092 (1971).

Poisson, S. D., *Mémoire sur la Théorie du Magnétisme*, Mémoires de l'Académie, Vol. V, p. 247.

Pomerantz, M., F. Dacol and A. Segmuller, *Phys. Rev. Lett.* **40**, 246 (1978); *Physics Today* **34**, 20 (1981).

Posfai, M., *et al.*, *Science* **280**, 880 (1998).

Priestley, J. B., *History of Electricity*, London, 1775, p. 86.

Privman, V. and M. E. Fisher, *Phys. Rev.* **B30**, 322 (1984).

Prutton, M., *Thin Ferromagnetic Films*, Butterworth, Washington, 1964.

Puga, M., *Phys. Rev. Lett.* **42**, 405 (1979).

Reed, P., *J. Phys.* **C12**, L799 (1979).

Reitz, J. and M. B. Stearns, *J. Appl. Phys.* **50** (3), 2066 (1979).

Riedel, E. K. and F. J. Wegner, *Phys. Rev.* **B9**, 294 (1974).

Richmond, P. and G. Rikayzen, *J. Phys.* **C2**, 528 (1969).

Rogiers, J., T. Lookman, D. D. Betts and C. J. Elliot, *Canad. J. Phys.* **56**, 409 (1978).

Ruderman, M. A. and C. Kittel, *Phys. Rev.* **96**, 99 (1954).

Rushbrooke, G. S., *J. Math. Phys.* **5**, 1106 (1964).

Russell, J. S., *Proc. Roy. Soc. Edinburgh* (1844), p. 319.

Saarloos, W. and D. Kurtze, *J. Phys.* **A17**, 1301 (1984).

Sachdev, S., *Phys. Rev.* **B68**, 064419 (2003).

Sachdev, S., C. Buragohain and M. Voijta, *Science* **286**, 2479 (1999).

Saenz, A. W. and R. O'Rourke, *Rev. Mod. Phys.* **27**, 381 (1955).

Satija, S., G. Shirane, Y. Yoshizawa and K. Hirakawa, *Phys. Rev. Lett.* **44**, 1548 (1980).

Schaefer, H. F. (ed.), *Methods of Electronic Structure Theory*, Plenum, New York, 1977.

Schilling, R., *Phys. Rev.* **B15**, 2700 (1977).

Schilp, P. A., (ed.), *Albert Einstein: Philosopher-Scientist*, Vol. 1, Harper & Row, New York, 1959, p. 9.

Schlottmann, D., *Int. J. Mod. Phys.* **B11**, 355 (1997).

Schrödinger, E., *Ann. Phys.* **79** (4), 734 (1926).

Schrieffer, J. R. and P. A. Wolff, *Phys. Rev.* **194**, 491 (1966).

Schultz, T., D. Mattis and E. Lieb, *Rev. Mod. Phys.* **36**, 856 (1964).

Schulz, H. J., *Int. J. Mod. Phys.* **B5**, 57 (1991).

Schwinger, J., "On Angular Momentum," U.S. Atomic Energy Commission Report NYO-3071 (1952), reprinted in L. Biedenharn and H. Van Dam (eds.), *Quantum Theory of Angular Momentum*, Academic Press, New York, 1965.

Shastry, B. S., *Phys. Rev. Lett.* **56**, 1529 and 2453 (1986).

Sherrington, D., *Phys. Rev. Lett.* **41**, 1321 (1978).

Sherrington, D. and S. Kirkpatrick, *Phys. Rev. Lett.* **35**, 1792 (1975); *Phys. Rev.* **B17**, 4384 (1978).

Sherwood, R. *et al.*, *J. Appl. Phys.* **30**, 217 (1959).

Simon, B., *Phys. Rev. Lett.* **44**, 547 (1980).

Slater, J. C., *Phys. Rev.* **34**, 1293 (1929).

Slater, J. C., *Phys. Rev.* **35**, 509 (1930).

Slater, J. C., *Phys. Rev.* **49**, 537, 931 (1936).

Slater, J. C., *Phys. Rev.* **52**, 198 (1937).

Slater, J. and G. Koster, *Phys. Rev.* **94**, 1498 (1954).

Smith, A., J. Janak and R. Adler, *Electronic Conduction in Solids*, McGraw-Hill, New York, 1967.

Smith, F., *Phys. Rev. Lett.* **36**, 1221 (1976).

Sommerfeld, A., *Electrodynamics*, Academic Press, New York, 1952.

Sompolinsky, H., *Phys. Rev. Lett.* **47**, 935 (1981).

Soohoo, R., *Magnetic Thin Films*, Harper & Row, New York, 1965.

Stanley, H. E., *Introduction to Phase Transitions and Critical Phenomena*, Oxford Univ. Press, Oxford, 1971.

Stanley, H. E. and T. A. Kaplan, *Phys. Rev. Lett.* **17**, 913 (1966).

Stauffer, D., *Int. J. Mod. Phys.* **C10**, 931 (1999).

Stearns, M. B., *J. Magnetism Magn. Mater.* **5**, 167 (1977).

Stearns, M. B., *Phys. Today* (April 1978), pp. 34–39.

Steiner, M., J. Villain and C. Windsor, *Adv. Phys. (Philos. Mag. Suppl.)* **25**, 87 (1976).

Stone, M., in *Bosonization*, M. Stone, ed., World Scientific, 1994.

Stoner, E. C., *Magnetism and Matter*, Methuen, London, 1934, p. 100.

Stoner, E. C., *Proc. Roy. Soc.* **169A**, 339 (1939).

Stoney, G. J., *Trans. Roy. Dub. Soc.* **4**, 583 (1891).

Stuart, R. and W. Marshall, *Phys. Rev.* **120**, 353 (1960).

Su, G. and M. Suzuki, "Josephson-like effect in a spin valve," *Mod. Phys. Lett.* **B16**, 711 (2002).

Sushkov, O. P., *Phys. Rev.*, **B68**, 094426 (2003).

Sykes, M., D. Gaunt, J. Essam and C. Elliott, *J. Phys.* **A6**, 1506 (1973).

Sykes, M., D. Gaunt, J. Essam and D. Hunter, *J. Math. Phys.* **14**, 1060 (1973).

Sykes, M., D. Gaunt, J. Essam, B. Heap, C. Elliott and S. Mattingly, *J. Phys.* **A6**, 1498 (1973).

Sykes, M., D. Gaunt, J. Martin, S. Mattingly and J. Essam, *J. Math. Phys.* **14**, 1071 (1973).

Sykes, M., D. Gaunt, S. Mattingly, J. Essam and C. Elliott, *J. Math. Phys.* **14**, 1066 (1973).

Sykes, M., D. Hunter, D. McKenzie and B. Heap, *J. Phys.* **A5**, 667 (1972).

Sykes, M. F., *J. Math. Phys.* **2**, 52 (1961).

Sykes, M. and M. Fisher, *Physica* **28**, 919, 939 (1962).

Sykes, M., J. Essam and D. Gaunt, *J. Math. Phys.* **6**, 283 (1965).

Sykes, M., D. Gaunt, P. Roberts and J. Wyles, *J. Phys.* **A5**, 624, 640 (1972).

Sykes, M., *et al.*, *J. Phys.* **A5**, 640 (1972).

Taggart, G. B., *Phys. Rev.* **B20**, 3886 (1979).

Takhtajan, L. A., *Phys. Lett.* **64A**, 235 (1977).

Taylor, D. R., *Phys. Rev. Lett.* **42**, 1302 (1979).

Thiele, A., *Bell. Syst. Tech. J.* **48**, 3287 (1969).

Thiele, A., *et al.*, *Bell Syst. Tech. J.* **50**, 711, 725 (1971).

Tjon, J. and J. Wright, *Phys. Rev.* **B15**, 3470 (1977).

Tjon, T. A., *Phys. Rev.* **B2**, 2411 (1970).

Tobochnick, J. and G. V. Chester, *Phys. Rev.* **B20**, 3761 (1979).

Toulouse, G., *Commun. Phys.* **2**, 115 (1977).

Uhlenbeck, G. E. and S. Goudsmit, *Naturwiss.* **13**, 953 (1925).

Uhlenbeck, G. E. and S. Goudsmit, *Nature* **117**, 264 (1926).

Utiyama, T., *Progr. Theor. Phys.* **6**, 907 (1951).

Van Vleck, J. H., *The Theory of Electric and Magnetic Susceptibilities*, Oxford, 1932, p. 104.

Van Vleck, J. H., *Nuovo Cimento* **6**, (Ser. X, Suppl. 3) 857 (1957).

Varma, C. M. and Y. Yafet, *Phys. Rev.* **B13**, 2950 (1976).

Venuti, L., *et al.*, *Int. J. Mod. Phys.* **B16**, 1363 (2002), *inter alia*.

Villain, T., *J. de Phys.* **35**, 27 (1974).

Villain, J., *J. Physique* **36**, 581 (1975).

Walker, L. R. and R. E. Walstedt, *Phys. Rev. Lett.* **38**, 514 (1977); *Phys. Rev.* **B22**, 3816 (1980).

Walker, L. R., *Phys. Rev.* **116**, 1289 (1959).

Wannier, G. H., *Phys. Rev.* **52**, 191 (1937).

Wannier, G. H., *Phys. Rev.* **79**, 357 (1950).

Weng, C.-Y., R. Griffiths and M. Fisher, *Phys. Rev.* **162**, 475 (1967).

Weiss, P., *J. Physique* **6**, 661 (1907); *Proc. of 6th Solvay Congress*, 1930, Gauthier-Villars, Paris, 1932, pp. 281ff.

Weiss, P., *J. de Phys.* **6** (4), 661 (1907).

Whittaker, E., *A History of the Theories of Aether and Electricity*, Vol. II, Harper & Row, New York, 1960.

Wiegmann, P. B., *JETP Lett.* **31**, 364 (1980).

Wiegmann, P. B., *Phys. Lett.* **80A**, 163 (1980).

Wiegmann, P. B., *et al.*, *Phys. Lett.* **81A**, 175, 179 (1981).

Wigner, E. P., *Trans. Faraday Soc.* **205**, 678 (1938).

Wilson, K. G., "Solution of the spin 1/2 Kondo Hamiltonian," in *Collective Properties of Physical Systems*, Proc. of the 24th Nobel Symposium, B. Lundqvist and S. Lundqvist, eds., Academic Press, New York, 1973.

Wilson, K. G., *Rev. Mod. Phys.* **47**, 773 (1975).

Wilson, K. G. and J. Kogut, *Phys. Reports* **12**, 75–200 (1974).

Wohlfarth, E. and J. Cornwell, *Phys. Rev. Lett.* **7**, 342 (1961).

Wolf, W. P., *Rep. Prog. Phys.* **24**, 212 (1961).

Wolff, P. A., *Phys. Rev.* **124**, 1030 (1961).

Wortis, M., *Phys. Rev.* **132**, 85 (1963).

Wortis, M., *Phys. Rev.* **138A**, 1126 (1965).

Wu, J. and D. C. Mattis, *Phys. Rev.* **B67**, 224414 (2003).

Yamada, T., *Prog. Theor. Phys. Jpn.* **41**, 880 (1969).

Yamazaki, H., *J. Appl. Phys.* **64**, 5391 (1988); "Fractal properties in magnetic crystal," in *Nonlinear Phenomena and Chaos in Magnetic Materials*, P. E. Wigen, ed., World Scientific, 1994, pp. 191ff.

Yang, C. N., *Phys. Rev.* **85**, 809 (1952).

Yang, C. N. and T. D. Lee, *Phys. Rev.* **87**, 404 (1952).

Yang, C. N. and C. P. Yang, *Phys. Rev.* **150**, 321 (1966).

Yeomans, J. M., *Statistical Mechanics of Phase Transitions*, Clarendon Press, Oxford, 1992.

Yosida, K., *Phys. Rev.* **106**, 893 (1957).

Yosida, K., *Phys. Rev.* **147**, 223 (1966).

Yosida, K. and A. Yoshimori, in *Magnetism*, Vol. 5, G. Rado and H. Suhl, eds., Academic Press, New York, 1973, p. 253.

Young, D. P. *et al.*, *Nature* **397**, 412 (1999).

Zhang, X. Y., E. Abrahams, and G. Kotliar, *Phys. Rev. Lett.* **66**, 1236 (1991).

Zittartz, J., *Z. Physik* **B40**, 233 (1980).

Index

Exchange bias

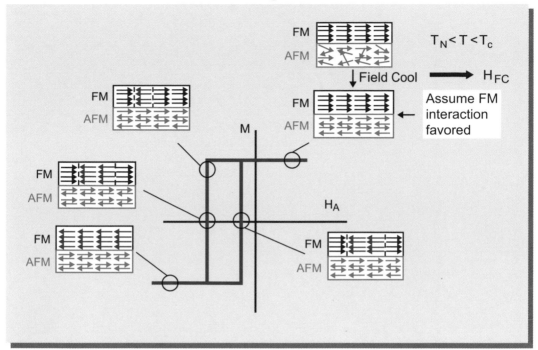

This picture illustrates "Exchange Bias": the effects of an antiferromagnetic substratum on the switching of a ferromagnet. As shown, the hysteresis loop becomes quite asymmetric. The mechanism for this broken symmetry is quite sturdy but poorly understood at the present time. [Figure courtesy Prof. Ivan K. Schuller, UCSD]